T0205735

Byron W. Hanks (Editor)

Proceedings of the 14th International Meshing Roundtable

Byron W. Hanks (Editor)

Proceedings of the 14th International Meshing Roundtable

With 326 Figures

 Springer

Byron W. Hanks
Computer Modeling and Sciences Dept.
Sandia National Laboratories
P.O. Box 5800
Albuquerque, NM 87185
USA
bwhanks@sandia.gov

ISBN-10 3-642-42134-2 Springer Berlin Heidelberg New York
ISBN-13 978-3-642-42134-1 Springer Berlin Heidelberg New York

Springer is a part of Springer Science+Business Media
springeronline.com

© Springer-Verlag Berlin Heidelberg 2005
Softcover re-print of the Hardcover 1st edition 2005

Typesetting: Dataconversion by authors
Final processing by PTP-Berlin Protago-TEX-Production GmbH, Germany
Cover-Design: design & production, Heidelberg
Printed on acid-free paper   89/3141/Yu – 5 4 3 2 1 0

Reviewers

Name	Affiliation
Aftosmis, Michael	NASA Ames, USA
Baker, Timothy	Princeton U., USA
Berkhahn, Volker	U. of Hanover, Germany
Bern, Marshall	Palo Alto Research Center, USA
Blacker, Ted	Sandia Nat. Labs, USA
Brewer, Michael	Sandia Nat. Labs, USA
Carbonera, Carlos	Jostens, Inc., USA
Cheng, Siu-Wing	Hong Kong U. of Science and Tech., Hong Kong
Chew, Paul	Cornell U., USA
Chou, Jin	Lawrence Livermore Nat. Labs, USA
Corney, Jonathan	Heriot Watt U., Scotland
Despres, Bruno	Com. Engergie Atomique, France
Dey, Tamal	Ohis State U., USA
Eiseman, Peter	PDC, USA
Fenwick, Liz	U. of California, San Diego, USA
Freitag, Lori	Lawrence Livermore Nat. Labs, USA
Frey, Pascal	Paris-6 U., France
Gopalsamy, Sankarappan	U. of Alabama, Birmingham., USA
Gopalsamy, Craig	Harvard U., USA
Haimes, Robert	M.I.T., USA
Hanks, Byron	Sandia Nat. Labs, USA
Hecht, Frederic	Paris-6 U., France
Helenbrook, Brian	Sandia Nat. Labs, USA
Hjelmfelt, Eric	Altair Eng., USA
Ito, Yasushi	U. of Alabama, Birmingham, USA
Jiao, Xiangmin	U. of Illinois, U-C., USA
Knupp, Patrick	Sandia Nat. Labs, USA
Kraftcheck, Jason	U. of Wisconsin, USA
Labelle, Francois	U. of California, Berkeley, USA
Lipnikov, Konstantin	Los Alamos Nat. Labs, USA
Loubere, Raphael	Los Alamos Nat. Labs, USA
Marzouk, Youssef	Sandia Nat. Labs, USA
Meyers, Ray	Elemental Tech., USA
Nandihalli, Sunil	PDC, USA
Ollivier-Gooch, Carl	UBC, Canada
Owen, Steve	Sandia Nat. Labs, USA
Pav, Steven	U. of California, San Diego, USA
Pebay, Philippe	Sandia Nat. Labs, USA
Phillips, Todd	CMU, USA
Pirzadeh, Shahyar	NASA Langley Research Center, USA
Rajagopalan, Krishnakumar	PDC, USA

Introduction

The papers in this volume were selected for presentation at the 14th International Meshing Roundtable, held September 11-14, 2005 in San Diego, CA, USA. The conference was started by Sandia National Laboratories in 1992 as a small meeting of organizations striving to establish a common focus for research and development in the field of mesh generation. Now after 14 consecutive years, IMR has become recognized as an international focal point annually attended by researchers and developers from dozens of countries around the world.

The 14th International Meshing Roundtable consists of technical presentations from contributed papers, keynote and invited talks, short course presentations, and a poster session and competition. The Program Committee would like to express its appreciation to all who participate to make the IMR a successful and enriching experience.

The papers in these proceedings were selected from among 46 submissions during the Program Committee meeting that took place in San Diego on April 29, 2005. Based on input from peer reviews, the committee selected these papers for their perceived quality, originality, and appropriateness to the theme of the International Meshing Roundtable. The Program Committee would like to thank all who submitted papers. We would also like to thank the colleagues who provided reviews of the submitted papers. The names of the reviewers are acknowledged on the previous pages.

As Program Chair, I would like to extend special thanks to the Program Committee and to the Conference Coordinators for their time and effort to make the 14th IMR another outstanding conference.

Byron Hanks
Sandia National Laboratories
14th IMR Chair

14th IMR Conference Organization

Committee:

Byron Hanks (Chair)
Sandia National Labs, Albuquerque, NM
bwhanks@sandia.gov

Tim Baker
Princeton University, Princeton, NJ
baker@tornado.princeton.edu

Steven Pav
University of California, San Diego
spav@ucsd.edu

Philippe Pebay
Sandia National Labs, Livermore, CA
pppebay@ca.sandia.gov

Krishnakumar Rajagopalan
Program Development Company, White Plains, NY
Krishna@gridpro.com

Alan Shih
University of Alabama at Birmingham, Birmingham, AL
ashih@uab.edu

Reza Taghavi
Simulation Works, St. Paul, MN
taghavi@siw.com

Coordinators:

Lynn Janik Washburn
Sandia National Labs, Albuquerque, NM
lajanik@sandia.gov

Aubrey Blacker
Sandia National Labs, Albuquerque, NM
aablack@sandia.gov

Web Designer:

Bernadette Watts
Sandia National Labs, Albuquerque, NM
bmwatts@sandia.gov

Web Site:

http://www.imr.sandia.gov

Contents

Session 1A

Session 1B

Design and Implementation of a Corporate Mesh Object

John T. Svitek, Wa Kwok, Joseph R. Tristano

{john.svitek,wa.kwok,joe.tristano}@ansys.com

ANSYS, Inc

ABSTRACT

Today, finite element technologies allow engineers to analyze complex assemblies and subsystems. With CPU power constantly increasing, it is not unreasonable to state that the engineer will hope to analyze the whole complex system, such as an entire automobile, in a single study. This study may include parameteric design, result animation, crash analysis, and so on. The traditional mesh data structure, which mainly serves a particular type of mesh algorithm, is far from enough to meet the challenges of tomorrow. This paper mainly focuses on storing, accessing, and manipulating mesh data within the vast scope of the analysis system for multiple purposes, such as meshing generators, solvers, pre- and post-processors, and so on. It details the decision-making put into the design of the ANSYS Corporate Mesh Object, the programming methods used to implement those designs, and future enhancements planned to meet ever-changing requirements.

1. INTRODUCTION

As machines have become more powerful and finite element analysis methods have become more elaborate, the amount of finite element data that must be managed and stored has grown to previously unimaginable levels. The vast amount of data produced by today's meshing technology needs to be readily available to several different users, accessing it in several different ways, all the while fitting in the small memory space afforded by a Windows-based PC. The data needs to be accessed quickly for graphical support, accurately for solving, but also easily for third-party users. In the face of these challenges, ANSYS Incorporated noticed the need to replace its existing (and very simplistic) mesh data structure with a much more flexible and powerful data structure, an ANSYS Corporate Mesh Object or ACMO.

1.1 Purpose and Previous Work

The main purpose of the ACMO as described in this paper is to act as a repository to store nodes and elements, whether created by a meshing algorithm or manually created by the user. This type of mesh object does not contain full mesh generation data such as the edge or face representations of each element, and the data is not expected to substantially change very often. Therefore, this object is not suited for use as a data structure for the actual meshing process, when nodes and elements are being generated, deleted, and changed continuously. Such data structures have been detailed extensively in the past, often coupled to a particular algorithm [4,5,6,7,8,10], or rigorously tailored to the output of a particular algorithm [3,9]. Rather, once a mesh has been successfully and completely created, the ACMO acts as a common location to store the data, and provides easy access to the data from a range of different users. Much previous research is available on the various ways of representing mesh data [1,2], though the ACMO attempts to also address the usability and accessibility of the data.

1.2 Background

In the ANSYS Workbench software, geometry is stored in a sophisticated object called the Part Manager (Fig. 1), which encapsulates the geometry data in easy to understand concepts such as "Assembly", "Part", "Body", "Face", "Edge", and "Vertex" as well as several other layers in between. An assembly is a collection of parts, which in turn is a collection of boundary representations. When a user asks Workbench to mesh geometry, each part in turn is meshed, and all these part meshes are then collected. In this way, a continuous mesh exists between bodies, but not between parts. Nodes are not duplicated at places where bodies share topology.

Unlike the Part Manager, the original mesh data structure used in the ANSYS Workbench software took a very simple approach to storing the nodes and elements created during meshing. After each part was meshed, the nodal coordinates were stored in an array, and the element's nodal connectivity was stored in an array, along with topological information. There was no real distinction between different bodies in the mesh, and one part mesh had no relationship whatsoever to another part mesh. Managing the various part meshes was the duty of the overall system; for example, a two part meshed assembly had two nodes numbered zero and it became the system's job to differentiate between those two nodes.

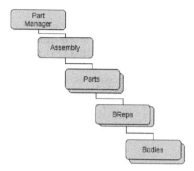

Fig. 1. Part Manager

1.3 Need for Change

Originally, the mesh data structure used in ANSYS Workbench was sufficient because there never was a part containing more than one body. The mesh was not exposed to many uses beyond graphics and solving. There was very little data actually stored in the mesh and any newly required data could be programmed in as necessary.

However, as the system around the mesh data structure matured, the mesh itself had trouble keeping up. The concept of a multiple-bodied part was introduced into the Part Manager, but the mesh was still stored part by part so that old code using the mesh would not need to be changed. The nodes and elements were stored exactly as before, so accessing them on a body-by-body basis became a time-consuming operation.

Furthermore, the mesh was not easily extensible. If a user wanted to store some data on each element, such as a value describing the volume of the element, a new array of values would need to be added to the data structure, along with all supporting implementation such as saving, resuming, copying, clearing, setting, and getting. If a user then wanted to store another value per element, such as a metric value, the entire process would have to be repeated.

As machines became more powerful, larger and larger meshes were being generated. Not long ago, one million nodes could be considered a large mesh, but a few short years later, ten million nodes were on the horizon. Such enormous amounts of data were putting strain on the mesh data structure, which did not have any kind of memory management system. Memory fragmentation became a very real and very serious concern as other uses of the same memory space (such as solvers) were unable to allocate enough large-block memory to operate.

New uses of the mesh were constantly being developed, and users often wanted to access the data in the mesh in a certain way. A user would need to query for an element based on its index on the part, or maybe its index on a particular body in that part. Further, the nodal connectivity of that element could be returned using nodal indices on either the part or the body. As a result, the user interface to the mesh became monstrous, so that every permutation of data access could be afforded to the user.

1.4 Future Challenges

With the need to integrate different mesh technologies into ANSYS Workbench, standardizing the mesh object became even more paramount. For example, ICEM CFD, Cadoe, and CFX all brought new technologies to the ANSYS product line, but they all brought different data representations that were not compatible with each other or with existing ANSYS software. In order to move data between the various technologies, a corporate mesh object that could be used by everyone was necessary.

2. DESIGN

2.1 Use Types

Before detailing the design process on the ACMO, it is important to illustrate the many different ways a single mesh could be used. (Fig. 2)

- **Mesher**: After the meshing process, the nodes and elements need to be stored into the mesh object. Meshing is done part-by-part, and one part has no bearing on or relationship to another part. Mesher only needs to add data to the mesh object.
- **Refiner**: An existing mesh can be refined one part at a time. The entire part mesh is removed from the assembly, and replaced with a new mesh after the refinement is finished. The refiner must be able to easily access all data in a part mesh, and to add data to the mesh object.
- **Graphics**: The mesh is drawn to the graphics screen body-by-body. In order to minimize its footprint in memory, graphics requires a reference to the mesh data, instead of a copy of the data. All nodes of a body are required, and nodal connectivity of elements must be returned such that it references the body-based indices of the nodes.
- **Preprocessor**: Data, such as beam orientation nodes and contact or surface effect elements are added by a separate preprocessor to the solver
- **Solver**: The solver requires all nodes and elements across the entire assembly. References to the data are necessary to minimize memory usage. All nodes and elements must be uniquely identified across all parts.
- **Postprocessor**: The post processor performs calculations on the mesh to report results on the mesh produced by the solver
- **Import/Export**: Specialty tools can build a mesh from legacy data such as NASTRAN, and convert legacy data between different representations. All nodes and elements must be uniquely identified. Data might be added to the mesh. The import/export operation exercises all possible permutations of data access. Speed, memory usage, and memory fragmentation are all very important.
- **SDK**: An openly published interface to the mesh object for third-party developers. Usability and a well-developed abstraction of the data are more important than speed or memory usage.

From this sampling of uses of the mesh, there are several clear general requirements for the ACMO. Mesh data must be accessed by a unique identifier, but also by a direct index to the data. For example, the first node in a mesh can be access by its index 0, or by a given identifier that is unique across the entire assembly. Also, some users of the mesh may access the node based on the part it exists on while another user may access it based on its location in the entire assembly. A node at index 0 on a certain part is not necessarily at index 0 on the assembly.

Also, different users of the mesh can use the mesh simultaneously. Therefore, the mesh object must be thread-safe since these users could exist on different threads. But even more challenging, one user could actually change the mesh while a different user is still using it. For example, the mesher could remesh (and therefore change) a part that had already been solved. The original solved part mesh can still exist in graphics and be available for perusal by the end user, and the newly meshed part must also be available. A single part would now exist twice in the assembly, and this condition would need to be automatically managed by the mesh object.

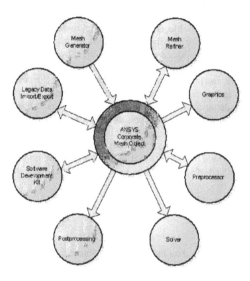

Fig. 2. Mesh Uses

2.2 Physical Design of the Mesh

With these cases in mind, the mesh object is designed to closely mirror the Part Manager in ANSYS Workbench, which manages geometry. An assembly mesh contains a collection of part meshes, each containing a collection of body meshes. Node and element identifiers are stored at the assembly, mapped to the physical location of those nodes and elements in the mesh. Nodal coordinates and topological information are stored on the part mesh because a node can be shared between multiple bodies. Elements are stored on the body mesh. To facilitate extensibility, an attribute mechanism that can attach an arbitrary data field to a node, an element, or an entire mesh is planned, so that any data that must be added to the mesh in the future could be done so easily.

A thin "configuration" layer between the assembly mesh and its collection of part meshes allows the assembly to change the collection of part meshes. By changing the assembly's configuration, different views of the same assembly are possible, without having to make an entire copy of the assembly.

2.3 Interface Design

Since the mesh will need to communicate across processes, a COM interface is necessary. The COM interface is a thin layer of code between the user and the core data object. However, the interface layer is completely separate from the actual core object and is not required for the actual mesh object to function. The internal core objects have no awareness of the COM interface layer; they are free to communicate with one another directly. In this way, the core objects can be used independent of the COM interface layer. This type of functionality facilitates unit-testing the mesh object, since it is trivial to write a driver program to create and operate the core objects without the general hassle of using COM.

Code for thread-safing the mesh object is placed in the COM object layer, which makes the actual implementation of the core objects more understandable and readable. Also, the COM object layer can be used to hide the internal storage of the data from the user to better abstract the data. For instance, there is no actual node object inside the mesh object. Rather, the concept of a node is made up from several data fields in the mesh object, such as the nodal coordinates stored in one array, the nodal topologies stored in another, and the elements, which use the node. A lightweight node object can be created in the COM interface layer which encapsulates all this data, allowing a very user-friendly view of the data to a third party. However, since the various bits of data that makes up the node must be collected and stored together, the advantage of user-friendliness comes at the expense of speed and memory usage.

2.4 Design Goals

The mesh object typically contains more data than any other object in a finite element analysis. With this in mind, the ACMO was written primarily to reduce memory fragmentation. Memory fragmentation can become a serious problem, which occurs when small chunks of memory are requested and then returned to the system. Over time, the largest free blocks of memory are repeatedly decimated in order to fulfill the user's request for smaller blocks, but these smaller blocks may not necessarily be rebuilt into larger blocks as they are returned to the system. Memory fragmentation can be minimized by preventing small memory allocations whenever possible using techniques such as pooling, where large amounts of small allocations are made as one large allocation. Pooling is used extensively in the ACMO.

Along with memory fragmentation, memory usage is a main concern when dealing with huge amounts of data. For example, on a 32-bit Windows PC, a single process is limited to 2-3 GB of memory, regardless of how much memory is actually available. At 10 million nodes, over 8% of usable memory may be utilized just to store the nodal coordinates, which is just a small fraction of all the data which is stored in the mesh, not to mention the memory requirements of other objects residing in the same memory space. To minimize memory usage, "lazy evaluations" are used throughout the ACMO. Lazily evaluated data is generated at the time of access, such as a node-to-element cross-reference, so it is not loaded into memory by default. Rather, the first time such data is requested by the user, it is calculated and stored into memory in its entirety. If the user never requests the data, it is never loaded into memory.

Finally, in some situations, the speed of accessing data is extremely important, even more so than memory issues. Any access to the mesh object during meshing would take only a minute fraction of the overall time of the meshing operation, and so the speed of the data access is not too important. On the other hand, when dealing with graphics, speed becomes a very real issue, whereby the access to the mesh data can be the limiting factor to the overall speed of drawing to the screen. Therefore, the data in the mesh is always organized in such a way to maximize speed when accessing it from the graphics system, even if it adversely affects speed when accessing the data from one of the users.

3. IMPLEMENTATION

3.1 Programming Conventions

As mentioned earlier, the mesh is comprised of a core group of objects and a COM interface layer. For instance, the top-level mesh object is known as an AnsAssemblyMesh, and a COM object known as an AssemblyMesh wraps it. The "Ans-" prefix identifies the object as an internal core object. Since the AssemblyMesh object does not have an "Ans-" prefix, it is understood to be the COM interface to an object named AnsAssemblyMesh.

3.2 Object Overview

The mesh is comprised internally of three main objects. The first, the AnsAssembly-Mesh, is a top-level object that contains all other objects. The middle-level object is the AnsPartMesh, and the lowest-level object is the AnsBodyMesh. An AnsAssemblyMesh contains a collection of AnsPartMeshes, which in turn contains a collection of AnsBody-Meshes. As mentioned earlier, each of these objects mirrors similar concepts in the ANSYS Part Manager, in which a geometric assembly is comprised of a collection of parts, and a part is made up of a collection of bodies. (Fig. 3)

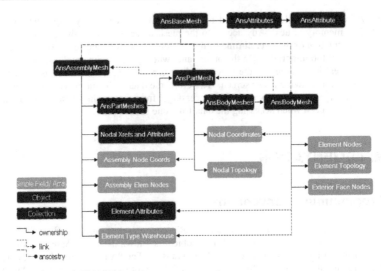

Fig. 3. Internal Object Model

Each of these core objects has a COM object that "wraps" it and acts as an interface. For each AnsAssemblyMesh, there exists a corresponding AssemblyMesh object. The COM object contains a reference to the core object. A user calls into the mesh object through these COM objects, which delegates the call into the core object. However, the core objects know nothing about the COM objects, and in fact can function without them. (Fig. 4)

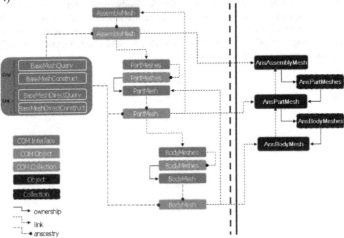

Fig. 4. COM Interface Model

The three mesh objects all derive from a single base mesh object, the AnsBaseMesh, which describes their common functionality. For example, a node can be accessed from a particular part, or from a body, or from the entire assembly, and so the AnsBaseMesh defines the ability to query for a node. Likewise, the three mesh interfaces all derive from a common interface, the BaseMesh. Since the functionality to query a node exists in the AnsBaseMesh, the corresponding interface to that functionality exists in the BaseMesh.

3.3 Identifiers

All data in the ACMO is given an identifier, which is either provided by the user or simply begins at 1 and accumulates as more data is added to the mesh. The identifiers are unique to a specific data type such that there can be a node identified as 1, and an element identified as 1, but never more than one node in the entire mesh identified as 1.

The identifiers are stored in the AnsAssemblyMesh in a grouped hash table, which maps the identifier to the data's physical location in the assembly and in its "group," usually the part in which the data exists. Entries in the grouped hash table are made in the form group g: [i, j, k], where i is the identifier, g is the index of the group in the table, j is the index of the identifier across all groups, and k is the index of the identifier in group g. For example, a typical entry in the hash table might look like "group 2: [1, 7, 5]," meaning identifier 1 is found at index 7 in the entire assembly, and at index 5 in its particular group. If this was a node, then the node identified as 1 is the eighth node in the assembly (located at index 7), and it is also the sixth node in the third part on the assembly. Using this design, if the groups of the hash table correspond to the parts in the assembly, the part meshes can be moved in the assembly mesh, and the hash table is easily manipulated to reflect the change.

When data is actually stored in the mesh object, a reference to the actual location of the data is used rather than the identifiers. For example, if an element were made up of nodes identified as 1, 2, and 3, which are located at indices 0, 1, and 2, then the element's nodal connectivity would be stored with 0, 1, and 2. This allows quick access to the nodal data without performing a lookup in the hash table.

3.4 Data Organization

Data stored in the mesh must be well organized in order to be accessed quickly with minimal memory overhead. Nodal coordinates and nodal topological information is stored in the part mesh, but it is ordered according to the node's classification and bodies. A node's classification is based on the elements that use the node. If a node only ever lies at the vertex of an element, it is considered a "corner" node and is placed at the beginning of the node list. Nodes that lie only on the edges of an element are considered "midside" nodes, and follow the corner nodes in the node list. Next stored in the node list are "multipurpose" nodes that lie on both vertices and edges of elements. Finally, "zombie" nodes, which are not used by any elements, are stored last. In this way, a user only interested in corner nodes can easily access the desired type without having to search through the entire node list.

Within the four categories of nodes, the node list is further organized based on the bodies that use them. Referred to as "exclusive" nodes, any nodes used only by elements that exist on the first body of the part mesh are stored at the beginning of the list, and so on. Any nodes shared by elements existing on more than one body are called "interface" nodes and are stored at the end of the node list following all exclusive nodes. (Fig. 5) Using this method, it is easy to determine the body on which an exclusive node exists, though an interface node necessarily requires more work.

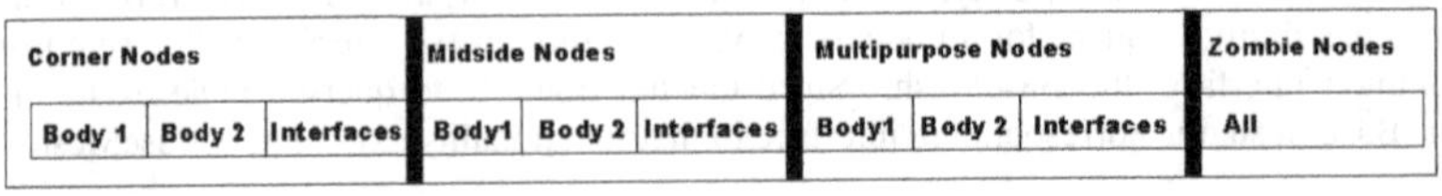

Corner Nodes			Midside Nodes			Multipurpose Nodes			Zombie Nodes
Body 1	Body 2	Interfaces	Body1	Body 2	Interfaces	Body1	Body 2	Interfaces	All

Fig. 5. Part-based Node Organization

Element connectivity and topological information is stored on the body. Elements of the same "type" are grouped together. An element type is based on the shape of an element, whether or not it has midnodes along its edges, and any arbitrary data assigned by the user. An element shape can be general or concrete. A concrete shape, such as a triangle or tetrahedron, has a known number of nodes making up the element and the number is always the same. In a general shaped element, the number of nodes is unknown; each element of that particular type is unique. By abstracting the concept of an element type into its own object, new types can easily be added to the mesh without additional programming.

Some nodes and elements transcend the boundaries of parts and bodies. For instance, a contact element may exist between two parts, and is used to identify the contact of the two parts to the solver. Rather than being generated during a normal mesh generation, a contact element is generally added to the mesh as a preprocess to the solver. Since the contact element does not exist on any one body, it is considered "independent," and is stored in assembly mesh, though it might reference non-independent nodes.

3.5 Memory Pools

Whenever possible, the mesh utilizes memory pools to combine many small allocations into one large allocation. As just stated, since the number of nodes in a concrete shaped element is known, the exact amount of memory required to store all element connectivity can be easily determined. For instance, when storing 1,000 tetrahedron, each with 20 nodes (which are referenced using one long each), allocating one long array of length 20,000 can prevent fragmentation of available memory that may occur by allocating 20 longs 1,000 times.

3.6 Decoupled Data Versus Committed Data

Organizing mesh data as described above can be an expensive operation, and in some cases, the organization is impossible until all data has been added to the mesh. Also, the exact length of memory pools cannot be determined until all data has already been put into the mesh. Therefore, data can exist in the mesh in two distinct states: decoupled or committed.

As data is added into the mesh, it is considered to be in a temporary "decoupled" state. When decoupled, the data is not organized in any way. Querying for decoupled data based on its unique identifier requires an expensive linear search through the mesh. However, the data is only stored in this state until the user is ready to "commit," or finalize all changes to the mesh. When the mesh object is committed, the data is sorted and moved to permanent memory pools. Also, the internal structure of the data is changed to afford the quickest possible access to the user. For instance, the unique identifiers of the nodes are stored to describe a decoupled element's nodal connectivity. However, in a committed element, the indices of those nodes are stored instead. Small changes to committed mesh data happen immediately, such as changing the value of a node's coordinates, but large changes, such as adding data or removing data, is deferred until the user commits the mesh, allowing the user to determine exactly when the speed hit caused by organizing the data occurs.

Storing data quickly and efficiently is the main goal of the decoupled state, rather than allowing quick access. Since the exact amount of data being added to the mesh may not be known, granularized memory pools can be utilized to store the data. The first time memory is requested to store data, enough memory is allocated to store many instances of the data, instead of just one instance. For example, the first time a user adds a node to the mesh, instead of allocating enough memory to store one node, enough memory to store one thousand nodes may be allocated. Once all of this memory has been used, enough memory for two thousand more nodes may be allocated. The exact granularities used are left to the user, allowing the memory allocation to be fine-tuned, and if the exact amount of data being added to the mesh is known in advance, then that amount can be used as the granularity, preventing any unnecessary memory reallocations.

Since data in a decoupled state is not tied to the assembly mesh (it does not use the assembly level hash tables to map identifiers to locations in the mesh), the data can be moved between assemblies. An entire part can be decoupled, basically a reverse-commit. The part can then be removed safely and easily from one assembly and added to another. Once in the other assembly, the data can be recommitted and reorganized as necessary, and the second assembly's hash tables would be updated with the new data's unique identifiers.

3.7 Filtered Data Access

The mesh object affords many different types of access to the same data through the same interfaces. To accomplish this, all method calls to the mesh object are filtered. A typical method declaration looks like this:

```
HRESULT GetNode( long     id,
                 ULONG    ulAppliedFilter,
                 ULONG    ulFilter0,
                 ULONG    ulFilter1,
                 _sNode *piNode );
```

where ULONG is an unsigned long, and _sNode is a structure containing nodal information such as coordinates.

The three ULONG filter values can affect both the input and output of the method. The ulAppliedFilter is made up by bitwise OR-ing the desired filters together. Each individual filter is simply an integer with exactly one bit set to one and all other bits set to zero. When two individual filters are bitwise OR-ed, the resultant integer has exactly two non-zero bits. Some filters, called unvalued filters, can work solely on their own and require no other input from the user. For example, the default behavior of the GetNode function treats the id input as the user-defined unique identifier of the node. By applying the DIRECT_ACCESS filter to the method call, the id input is instead treated as the index of the node at the level of the mesh from which it is queried. That is, if the GetNode call were made from assembly mesh filtered with DIRECT_ACCESS, then the id input would be treated as the index of the node on the assembly. If the same call were made from a part mesh, the id input would instead be the index of the node on that particular part. The same method call can behave differently based on the filters applied to it or based on the level of the mesh it is called from. However, if no filter were applied to the GetNode method, it would behave exactly the same no matter what level of the mesh it was called from, since the id input would be treated as a unique identifier, which is the same throughout the mesh.

Some filters require further input from the user, such as BY_PART. This filter, when used in conjunction with the DIRECT_ACCESS filter, instructs the method to treat the input as an index on the given part. In this case, the id of the desired part would be provided in the ulFilter0 field. An example use might be:

```
HREULT  hr  =  piMeshObject->GetNode(  5,  DIRECT_ACCESS  |
BY_PART, 3, 0, &iNode );
```

In this case, the user is querying for the node at index 5 on part 3. Since the user only used one valued filter (BY_PART), the ulFilter1 is unused. Any number of unvalued filters, and up to two valued filters, can be applied together to any particular method call.

Some examples of unvalued filters:

DIRECT_ACCESS	(1L<<0)
DIRECT_RETURN	(1L<<1)
NO_ALLOC	(1L<<2)
ASSEMBLY_ACCESS	(1L<<3)
ASSEMBLY_RETURN	(1L<<4)
PART_ACCESS	(1L<<5)
PART_RETURN	(1L<<6)
BODY_ACCESS	(1L<<7)
BODY_RETURN	(1L<<8)
FORCE_COPY	(1L<<9)
NO_CLEANUP	(1L<<9)
NO_MIDSIDES	(1L<<10)
NO_ERRORCHECK	(1L<<11)
EMPTY_RETURN	(1L<<12)
INDEPENDENT	(1L<<13)
DECOUPLED	(1L<<14)
DESCRIPTION_ACCESS	(1L<<15)

Some valued filters:

BY_PART	(1L<<24)
BY_BODY	(1L<<25)
BY_PID	(1L<<26)
BY_ATTRIBUTETYPE	(1L<<27)
BY_ATTRIBUTEVALUE	(1L<<28)
USE_OPTION	(1L<<29)

In these examples, the individual filters are created by shifting the non-zero bit by a unique amount. DIRECT_ACCESS does not shift the non-zero bit, so it remains as integer 1. PART_ACCESS shifts the non-zero bit 5 times to the left, and is binary 100000 or decimal 32. These two values can be bitwise OR-ed together (DIRECT_ACCESS | PART_ACCESS) to binary 100001, or decimal 33. Using this implementation, 32 unique filters can be created, and they can be used together in any combination.

Completely unfiltered methods to the mesh object also exist, mirroring each of the normal filtered methods. For example, the unfiltered version of GetNode looks like this:

```
HRESULT GetNodeEZ( long     id,
                   _sNode *piNode );
```

In this case, the input id is always treated as a unique identifier rather than an index, because that is the default behavior of the GetNode method. In this way, code that only uses the default behavior of the mesh is made much cleaner.

The concept of filtered methods works well, but can be improved upon. Most users of the mesh always use the exact same filters for all method calls. To accommodate, in the future, filtered methods will be removed in favor of filtered mesh objects, discussed below in Section 4.3.

3.8 Attribute Mechanism

Attributes are used to tag nodes, elements, or entire meshes with arbitrary data. They are also used to define element types. By utilizing an attribute mechanism, a user can easily add new data to the mesh without performing any additional programming. In the ACMO, an attribute is defined with an identifier, an optional description, and a variant data value. The description makes access easier to the end user (by asking for attribute "volume" rather than attribute id 4 for example) at the expense of some memory to store the string. In order to keep memory fragmentation in check, attributes attached to similar data are stored together in granularized pools, and share strings (rather than making multiple copies of the same string). For instance, all attributes connected to nodes on a particular part are stored together in one pool; all attributes connected to the actual part itself are stored in a separate pool.

To reduce database sizes, attributes can also be treated as transient data that is not saved. Rather, it needs to be lazily evaluated when required.

Some attributes without fail are tagged to every bit of data in a mesh. For instance, the user may apply a "volume" attribute to every element. In this case, a meta-attribute is used to further reduce memory usage. Rather than storing only one variant data value, a meta-attribute stores an array of such values, one for each possible instance of the attribute. The identifier of the attribute and a reference to its description only need to be stored one time for a meta-attribute, instead of multiple times for normal attributes.

3.9 Memory Management

Since the mesh can be accessed in many different ways through the same methods (using filters), the same method call can sometimes cause a memory allocation and other times not. Any time possible, the mesh returns references to its data, rather than copies of the data, though this behavior can also be overridden using the FORCE_COPY filter. For example, if a user calls GetNodes on a part without filters, the mesh can return a pointer directly to its nodal coordinates. If the same call is made using the NO_MIDSIDES filter to only return corner nodes, the mesh might allocate memory for the output. The mesh uses standard HRESULT return codes such as S_OK for success and E_FAIL for general failure, but also uses a custom code S_ALLOC to signify a success but with memory allocation that must be freed by the caller.

The S_ALLOC code (defined as 512) alerts the user that memory has been allocated, but it cannot tell the user exactly what memory must be freed. Some method calls can return several different outputs simultaneously, and not all of them may need to be freed. Also, some methods may return "nested" allocations, when an allocated array contains an allocated array, such as in the case of the GetElements method without filters. By default, the mesh always uses the global identifiers of nodes and elements, even though internally those nodes and elements may be stored by reference using indices. If a user calls GetElements, the mesh will return the nodal connectivity using nodal identifiers, which will require a memory allocation. In this case, the mesh must allocate an array to store each element, and a pool to store the nodal connectivity of each element.

To ease the burden on the user, a simple and very lightweight memory manager is in place to handle memory allocations. For each and every memory allocation made, a reference to the allocated memory is added to a stack. If a method allocates three arrays, three memory references are pushed onto the memory stack. After calling a method that allocated memory, the user simply needs to call ReturnFunctionMemory, and any memory allocated by the last function call will be freed. In no memory was allocated, the call will be ignored. If a user calls another method that allocates memory, any references to memory left on the memory stack will be removed and pushed onto a queue for later handling. Users can also request to have memory removed from the memory manager, in which case the user becomes responsible for freeing the memory when it is no longer needed.

3.10 Multiple Configurations

Many times, multiple users of the mesh require simultaneous but conflicting access to the same data. For example, in the Workbench software, the end user can solve a mesh and view the results even while remeshing the original mesh. In this case, the same mesh must exist in multiple states. In the ACMO, configurations allow different "views" of the same assembly. The assembly mesh contains a configuration manager that can construct the various views and switch between them. Each configuration contains a list of part "instances", which are references to actual part meshes. In this way, the same part mesh can be shared between multiple configurations of the mesh. When the configuration needs to change, the configuration manager retrieves the necessary part meshes for the assembly, and then the hash tables that map global identifiers for the nodes and elements are rebuilt.

For instance, if a three-part assembly is meshed and the solved, the assembly mesh will contain one configuration with three part meshes. If the user then changes the mesh on the second part, the assembly mesh does not immediately remove the original part mesh that was remeshed, since the end user can still view the results of the original solve. At this point, the assembly will contain two configurations, but four part meshes. One configuration has an instance of part one $P1^1$, an instance of part two $P2^1$, and an instance of part three $P3^1$. The second configuration has instances of the same parts one and three as in the first configuration, but it has an instance of a completely different part two $P2^2$. As the assembly needs to change its view of the data, these parts are switched in and out of the assembly as required. For example, if the user asks to view the results of the solve, then the assembly switches to configuration one. If the user asks to view the current mesh, the assembly switches to configuration two. (Fig. 6) If the user resolves the mesh at this point, the original configuration is no longer required and is deleted. Since one part mesh no longer has any instances in the configuration manager, it too is deleted. The ACMO relies on accurate reference counting in the COM interface layer to determine when deletion can safely take place.

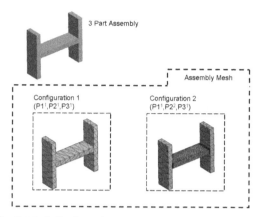

Fig. 6. Mesh Configurations

3.11 Thread-Safing

A final layer of complexity lies above the configuration manager: multiple threads accessing the mesh data simultaneously. A meshing process can be writing mesh data into one configuration of the object while a user is viewing another configuration. In the midst of the writing process, the user can change the configuration at will, and all the while the ACMO must be protected from access conflicts that can occur if the same memory is being read and written at the same time, not to mention that the physical structure of the mesh is changing while data access is taking place. To prevent such a catastrophe, the mesh object uses a simple technique utilizing a variation of semaphores called a critical section. A critical section prevents a thread from accessing an object when another thread has requested exclusive access to the same object. The mesh uses an extremely lightweight object that locks the critical section when the object is created, and releases the lock when the object is destroyed. In this way, it becomes a simple matter to create an instance of this critical section object at the beginning of every call into the mesh object in the COM layer. When the call finishes, the object is automatically released from memory and the critical section is unlocked. However, when using this method of thread-safing, only one operation can be performed on the mesh at any given time, so during a long operation such as committing mesh data, access to the mesh by other threads will be completely blocked until the operation finishes. The advantages of this method, such as ease of programming and inherent stability, offset the disadvantage of one thread needing to wait for another thread to completely finish accessing the data.

Since threads may be using different configurations in the ACMO, the mesh stores a map of thread ids to configuration ids, and every call into the mesh checks the calling thread immediately after creating the critical section. If the thread has changed since the last access to the mesh, then the current configuration is automatically changed. The calling system only needs to set the required mesh configuration once instead of setting it before every single call into the mesh object.

As an example, consider the situation where a user is refining a mesh that has already been solved. The refinement process starts on a second thread, and a new configuration is created in the ACMO that is tied to the refinement thread. While refinement runs, the primary GUI thread can also access the mesh object. If the user requests to view the results of the solve, the GUI thread will try to call into the mesh object, which will first attempt to lock the critical section. If the refinement thread is currently accessing the mesh, the GUI thread is forced to wait until the method call on the refinement thread finishes. At that time, the GUI thread can lock the critical section, forcing the refinement thread to wait if it attempts to call into the ACMO. Next, the mesh object checks the thread id against the last stored thread id. Since the id of the GUI thread differs from the stored thread (because the last access to the mesh object was made by the refinement thread), the ACMO changes the configuration of the mesh automatically. The GUI's method call is then allowed to finish. When the method call completes, the critical section wrapper is destroyed, and the actual critical section is automatically unlocked. If the refinement thread then accesses the ACMO, the entire process is repeated. (Fig. 7)

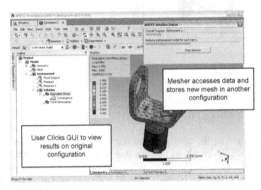

Fig. 7. Multiple Thread Access

4. FUTURE WORK AND ENHANCEMENTS

4.1 Part Spooling

As meshes grow ever larger in size, new ways of reducing memory usage must be pursued. The single largest memory gain may be found in the complete removal of parts from memory. A part that is not instanced by the current configuration of the assembly mesh can be spooled off to disk and freed from memory. The part can remain on disk until such time as it is needed. The configuration manager would then read the part from the disk back into memory. Though the time to switch between configurations would grow, the memory savings could be substantial. Along the same line, the implementation of memory map storage could also benefit the ACMO. A memory map acts like normal memory, except that it is written to disk instead of actual RAM. Access to the data is slowed considerably, but the fact that the disk is being used instead of physical RAM is transparent to the user.

4.2 Part Ghosting

Oftentimes, the meshes of several parts are exact copies of one another, but with a transform (a translation or rotation) applied. By storing the data only once and applying the transform when the data is accessed, memory usage can be greatly reduced at the expense of speed.

4.3 Filtered Interfaces to the Mesh

As mentioned briefly, many users always require the same type of access to the mesh object. The user may always use part-based indices when accessing nodes for instance. Instead of being burdened with always remembering to use the right combination of filters and always calling methods from the right level of the mesh (be it assembly, part, or body), a user can create a new interface to the mesh data which always acts the same way. As preliminary work, a FilteredBodyMesh interface to the mesh has been created. A user can create a FilteredBodyMesh from a normal BodyMesh interface to the mesh. The user can then apply various filters to the interface, which will always be respected when accessing the mesh data. This approach leads to much more understandable and readable code than the filtered methods currently implemented in the mesh, as well as allowing much more complex types of filtering.

4.4 Iterators

Iterator access may be the best solution to the memory management problem in the mesh, by removing memory allocations all together. When the user first accesses data in the mesh, instead of retrieving the actual data, the mesh would return an iterator, which points to the first instance of the data. Once the user queries the data being pointed to by the iterator, the iterator points to the next instance of the data. For example, if the user queries the mesh for the nodes lying along an edge, instead of building an array containing that information, the mesh would return an iterator pointing at the first node on the edge. Once the user has finished querying that node, the iterator moves on and points to the next node along the edge.

As a compromise between iterator access and array access to the mesh data, a bucket iterator can be used to retrieve the data in chunks, the size of which is controlled by the user. As in the previous example, if the user accesses nodes along an edge using an iterator with a bucket sized for three nodes, then the first three nodes along the edge would be retrieved and stored in the iterator. Once the user has queried those three nodes, then the next three nodes on the edge would be stored in the iterator.

5. CONCLUSION

As meshes become larger and the systems surrounding mesh data structures become more diverse and complicated, the ability to efficiently store and easily access mesh data becomes more and more vital. The ACMO strikes a reasonable balance between several conflicting priorities, such as speed of data access, memory usage, and memory fragmentation, by identifying the many users of the data and how those users interplay with one another. Further, the ACMO allows users to query for data in many ways, but through the exact same interface. Also, the ACMO is designed to be easily expanded without additional programming.

REFERENCES

1. Beall, Mark W, Shephard, Mark S "A General Topology-Based Mesh Data Structure", International Journal for Numerical Methods in Engineering, John Wiley & Sons, Ltd., Vol 40, Num 9, pp.1573-1596, May 1997

2. Garimella, Rao V "Mesh data structure selection for mesh generation and FEA applications", International Journal for Numerical Methods in Engineering, John Wiley & Sons, Ltd., Vol 55, Num 4, pp.451 - 478, October 2002

3. George, Paul-Louis and Houman Borouchaki "Delaunay Triangulation and Meshing: Application to Finite Elements", Hermes, pp.311-315, 1998

4. Hitchsfeld, N "Algorithms and Data Structures for Handling a Fully Flexible Refinement Approach in Mesh Generation", Proceedings, 4th International Meshing Roundtable, Sandia National Laboratories, pp.265-276, October 1995

5. Jia, Xiangmin, Herbert Edelsbrunner and Michael T Heath "Mesh Association: Formulation and Algorithms", Proceedings, 8th International Meshing Round-table, South Lake Tahoe, CA, U.S.A., pp.75-82, October 1999

6. Karamete, B K, T Tokdemir and M Ger "Unstructured Grid Generation and A Simple Triangulation Algorithm For Arbitrary 2-D Geometries Using Object Oriented Programming", International Journal for Numerical Methods in En-gineering, Wiley, Vol 40, pp.251-268, 1977

7. Kwok, W, K Haghighi and E Kang "An Efficient Data Structure for the Ad-vancing Front Triangular Mesh Generation Technique", Communications in Numerical Methods in Engineering, Vol 11, pp.465-473, 1995

8. Mobley, A, J Tristano, and C Hawkins "An Object Oriented Design for Mesh Generation and Operation Algorithms", Proceedings, 10th International Mesh-ing Roundtable, Newport Beach, CA, U.S.A., pp.179-183, October 2001

9. Noel, F J C Leon and P Trompette "A Data Structure Dedicated to an Integrated Free-form Surface Environment", Computers and Structures, Pergammon, Vol 57, Num 2, pp.345-355, 1995

10. Remacle, J F , B Karamete, M Shepard "Algorithm Oriented Mesh Database", Proceedings, 9th International Meshing Roundtable, Sandia National Labora-tories, pp.349-359, October 2000

Interface Reconstruction in Multi-fluid, Multi-phase Flow Simulations

Rao V. Garimella[1], Vadim Dyadechko[2], Blair K. Swartz[3], and Mikhail J. Shashkov[4]

[1] Los Alamos National Laboratory, Los Alamos, NM 87545 rao@lanl.gov
[2] Los Alamos National Laboratory, Los Alamos, NM 87545 vdyadechko@lanl.gov
[3] Los Alamos National Laboratory, Los Alamos, NM 87545 bks@lanl.gov
[4] Los Alamos National Laboratory, Los Alamos, NM 87545 shashkov@lanl.gov

Summary. An advanced Volume-of-Fluid or VOF procedure for locally conservative reconstruction of multi-material interfaces based on volume fraction information in cells of an unstructured mesh is presented in this paper. The procedure employs improved neighbor definitions and topological consistency checks of the interface for computing a more accurate interface approximation. Comparison with previously published results for test problems involving severe deformation of the materials (such as vortex-in-a-box problem) show that this procedure produces more accurate results and reduces the "numerical surface tension" typically seen in VOF methods.

1 Introduction

Hydrodynamic simulations of flows involving multiple fluids and/or multiple phases are an important research area with many applications such as droplet deposition, sandwich molding processes, underwater explosions, mold-filling in casting, simulations of micro-jetting devices, etc.

A very important feature of multi-fluid, multi-phase flow simulations is the interface between materials and phases, and it is often crucial to follow such interfaces at each step of the simulation. Lagrangian simulations (where the mesh deforms with the material) automatically maintain interfaces, but fail if the mesh is excessively deformed and the interface topology changes. On the other hand, Eulerian simulations (where the material moves through a stationary mesh) but often require special procedures to keep track of the interfaces in the flow.

In general, there are three broad categories of methods to track interfaces in hydrodynamic simulations – front tracking[GLIM98a], level set methods[OSHE01a, SUSS98a] and interface reconstruction [YOUN82a, RIDE98a]. Front tracking methods advect marker points on an initial interface with the flow so that a continuous, piecewise smooth interface approximation is known at each time step. In general, in this method, the global topology of the interface is fixed at the initial and not changed during the simulation. This is obviously disadvantageous for flows in which materials coalesce or fragment. Level set methods model the interface as the zero contour line of a distance function from mesh points to the interface. Level set methods model complex interface topology relatively easily but, in general, do not conserve material volumes very well. Volume tracking or Volume-of-fluid (VOF) methods compute an interface approximation at each time step using volume fraction information in mesh cell. Volume tracking methods can be accurate but the local topology of the interface in a cell is quite limited (such as a single line segment) making it difficult to capture subcell details such as thin filaments.

In this paper, an improved volume tracking technique is described for reconstructing complex material interfaces in unstructured meshes. The method currently allows one arbitrarily aligned linear interface segment per cell and thereby belongs to a venerable class of interface approximations that goes back at least to Debar [DEBA74a] and Youngs [YOUN82a, YOUN84a]. More recently, this class of interface approximations have been referred to as PLIC or Piecewise Linear Interface Calculation by Rider and Kothe [RIDE98a]. The method described here incorporates several new techniques designed to make the reconstruction method more accurate (generally second-order), rapid, and robust. These include the careful selection and use of interface neighbor cells, and a topological consistency checking and repair algorithm for the interface that is designed to minimize fragmentation of the material regions being reconstructed.

2 Interface Reconstruction Procedure

2.1 Overview

The primary problem being addressed in this research is the reconstruction of a piecewise linear interface given the volume fractions of different materials in the cells of an unstructured mesh. Cells filled with only one material are referred to as *pure* cells, and cells with some amount of two or more materials are referred to as *mixed* cells. Each mesh cell may contain at most one linear segment representing the interface. The reconstruction procedure must conserve the different cellular material volumes, make the interface as continuous as possible, and avoid non-physical holes and fragments in the material regions.

The main steps of the interface reconstruction procedure are:

1. *Interface Estimation:* A rough estimate of the interface is constructed using the volume fraction data specified on cells.

2. *Interface Smoothing:* Interface segments are adjusted taking into account other interface segments in the neighborhood so that the resulting interface is as smooth as possible. For example, straight line interfaces are typically recovered by this interface smoothing step.

3. *Interface Topology Repair:* The interface segments are adjusted so that they satisfy topological consistency checks (described below). This operation tries to ensure that the reconstructed material regions are continuous and do not have holes or fragments where the physics does not dictate it.

4. *Constrained Interface Smoothing:* The alterations made to the interface in the repair step are smoothed, with the constraint that topological consistency of any vertex cannot be destroyed.

2.2 Interface Estimation

Material interfaces are general curves in 2D and general surfaces in 3D. The interface normal with respect to a material is defined as the outward normal to the interface between this and another material. If an estimate of the interface normal is available at some point in the cell, then the interface can be approximated as a line segment perpendicular to this normal, properly positioned to cut off the right amount of material in the cell.

Like most volume tracking methods, this research assumes that there is a continuous and differentiable function whose values are given as volume fraction data at the centroid of each cell. This function is often referred to as a volume fraction function[5]. Then, the gradient of this function is approximated in a mixed cell using the discrete values of the function in the neighborhood of the cell using Green's theorem in the plane [6] (cf. [SHAS96a]). The estimate for the interface normal is then taken as the negative of the gradient of the function.

Once an interface normal estimate is obtained in a cell, the interface approximation in that cell is simply a straight line perpendicular to the normal. This line segment, however, must still be positioned correctly in the cell to subdivide it in the specified proportions. Fortunately, this is an easy problem involving quadratic splines lending itself nicely to a closed form solution [SWAR89a].

This estimate of the interface normal is quite crude due to the imprecisely defined nature of the volume fraction function, because of the character of its rapid local variability (e.g., the function steepens as the mesh size decreases), and because the data around the mixed cell is often too sparse to give an accurate estimate (there are typically far more pure cells than mixed cells, particularly in the normal direction). Therefore, this calculation rarely recovers the exact interface, even if the interface is just a straight line. Although some researchers have advocated concentrating on this step to make it as accurate as possible, that is not the approach taken here. Instead, the interface normal calculation from this step is only taken to be a rough estimate that is improved by a procedure referred to in this research as interface smoothing (described in the following section).

2.3 Interface Smoothing

Interface smoothing is defined in this paper as the process of making the piecewise linear reconstruction of the interface, with one line segment per cell, as accurate and smooth as possible. This means that the procedure tries to minimize the change in

---

[5] Most articles on this subject have not clarified what exactly this function means; it is often assumed to be a discontinuous function specific to each material that is unity on one side of the interface and null on the other. In this case, it is also unclear what is meant by taking the gradient of such a function. The notion used here is that the volume fraction function is akin to a color function that indicates the average color in a small window that is moved around the mesh particularly in a normal direction to the interface. Such a function is still an imprecise function which varies with the shape and orientation of the window, and steepens with decreasing size of the window. See [SWAR89a] for a detailed exposition of this topic.

[6] Some alternate methods such as the least squares construction of Barth [BART89a] are presented in [RIDE98a, KOTH96a].

interface normals from one mixed cell to the next. Such a procedure would recover the exact interface if the volume fraction data came from straight line material boundary.

The interface smoothing procedure in this paper derives from work by Mosso and Swartz [MOSS97a] in which they described an iterative interface smoothing algorithm based on finding the straight line that subdivides two non-overlapping cells according to an arbitrary volume fraction specified in each cell. The uniqueness and existence of such a line (and a plane in 3D) are discussed in detail in an earlier paper by Swartz [SWAR89a]. The Mosso-Swartz procedure recovers a straight line interface in a general unstructured mesh. Second order accuracy for curved interfaces is expected with their algorithm except in unusual geometric circumstances – for example, one should avoid just barely cutting off the corner of one or more cells. Rider and Kothe [RIDE98a] also propose an algorithm for interface smoothing based on the kernel of the Swartz's work [SWAR89a], and Swartz and Mosso's work [MOSS97a].

More specifically, consider two cells in a two material problem for which the volume fractions of the materials and an initial guess for the interface segments are known in each cell. Then the (quadratically convergent) iterative algorithm to find the common line that subdivides the two cells in the right proportions, as described in [SWAR89a], is as follows:

- Connect the centers of the interface segments in the two cells by a common line segment.
- Use the perpendicular to this line as the estimate for the pairwise interface normal for both cells.
- Move the interface segment in this normal direction in each cell to rematch the volume fraction associated with that cell.
- Repeat until convergence.

The above procedure forms the basic building block for the interface smoothing algorithm of this paper. The interface smoothing procedure visits each mixed cell in the mesh and carries out the following steps:

- Obtain the mixed cell neighbors of the cell.
- For each neighbor, find the normal of the common line that subdivides the cell and the neighbor according to the specified proportions.
- Perform a weighted average of all such pairwise normals to obtain the smoothed interface normal for the cell.
- Adjust the position of the interface so that the cell is subdivided according to specified proportions.

If, for each cell, the weighting of the normals from its neighboring cells is the same, then no iteration is needed at the global level. In this implementation, however, the common line normal associated with a cell and a neighbor of the cell gets a higher weighting if the interface segments that currently exist in the two cells are probable continuations of each other. This makes the smoothing process order-dependent in some situations, and therefore, a few (2-3) global iterations may be necessary for convergence.

It is useful to point out a key difference between this interface reconstruction procedure and the Mosso-Swartz procedure [MOSS97a]. In the current procedure,

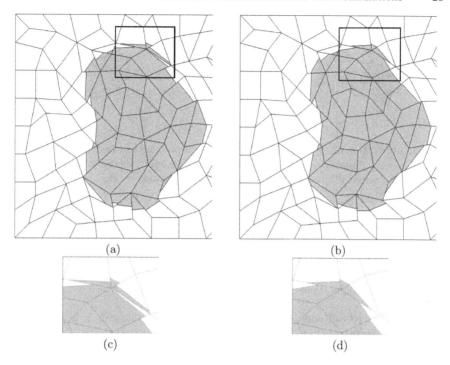

Fig. 1. Comparison of the reconstructed interface after (a) gradient estimation and (b) interface smoothing. Figures (c) and (d) show zoomed-in views of the interface in (a) and (b) respectively.

the pairwise normal of a given cell and each of its neighbors is iterated to convergence. Then, all such pairwise normals are averaged to give a final estimate for the interface normal in that cell. Global iterations are needed only to stabilize some changes as explained above. In the Mosso-Swartz algorithm, one iterative step is performed in a given mixed cell to improve the pairwise normal of the cell and each of its neighbors. These normals are averaged to get an improved estimate for the interface normal in that cell. Then the algorithm moves on to process another mixed cell. Therefore, it must reiterate this procedure over the whole interface until it converge to a final solution. This difference may account for our numerical observation that the current interface reconstruction procedure is quadratically convergent while the Mosso-Swartz algorithm is linearly convergent.

2.4 Interface Topology Repair

Consider a hypothetical procedure that recovers the exact interface curves on a given mesh. In such a case, it is evident that every vertex in the mesh will lie inside some material region or exactly on a material interface (considering external boundaries to be interfaces between a material and unmodeled space). This condition is referred to here as *topological consistency* of the interface. Even if the interface reconstruc-

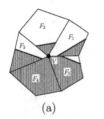 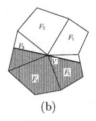 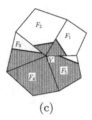

(a)   (b)   (c)

Fig. 2. Illustration of material classification and topological consistency for vertices with respect to interface reconstruction (a) Topologically inconsistent vertex (b) Vertex on a material interface (c) Topologically consistent vertex.

tion is approximate, it is reasonable to impose the topological consistency condition on it. Examples shown later demonstrate that imposing topological consistency on the interface results in material regions whose parts are more contiguous. This is invaluable in avoiding non-physical fragments and holes in the interface reconstruction process.

Given a mixed cell *and* an interface approximation in that cell, it is possible to classify each vertex of that cell as belonging to a particular material contained in that cell or as being on the interface between two or more materials. If the cell is a pure cell, then all of its vertices are classified with respect to that cell as belonging to the only material that the cell contains. From the vertex point of view, one can say that given a mesh and an approximate interface, every mesh vertex has a material classification with respect to each cell that uses this vertex, with the exception of cells where the vertex is on the interface.

For example, in Figure 2a, vertex V is considered to be "grey" (i.e. in the grey material) with respect to cells F_2, F_4 and F_5, but "white" with respect to cells F_1, F_2. Likewise, in Figure 2b, vertex V is considered to be "grey" with respect to cell F_4, "white" with respect to cells F_1 and F_2, and on the interface with respect to cells F_3 and F_5.

Definition 1. *Topological Consistency of Vertex*

A mesh vertex is considered to be topologically consistent in the context of interface reconstruction, if the vertex is classified as being in the same material with respect to all cells connected to it.

Using the above definition, vertex V is not topologically consistent in Figure 2a, is on the interface in Figure 2b, and is consistent in Figures 2c.

Definition 2. *Topological Consistency of an Interface Approximation*

An approximately reconstructed interface is said to be topologically consistent if every vertex in the mesh is topologically consistent or is on the interface.

Interface topology repair is then defined as the process of making an interface approximation topologically consistent.

The procedure used in this research for interface topological repair visits each mesh vertex, V, and checks its topological consistency. If the vertex is not topologically consistent and is not classified as being on the interface, then the procedure tries to make the vertex topologically consistent as described next:

- The cells, $\{C_i, i = 1, N_c\}$, connected to V are retrieved.
- The materials, $m_j, j = 1, N_m$, present in the cells, $\{C_i, i = 1, N_c\}$.
- For each material, m_j, the number of cells, N_j, in which V belongs to m_j is found.
- The materials are arranged in descending order of N_j.
- The most populous material around the vertex — the material with the maximum N_j — is chosen as the primary material. Cells in which V does not belong to the primary material are labeled secondary cells.
- The interface is adjusted in each mixed secondary cell so that the material classification of the vertex V in the cell changes to the primary material. This done by incrementally rotating the interface normal, readjusting its position to capture the right volume fraction and checking the material classification of V in that cell. Both positive and negative increments from 0 to 180 degrees are tried simultaneously to speed up the process.
- If all the secondary cells could not be repaired in this fashion, then the procedure is repeated with the next most populous material as the primary material.

The interface topological repair process modifies interface normals only as much as necessary to fix topological inconsistencies. Therefore, there may be some modified segments of the interface whose normals do not vary very smoothly with respect to their neighbors. A constrained interface smoothing procedure is reapplied to the interface to fix this problem. Unlike the first smoothing step, the constrained smoothing step takes care not to destroy the topological consistency of any vertex.

3 Tests

In this section, some test results are presented to demonstrate the accuracy and robustness of the interface reconstruction procedure.

First, results of interface reconstruction on some static examples are presented. In these examples, known analytical shapes are used to create volume fraction data in polygonal meshes and the interface is reconstructed.

Next, results of interface reconstruction in a dynamic setting similar to a fluid flow problem are presented. In these dynamic examples, a known shape is superimposed on a mesh to get the initial volume fraction data which is then advected using an analytical velocity field. The interface is reconstructed at each time step, and the quality of the interface is evaluated at the final step.

3.1 Static Tests

The first example, shown in Figure 6a, is a simple test to ensure that interface reconstruction procedure recovers a straight line exactly on a general unstructured mesh. Figure 6b shows the reconstruction of a circular interface on an unstructured triangular/quadrilateral mesh.

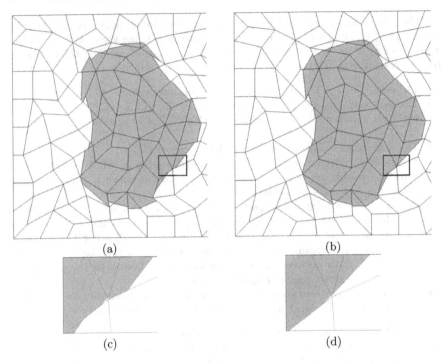

Fig. 3. Comparison of the reconstructed interface after (a) interface smoothing and (b) interface topology repair. Figures (c) and (d) show zoomed-in views of the interface in (a) and (b) respectively.

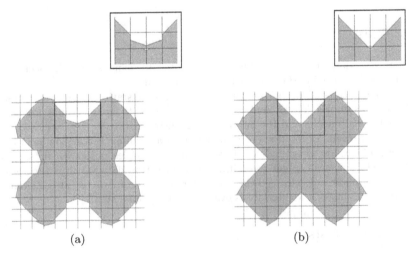

Fig. 4. Example showing elimination of gaps via interface topology repair.

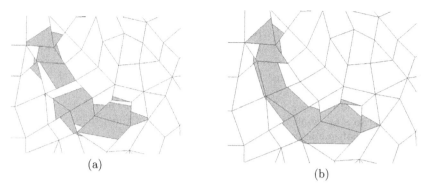

Fig. 5. Example showing reduced fragmentation via interface topology repair.

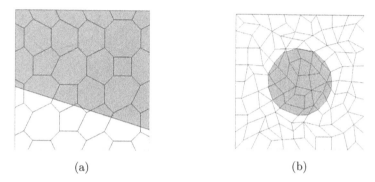

Fig. 6. Static Interface Reconstruction Tests: (a) Reconstruction of a straight line on an unstructured polygonal mesh (b) Reconstruction of a circular interface on an unstructured triangular/quadrilateral mesh

3.2 Advection

The advection scheme described below is primarily developed for the incompressible two-phase fluid flow. In this case, it is sufficient to follow the evolution of single material further referred to as a *reference* one.

It is assumed that a *solenoidal* velocity field is given analytically and no restriction is placed on the Courant number $CFL = v\,\Delta t/h$, where Δt is a time step, h is local mesh spacing, and v is the flow speed.

3.3 Advection of volumes

Given reference volume fractions in mesh cells at $t_{k-1} = \Delta t\,(k-1)$, one can calculate the reference volume volumes at $t_k = \Delta t\,k$ as described below.

First, the vertices of a polygonal cell C_i are tracked back along the streamlines by means of the 4-th order Runge-Kutta scheme and then connected by the straight segments. This results in a polygon that is considered to be the *Lagrangian prototype* $C_{i,k-1}$ of cell C_i (see Fig. 7).

Using a polygon intersection routine[7], we find the intersections of $C_{i,k-1}$ with all covered cells at $t = t_{k-1}$. By partitioning the bounding box of the entire computational domain into 2D array of rectangular bins and presorting all the mesh cells among these bins based on their centroid location, one can significantly accelerate the search of the cells covered by the prototype. Whenever $CFL \leqslant 1$, it is sufficient to intersect the prototype only with C_i and its direct neighbours. Moreover, if $CFL \leqslant 1$ and C_i along with all its neighbours is empty at $t = t_{k-1}$, then it is guaranteed to stay empty the next discrete moment of time; there is no need to perform any polygon intersection.

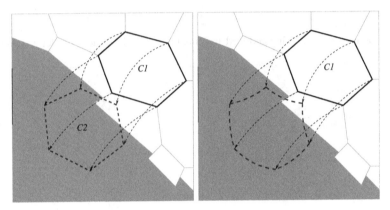

Fig. 7. Pseudo-Lagrangian (left) and the true Lagrangian prototypes of a cell.

In a *linear* velocity field, which possess the property of preserving straight lines, the algorithm employed for tracking cells back in time results in *true Lagrangian prototypes*. The local area defect $|C_{i,k-1}| - |C_i|$, introduced in this case by approximate integration, is $O(h\,\Delta t^{p+1})$ for a p-th order scheme.

In general, a solenoidal velocity field does not preserve straight lines and the Lagrangian prototype of a polygonal cell is not guaranteed to be a polygon. Therefore, the polygon $C_{i,k}$ is not the *true Lagrangian prototype* of C_i and introduces a local area defect of $O(h^3 \Delta t)$ due to the fact that we ignore the curvature of the prototype edges.

Any significant local area defect may eventually cause the advection algorithm to halt. Indeed, if the volume of the reference phase $\tilde{m}_{i,k-1}$ enclosed in the prototype $C_{i,k-1}$ exceeds the cell capacity $|C_i|$, we are in trouble. Another, less critical situation occurs when the prototype, being filled with the reference material, happens to have the area $|C_{i,k-1}|$ smaller than $|C_i|$. In this case the cell becomes mixed, even though its prototype was *pure* (contained only one fluid phase).

In order to fix these flaws we use a post-advection *repair* procedure. For every cell C_i, $i = \overline{1, N}$ we specify the lower and the upper bounds $\underline{v}_{i,k}$, $\overline{v}_{i,k}$ of the reference fraction volume $v_{i,k}$ allowed:

---

[7]Due to the COnservative REmapper (CORE) library by M. Staley [STAL04].

$$\begin{aligned}
\underline{v}_{i,k} &= & \overline{v}_{i,k} = 0 & \quad \text{if the prototype is empty,} \\
\underline{v}_{i,k} &= & \overline{v}_{i,k} = |C_i| & \quad \text{if the prototype is full,} \\
\underline{v}_{i,k} &= 0, & \overline{v}_{i,k} = |C_i| & \quad \text{otherwise,}
\end{aligned}$$

and then force each volume $v_{i,k}$, $i = \overline{1,N}$ to fit in:

for each cell C_i, $i = \overline{1,N}$ **do**
    if C_i is *overfilled* $(\overline{v}_{i,k} < v_{i,k})$ **then**
        try to redistribute the excess between the *non-overfilled* neighbors
        while there are still some leftovers **do**
            redistribute them among the next layer of the surrounding cells
        end do
    else if C_i is *underfilled* $(v_{i,k} < \underline{v}_{i,k})$ **then**
        try to borrow the lack from the *non-underfilled* neighbors
        while there is still some lack of material **do**
            borrow it from the next layer of the surrounding cells
        end do
    end if
end do

Due to the local nature of the volume defect, the redistribution usually involves only direct neighbours of the cell. Therefore, the complexity of the whole repair step comes to the total of $O(N)$.

3.4 Dynamic Tests

Rider and Kothe argued in their paper [RIDE98a] that simple translation and rotation tests are inadequate to predict the performance of interface reconstruction algorithms in real flow simulations which often involve stretching and shearing of materials. They proposed a test with an analytical velocity field (based on [BELL89a] and [LEVE96a]) which sets up a single vortex that stretches a circular region into a spiral with several full rotations and reverses back to a circle. The circle is of radius 0.15, placed at (0.5,0.75) in a unit square whose lower left corner is at (0.0,0.0). The two indicators used to evaluate the performance of the interface reconstruction algorithm are how well the structure of the spiral is maintained at the point of maximum stretching, and how close to the original shape the material is, when the flow is fully reversed.

The forward motion of the vortex is defined by the stream function:

$$\Psi = \frac{1}{\pi} sin^2(\pi x) sin^2(\pi y), \tag{1}$$

from which we get the divergence free velocity field

$$u_x = -\frac{\partial \Psi}{\partial y} = -sin^2(\pi x) sin(2\pi y) \tag{2}$$

$$u_y = \frac{\partial \Psi}{\partial x} = sin(2\pi x) sin^2(\pi y) \tag{3}$$

This velocity is reversed in time by multiplying the components with the term $cos(\pi t/T)$, where $T/2$ is the point of maximum stretching [RIDE98a, LEVE96a].

Figure 8 shows the results of the vortex test for three successively finer meshes at the point of maximum stretch and after full reversal of the flow. As earlier researchers have observed, volume tracking methods tend to exhibit break-up of the tail of the vortex if the mesh is not sufficiently refined. This is evident in the 32x32 rectangular mesh examples but largely disappears as the mesh is refined to 128x128 cells. The primary reason for this behaviour is that the method assumes that one line segment per cell is sufficient to represent the interface whereas the procedure generally needs at least two disjoint line segments per cell to represent the vortex tail as it thins out in a coarse mesh. Failure to capture this structure leads to ever accumulating errors in the interface approximation (but not volume conservation), and to eventual disintegration of the vortex structure. It must be noted that, in spite of some break-up of the vortex tail, the results presented here show less fragmentation than those presented in some earlier works ([RIDE98a],[ENRI02a]).

4 Conclusions

A volume tracking interface reconstruction for multi-material, multi-phase hydrody-namic simulations was presented in this paper. The procedure incorporates several new techniques to make the reconstructed interface more accurate, smooth and consistent. Results of simulations indicate that the procedure performs better than previous techniques on difficult problems. Extensions to handle more than two ma-terials per cells are planned.

5 Acknowledgements

This research was performed at Los Alamos National Laboratory operated by the University of California for the US Department of Energy under contract W-7405-ENG-36. The authors also acknowledge the support of the ASC program at Los Alamos National Laboratory. Los Alamos National Laboratory strongly supports academic freedom and a researcher's right to publish; as an institution, however, the Laboratory does not endorse the viewpoint of a publication or guarantee its technical correctness. The submitted manuscript has been authored by a contractor of the United States Government under contract. Accordingly the United States Government retains a non-exclusive, royalty-free license to publish or reproduce the published form of this contribution, or allow others to do so, for United States Government purposes.

[GLIM98a] J. Glimm, J. Grove, X. L. Li, K.-M. Shyue, Y. Zeng, and Q. Zhang. Three-dimensional front tracking. *SIAM Journal of Scientific Computing*, 19:703–727, May 1998.

[OSHE01a] S. Osher and R. P. Fedkiw. Level set methods: An overview and some recent results. *Journal of Computational Physics*, 169(2):463–502, May 2001.

[SUSS98a] M. Sussman, E. Fatemi, P. Smereka, and S. Osher. Improved level set method for incompressible two-phase flows. *Computers & Fluids*, 27(5-6):663–680, Jun-Jul 1998.

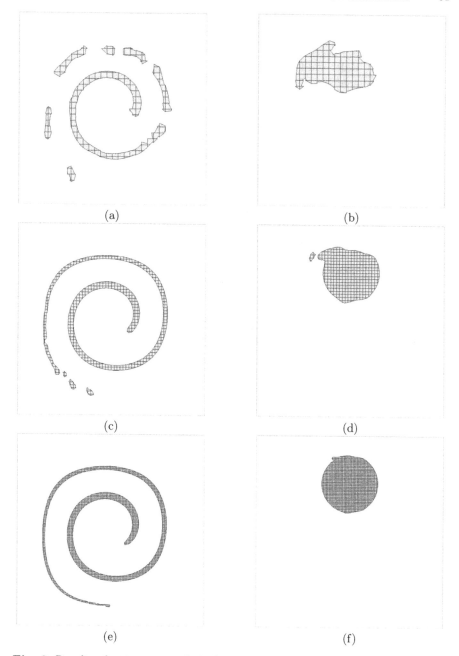

(a)

(b)

(c)

(d)

(e)

(f)

Fig. 8. Results of a time reversed single vortex simulation for different rectangular grids; (a) 32x32 grid at t=4 (maximal stretch) (b) 32x32 grid at t=8 (fully reversed flow) (c) 64x64 grid, t=4 (d) 64x64 grid, t=8 (e) 128x128 grid, t=4 (f) 128x128 grid, t=8

[YOUN82a] D. L. Youngs. Time-dependent multi-material flow with large fluid distortion. In K. W. Morton and M. J. Baines, editors, *Numerical Methods for Fluid Dynamics*, pages 273–285. Academic Press, 1982.

[RIDE98a] W. J. Rider and D. B. Kothe. Reconstructing volume tracking. *Journal of Computational Physics*, 141:112–152, 1998.

[DEBA74a] R. DeBar. Fundamentals of the KRAKEN code. Technical Report UCID-17366, Lawrence Livermore National Laboratory, 1974.

[YOUN84a] D. L. Youngs. An interface tracking method for a 3d eulerian hydrodynamics code. Technical Report 44/92/35, AWRE, 1984.

[SWAR89a] B. Swartz. The second-order sharpening of blurred smooth borders. *Mathematics of Computation*, 52(186):675–714, Apr 1989.

[BART89a] T. J. Barth and D. C. Jespersen. The design and application of upwind schemes on unstructured meshes. In *Proceedings of the 27th Aeroscience Meeting*, Reno, NV, Jan 1989. AIAA.

[KOTH96a] D. B. Kothe, W. J. Rider, S. J. Mosso, J. S. Brock, and J. I. Hochstein. Volume tracking of interfaces having surface tension in two and three dimensions. In *Proceedings of the 34th Science Meeting*, Reno, NV, Jan 1996. AIAA.

[SHAS96a] M. J. Shashkov. *Conservative Finite-Difference Methods on General Grids*. CRC Press, Boca Raton, FL, 1996.

[MOSS97a] S. J. Mosso, B. K. Swartz, D. B. Kothe, and R. C. Ferrell. A parallel, volume-tracking algorithm for unstructured meshes. In P. Schiano, A. Ecer, J. Periaux, and N. Satofuka, editors, *Parallel Computational Fluid Dynamics: Algorithms and Results Using Advanced Computers*, pages 368–375. Elsevier Science B.V., 1997.

[STAL04] M Staley. Core, conservative remapper. Technical Report LAUR-04-8104, Los Alamos National Laboratory, 2004.

[BELL89a] J. B. Bell, P. Colella, and H. M. Glaz. A second-order projection method of the incompressible navier-stokes equations. *Journal of Computational Physics*, 85:257–283, 1989.

[LEVE96a] R. LeVeque. High-resolution conservative algorithms for advection in incompressible flow. *SIAM Journal of Numerical Analysis*, 33:627–665, 1996.

[ENRI02a] D. Enright, R. Fedkiw, J. Ferziger, and I. Mitchell. A hybrid particle level set method for improved interface capturing. *Journal of Computational Physics*, 183:83–116, 2002.

Identifying and Meshing Thin Sections of 3-d Curved Domains

Luzhong Yin, Xiaojuan Luo, Mark S. Shephard

Scientific Computation Research Center, Rensselaer Polytechnic Institute, Troy, NY 12180.
lyin@scorec.rpi.edu; xluo@scorec.rpi.edu; shephard@scorec.rpi.edu

Summary: Realization of the full benefits of variable p-version finite elements requires the careful construction of prismatic elements in thin sections. This paper presents a procedure to automatically isolate the thin sections using the points on an approximate medial surface computed by an octree-based algorithm. Using the pairs of triangles associated with medial surface (MS) points, in conjunction with adjacency, classification and distance information, sets of surface triangles that are on opposite face patches in thin sections are identified. Mesh modifications are then executed to match the surface triangulations on the opposite face patches such that prismatic elements can be generated without diagonal edges through the thickness directions.

Keywords: thin sections, medial surface, prismatic elements

1. Introduction

Historically, the methods used to analyze thin sections involved applying deformation assumptions to the 3-D elasticity equations allowing the problem dimensionality to be reduced [1]. The application of such methods requires a reduced dimensional domain model. Application of these methods requires the identification of the thin sections and then the application of model dimension reduction on those portions of the domain [2]. Handling the interconnection between two-dimensional reduced elements to fully three-dimensional solid elements is another source of difficulty [3].

Since the assumptions corresponding to those deformation models are equivalent to allowing only low order deformation modes in the thickness direction, an alternative is to apply full three-dimensional model discretized with p-version finite elements with low polynomial order through the thickness [4, 5, 6]. Tetrahedral meshes cannot effectively be used to implement the appropriate low order deformation modes through the thickness due to the presence of the through the thickness diagonals. Therefore a mesh that contains a single element through the thickness without through the thickness diagonals is needed.

The automatic generation of meshes for general 3-D domains with such elements (prism of hexahedra) in the thin sections is not a straightforward process, particularly in the case where adaptive p-version finite element methods are applied that will require large curved elements of high polynomial order in the other directions. This paper reports on the status of efforts to development mesh generation procedures aimed at producing the desired p-version curved finite elements for models containing thin sections.

A key challenge to properly mesh the thin sections in general geometrical models is to identify and isolate thin sections from the rest of the domain. Most of efforts on meshing a thin section use a priori information on the thin section [7]. One approach identified as appropriate for identifying thin sections is the medial surface [2, 8]. The medial surface is the locus of the center of a maximum sphere rolls around the interior of the model. It is the set of interior points that is equidistant to more than one points on the model boundary. The medial surface can provide the following information on the region's geometry and topology [9]:

- Indication of local feature size (or local thickness) by the distance from a medial surface point to its closest boundary points
- Information on 'opposite' boundary points by relating the closest boundary points to a medial surface point.

The medial surface has been used to partition the geometry model into easily meshable subregions by several authors to generate volume mesh, for example [9, 10, 11]. Other mesh-related applications of medial surface include construction of three-dimensional anisotropic geometric metrics for geometric adaptation [12]. In the present procedure we determine and use a limited number of points on an approximation medial surface to identify the thin sections and generate prismatic layer mesh through thickness.

Section 2 presents the criteria to identify pairs of opposite triangles on the thin sections based on points on the medial surface for a classified surface triangulation of the model and gives an octree-based algorithm for their determination. Section 3 discusses the procedure that given those pairs of triangles determines any missing thin section triangles and isolates the thin sections. Section 4 considers the procedures for then meshing the thin sections. A general volume mesh generator is applied to fill the remaining domains for p-version adaptive analysis. The geometric approximation required by the p-version finite elements is achieved by applying a curving procedure in [13, 14]. Section 5 presents an example to show the benefits of above meshing procedure in p-version analysis.

1.1 Nomenclature

M_i^d the i-th mesh entity of dimension d, d=0 for vertex, d=1 for an edge, d=2 for a face.

$\overline{M}_i^d$ the closure of the i-th mesh entity of dimension d = 1 or 2.

G_i^d the i-th entity of dimension d in geometry model.

[classification symbol used to indicate the associated of one or more mesh entities with an entity in geometry model [15].

O_i the i-th octant.

O_i^d the i-th octant entity of dimension d, d=0 for octant vertex, d=1 for oc-tant edge.

E_i^0 the i-th medial surface point.

$p(\bullet)$ single closest point from entity $\bullet$ of dimension 0 to surface mesh.

$p_i(\bullet)$ the i-th closest point from entity $\bullet$ of dimension 0 to surface mesh.

2. Determination of Medial Surface Points and Associated Triangle Pairs

2.1 Criteria to Define Thin Section Triangle Pairs

The definition of a thin section is closely related to size of and order of the elements in the mesh. The geometric characteristic for a thin section is the dimension through the thickness is far less than the "in-plane" dimensions. We identify the thin sections using a surface triangulation of the model. The basic idea is to find pairs of triangles on "opposite model faces" that are close to each other relative to their size, thus indicating they are within a thin section.

A point on the medial surface can provide the local thickness [9] by the distance to its closest boundary points and 'opposite' boundary points by relating the closest boundary points. Therefore, the concerned pair of the opposite triangles can be defined based on a medial surface point as fol-lows.

A pair of triangles M_i^2 and $M_{i'}^2$ is candidate *thin section triangle pair* if there exist a pair of closest boundary points P_1 and P_2 from a medial surface point E_i^0, such that $P_1 \subset \overline{M_i}^2$ and $P_2 \subset \overline{M_{i'}}^2$, where the P_1 and P_2 have following properties:

(1) The ratio of thickness (defined as the diameter of the maximum inscribed sphere associated with E_i^0) to the average size of M_i^2 and $M_{i'}^2$ is smaller than a default value, for example 1/3 of the average edge length of the element.

(2) The angle formed by the outward normal to M_i^2 and $M_{i'}^2$ is close to π.

The situation of the medial surface point defined by conditions of (1) and (2) is shown in Fig. 1.

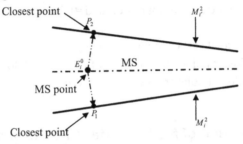

Fig. 1. A thin section triangle pair identified by a medial surface point E_i^0

A candidate thin section triangle pair is further processed to ensure that all points on their closures meet those conditions. The key step to identify the thin section triangle pairs is to calculate the points on the medial surface of the classified surface triangulation. We use octree to calculate the medial surface points with the goal of identifying most, but not all, triangles in the thin sections.

2.2 Octree-Based Algorithm to Compute the Medial Surface Points

The medial surface points are calculated for a classified surface triangulation. The classification information of the surface triangulation is used to ignore the medial surface branches of the triangulated model that do not exist in the smooth curved model. That is, the two closest points of a medial surface point on two adjacent triangles that are classified on one C^1 continuous model face will be ignored in the calculations. From the property that a closed geometrical model has a closed set of medial surfaces [16], a medial octant tracing algorithm was constructed. In this algorithm, medial octants are defined as octants that intersect medial surface as O_1 and O_2 shown in Fig. 2. The steps of the tracing algorithm are as follows.

- Construct octree by inserting surface mesh entities into boundary octants.
- Determine an octant with an edge that intersects the medial surface.
- Resolve all intersections of that octant edge by a traversal algorithm.
- Continue the traversal on the other edges until all intersections are resolved.
- Move to neighboring octants of the intersection points to process their other octant edge/medial surface intersections.

The above tracing procedure can be illustrated by Fig. 2. Suppose by the traversal algorithm discussed below, we resolve the intersections shown as E_1^0, E_2^0 and E_3^0 on the edges of O_1. After that we move to the next medial octant O_2 adjacent to E_3^0 to process the other intersection points on all the edges of O_2. The procedure will repeat until no new neighbor medial octant can be found.

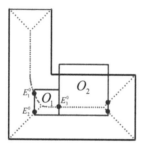

Fig. 2. Move from one resolved medial octant O_1 to the next octant O_2 adjacent to a known medial surface point E_3^0 to calculate other new intersection locations.

To control the medial octant size, before calculating intersections, recursively subdivide the neighbor medial octants to be no more than one level different. Further subdivide the medial octant to the same order of the size of surface triangles that are closest to the octant vertices.

The goal of the algorithm is to determine the intersection between an octant edge and the medial surface. An octant edge can have multiple intersections. To determine those intersections, we employ the relationship among Voronoi regions and medial surface [8] for polyhedron. The medial surface of a convex polyhedron is identical to its Voronoi diagram which is defined as the boundaries of the Voronoi regions. If a polyhedron is non-convex, the Voronoi diagram is a superset of medial surface. In the presence of concave edges whose interior dihedral angles $w_i > \pi$, the Voronoi region associated with each of concave edges is a particular subregion bounded by a portion of the medial surface and planes perpendicular to the boundary planes at the concave edge and intersect the medial surface. In this case, the difference between the medial surface and Voronoi diagram is that medial surface does not include planes of the Voronoi diagram incident at the concave edges, shown as the dash lines in Fig. 3, where the dot lines is medial surface.

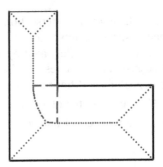

Fig. 3. Relationship among Voronoi regions, Voronoi diagram and medial surface.

Taking into account non-convex region cases, an octant edge with bounding vertices closest to two different surface entities (i.e. in two different Voronoi regions associated with the two entities) that are not adjacent to one re-entrant edge or corners, intersects the medial surface. The closest point $p(O_1^0)$ determines the Voronoi region that the vertex O_1^0 is in. That is, if $p(O_1^0)$ is found on a surface entity, then O_1^0 is in the Voronoi region associated with that surface entity. The current octree is employed to determine the closest point information. The procedure to resolve the intersections is as follows:

Assume there is just one intersection on the edge bounded by O_1^0 and O_2^0. Using the equidistant condition to the boundary entities whose associated Voronoi regions O_1^0 and O_2^0 are in, the location of the intersection can be found for the parameter t in the interpolation formulation

$$\mathbf{E}_i^0 = (1-t)\mathbf{O}_1^0 + t\mathbf{O}_2^0 \qquad (1)$$

where the bold letters denotes the location vectors at the corresponding vertices. Note a Voronoi region in a polyhedron could be associated with face, concave edge or concave corner. For the case that two Voronoi regions are associated with two faces as shown in Fig. 4, t is given as

$$t = \frac{\left(\mathbf{O}_2^0 - \mathbf{O}_1^0\right) \cdot \mathbf{n}_2 + d_2 - d_1}{\left(\mathbf{O}_1^0 - \mathbf{O}_2^0\right) \cdot \mathbf{n}_1 + \left(\mathbf{O}_2^0 - \mathbf{O}_1^0\right) \cdot \mathbf{n}_2} \qquad (2)$$

where d_1 and d_2 are distances from O_1^0 and O_2^0 to their closest points on boundary, $\mathbf{n}_1$ and $\mathbf{n}_2$ are unit vectors from O_1^0 and O_2^0 to their closest points, respectively. Note that Eqn (2) is also valid when O_1^0 and O_2^0 are on the boundary, which leads to d_1 and d_2 equal to zero, and $\mathbf{n}_1$ and $\mathbf{n}_2$ are in the outward normal directions to the faces where O_1^0 and O_2^0 are on.

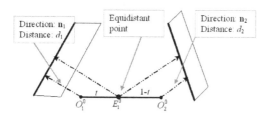

Fig. 4. Medial surface point calculation

If $p(O_1^0)$ or $p(O_2^0)$ is on concave edge or concave vertex, the equidistant condition leads to a quadratic equation to solve the parameter t in Eqn (1).

After getting the assumed intersection in Eqn (1), we request its closest points on the boundary. If multiple closest points are returned there is a single intersection. If a single closest point is returned there are multiple intersections, in which case subdivide the edge at that location and repeat until the intersections are resolved.

The efficiency of the above algorithm is illustrating using Fig. 5. Given the octant edge whose bounding vertices are O_1^0 and O_2^0, we obtain the closest points on the boundary to O_1^0 and O_2^0 and an assumed medial surface intersection point E_j^0 is found. The closest point to E_j^0 is a single point $p(E_j^0)$, thus indicating the point is not on the medial surface. After subdivide the edge at E_j^0, we obtain the correct intersection points, one on each sub-edge bounded by (O_1^0, E_j^0) and (E_j^0, O_2^0).

It is noted the tracing algorithm can start from a convex model edge whose interior dihedral angle is $w_i < \pi$, or by a medial surface point calculated on a ray in the direction of the normal to a surface triangle to the interior of model by the above edge intersection algorithm. Operations to calculate medial surface points are not applied to octant edges external to the model.

Fig. 6 shows a result of the above algorithm for an example. Fig. 6(b) shows the medial octants that are traversed by the above tracing algorithm. The medial octant sizes are refined to match surface mesh size.

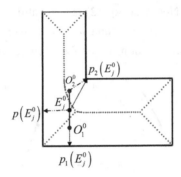

Fig. 5. Multiple intersection on the edge bounded by O_1^0 and O_2^0

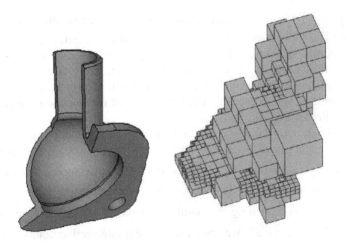

Fig. 6. (a) Model; (b) traversed medial octants.

3 Defining Thin Sections

The medial surface point calculations provide a set of unorganized and incomplete thin section triangle pairs. We organize the thin section triangles in the sets using knowledge of which model faces they are classified on. After the initial sets are constructed, the missing thin section triangles are determined. The procedure has three steps:
- Collect starting triangle sets using classification information.
- Complete the triangle sets to define thin section surface patches
- Construct the loops for each thin section surface patch and match the loops on the boundary of opposite thin section patches.

3.1 Collect Starting Thin Section Triangle Sets

Given medial surface point E_i^0, introduce

$$\left| E_i^0 \right|_{\bullet} = \begin{cases} 1 & thin \\ 0 & not\ thin \end{cases} \tag{3}$$

to indicate when E_i^0 defines a thin section triangle pair. $\left| E_i^0 \right|_{\bullet} = 1$ indicates the medial surface point is associated with thin triangle pair and $\left| E_i^0 \right|_{\bullet} = 0$ means they are not part of a thin section, see Section 2.1. Denote the triangle pair as

$$\left[E_i^0 \right]_{\bullet} = \left\{ M_k^2, M_{k'}^2 \right\}. \tag{4}$$

With the above notations, a starting thin section triangle set is defined as

$$\hat{G}_j^2 = \left\{ M_i^2 \middle| M_i^2 \left[\ G_j^2 \ and \ M_i^2 \in \left[E_i^0 \right]_{\bullet} and \left| E_i^0 \right|_{\bullet} = 1 \right\} \tag{5}$$

Note that each $\hat{G}_j^2$ is uniquely associated with model face G_j^2. For this unique association, the identity tags of "opposite" sets for $\hat{G}_j^2$ can be recorded during the construction of $\hat{G}_j^2$. Generally, $\hat{G}_j^2$ may have one or more opposite sets denoted as $opp(\hat{G}_j^2)$. A simple example in Fig. 7 shows $\hat{G}_1^2 = \left\{ M_a^2, M_b^2 \right\}$, $\quad opp(\hat{G}_1^2) = \left\{ \hat{G}_3^2 \right\}$; $\quad \hat{G}_2^2 = \left\{ M_c^2, M_d^2, M_e^2 \right\}, opp(\hat{G}_2^2) = \left\{ \hat{G}_3^2 \right\}$ and $\hat{G}_3^2 = \left\{ M_{a'}^2, M_{b'}^2, M_{c'}^2, M_{d'}^2, M_{e'}^2 \right\}, \ opp(\hat{G}_3^2) = \left\{ \hat{G}_1^2, \hat{G}_2^2 \right\}$.

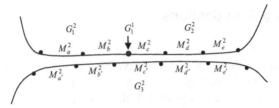

Fig. 7. An example to demonstrate the starting triangle sets

Note sets at this point may have to be later split or merge to represent a thin section surface patch.

3.2 Determining the Missing Thin Section Triangles

The majority of thin section triangles are identified by the medial surface points in the tracing algorithm but some are missed, see Figure 8 (a), where the dark shaded triangle faces are the identified thin section triangles.

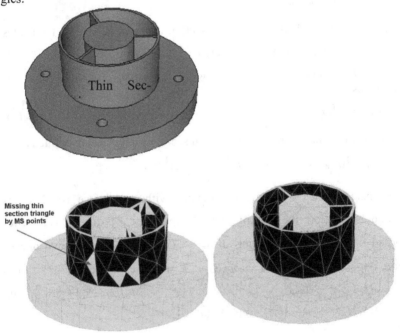

Fig. 8. (a) Geometry model with thin section; (b) Thin section triangles obtained by medial surface points; (c) The completed thin section surface patches.

To determine whether a missing triangle M_e^2 on G_j^2 belongs to $\hat{G}_j^2$, local thickness h_e at M_e^2 is compared with the local thickness h_i at an edge adjacent triangle M_i^2 that is inside $\hat{G}_j^2$. If $|h_e - h_i|/h_i$ is smaller than a default value, place M_e^2 in $\hat{G}_j^2$. The local thickness h_e at M_e^2 is obtained by searching for the closest point on the triangles classified on the model faces that are known to be opposite to G_j^2. The triangle $M_{e'}^2$ that the closest point is on is defined as opposite triangle to M_e^2. Also place $M_{e'}^2$ in the set that is opposite to $\hat{G}_j^2$ if it is not there already. Note, $M_{e'}^2$ must be in the neighborhood of $M_{i'}^2$ which is opposite to M_i^2. The search is a local operation. Fig 8 shows an example before (Fig 8 (b)) and after completing the thin section triangle sets (Fig. 8(c)). In Fig 8(b), the faces in black are determined by medial surface points and those in gray on the thin sections are missing surface triangles. Fig. 8(c) shows all the thin section triangles are recovered shown in black. Note the boundary edges of $\hat{G}_j^2$ can also be identified as those used by only one triangle in the set. We denote the set of boundary edges of $\hat{G}_j^2$ as $\partial\hat{G}_j^2$.

3.3 Construct the Boundary Loops on Thin Section Surface Patches

To complete the definition of the thin section surface patches opposite each other, the loops on the boundary of surface patches have to be matched. The process can lead to the need to split surface patches. Figure 9 (a) shows an example that has three thin section surface patches, where $\hat{G}_3^2$ is opposite to $\hat{G}_1^2$ and $\hat{G}_2^2$ with each loop on each of the sets. In this case, splitting $\hat{G}_3^2$ to form two loops on $\hat{G}_3^2$ to match the loops on $\hat{G}_1^2$ and $\hat{G}_2^2$ is needed as shown in Fig. 9 (b). Note the loops in $\hat{G}_1^2$ and $\hat{G}_2^2$ cannot be merged to form one loop since the model edge must be used in the volume mesh generation.

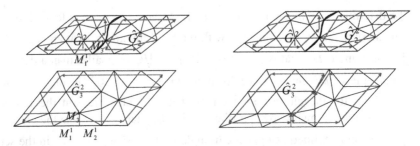

Fig. 9. (a) Loops on thin section surface patches (b) Opposite loops

The procedure to create the loop L_k and its opposite loop $L_{k'}$ can be started from any $M_f^1 \in \partial \hat{G}_j^2$ which looks for its opposite edge $M_{f'}^1 \in \partial \hat{G}_{j'}^2$ through the opposite triangles in following steps:

- If $M_f^1 \in \partial \hat{G}_j^2$ can find its opposite edge $M_{f'}^1 \in \partial \hat{G}_{j'}^2$, the procedure continues to process the next edge $M_d^1 \in \partial \hat{G}_j^2$ adjacent to M_f^1. For example, in
Fig. 9(a), the edge $M_r^1 \in \partial \hat{G}_1^2$ is found opposite to $M_1^1 \in \partial \hat{G}_3^2$. The next
edge $M_2^1 \in \partial \hat{G}_3^2$ adjacent to M_1^1 is processed.

- If $M_d^1 \in \partial \hat{G}_j^2$ cannot find its opposite edge on $\partial \hat{G}_{j'}^2$, then from the edge
$M_{d'}^1 \in \partial \hat{G}_{j'}^2$ adjacent to $M_{f'}^1 \in \partial \hat{G}_{j'}^2$, its opposite edge is found from the interior edges of $\hat{G}_j^2$ adjacent to M_f^1. Fig. 9 shows a such case, where
$M_2^1 \in \partial \hat{G}_3^2$ cannot find an opposite edge on $\partial \hat{G}_1^2$, therefore, from $M_3^1 \in \partial \hat{G}_1^2$
adjacent to M_r^1, the interior edge M_3^1 is found opposite to it. The proce-
dure leads to split thin section surface patches from Section 3.2 as
shown in Fig. 9(b).

In the above procedure, the interior edge refers to the edge not belonging to $\partial \hat{G}_j^2$ but on the closure of the triangles in $\hat{G}_j^2$.

4. Meshing Thin Sections

The thin section information obtained from Section 3 is characterized by a pair of opposite thin section surface patches with paired closed opposite loops as boundaries. To generate structured prismatic elements without long diagonal edges through the thickness the opposite triangulation sets for thin section surface patch must be topologically matched and be geometrically similar.

4.1 Overall Algorithm

The overall procedure to mesh each thin section consists of the following steps:
- Apply local mesh modifications to match the thin section boundaries
- Delete the surface triangulation of one triangle set
- Copying the remaining triangulation to its opposite model face
- Connect the matched triangulations to form prismatic elements

4.2 Boundary Matching

The procedure to match the boundaries of the triangle sets for a thin section can be divided into two continuous operations. First, apply split or collapse operations to modify the mesh topology to ensure the mesh edges in each paired opposite loops are one-to-one matched. Second, the desired target locations for the vertices in the loops are computed and local mesh modifications as in [13] are applied to incrementally move the vertices towards to the target locations.

4.2.1 Topological Matching

For each pair of opposite loops, the process begins to traverse one loop through vertex adjacency information from one selected starting mesh edge and vertex to match the topological configuration to its opposite loop. Split and/or collapse operations are used to keep a loop L_i one-to-one matching with its opposite loop $L_{i'}$. The procedure starts from a vertex with lowest dimensional classification.

For each mesh edge M_i^1 in the closed loop L_i, retrieve the attached edges $\{M_j^1\}$ in the opposite closed loop $L_{j'}$. Let

$$|M_j^1|^* = \begin{cases} 1, assigned \\ 0, unassign \end{cases} \tag{4}$$

be the operation to determine whether the mesh edge M_j^1 already associates with one mesh edge in loop L_i. If only one mesh edge M_j^1 is attached and $|M_j^1|^* = 0$, let M_j^1 associate with M_i^1 and update $|M_j^1|^* = 1$ and continue to next mesh edge M_{i+1}^1. Otherwise, either split or collapse is applied to produce or eliminate mesh edges to maintain the one-to-one matching. As examples, Figure 10 shows that a split operation is applied on mesh edge

M_1^1 to produce one more edge M_2^1 to obtain the matched pair edges (M_1^1, M_1^1) and (M_2^1, M_2^1) then update $|M_2^1|=1$. Considering the mesh vertices will be moved in the next step, the split operation applied currently does not snap the new introduced vertices to the model boundary.

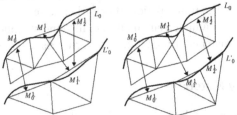

Fig. 10. Split operation to assign one-to-one match for mesh edge M_2^1

4.2.2 Target Location for Vertices in the Opposite Model Face

To achieve the geometrical similarity between the two triangle sets for a thin section, each vertex M_i^0 in one triangle set need to compute an appropriate location for its matched vertex M_i^0 on the opposite model face. The target location for M_i^0 is obtained by first computing the closest point P_i^0 on the mesh face M_i^2 classified on the opposite model face to M_i^0, and then, projecting P_i^0 to the model face by the model parameters of P_i^0 determined by the interpolation of P_i^0 on the triangle face M_i^2.

4.2.3 Incrementally Move Vertices on the Thin Section Boundary

The movement of the mesh vertices on the thin section boundary loop L_i can cause the surface mesh to become invalid. Fig 11 shows an example where two triangle faces marked as shaded in Fig 11(b) become invalid because of moving vertices M_2^0 to its target locations. This problem is avoided by applying the following procedure [13]:

- Put all of the vertices with attached target locations into a list
- Traverse the list and deal with one vertex at each step
- If the vertex moves to the target location without causing any problem, move it and remove it from the list. Otherwise, apply local mesh modi-

fications to correct the invalid elements. Remove the vertex from the list (See reference [13]).

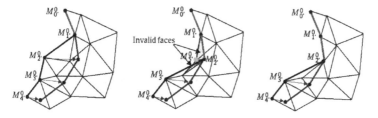

Fig. 11. (a) Move M_0^0 and M_1^0. (b) Move M_2^0, invalid faces marked as shaded. (c) Collapse M_5^0 to M_2^0.

4.3 Surface Triangulation Matching

The surface triangulation matching between the two triangle sets for a thin section is achieved through the triangulation deletion of one triangle set and the copying of the remaining triangle set to the opposite model face. The key technique is the location computation of the copied vertices on the opposite model face. This is accomplished by the interpolation strategy discussed in Section 4.2

4.4 Volume Mesh Generation

With the topologically and geometrically matched surface triangulation for thin sections, the volume mesh generation procedure constructs prismatic elements by directly connecting each paired triangles classified on the two opposite model faces. Because the generalized volume mesh generator requires the exposure mesh faces to be triangles, one layer of pyramid elements are added neighboring to the interior quadrilateral faces of the structured prismatic elements.

4.5 Examples

Two example models with thin sections are given in Figure 12. The first row shows the input surface triangulation for the two models. The second row shows just the isolated opposite thin section surface patches after loop construction and loop matching. As can be seen from the figures the two surface meshes are not yet matched. The bottom row shows the final meshes where the thin section meshes have been matched, the prismatic thin section elements created, volume mesh completed and the mesh properly curved to the domain boundary.

5. p-Version Analysis Example for Model with Thin Sections

In this section, a structural part with thin sections shown in Figure 13 is analyzed using p-version method. Due to the symmetry of the problem, only one half of domain is analyzed by assigning properly symmetric boundary conditions. For this particular model, there are two portions of the domain can be regarded as thin sections with thickness 0.5 (marked in Figure 13) determined by the dimensions of the model.

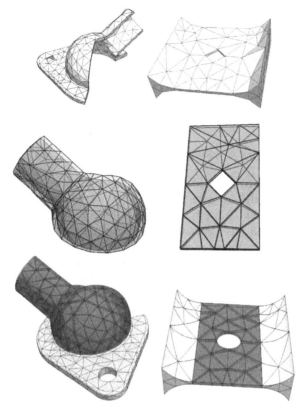

Figure 12: Examples meshes for models with thin sections: Surface triangulations (top). Thin sections (middle). Curved meshes with prismatic elements (bottom).

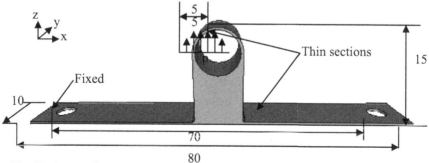

Fig. 13. Geometric model of the structural part with thin sections

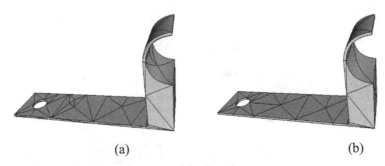

(a)   (b)

Fig. 14. Two meshes for one quarter of the domain

Figure 14 shows two curved meshes for the one half of the domain. The mesh in Figure 14(a) is generated by the automatic tetrahedral volume mesh generators and the mesh in Figure 14(b) is generated using the procedure presented in this paper that has mixed topological elements. The two thin section structures are meshed with prismatic elements without diagonal edge through the thickness directions in Figure 14(b) comparing to the all tetrahedral element mesh. Table 1 presents the summary of these two meshes and the statistic indicates that number of elements for mixed topological mesh has been reduced almost 50% comparing to the all tetrahedral mesh.

Table 1. The comparision between the two meshes

	Regions			Faces		Edges	Vertices
All tetraheral	131			328		264	67
Mixed topology	66			197		194	63
	Tet	Prism	Pyramid	Tri	Quad		
	38	24	4	149	48		

The material is assumed to be linearly elastic with Young's modulus E = 200,000PA and Poission's ratio v=0.33. The quadratic z direction pressure $p = 1.-(x/4.5)^2$ is applied on the inner circular surface that will cause the bending of the thin sections. The problem is analyzed using:

- Uniform p-version method that varies the polynomial from p=2 to p=8.
- Adaptive p-version method that enriches the polynomial independently at each coordinate directions of the elements.

The implicit element residual posterior error estimator [17] is applied to evaluate the solution accuracy in relative error in energy norm e_r as

$$e_r = \sqrt{\frac{\|u - u_X\|}{\|u\|}}\% = \sqrt{\frac{\|e\|}{\|u\|}}\% = \sqrt{\frac{\sum_k \eta_k}{\|u\|}}\%$$

(5)

where u, u_X are the exact solution and finite element solution of the problem. $e = u - u_X$ defines the error and η_k is the elemental error estimator [17]. $\|u\|$ is the exact strain energy. Since the true exact value is not known, we adopt the value 53.7761 computed from StressCheck [18] on a mesh with 3081 tetrahedral regions at polynomial order 8. Table 2 presents the analysis results with respect to the number of degree of freedom for uniform p-version method and Table 3 for adaptive p-version method.

Table 2: Uniform p-version analysis results

p-order	All tetrahedral mesh		Mixed topology mesh	
	Dof	e_r	Dof	e_r
2	890	60.42	710	47.74
3	2566	28.86	1699	15.39
4	5586	17.78	3380	12.20
5	10343	11.61	5939	7.39
6	17230	7.76	9574	4.13
7	26640	5.56	14483	2.16
8	38966	4.15	20864	1.57

Table 3: Adaptive p-version analysis results

Iteration step	All tetrahedral mesh		Mixed topology mesh	
	Dof	e_r	Dof	e_r
1	890	60.42	710	47.74
2	2635	24.02	1925	14.35
3	2837	22.08	2145	10.46
4	3371	17.06	3536	8.87
5	4129	11.19	5767	4.36
6	4432	9.02	6768	1.97

Results show that,

- In case of the uniform p-version method, the mixed topology mesh can obtain more accurate solution (1.57%) than the all tetrahedral mesh (4.15%) with only 54% of the degrees of freedom.
- In case of the adaptive p-version method for the all tetrahedral mesh, the adaptive analysis stops after 6 iteration steps with an unsatisfied solution accuracy level (9.02%) because some elements reached the polynomial limitation p=8.
- For the mixed topology mesh, the adaptive p-version method uses only 47% of the degrees of freedom than used for uniform p-version method to achieve an solution accuracy that is slightly better. The highest polynomial order at the last step is p=7 for the elements next to the fixed cylinder hole.

6. Closing Remarks

The paper presented a procedure to automatically isolate and mesh thin sections of 3-D solid models with prismatic elements for directional p-version finite element analysis. Key ingredients of the procedure are:

- Construction of an octant tracing algorithm to calculate a limited number of medial surface points to define the thin section triangle pairs,
- A strategy to organize the thin section triangle pairs to define thin section face patches that are opposite to each other,
- A procedure to generate prismatic volume mesh by copying one side of surface mesh to the other side and connecting the corresponding opposite vertices.

Uniform p-version analysis results clearly demonstrated that the mixed topology mesh generated by the presented procedure for model with thin section can obtain more accurate solution substantially fewer degree of freedom than the mesh with all tetrahedral elements.

7. Acknowledgement

This work was supported by the National Science Foundation through SBIR grant number DMI-0132742.

References

1. J.N. Reddy, Theory and analysis of elastic plates. 1999, Taylor & Francies.
2. R.J. Donaghy, C.G Armstrong, M.A. Price (2000) Dimensional Reduction of surface models for analysis. Eng. with Computers 16: 24-35.
3. RL Actis, SP Engelstad, BA Szabo (2002) Computational requirements in design and design certification, Collection of Technical Papers - AIAA/ASME/ASCE/AHS/ASC Structures, Structural Dynamics and Materials Conference, pp. 1379-1389.
4. Duster, H. Broker, E. Rank (2001) The p-version of the finite element method for three dimensional curved thin walled structures, Int. J. Numer. Methods Engrg, 52: 673-703.
5. C.A. Duarte, I. Babuska (2002) Mesh-independent p-orthotropic enrichment using the generalized finite element method, Int. J. Numer. Methods Engrg, 55: 1477-1492.
6. S. Dey, M.S. Shephard, J. E. Flaherty (1997) Geometry representation issues associated with p-version finite element computation, Comput. Methods. Appl. Mechanics. Engrg., 150: 39-55.
7. C.K. Lee, Q.X. Xu (2005) A new automatic adaptive 3D solid mesh generation scheme for thin-walled structures. Inter. J. Numer. Meth. Engng 62: 1519-1558.
8. N. M. Patrikalakis, H. N. Gursoy (1990) Shape interrogation by medial axis transform, ASME Advances in Design Automation, 16th design automation conf., Chicago, DE-vol 23, 77-88.
9. P. Sampl (2001) Medial axis construction in three dimensions and its application to mesh generation. Engineering with Computers 17: 234-248.
10. M. A. Price, C. G. Armstrong, M. A. Sabin (1995) Hexahedral mesh generation by medial surface subdivision: I. Solids with convex edges, Int. J. Numer. Meth. Eng, 38, 3335-3359
11. M. A. Price, C. G. Armstrong (1997) Hexahedral mesh generation by medial surface subdivision: part II. Solids with flat and concave edges, Int. J. Numer. Meth. Eng, 40, 111-136
12. K-F Tchon, M. Khachan, F. Fuibault, R. Camarero (2005) Three-dimensional anisotropic geometric metrics based on local domain curvature and thickness, Computer-Aided Design 37: 173-187
13. X. Li, M.S. Shephard, M.W. Beall (2003) Accounting for curved domains in mesh adaptation, Int. J. Numer. Methods Engrg, 58: 247-276.
14. X.J. Luo, M.S. Shephard, R.M. Obara, R. Nastasia, M.W. Beall (2004) "Automatic p-version mesh generation for curved domains", Engrg. Comput. 20: 265-285.

15. M.W. Beall and M.S. Shephard (1997) A general topology-based mesh data structure, Inter. J. Numer. Meth. Engng 40: 1573-1596.
16. A. Lieutier (2004) Any open bounded subset of r^n has the same homotopy type as its medial axis, Computer-Aided Design 36: 1029-1046.
17. M. Ainsworth and J.T. Oden (2000) A posteriori error estimation in finite element analysis, A wiley-interscience publication.
18. StressCheck User's Manual (2005)

A COMPUTATIONAL FRAMEWORK FOR GENERATING SIZING FUNCTION IN ASSEMBLY MESHING

William Roshan Quadros[*], Ved Vyas[*], Mike Brewer[+], Steven James Owen[+], Kenji Shimada[*]

[*]Dept. of Mechanical Engineering, Carnegie Mellon University, Pittsburgh, PA, 15213, USA

[+]Sandia National Laboratories[1], Albuquerque, NM, 871850, USA

ABSTRACT

This paper proposes a framework for generating sizing function in meshing assemblies. Size control is crucial in obtaining a high-quality mesh with a reduced number of elements, which decreases computational time and memory use during mesh generation and analysis. This proposed framework is capable of generating a sizing function based on geometric and non-geometric factors that influence mesh size. The framework consists of a background octree grid for storing the sizing function, a set of source entities for providing sizing information based on geometric and non-geometric factors, and an interpolation module for calculating the sizing on the background octree grid using the source entities. Source entities are generated by performing a detailed systematic study to identify all the geometric factors of an assembly. Disconnected skeletons are extracted and used as tools to measure 3D-proximity and 2D-proximity, which are two of the geometric factors. Non-geometric factors such as user-defined size and pre-meshed entities that influence size are also addressed. The framework is effective in generating a variety of meshes of industry models with less computational cost.

Keywords: Assembly meshing, finite element mesh sizing function, skeleton, and pre-mesh

---

[1] Sandia is a multiprogram laboratory operated by Sandia Corporation, a Lockheed Martin Company, for the United States Department of Energy under Contract DE-AC04-94AL85000

The submitted manuscript has been authored by a contractor of the United States Government under contract. Accordingly the United States Government retains a non-exclusive, royalty-free license to publish or reproduce the published form of this contribution, or allow others to do so, for United States Government purposes.

1. INTRODUCTION

Most automatic, unstructured finite element (FE) meshing algorithms do not recognize the geometric complexity and other non-geometric factors in assembly meshing. Hence, it is difficult to generate a desired 3D mesh of assemblies in one step, and it is worthwhile to split the meshing process into two steps: (1) The analysis of the geometric complexity of the input assembly and other non-geometric factors, and the generation of functions that provide element size, anisotropy, and orientation information; (2) Generation of a FE mesh using the element size, shape and orientation information available from the first step. In this paper, the objective is to analyze the geometric complexity and other non-geometric factors for generating element sizing function for assemblies. This paper does not address element anisotropy and orientation functions.

Sizing function plays a crucial role in mesh generation and in finite element simulation. Meshing assembly models is of increasing importance as simulations are more routinely performed at the system level rather than the part level. Fig. 1 shows a uniform mesh and a graded mesh of a system or assembly. The uniform mesh consists of 39,165 elements and the graded mesh consists of only 17,210 elements, with fine mesh at the holes (see Fig. 1(b)). This shows that with a proper sizing function, a high quality FE mesh with a fewer number of elements can be obtained. With a high-quality mesh, more accurate FE analysis results can be obtained. With a proper sizing function, the number of elements can be greatly reduced; therefore, memory usage and computation time can be greatly reduced during analysis. Consequently, there is great demand for the automatic generation of proper mesh sizing functions for assemblies.

A computational framework must automate the process of generating the mesh sizing function while meeting industry requirements. Automating the sizing function generation is important because manual specification of the sizing information is tedious and time-consuming, and for complex CAD models it may not be practical to specify the size manually. At the same time, the framework should provide the user with options/controls in generating a variety of meshes to satisfy the industry needs. The framework should consider both geometric and non-geometric factors that influence mesh sizing. The framework should generate one common sizing function for all the geometric entities of the assembly. This common sizing function must provide consistent element sizing across 1D, 2D, and 3D entities of the assembly. This framework is independent of the meshing algorithms that it is used in conjunction with. The next section gives a brief review of the literature on previous work in mesh sizing.

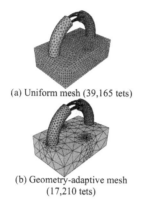

(a) Uniform mesh (39,165 tets)

(b) Geometry-adaptive mesh
(17,210 tets)

Fig. 1. Uniform and Graded Assembly Meshes

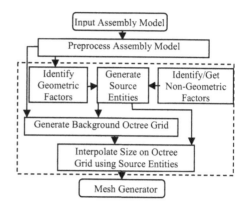

Fig. 2. Schematic Diagram of the Framework

2. LITERATURE REVIEW

This section discusses previous meshing approaches with more emphasis on sizing. Many of the previous meshing algorithms incorporate some sort of element/nodal spacing control; however, sizing is integrated with the meshing process in many cases. Also, consideration has been given to mesh sizing for a single part, not an assembly.

In the past, background meshes have been used as a mechanism for storing sizing function. In early advancing front methods [1], a background mesh consisting of simplicial elements (triangles) was manually constructed using sample points. The generation of a background mesh was later automated by generating Constrained Delaunay Triangulation (CDT) of a set of vertices. Cunha et al. [2] automate the placement of the background mesh nodes on the curves, followed by placement on surfaces, using curvature and proximity. Proximity is determined based on the distance between the facets and the nodes. Measuring proximity in this manner is a combinatorial problem and is generally time-consuming and less accurate. Owen and Saigal [3] use a natural-neighbor interpolation method on a background mesh to alleviate the abrupt variations in target mesh size. In their approach, the sizing function is critically dependent upon node placement in an initial background mesh.

One disadvantage of the background tri/tet mesh is that, while calculating the mesh size at a point, finding the tri/tet that contains the point is expensive. An alternative for storing mesh sizing function is the background grid. Pirzadeh [4] uses a uniform Cartesian grid to store the mesh sizing function. However, a uniform grid is not suitable for capturing large gradients of sizing function, and it consumes a large amount of memory. Another class of background grids that have overcome the uniformed grid limitations are non-uniform hierarchical grids called the "Quadtree" and "Octree" [5, 6]. The size of quadtree/octree cells (squares/cubes) depends on the subdivision of the bounding box, which is governed by the user-specified spacing function or a balance condition for the tree. The drawback of this approach is that quadtree and octree are orientation sensitive, and it is difficult to control the sizing gradient. Zhu et al. [7] use octree as a background overlay grid to store mesh sizing, rather than for the primary purpose of meshing. Their refinement is not directly based on the geometry of the domain, but rather on the sizing function gradient. Zhu [8] has extended the background overlay grid approach to consider premeshed geometric entities.

Researchers have also looked at limiting the gradients of sizing functions. Borouchaki et al. [9] present a corrective procedure to control the size gradation. In their method, gradients of a discretized size function are limited by iterating over the edges of a background mesh and updating the size function locally for neighboring nodes. Persson [10] proposes a method for limiting the gradients in a mesh size function by solving a nonlinear partial differential equation on the background mesh.

Another class of meshing approach relevant to this paper uses geometric skeletons, in particular medial axis transform (MAT) [11] for geometry-adaptive meshing. Researchers [12] use the radius function of MAT to control nodal spacing on the boundary and interior of a 2D domain while generating a triangular mesh adaptively. Quadros et. al. use the medial axis to generate adaptive quadrilateral meshes on surfaces by varying the width of the tracks using the radius function [13]. They also use skeletons in generating the mesh sizing function for surfaces [14] and single solids [15], by considering only geometric factors.

3. PROBLEM STATEMENT

Given an assembly A in R^3 and the bounds on mesh size, d_{min}, and d_{max}, and the upper bound on the discrete gradient α. Develop a framework for *automatically generating the mesh sizing function s, based on both geo-metric and non-geometric factors, such that,*

1. *Mesh size d = s (**p**) where point **p**(x,y,z) ∈ A and $d_{min} \leq d \leq d_{max}$*

2. *For any two grid nodes $**n_1**$, $**n_2**$ ∈ **O**, where **O** is the background*

   octree grid of A

$$| s(**n_1**.\text{coord}()) - s(**n_2**.\text{coord}()) | \leq \alpha || **n_1**.\text{coord}() - **n_2**.\text{coord}() ||$$

4. OVERVIEW OF THE FRAMEWORK

The schematic diagram of the computational framework for generating the mesh sizing function of assemblies is shown inside the dotted lines in Fig. 2. The input assembly consists of many single parts, which are repre-sented in B-rep format (e.g. ACIS sat file, supported by many commercial geometric modelers). The input assembly is preprocessed before passing it to the mesh sizing framework. Then source entities are generated based on the geometric and non-geometric factors to control the size. The source entities provide sizing information, and they are associated with a firmness level to facilitate the overriding of sizes among the different factors. In Section 6.1, a systematic study is performed to identify the geometric fac-tors that must be considered to completely measure the geometric com-plexity of an assembly. Section 6.2 discusses the non-geometric factors that influence the mesh size. Next, to store the sizing function, a back-ground octree grid is generated, instead of a background mesh, by using both geometric data and source entities (detailed in Section 7). Because a point containment test is expensive in a background mesh, a background grid is used, with a time complexity of $O (\log N)$, where N is the number of leaf cells. The final step is to interpolate the sizing function over the background grid using the source entities (detailed in Section 8). The in-terpolation scheme blends the size smoothly while respecting the firmness levels of geometric and non-geometric factors. During mesh generation, the target mesh size at a point is interpolated using the size stored at the grid-nodes of the octree cell containing that point.

5. PREPROCESS ASSEMBLY MODEL

In the preprocessing stage, an input assembly is made ready for the sizing function module. First, the assembly model is repaired, or healed, if necessary. To obtain a conformal mesh, imprint and merge operations are performed on the assembly model. Note that element sizes are affected by the imprint and merge operations as they change the geometry and topology of a domain at the interfaces. In Fig. 3, at 'A', the circular curve of the hole comes in close proximity with another curve, edge of the cube base. At 'B', the cube shown in green shares a common square interface surface with the circular base, shown in red. As the thickness of the circular base is much less than that of the cube, a fine mesh is required, even at the cube near the interface.

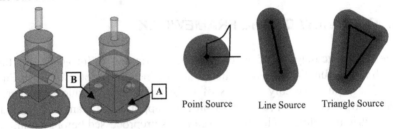

Point Source   Line Source   Triangle Source

Fig. 3. Before and After Imprint and Merge Operations

Fig. 4. Point, Line, and Triangle Source Entities

6. GENERATE SOURCE ENTITIES BASED ON GEOMETRIC AND NON-GEOMETRIC FACTORS

A source entity represents the size and gradient at a location due to a geometric or non-geometric factor. Fig. 4 shows source entities that are defined by: coordinate $c[]$; size $s[]$; scope $scp[]$; and local sizing function f. A source entity can be a point, a line segment, a triangle, or a tetrahedron, defined by $c[]$, $s[]$, and $scp[]$ (See Fig. 4). The local sizing function f can be constant (CONST), linear (LINEAR), geometric progression (GEOM), exponential function (EXP), etc. Source entities are also associated with a firmness level, l, and a list of incident geometric entities, L. The influence of source entities will be restricted to only the geometric entities present in the list of incident entities, L. The three levels of firmness are LIMP, SOFT, and HARD. The size is imposed only on the geometric entities of L. By controlling the size, scope, local sizing function, and firmness level, a variety of meshes can be generated.

6.1. Generate Source Entities Based on Geometric Factors of an Assembly

As it is difficult to analyze the geometric complexity of an assembly, A, at once, first the assembly A, embedded in $\mathfrak{R}^3$, is decomposed into disjoint subsets; then the geometric complexity of each subset is analyzed in reference to the FE mesh generation. This decomposition of an assembly into disjoint subsets is for the purpose of theoretical analysis only.

6.1.1. Disjoint Subsets of an Assembly

Let an assembly A contain N solids, S_i, where i = 1, 2 … N, which do not intersect. Here it is assumed that the imprinted and merged assembly A contains curvature-continuous interfaces and bounadries with no degenerate entities.

Fig. 5 shows the anatomy of A, where interior, boundary, and interface are denoted by "in", "bnd", and "inf", respectively. The root of the tree is assembly A, and the leaf nodes are the disjoint subsets, which are shown in rectangular blocks. Fig. 5 shows the disjoint subsets of assembly A, the interior of each solid, $in(S_n)$, the interior of each boundary surface, $in(F^*_m)$, the interior of each interface surface, $in(F^{**}_l)$, the interior of each boundary curve, $in(C^*_p)$, the interior of each interface curve, $in(C^{**}_q)$, and vertices, V_r. As the subsets are disjoint, the geometric complexity of each subset is independent of the others.

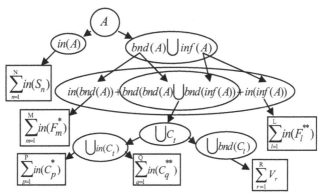

Fig. 5. Disjoint Subsets of an Assembly

6.1.2.   Geometric Factors for Each Disjoint Subset

The geometric complexity of each disjoint subset is analyzed in reference to FE meshing to determine the geometric factors [14, 15]. The only geometric factor essential to capture the complexity of the interior of a solid, $in(S)$, is 3D-proximity. 3D-proximity implies proximity between the vertices, curves, and surfaces that bound $in(S)$, which is a global measure of the geometric complexity of $in(S)$. The geometric factors which capture the complexity of the interior of a boundary surface, $in(F^*)$, or interface surface, $in(F^{**})$, are the 2D-proximity between its boundary curves and vertices, and curviness. 2D-proximity is a global measure, and curviness is a local measure of geometric complexity. The geometric factors of the interior of a boundary curve, $in(C^*)$ or interface curve, $in(C^{**})$, are the 1D-proximity between end vertices of a curve, curviness and twist [14]. As interior of a curve is a 1D entity embedded in 3D space, these three geometric factors are required in order to completely measure its complexity. No geometric factors are required for a vertex V; a FE node at each vertex is sufficient to represent the vertex in a FE model.

In the beginning of Section 6.1.1 it was assumed that curves and surfaces were curvature continuous, but cusps may exist (such as the tip of cone), and sharp bends in the curves and surfaces, where tangent and curvature vector are not well defined. These points, curves, and regions should be considered as hard points, curves, and regions [14]. These hard entities are not addressed in the remainder of this paper.

6.1.3.   Tools to Measure Geometric Factors of Each Subset

The following paragraphs discuss the tools needed to measure the geometric factors of each subset of A. In this paper, disconnected skeletons are proposed as the tools to measure the proximity in $in(A)$, $in(F^*)$, and $in(F^{**})$, more accurate tools than previous methods used by researchers.

A 3D-skeleton of an assembly is extracted and used as the tool to measure 3D-proximity in $in(A)$. On the PR-octree, the 3D-skeleton of an assembly is extracted by propagating a wave front from the boundary surfaces and the interface surfaces towards the interior of an assembly, similar to that of extracting the 3D-skeleton of a single solid [16]. The three important phases of the wave propagation are: initiating the initial wave front (see Fig. 6(a)), propagating the wave front, and terminating the wave. Note that in Fig. 6(a), the wave is propagated in both directions at the interfacial surfaces, whereas at the boundary surfaces the wave is only propagated inward. The skeleton points are generated where the wave terminates. The distance traveled (also called the radius) by the wave at the skeleton points measures 3D-proximity. In Fig. 6(b), the skeleton points where the wave has traveled the least are shown in red and the furthest in blue.

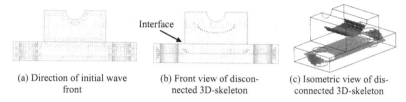

(a) Direction of initial wave front

(b) Front view of discon-nected 3D-skeleton

(c) Isometric view of dis-connected 3D-skeleton

Fig. 6. Initial Front and 3D-skeleton

A disconnected skeleton of a surface provides local thickness informa-tion, and it is used as a tool to measure the proximity in $in(F^*)$ and $in(F^{**})$. A disconnected 2D-skeleton of the surfaces is generated by com-bining the concept of medial axis transform (MAT) and chordal axis trans-form (CAT) [14]. The skeleton generated using this method is computa-tionally efficient and is sufficiently accurate for the purpose of this work.

Fig. 7 shows how a skeleton effectively measures changes in surface complexity before and after imprint operations. Fig. 7(a) shows the whole assembly model and the magnified view of a bolt and a nut. The 2D-skeleton of the top surface of the flange before the imprinting is shown in Fig. 7(b). Note that after imprinting the circular base of the hexagonal bolt-head on the top surface of the flange, a thin circular ring appears at the holes (Fig. 7(c)). Thus the 2D-skeleton measures the proximity between the outer and inner circles of these thin circular rings.

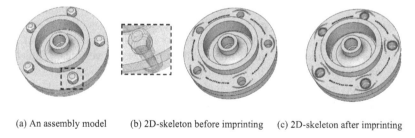

(a) An assembly model

(b) 2D-skeleton before imprinting

(c) 2D-skeleton after imprinting

Fig. 7. 2D-skeleton Before and After Imprint

Other tools used to measure other geometric factors mentioned in Sec-tion 6.1.2 are briefly discussed here. The curviness in $in(F^*)$ and $in(F^{**})$ is measured using the minimum principal radius of curvature[2, 3, 14, 17].

The proximity between the end vertices in $in(C*)$ and $in(C**)$ is measured using the length of a curve. The curviness of $in(C*)$ and $in(C**)$ is measured using the curvature of a curve [2, 7, 14]. The twist of $in(C*)$ and $in(C**)$ is measured using torsion [14].

6.1.4. Generate Source Entities using the Tools

Source entities are generated using the tools for controlling the size due to geometric factors. The source entities cover the disjoint subsets in order to reflect the influence of the geometric factors. All the source entities generated based on the geometric factors will be of LIMP firmness level, to give high precedence to non-geometric factors. The following paragraphs explain how to generate the source entities using skeletons and other tools.

A 3D-skeleton is converted into point sources with a size and local sizing function given by Equation (1). The *num_of_layers* controls the number of layers of finite elements across the thickness. A LINEAR local sizing function is used with *end_factor* = 0.1 to have a slightly larger size at the interior of $in(S)$. The coordinate, **c**, and the scope, *scp*, of a source point is set equal to the skeleton point coordinate and the radius, R, respectively. Note that size due to skeleton-based source points are interpolated over the entire grid to cover $in(A)$.

$$s = \frac{2R}{num\_of\_layers}, f(r) = s - \frac{r}{scp} \times (s \times end\_factor) \tag{1}$$

Triangle sources can be generated using the 2D-skeleton points to control size over $in(F*)$ and $in(F**)$. The 2D-skeleton radius can be interpolated over the facets (obtained from the geometry engine) to generate triangle sources (Equation 1). A GEOM function is used as local sizing function f, as given in Equation (2) [8]. In Equation (2), r represents the distance between a point **p** (inside the scope) and its projection **p'** on the triangular source. Growth factor g is set to 1.2, and s_0 is the size at **p'**, which is calculated using linear interpolation of sizes on vertices. Thus the size on a source entity is influenced by sizes on the vertices; the sizes at surrounding regions inside the incident entities are controlled by the growth factor. Note that at interfaces $in(F**)$, the size is radiated inside the incident entities L.

$$f(r) = s_0 \times g^n, n = \frac{\ln\left(\frac{r(g-1)}{s_0} + g\right)}{\ln(g)} \tag{2}$$

The GEOM function is used at both the triangle and the line sources (obtained from facets) on the surfaces and curves, respectively, to radiate the size into incident entities. In $in(F^*)$ and $in(F^{**})$, triangular surface curvature-based sources are generated using minimum principal radius of curvature and maximum spanning angles [2, 14, 17] (Fig. 9(c)). In $in(C^*)$ and $in(C^{**})$ line sources are generated using the tools used for measuring geometric factors of curves [14] (Fig. 9(d)).

6.2. Generate Source Entities Based on Non-Geometric Factors

The following paragraphs discuss non-geometric factors influencing mesh size during simulation of complex assemblies. The firmness level of the source entities can be set depending on the requirement.

User-Defined Size: Source entities are also generated based on the user's input. Experienced users in industry may prefer specific sizes at certain regions based on their knowledge and experience. The framework can handle user-specified sizes on solids, surfaces, curves, and vertices. Users can control the size by specifying size, scope, and local sizing function. Facets of the specified surfaces and curves are extracted, then, based on the user's input, triangle, line, and point sources are generated. The firmness level is set to SOFT to override the sizing functions of geometric factors.

Pre-Meshed Entities: The source entities due to pre-meshed entities are generated to obtain a smooth transition in element size at the pre-meshed entities. Pre-meshed entities often appear while meshing assembly models; they could be surfaces or curves. The 1D/2D elements of the pre-meshed entities are converted into line/triangle sources. Edge lengths of the 1D/2D element are used as size on the source entities, and the growth factor for the GEOM function is set to 1.2. The firmness level is set to SOFT.

Meshing Scheme: Some meshing schemes require a specified number of elements/intervals on the boundary curves. For example, quad meshing algorithms require an even number of 1D elements overall on the boundary curves of a surface. Also, in mapped surface meshing schemes the same number of intervals is required on pairs of opposite curves. These constraints can be imposed by artificially generating the line sources on a curve and by setting the firmness level to HARD.

Previous Analysis Results: Previous analysis results can be incorporated into the sizing function by generating source entities. For example, in a region of stress concentration, the size of source entities can be calculated based on the stress; local sizing function can be controlled based on the stress gradients, which can then be used in remeshing a model for the next iteration of analysis.

Boundary Conditions: At a boundary condition, the elements' sizes can be controlled before running the analysis by generating source entities using the domain knowledge of the analysis. For example, consider a point load boundary condition, which is frequently encountered in structural/solid mechanics problems. The element size can be controlled using a suitable local sizing function based on radial stress distribution.

7. GENERATE BACKGROUND OCTREE GRID

The background octree grid of an assembly is first generated using geometric information and then refined based on source entities. A PR-Octree is generated using graphics facets because it captures geometric features and provides a suitable lattice for storing a sizing function[15]. The vertices and centroids of facets of an assembly model are given as input point list while generating a PR-Octree starting from the bounding box of the assembly. The graphics facets capture the boundary curvature and small features, and hence small-sized facets exist at high curvature regions and at fine features, which results in finer cells in those regions. The PR-Octree is further subdivided based on the size and sizing gradient of source entities. The cells are further subdivided until only one level of depth difference is maintained between the adjacent cells; this ensures a smooth transition in grid cells for storing the sizing function.

The octree cells intersecting with the boundary facets are efficiently identified using the separating axis theorem [18], and these boundary cells and their nodes are colored gray. Then inner nodes and cells are colored black. Octree cells and nodes lying outside the assembly are removed to reduce memory usage and computational time.

8. INTERPOLATE SIZE ON BACKGROUND GRID USING SOURCE ENTITIES

Interpolating mesh size over the background octree grid using source entities is discussed in two steps. Section 8.1 explains how to interpolate sizing function due to a single factor, geometric or non-geometric. In Section 8.2, the blending of the sizing functions, due to different factors based on firmness level, is explained.

8.1. Interpolate Sizing Function of a Factor

First, every source entity is linked with the grid-nodes that fall inside its scope. The linking process starts from a set of initial nodes of the octree cells that intersect with the source entity. The grid-nodes are visited in breadth-first traversal, starting with the initial nodes, until all the grid-nodes contained inside its scope are linked with the source entity.

The size at a grid node n, due to a factor k, is interpolated by taking the weighted sum of the sizes determined by the local sizing function f of m source entities, linked to that grid-node, as given in Equation (3). Here, r_i is the distance between the ith source entity and grid-node n, and $f_i(r_i)$ is the size determined by the local sizing function of the ith source entity. k could be any geometric or non-geometric factor. This ensures that a source entity that is closer and has a smaller size has a greater influence on the grid-node.

$$s_k = \sum_{i=1}^{i=m} f_i(r_i) \times \left(\frac{W_{i\_dist} + W_{i\_size}}{2} \right), \ W_{i\_dist} = \frac{\frac{1}{r_i^2}}{\sum_{j=1}^{j=m} \frac{1}{r_j^2}}, \ W_{i\_size} = \frac{\frac{1}{f_i(r_i)^2}}{\sum_{j=1}^{j=m} \frac{1}{f_j(r_j)^2}} \tag{3}$$

8.2. Blend Sizing Functions of Geometric and Non-Geometric Factors

This section gives details of blending different sizing functions. The firmness level is used in overriding a sizing function of one factor with that of another. All the sizing functions due to geometric factors have a firmness level of LIMP, and the sizing functions due to non-geometric factors have firmness level of either SOFT or HARD, depending on the firmness level of a source entity (see Section 6.2).

As all the geometric factors have LIMP firmness, the final size s at each grid-node is calculated as given in Equation (4), where k represents different geometric factors. w are the weights, which are initially set to 1.0. These weights can be controlled to achieve variation in meshes. The *scale* is used to control the overall coarseness of the mesh.

$$s = \min\{s_k \cdot w_k\} \cdot scale \tag{4}$$

The sizing function generated by combining all the geometric factors can be overridden by the sizing function of non-geometric factors, as the non-geometric factors have a higher firmness level. The maximum size, d_{max}, and minimum size, d_{min}, are enforced by trimming the sizing function stored on the background grid.

Smoothing techniques are used to alleviate the abrupt gradients caused by combining the sizing functions of the geometric factors, and by overriding the sizing function due to non-geometric factors. The concepts of digital filters used in smoothing 2D images are used here in smoothing the mesh sizing function stored on the octree. First two iterations of median filter are used to remove the sharp gradients, and then a modified mean filter is used iteratively to smooth the gradients. During mesh generation, trilinear interpolation is used to calculate the target mesh size at $\mathbf{p}$, using the size at the grid-nodes of the cell containing $\mathbf{p}$. Thus the target mesh size is calculated in $O(max\_depth)$, where max_depth is the maximum depth of the octree.

9. RESULTS AND DISCUSSION

The proposed framework has been implemented in C++ within CUBIT, a finite element mesh generation toolkit by Sandia National Laboratories. The proposed framework has been tested on industrial assembly models and results obtained on three such models are shown in Fig. 8 to Fig. 10 (original models courtesy of Ansys, Inc.).

Fig. 9 shows the components of the proposed framework by considering only geometric factors in a twelve-volume assembly. Fig. 9(a) and Fig. 9(b) show 16,037 3D-skeleton points and 16,474 2D-skeleton points, which are used as tools to measure 3D-proximity and 2D-proximity, respectively. Fig. 9(c) and Fig. 9(d) show 7,405 triangle sources and 5,320 line sources on surfaces and curves, respectively. Fig. 9(e) shows the mesh size on the background octree grid of $min\_depth = 5$ and $max\_depth = 7$ (as in all three models) using a color scale. Fig. 9(f) shows the geometry-adaptive mesh containing 83,584 tet elements.

Fig. 8(b) shows a mesh generated by combining both the geometric factors and sizing based on a pre-meshed surface (see Fig. 8 (a)) in a four-volume assembly. The number of tris in the pre-mesh and average edge length of the tris were 787 and 2.00 respectively. 787 triangle sources were generated using the tris of the pre-meshed surface with sizes calculated based on the edge lengths of tris. A GEOM local sizing function with growth factor = 1.2 was used. The sectional view in Fig. 8(b) shows a smooth transition in the size of the tet elements at the pre-meshed surface. The tet mesh shown in Fig. 8(b) consisted of 44,462 tets.

Fig. 10 shows the meshes generated by combining the geometric factors and a user-defined size in a two-volume assembly. Fig. 10(a) contains 68,325 tets that are generated using geometric factors only. Fig. 10(b) contains 73,289 tets that are generated by incorporating user-defined size specified on the top surface. The user has requested four layers of tet elements of size 2.5. Triangle sources were generated by extracting the ACIS facets with size = 2.5, scope = 10.0, and a CONST local function.

For the assembly in Fig. 9, the maximum and minimum sizes at the grid nodes were 37.20 and 1.29 respectively, which were within the user specified bounds d_{max} = 40, and d_{min} = 1.0. Similarly, maximum and minimum sizes were 15.05 and 1.32, respectively, for Fig. 8. In Fig. 9, maximum and minimum sizes were 40.68 and 1.38, respectively.

Table 1 shows the computational time taken by the components of the framework in generating sizing function for the assembly models shown in Fig. 8 to Fig. 10. The timings were measured in e*machines* M6811 notebook. From Table 1 it is observed that interpolation has taken the most time in all three models. Note that octree generation time in assemblies shown in Fig. 8, Fig. 9, and Fig. 10 is proportional to the number of graphics facets extracted (2,927, 12,161, and 7,346 facets, respectively). There is a strong correlation between number of source entities and computational time for interpolation. The number of source entities in the three models were 11,211, 45,236, and 44,957.

Table 2 shows that the approach used in extracting disconnected 3D and 2D skeletons are computationally efficient. The 2D- and 3D-skeleton times are accompanied by their generation times relative to the total source entity generation times for each assembly. The 2D-skeleton times are much larger than the 3D-skeleton times due to a larger number of surfaces than volumes in the models. For example, the model in Fig. 10 contains only two volumes but has 186 surfaces.

Table 1. Computational Time (sec) in Generating Mesh Sizing Function

	Figure 8	Figure 9	Figure 10
Octree Generation	1.689	4.438	3.999
Source Entity Generation	0.702	2.547	2.501
Interpolation	2.188	7.453	9.000
Total	4.579	14.438	15.500

Table 2. Computational Time (sec) in Generating Disconnected Skeletons

	Figure 8	Figure 9	Figure 10
3D-Skeleton	0.062 (8.9%)	0.234 (9.2%)	0.282 (11.3%)
2D-Skeleton	0.312 (44.4%)	1.094 (43.0%)	0.782 (31.3%)

10. CONCLUSION

In this paper a computational framework for generating a mesh sizing function for assembly meshing is proposed. The framework generates a sizing function by considering both geometric and non-geometric factors. A systematic study has been performed to determine the geometric factors that influence the mesh size. Disconnected 3D and 2D skeletons are extracted and used for measuring the proximity. The proposed framework is computationally efficient and it has been tested on many industry models to verify the effectiveness.

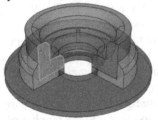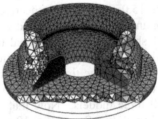

(a) Assembly model with a pre-meshed surface   (b) Sectional view of a tet mesh generated using
                                                  geometric factors and a pre-meshed entity

Fig. 8. Mesh Sizing due to Geometric and Non-geometric Factors (pre-meshed entity)

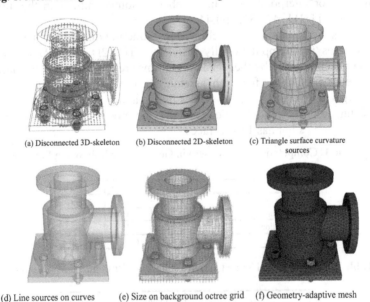

(a) Disconnected 3D-skeleton   (b) Disconnected 2D-skeleton   (c) Triangle surface curvature
                                                                  sources

(d) Line sources on curves   (e) Size on background octree grid   (f) Geometry-adaptive mesh

Fig. 9. Components of the Framework

(a) Graded mesh based on only geometric factors

(b) Graded mesh based on geometric factors and a user defined size

(c) Enlarged view of the mesh at a user-defined surface

Fig. 10. Mesh Sizing due to Geometric Factors and a Non-Geometric Factor (user defined size)

REFERENCES

[1]   R. Lohner and P. Parikh, "Generation of Three-Dimensional Unstructured Grids by the Advancing Front Method," *AIAA-88-0515*, 1988.

[2]   A. Cunha, S. A. Canann, and S. Saigal, "Automatic Boundary Sizing For 2D and 3D Meshes," *AMD Trends in Unstructured Mesh Generation, ASME*, vol. 220, pp. 65-72, 1997.

[3]   S. J. Owen and S. Saigal, "Neighborhood Based Element Sizing Control for Finite Element Surface Meshing," *Proceedings, 6th International Meshing Roundtable*, pp. 143-154, 1997.

[4]   S. Pirzadeh, "Structured Background Grids for Generation of Unstructured Grids by Advancing-Front Method," *AIAA*, vol. 31, 1993.

[5]   W. C. Tracker, "A Brief Review of Techniques for Generating Irregular Computational Grids," *Int. Journal for Numerical Methods in Engineering*, vol. 15, pp. 1335-1341, 1980.

[6]   M. S. Shephard, "Approaches to the Automatic Generation and Control of Finite Element Meshes," *Applied Mechanics Review*, vol. 41, pp. 169-185, 1988.

[7]   J. Zhu, T. Blacker, and R. Smith, "Background Overlay Grid Size Functions," *Proceedings of 11th International Meshing Roundtable*, pp. 65-74, 2002.

[8]   J. Zhu, "A New Type of Size Function Respecting Premeshed Entities," *12th International Meshing Roundtable*, 2003.

[9]   H. Borouchaki and F. Hecht, "Mesh Gradation Control," *6th International Meshing Roundtable*, 1997.

[10] P.-O. Persson, "PDE-Based Gradient Limiting for Mesh Size Functions," *Proceedings, 13th International Meshing Roundtable*, pp. 377-388, 2004.

[11] H. Blum, "A Transformation for Extracting New Descriptors of Shape," *Models for the Perception of Speech and Visual Form Cambridge MA The MIT Press*, pp. 326-380, 1967.

[12] V. Srinivasan, L. R. Nackman, J. M. Tang, and S. N. Meshkat, "Automatic Mesh Generation using the Symmetric Axis Transformation of Polygonal Domains," *Proc. IEEE*, vol. 80(9), pp. 1485-1501, 1992.

[13] W. R. Quadros, K. Ramaswami, F. B. Prinz, and B. Gurumoorthy, "Automated Geometry Adaptive Quadrilateral Mesh Generation using MAT," *Proceedings of ASME DETC*, 2001.

[14] W. R. Quadros, S. J. Owen, M. Brewer, and K. Shimada, "Finite Element Mesh Sizing for Surfaces Using Skeleton," *Proceedings, 13th International Meshing Roundtable*, pp. 389-400, 2004.

[15] W. R. Quadros, K. Shimada, and S. J. Owen, "Skeleton-Based Computational Method for Generation of 3D Finite Element Mesh Sizing Function," *Engineering with Computers*, 2004.

[16] W. R. Quadros, K. Shimada, and S. J. Owen, "3D Discrete Skeleton Generation by Wave Propagation on PR-Octree for Finite Element Mesh Sizing," *ACM Symposium on Solid Modeling and Applications*, 2004.

[17] S. J. Owen and S. Saigal, "Surface Mesh Sizing Control," *International Journal for Numerical Methods in Engineering*, vol. 47, pp. 497-511, 2000.

[18] D. Eberly, "Intersection of Convex Objects: The Method of Separating Axes," *Magic Software, Inc.*, 2003.

Unstructured Computational Meshes for Subdivision Geometry of Scanned Geological Objects

Andrey A. Mezentsev[1] Antonio Munjiza[2] and John-Paul Latham[1]

[1] Department of Earth Sciences and Engineering, Imperial College London, London, UK (A.Mezentsev),(J.P.Latham)@imperial.ac.uk
[2] Department of Engineering, Queen Mary University of London, London, UK A.Munjiza@qmul.ac.uk

Summary. This paper presents a generic approach to generation of surface and volume unstructured meshes for complex free-form objects, obtained by laser scanning. A four-stage automated procedure is proposed for discrete data sets: surface mesh extraction from Delaunay tetrahedrization of scanned points, surface mesh simplification, definition of triangular interpolating subdivision faces, Delaunay volumetric meshing of obtained geometry. The mesh simplification approach is based on the medial Hausdorff distance envelope between scanned and simplified geometric surface meshes. The simplified mesh is directly used as an unstructured control mesh for subdivision surface representation that precisely captures arbitrary shapes of faces, composing the boundary of scanned objects. CAD model in Boundary Representation retains sharp and smooth features of the geometry for further meshing. Volumetric meshes with the MezGen code are used in the combined Finite-Discrete element methods for simulation of complex phenomena within the integrated Virtual Geoscience Workbench environment (VGW).

Key words: laser scanning, unstructured mesh, mesh simplification, subdivision surfaces

1 Introduction

Recent developments in the Finite Element method (FEM) and advances in power of affordable computers have broadened the FEM application area to simulation of complex coupled phenomena in natural sciences, geology, biology and medicine in particular [Zienkiewicz]. The formulation of the combined Finite-Discrete element method (FEM-DEM) in the nineties [Munjiza] has established a connection between the continuous and discrete modeling of complex coupled phenomena. Such a formulation opens a possibility for development of integrated Virtual Prototyping Environments (VPE) in natural sciences, similar the VPE found in engineering [Latham].

VPE is typically a unification of highly inhomogeneous interacting computational components, representing the models on different levels of mathematical abstraction. For the success of VPE in natural sciences it is highly desirable to provide unified means for model representation on different levels of the models abstraction: the so-called micro and macro levels (sometimes also addressed as Mechanics of Continua and Discontinua) [Munjiza],[Latham]. For a micro level of simulation the model approximates continuous fields of system variables and systems of partial differential equations form the mathematical model. On macro level of simulation discontinuous fields of systems variables are approximated by the model and mathematically are represented by systems of ordinary or differential-algebraic equations. The FEM-DEM method is a unique computational technology, which permits representation on different levels of modeling to be combined: both micro level and macro level. Methodologically it provides a unified framework for simulations within the framework of natural sciences VPE.

As a starting point for simulation, both the FEM and the DEM require domain discretisation into a set of geometrical simplicies - a mesh. For many natural sciences applications, and specifically in geology, the main problem, making the workflow very complex, is related to absence of fully automatic methods of geometry definition and meshing. Most of the natural objects, i.e. geological particles or bio-medical entities, are characterized by complex shape that can only be captured with sophisticated scanning equipment. With increasing robustness of scanning technology it is has become possible to use realistic point-wise scanned data to define natural object geometries for simulations. Unfortunately, the output from scanned data is not usable directly for meshing and there has been much recent research reported in the area of process automation (see, for example, [Bajaj] –[Xue]).

It should be stressed that geometry definition and downstream computational mesh generation are very application specific. Moving to a new application area generally requires development of a new geometric model with different parameters, meeting specific requirements of downstream applications. Importantly, most of the developed geometric formats do not fully address discretisation requirements from the point of view of the efficiency of organization and application in the VPE, pursuing rather conflicting requirements for geometric models. The present paper addresses this problem from the point of view of CAD/mesh integration for the Virtual Geoscience Workbench environment (VGW) in natural sciences, reflecting a growing shift from stochastic to deterministic models in geological simulations.

The rest of the paper is organized as follows. In Section 2 the automatic methods for geometric models derivation based on discrete data are discussed together with basic principles of subdivision surfaces. In Section 3 a new mesh simplification concept using medial Hausdorff distance is presented. In Section 4 numerical results are given, while Section 5 gives future work and conclusions.

2 CAD Definition from Discrete Scanned Data

With the development of new scanning technology it is now possible to create large data bases of point-wise data in different areas of science and engineering. Increased accuracy of scanning permits acquisition of data sets, containing millions of data points and precisely defining the shape of different objects. Unfortunately, this information cannot be directly used in the process of the computational model definition. The size of data sets dictates development of automatic conversion methods of scanned data to geometric and further to computational models. This problem has received in recent years a lot of attention in computer vision, computational geometry and mesh generation communities [Bajaj]–[Xue].

Typically, geometry of scanned objects is defined by the boundary, using different incarnations of the so-called Boundary Representation (BREP) model. The BREP model combines surfaces with elements of topology, organized in a tree form (see, for example, [Mezentsev]–[Lang]). Most popular choices for faces underlying representation in BREP are piecewise polyhedral meshes [Owen] or spline surfaces, approximating discrete data [Lang]. As scanned data sets contain redundant data points, not corresponding to the underlying geometric complexity of the objects, initial discrete data requires simplification in most cases. Scanned discrete representation also greatly differs from application area to application area, and it changes significantly with the scanning technology used, so it is necessary to define specifics of considered input data sets.

2.1 Specifics of Scanned Data and Geometry in the Geological Applications

The development of a highly automated method of converting scanned data to surface geometry is considered. It should work with clouds of points, organized as unstructured set of X, Y, Z coordinates, type of data is common for reverse engineering, image recognition and computer visualization problems [Bajaj],[Frey]. It does not have ordered sub-parallel sliced structure, frequently found in the tomography-based bio-medical applications [Cebral]. In the considered case, the data consists of a dense bounded noisy cloud of points, lying on the boundary of the domain (see, for example Fig. 1 and Fig. 2). The discrete data have the following features [Frey]:

1. Data may be very noisy
2. Data sets are very dense (Fig. 2)
3. Straightforward approaches (like Marching Cubes [Lorensen]) frequently introduce errors of polygonal approximation, the so-called staircase effects
4. Surface reconstruction algorithms are not targeted to produce computational meshes, so the quality of meshes is low.

Addressing the specifics of the objects under consideration, it could be observed, firstly, that geological particle geometry is constrained, but not limited to a single-connected domain. It is mostly convex with random combinations of smooth rounded C_1 and highly irregular regions with sharp C_0 edges. The set up of the problem has clearly different geometry requirements in many other application areas. Secondly, data is very noisy and it is likely to have isolated scanned data points largely off-set from the reconstructed surface. This feature requires special measures to be taken to insure stability of surface reconstruction and simplification.

The problem of surface reconstruction from extremely noisy data is far from solved and a number of research papers have been published recently, addressing specific types of smooth surfaces in certain application areas (see, for example [Kolluri]). However, none of the papers address geological geometry, which requires reconstruction for the complete hull without unresolved areas, i.e. lower part of the geometry.

Thirdly, specifics of the usage of the geometrical model in the VPE simultaneously require efficiency of the geometry storage, access, rendering and meshing for

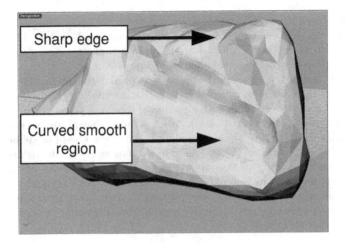

Fig. 1. Typical particle geometry as a combination of curved smooth and non-smooth sub-regions with sharp edges

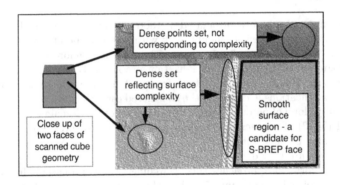

Fig. 2. Density of scanned points and geometric complexity of underlying geometry for two perpendicular faces of cube-type geometry

multiple particles at a time. Note that the VGW applications are designed to handle millions of free form particle objects with sharp and rounded features, similar to the example shown in Fig. 1.

2.2 Related Surface Reconstruction and Simplification Techniques

A comprehensive survey of recent surface reconstruction and simplification methods could be found in [Frey] and in [Kolluri]. For the sake of completeness a surface reconstruction method for the discrete data problems described above is outlined here. Typically, surface reconstruction is a two stage process, firstly a Delaunay tetra-hedrization is constructed for a set of scanned points. Secondly, a polygonal surface is extracted from volumetric discretisation using formal or heuristic techniques. In one of the most robust formal approaches, Boissonnant and Cazals [Boissonnat] success-fully applied natural neighbor interpolation for surface reconstruction. Taking into consideration, that for mostly convex configurations of geological scanned particle data (see, for example, Fig. 1) outward pointing normals are known from scanning device the aforesaid method could also be used in the proposed technique. Together with interpolation of signed distance functions, proposed by Hoppe et al. [Hoppe] this method permits reconstruction of triangulated surfaces, corresponding to the initial dense and largely redundant set of scanned points. However, for the discussed geological geometry, special measures should be taken to preserve distinctive sharp features of the geometry with the C_0 continuity. The next stage of the algorithm applies a mesh simplification algorithm, based on the discrete Hausdorff distance between initial scanned points set triangulation M and simplified triangulation set M_s. Let us recall, that firstly the so-called directional Hausdorff distance can be defined as follows:

$$h(M, M_s) = max_{m \in M} \ min_{n \in M_s} \| m - n \| \tag{1}$$

Here, h - the directed Hausdorff distance from M to M_s, will be small when every point of M is close to some point of M_s. The symmetric Hausdorff distance H will be as follows:

$$H(M, M_s) = max\{h(M, M_s), \ h(M_s, M)\} \tag{2}$$

However, the distances in (1) and (2) are rather fragile for a noisy scanned set. For example, a single point in M that is far enough from any point in M_s will cause h to be large. For better results, Hausdorff distances for geological scanned data requires re-formulation, reflecting possible presence of such points in the data or, alternatively, filtering of points prior to the mesh simplification may be used.

The proposed mesh simplification procedure involves iterative removal of redun-dant mesh nodes, not corresponding to the geometric complexity of the underly-ing surface. Typically, the node is removed from the mesh and resulting void is re-triangulated. Should the deviation δ (Fig. 3) of re-triangulation be within the tolerance envelope of the mesh, based on the directed Hausdorff distance H, node removal is successful. If not, the mesh node is associated with the surfaces geomet-ric complexity and should be retained (see Fig. 3 and Fig. 4). In our case the main challenge is to develop robust technique of simplification, suitable for noisy scanned data in geological applications.

In stand-alone computational applications optimization of the simplified mesh with respect to the requirements of simulation methods produces good results [Bajaj], [Karbacher], [Frey], [Cebral]. However, the necessity of having fast geom-etry visualization with different levels of smoothness and details requires develop-ment of a very specific geometric model for the VGW applications, equally efficient in computer graphics and simulation related (e.g. computational mesh generation) applications. This problem is discussed in the following section.

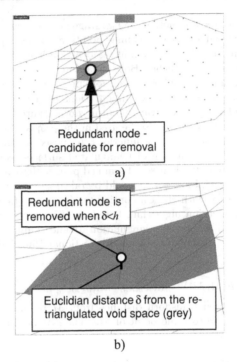

Fig. 3. Initial mesh (a), redundant node and re-triangulated simplified mesh (b)

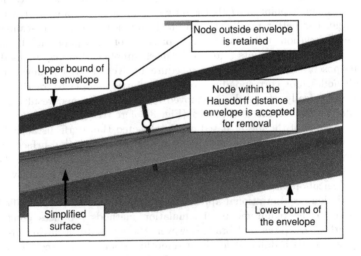

Fig. 4. Hausdorff distance envelope (upper and lower surfaces) for a given re-triangulated face, shown in Fig. 3 b), side view

2.3 Related Geometric Model

The BREP model is an efficient way of geometry definition for meshing (the topological tree of the BREP model is shown in Fig. 5). However, the BREP model should have so-called watertight properties, uniquely defining objects boundary without gaps or overlaps [Mezentsev], [Owen], [Beall]. Frequently, the classic BREP models based on splines also contain faces that are too small for quality mesh generation. The faces are unified to form bigger entities the so-called Super Elements (SE) and Constructive Elements (CE) during the BREP model meshing. From the mesh generation point of view, the best geometric model should contain only one face, covering the entire model.

Here the detailed discussion of the BREP model topology will not be given (see [Mezentsev] for details), only the concept of the subdivision BREP or S-BREP model is introduced. S-BREP is a new modification of the BREP model, with a full topological tree providing information on the hierarchy and mutual relationships of the models elements. The main building block of the S-BREP model is a face, which is based on a specific underlying surface definition. Most BREP models apply Non-Uniform Rational B-Spline (NURBS) curves and surfaces.

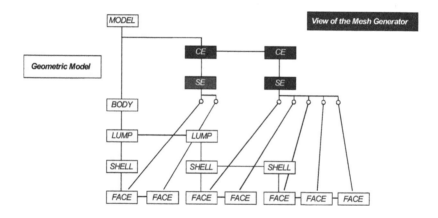

Fig. 5. Topological tree of classical BREP geometric model: CAD and mesh generation representations. Meshing typically requires coarsening of faces to unions - Super Elements (SE) and Constructive Elements (CE)

Alternatively, having both visualization and mesh generation in mind for our applications, the interpolating subdivision surface (see Zorin et al. [Zorin]) is used to define underlying face geometry in the S-BREP model. No trimming curves are required, as subdivision surface faces could be of arbitrary shape and topology. Previously interpolating subdivision surfaces have been used in mesh generation by Rypl and Bittnar [Rypl] and Lee [Lee1], [Lee2]. In [Lee1] Lee also has proposed a simplified topological model based on the subdivision surface geometry, which is further developed and generalized in present work. Let us recall main features of the subdivision surfaces.

The idea of interpolating subdivision is in representation of a smooth curve or surface as a limit of successive subdivisions of initial mesh. By starting with initial coarse mesh new positions of the inserted points are calculated according to predefined rules. In most cases local rules how to insert points (i.e. weighted sum of surrounding nodes coordinates [Zorin], [Dyn], [Loop]) and how to split the elements of the previous mesh are used. The resulting subdivision mesh will be smoothed out so the angles between adjacent elements will be nearly flattened. Eventually, after an infinite number of refinements, a smooth curve or surface in differential geometric sense can be obtained. For example, Fig. 6 a) shows a number of successive subdivisions for a curve. Initial coarse mesh (left, nodes are represented by hollow points) is refined by insertion of new nodes (shown as filled points) to obtain rather smooth curve representation (right).

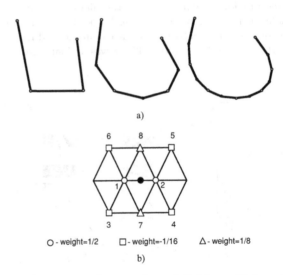

Fig. 6. Principles of subdivision: a) successive subdivisions of interpolating subdivision curve b) Butterfly subdivision scheme

The advantages of subdivision algorithms are that the schemes are local and surface representation will be good enough for most applications after a small number of refinement steps. Moreover, surface at any point can be improved arbitrarily by applying more local refinements. As a basis for the S-BREP geometry the so-called interpolating Butterfly scheme is used, proposed initially by Dyn et al. [Dyn]) and later modified by Zorin et al. [Zorin]. The scheme could be applied for an arbitrary connectivity pattern of triangular mesh and uses eight points of the coarse level (Fig. 6 b), hollow points, triangles and quads) to compute position of the node on the new level of refinement (filled point, regular case). Note that the position of nodes on the previous level is retained. The following formula is used for computation of the regular node position:

$$X_p = \frac{1}{2}(X_1 + X_2) - \frac{1}{16}(X_3 + X_4 + X_5 + X_6) + \frac{1}{8}(X_7 + X_5) \qquad (3)$$

For nodes with valence different from six (extraordinary internal and external nodes) different subdivision rules with different weights are applied. A complete set of rules for the modified Butterfly scheme could be found in [Zorin].

Interpolating subdivision surface is a generalization of spline surfaces for control net of arbitrary topology. Modified Butterfly scheme gives in the limit a C_1- continuous surface and tangent vectors could be computed at any point of the surface. With reference to the triangular surface meshes considered in this study, it is also possible to apply the Loop scheme [Loop]. However, the modified Butterfly scheme produces better results on sharp corners without special topological features, producing only minor smoothing. Note that our approach models C_0 features of geometry using the tagging process, as proposed by Lee [Lee1]. For smooth regions of particle geometry the simplified geometric mesh is directly used as a control mesh of interpolating subdivision surface, sharp edges of particles are represented by discontinuities between subdivision faces.

3 Proposed Method

As it was identified in Section 2, for efficient geometry definition and storage within VGW a new definition of the Hausdorff distance for automatic scanned mesh simplification is required. It is also shown here, how the S-BREP model is constructed and utilized for mesh generation.

3.1 Medial Hausdorff Distance

For a noisy initial set of scanned points it is difficult to apply classical formulation of the Hausdorff distance (1) for initial data simplification. In pattern recognition, the modified *medial* Hausdorff distance has been introduced and successfully used (see, for example, [Chetverikov]) for pattern recognition on noisy data:

$$h^f(M, M_s) = f^{th}_{m \in M} \, min_{n \in M_s} \parallel m - n \parallel \qquad (4)$$

where $f^{th}_{m \in M}$ denotes the $f - th$ quantile value of $g(x)$ over the set X for a value of f that is between zero and one. When $f = 0.5$ the so-called medial Hausdorff distance [Chetverikov] is received, which is used in our method. This measure generalizes the directed/symmetric Hausdorff distance measure, by replacing the maximum with a quantile. Medial Hausdorff distance is computed in a discrete form, similar to discrete computations in traditional formulation, proposed by Borouchaki in [Borouchaki]. Applied on a patch-wise basis, proposed modified formulation of the Hausdorff distance permits to perform filtering of the noisy data for geological and engineering applications, as demonstrated below.

3.2 Automatic Patching

Next step involves construction of the medial Hausdorff envelope on both sides of the dense scanned mesh M, as discussed in Section 2.2 and shown in Fig. 4. Together with smoothes criterion (similar to one formulated by Frey in [Frey]) discrete normal deviation of the simplified mesh M_s is ensured to be changing smoothly. It appears, that this method works well, as its curvature modification was also successfully used in [Karbacher]. In the proposed approach normal deviation of triangles is used to automatically define patching of the whole geometric model to individual faces. For example, in Fig. 2 the smooth region on the right (outlined by the bold line) is represented by a single face, defined by underlying subdivision surface. The face is bounded by edges, naturally represented by subdivision curves. Simplified surface mesh (Section 2.2 with modifications, discussed in Section 3.1) is used directly as a control mesh for the construction of the S-BREP faces.

3.3 Computational Mesh Generation for the FEM-DEM

The problem of computational mesh generation on subdivision geometry requires further clarification. Though it is mostly reduced to the problem of surface mesh generation on the S-BREP geometric model it is worth mentioning, that *volumetric constrained* mesh generation on subdivision faces is not yet established. The methods of *surface meshing on subdivision geometry* are mainly discussed in the literature (i.e. [Rypl], [Lee1]).

In the proposed technique, two main approaches to surface meshing on the S-BREP geometry have been used for the further volumetric computational mesh generation. The first approach uses subdivision surface mesh on different levels of refinement for further constrained 3D mesh generation inside the domain, bounded by subdivision faces. The second approach to surface meshing is based on the method, proposed by Lee [Lee2] and uses parameterization of the limiting subdivision faces in the S-BREP model. It should be stressed once again, that in the S-BREP no trimming curves are required and each face is naturally bounded by the limiting boundary curves. For each face in the second approach a corresponding parametric mesh is generated, defined on the parametric $u-v$ space. In the simplest case, a parametric mesh can be obtained by face projection on one of the coordinates planes, but in more complex geometric configurations faces splitting to non-overlapping sub-domains is required. Having obtained parameterization, it is possible now to generate mesh in physical space of the S-BREP face, using traditional parametric surface mesh generation algorithms. See [Mezentsev1] for details.

MezGen unstructured Delaunay mesh generator has been applied for volumetric meshing (Mezentsev,[Mezentsev1]). Following the method, formulated in [Weatherill], the boundary recovery process in MezGen code is addressed in two main phases: an edge recovery phase and a face recovery phase. The main advantage of the MezGen approach is that rather than attempting to transform the tetrahedrons to recover edges and faces as proposed by George [George], the nodes are inserted directly into the triangulation when the edge or facet of subdivision triangulation cuts non-conforming tetrahedrons. This process temporarily adds additional nodes to the surface. Once the surface facets have been recovered, additional nodes that were inserted to facilitate the boundary recovery are deleted and the resulting local void remeshed. It is important, that in this method, no parameterization is introduced and all the operations are performed in physical space of the model. Straightforward formulation makes the method very robust, but potential pitfalls are related to the quality of direct usage of subdivision surface facets as computational meshes. That could affect the quality of the mesh for complex faces with geometrical constraints, similar to the example in Section 4.3.

3.4 Algorithm Overview

In summary, the proposed algorithm for derivation of CAD and corresponding FEM/DEM computational models for the cloud-type discrete scanned data will be as follows:

1. Acquire discrete (point-wise) data defining object boundary from scanning device. Using current scanning technology, normal orientation is known.

2. Perform Delaunay tetrahedrization of the dense point-wise data set, using any of the appropriate algorithms, for example [Frey].

3. Reconstruct triangular surface mesh, defining scanned object, using natural neighbor interpolation method of the signed distance to the tangent hyperplanes, following [Boissonnat] and [Karbacher].

4. Simplify dense surface mesh to sparse geometric BREP mesh, using edge swapping and edge collapsing operations within the tolerance envelope of the surface, based on the modified median Hausdorff distance measure.

5. Extract smooth sub-regions (geometric faces) of the simplified surfaces mesh, using normal deviation criterion.

6. Generate faces of the S-BREP CAD model using obtained simplified mesh as a control mesh for interpolating subdivision surfaces.

7. Store S-BREP CAD model in the database of the VGW for further use.

8. Retrieve S-BREP CAD model from the data base and perform limiting subdivision to the required level of surface approximation.

9. Mesh refined subdivision CAD using constrained Delaunay methods.

It is important that based on this formulation a robust and flexible model both for visualization and computational modeling with the FEM/DEM methods is defined.

4 Numerical Results

The present implementation for the automated extraction of the described CAD model from noisy scanned cloud-of-point type data is based on the modification of the MezGen unstructured Delaunay mesh generation code [Mezentsev1] to the requirements of the VGW. VGW is a computer environment that enables the user to automate model creation for complex simulation scenarios for geosciences applications. The core component of the VGW is a virtual information space (specialized data base), unifying inhomogeneous components of different computational modules. The VGW is designed using the Object Oriented programming paradigm. This work establishes pre-processing and geometry visualization components of the VGW: scanning, data simplification and automatic S-BREP CAD creation for mesh generation.

4.1 Ellipsoid Geometry

Simple smooth point-wise ellipsoid geometry is presented in Fig. 7, a). The data was scanned with the Optix 300SE scanner (3DD Digital Corporation). Specifics of scanning process dictate separate scanning of the upper and lower parts of the model that results in reduced model accuracy at the equator regions. This draw back is related to the obvious limitations of the technology, relying on the rotary table for scanned object movement.

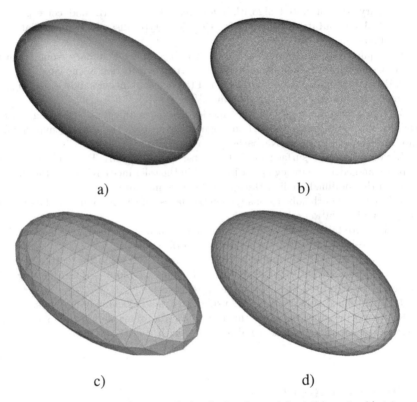

a)

b)

c)

d)

Fig. 7. Ellipsoid model: a) scanned cloud of points with visible noise b) triangulated dense cloud of points c) simplified surface mesh, used as a control mesh for subdivision surface d) computational isotropic 3D mesh on S-BREP CAD model

This test data was used to check the paradigm of automatic dense mesh (shown on Fig. 7, b)) simplification for the S-BREP limiting subdivision surface control mesh (shown on Fig. 7, c)). It appears, that for imperfect (shown on Fig. 7, a)) data set, rather noisy at the equator regions, of the scanned ellipsoid geometry, proposed method robustly captures a one-face subdivision surface in the S-BREP CAD format. The control mesh (Fig. 7, c)) has rather regular structure, simplifying application of the modified Butterfly subdivision scheme. Fig. 7 d) depicts isotropic volumetric computational mesh, generated by MezGen on the parameterized S-BREP face.

4.2 Geological Particle

Fig. 8 gives an example of the *volumetric* computational mesh (generated by the MezGen mesh generation code) on the limited subdivided S-BREP geometry of scanned particle. Our method automatically extracted 10 S-BREP faces in the form of the interpolating subdivision surfaces from a simplified subdivision mesh (shown in bold lines on Fig. 8). Only one operator intervention has been required in the process of automated face definition. Unfortunately, for free-form geometric models obtained from scanned data, minimal intervention will be required, especially for irregular rock particles with geometrically smooth and discontinuous regions. The operator driven decisions are taken when automatic algorithms fail to define CAD faces topology in one case of sharp transition between faces (Fig. 8).

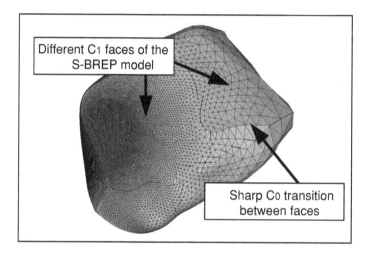

Different C_1 faces of the S-BREP model

Sharp C_0 transition between faces

Fig. 8. Volumetric computational MezGen mesh on the limit S-BREP (subdivision surface) model of a scanned geological particle, precisely capturing the shape of object

Application-specific computational mesh refinement is also straightforward and mesh refinement shown in Fig. 8 and Fig. 9 illustrate anticipated contact area for accurate resolution of the contact forces in the FEM-DEM simulation [Munjiza].

Fig. 9 depicts volumetric computational mesh with the same level of refinement on coarse (non-limiting) subdivision geometry, practically corresponding to subdivision face control mesh after stage 4 of the described algorithm (see Section 3.4 for details).

It can be observed that both S-BREP models in Fig. 8 and Fig. 9 represent the same shape on different levels of subdivision, retaining nearly automatically extracted S-BREP faces of the geometry. This example proves applicability of the proposed method of geometry definition for meshing applications. It establishes usability of the S-BREP model with C_1 continuity of subdivision faces, combined

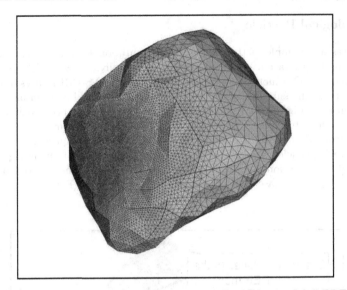

Fig. 9. Volumetric computational MezGen mesh on a coarsened S-BREP model of a scanned geological particle. Coarse facets of underlying subdivision surface are shown in bolder lines, while the quality of computational mesh is maintained (smallest dihedral angle is 11.7 degrees)

with C_0 continuity of sharp features across surface faces of free form geological particles. The other interesting observation is sharp decrease of the CAD modeling time, required for preparation of the geometric model. Fot the particle shown in Fig. 8 and Fig. 9 using traditional method of CAD definition with the Rhino3D package (Robert McNeel and Associates) it takes from 14 to 17 hours (depending on the operator qualification) to produce the NURBS CAD geometry. The proposed automated geometry definition and meshing method reduces this time to 2.25 hours. In terms of CAD model storage requirements new format tends to provide two times more efficient solution as compared to the Rhino3D models in ACIS (Spatial Inc.) file format.

4.3 Smooth Surface with Irregular Boundary

Earlier references related to application of interpolating subdivision surfaces for CAD modeling and computational mesh generation (e.g. [Rypl], [Lee2]) reveal certain limitations for CAD complexity, i.e. most test models were limited to a simple topology of the surface boundary. In our opinion, this is attributed to the simplified structure of applied topological model. Further development of the S-BREP overcomes this limitation, as discussed below.

In the presented example, interpolating subdivision boundary curve for the S-BREP faces is rather complex, forming highly irregular (fractal-like) boundaries. Fig. 10 a) gives scanned free form model of a surface with a regular boundary bounded by a simple four-curve outer loop. Fig. 10 b) and c) presents the same face geometry with complex irregular internal cut. Inner boundary loop is composed

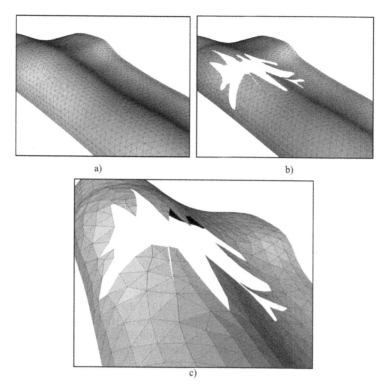

Fig. 10. Regular a) and irregular b) boundaries of free form S-BREP geometry with zoom-in on coarser level of subdivision – c)

from 14 subdivision curves, automatically connected by inserted crease extraordinary boundary node of the subdivision curve [Zorin], providing both C_0 and C_1 continuity regions on the inner boundary. This process is somewhat similar to the known effect of knot insertion into NURBS curve, creating a C_0 continuity between segments of a single spline. Given example shows, that proposed geometric model can easily integrate multiply connected domains with highly irregular boundary loops.

5 Conclusions

A new highly automated method for geometry definition and meshing of complex objects has been proposed. It is based on the methods of natural neighbors interpolation and signed distance function and a new application of the median Hausdorff distance (4) as a distance functions for automatic simplification of the scanned objects. A new concept of the S-BREP, the Boundary representation model with interpolating subdivision surfaces is introduced and implemented. The new model provides a robust alternative to the existing geometric models. It combines adaptive resolution of the geometry with automatic methods for defining faces, based on point wise data acquired from laser scanning technology.

A number of examples of different complexity show applicability and efficiency of the approach to the problem of automatic CAD definition and generation of computational meshes for noisy scanned data.

However, the problem is very complex and far from solved. More effort is required for the development of the theoretical aspects of automatic normals definition for non-convex geometry, regularization of the mapping between initial (scanned) dense and simplified surface meshes for noisy data. Further research in this area will improve the automatic extraction of faces and appropriate subdivision for VGW code development.

Acknowledgements

This work has been carried out under the Grant Virtual Geoscience Workbench, supported by the Engineering and Physical Sciences Research Council of the United Kingdom.

[Zienkiewicz] Zienkiewicz O, Taylor R (2000) The finite element method, Volume 1. Butterworth-Heinemann, Oxford

[Munjiza] A. Munjiza (2004) The combined discrete finite element method. Wiley, Chichester

[Latham] Latham J-P, Munjiza A (2004) The modeling of particle systems with real shapes. Phil Trans Royal Soc London 362:1953–1972

[Bajaj] Bajaj C, Bernardini F, Xu G (1996) Reconstructing surfaces and functions on surfaces from unorganized three-dimensional data. Algorithmica 19:362–379

[Karbacher] Karbacher S, Laboureux X, Schon N, Hausler G (2001) Processing range data for reverse engineering and virtual reality. In: Proceedings of the Third International Conference on 3-D Digital Imaging and Modeling (3DIM'01). Quebec City, Canada, 270–280

[Boissonnat] Boissonnat J-D, Cazals F (2002) Smooth surface reconstruction via natural neighbour interpolation of distance functions. Comp Geo 22:185–203

[Frey] Frey P (2004) Generation and adaptation of computational surface meshes from discrete anatomical data. Int J Numer Meth Engng 60:1049–1074

[Xue] Xue D, Demkowicz L, Bajaj C (2004) Reconstruction of G1 surfaces with biquartic patches for HP Fe Simulations. In: Proceedings of the 13th International Meshing Roundtable, Williamsburg

[Mezentsev] Mezentsev A, Woehler T (1999) Methods and algorithms of automated CAD repair for incremental surface meshing. In: Proceeding of the 8th International Meshing Roundtable, South Lake Tahoe

[Owen] Owen S, White D (2003) Mesh-based geometry. Int J Numer Meth Engng 58:375–395

[Lang] Lang P, Bourouchaki H (2003) Interpolating and meshing 3D surface grids, Int J Numer Meth Engng 58:209–225

[Cebral] Cebral J , Lohner R (1999) From medical images to CFD meshes. In: Proceeding of the 8th International Meshing Roundtable, South Lake Tahoe

[Lorensen] Lorensen W, Cline H (1987) Marching Cubes: a high resolution 3D surface reconstruction algorithm. Comp Graph 21(4):163–169

[Kolluri] Kolluri R, Shewchuk J, O'Brien J (2004) Spectral surface reconstruction from noisy point clouds. In: Eurographics Symposium on Geometry Processing

[Hoppe] Hoppe H, DeRose T, Duchamp T, McDonald J, Stuetzle W (1992) Surface reconstruction from unorganized points. Comp Graph 26(2):71–78

[Beall] Beall M, Walsh J, Shepard M (2003) Accessing CAD geometry for mesh generation. In: Proceeding of the 12th International Meshing Roundtable

[Zorin] Zorin D, Schroder P, Sweldens W (1996) Interpolating subdivision with arbitrary topology. In: Proceedings of Computer Graphic, ACM SIGGRAPH'96, 189–192

[Rypl] Rypl D, Bittnar Z (2000) Discretization of 3D surfaces reconstructed by interpolating subdivision. In: Proceeding of the 7th International Conference on Numerical Grid Generation in Computational Field Simulations, Whistler

[Lee1] Lee C (2003) Automatic metric 3D surface mesh generation using subdivision surface geometrical model. Part 1: Construction of underlying geometrical model. Int J Numer Meth Engng 56:1593–1614

[Lee2] Lee C (2003) Automatic metric 3D surface mesh generation using subdivision surface geometrical model. Part 2: Mesh generation algorithm and examples. Int J Numer Meth Engng 56:1615–1646

[Dyn] Dyn N, Levin D, Gregory J (1990) A butterfly subdivision scheme for surface interpolation with tension control. In: Proceedings of Computer Graphic, ACM SIGGRAPH'90, 160–169

[Loop] Loop C (1987) Smooth subdivision surface, based on triangles. MA Thesis, University of Utah, Utah

[Chetverikov] Chetverikov D, Stepanov D (2002) Robust Euclidian alignment of 3D point sets. In: Proc. First Hungarian Conference on Computer Graphics and Geometry, Budapest, 70–75

[Borouchaki] Borouchaki H (2000) Simplification de maillage basee sur la distance de Hausdorff. Comptes Rendus de l'Académie des Sciences, 329(1):641–646

[Mezentsev1] Mezentsev A (2004) MezGen - unstructured hybrid Delaunay mesh generator with effective boundary recovery. http://cadmesh.homeunix.com/andrey/Mezgen.html

[Weatherill] Weatherill N, Hassan O (1994) Efficient three-dimensional Delaunay triangulation with automatic point creation and imposed boundary constraints. Int J Numer Meth Engng 37:2005–2039

[George] George P, Hecht F, Saltel E (1991) Automatic mesh generator with specified boundary. Comp Meth Appl Mechan Engng 92:269–288

Automatic Near-Body Domain Decomposition Using the Eikonal Equation

Yuanli Wang[1], Francois Guibault[2], and Ricardo Camarero[3]

[1] École Polytechnique de Montréal, C.P. 6079, Succ. Centre-ville,
Montréal (QC) H3C 3A7, Canada yuan-li.wang@polymtl.ca
[2] École Polytechnique de Montréal, C.P. 6079, Succ. Centre-ville,
Montréal (QC) H3C 3A7, Canada francois.guibault@polymtl.ca
[3] École Polytechnique de Montréal, C.P. 6079, Succ. Centre-ville,
Montréal (QC) H3C 3A7, Canada ricardo.camarero@polymtl.ca

1 Introduction

The purpose of this research is to generate high-aspect-ratio cells for structured grids in the vicinity of boundaries for wall dominated phenomena such as viscous layers in aerodynamic applications. When such domains are discretized for complex geometries, it is important that the grid fits the boundaries well, and this becomes even more difficult to achieve with strongly curved boundaries. Domain decomposition simplifies this problem by subdividing the domain into multiple blocks and makes the task of meshing complex geometries more manageable.

There are many benefits to using a multi-block strategy. It is possible to control the orthogonality and grid quality more precisely within smaller blocks, and it allows different mesh types and generation techniques in each individual block. Also, parallel algorithms can be embedded in the mesh generation algorithm, where different blocks can be assigned to different processors, thus greatly improving the efficiency of the process.

The present paper introduces an approach to automatically decompose an arbitrary complex domain into face-matching multi-blocks emanating from the boundaries. The subdivision procedure is based on the creation of offset surfaces which closely fit their geometries instead of arbitrary planes or surfaces. This fitting is achieved through a weak solution of the *Offset Distance Function*, which is a variation of the Eikonal equation. Computing the normal directions in this manner insures that they do not cross, and the resulting propagation avoids self-intersections which arise from direct construction methods.

Thus this method can be applied on an arbitrary complex domain, i.e., a concave, or a convex shape, it may have sharp corners, or even be multi-connected. In addition, the proposed method transports the original parameterization to the propagated surface which allows rigorous matching of block faces. The geometric and topological configurations are thus well defined which allows to increase the automation level without user intervention.

This paper is arranged as follows: Section 2 gives a brief survey for multi-block strategies, and the methodology used in this work. Section 3 introduces the techniques used for the computation of the front propagation, and the proposed mathematical foundation, based on a variation of the Eikonal equation, is reviewed together with the numerical scheme used to solve this form of equation. Section 4 describes the application of this method to complex industrial configurations. Finally, Section 5 summarizes the presentation.

2 Multi-Block Generation Methodology

The numerous domain decomposition methods developed over the years were devised for entire geometric domains. The present work aims to generate viscous grids, thus we will limit the approach to generating multi-blocks around solid boundaries.

One early strategy is to use unstructured meshing methods where triangular sub-domains are generated by applying the Delaunay technique [Bergman1990, Cordova1992], and then transform the triangles into quadrilaterals by removing the appropriate edges or subdividing each triangle into three quadrilaterals that are used as blocks [Bergman1990]. Advancing-front techniques have also been used to decompose the domain directly into quadrilateral [Schonfeld1991], or coarse hexahedra cells [Kim1995] for 3D problems.

Piperni and Camarero developed a domain decomposition method [piperni2003] whose basic idea is to topologically map the original domain into a rectilinear polygon. It is then decomposed into prime rectangles which are finally mapped back to the geometric domain to generate quad (2D) or hexahedra (3D) blocks. Park and Lee suggested a hyper cube++ concept, in which complex geometry is first transformed into parametric domain, after decomposing the parametric domain, these decomposing informations are mapped back to physical space [ParkLee1998].

Guibault proposed a domain decomposition method, based on a direct propagation of the boundaries using Eqn. (2), aimed at generating multiple blocks around geometric boundary area [PHD-guibault]. Topologically, each block is equivalent as a cube template. The blocks are obtained by first generating surface grids on the boundaries to be offset, then propagating these surface grid points along their normal directions. Finally, the relative topological connectivities are constructed by sequentially connecting propagated points into edges, faces and volumes. The input and output files include both geometric information, (points, curves and surfaces), and topological information (vertices, edges, faces and volumes).

The advantage of this method is that it enforces parametric consistency between the original boundary, and its offset counterpart. All the points located on the original front are propagated along their local normal directions, thus all the points on the original front and its offset share a one-to-one parametric mapping. The weakness of this method is that self-intersection eliminating algorithms are difficult to implement and may fail for severely concave regions in 3D .

The present multi-block generation strategy is directly inspired from [PHD-guibault]. The major difference lies in the computation of the normal direction, $\mathbf{n}$. Instead of using the boundary's geometry directly, it is calculated through the use of a distance function ϕ field. This is obtained by the weak solution of the Offset Distance Equation, rather than using the coordinates of neighboring points.

$$\mathbf{n} = \frac{\nabla \phi}{|\nabla \phi|} \tag{1}$$

This property guarantees that local normal directions will not intersect during propagation, thus self-intersections are naturally avoided by this formulation. As no additional care is required to detect or eliminate collisions, this method can be applied on arbitrary complex domains. These can be closed or even open domains (as in through flow applications) by the addition of special boundary planes (Section 4).

3 Front Propagation Model

In dealing with surface offset or propagation, two important problems arise, for which no perfectly satisfactory solution has yet been proposed. The first problem relates to the potential self-intersection of the front as it grows from the original surface as shown in Fig. 1. *Local* self-intersection may occur during propagation when the offset distance is greater than the local curvature radius in concave regions. *Global* self-intersection, on the other hand, arises when the distance between two distinct points on the curve or surface reaches a local minimum.

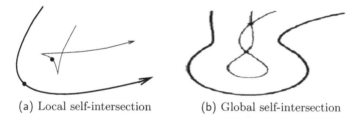

(a) Local self-intersection   (b) Global self-intersection

Fig. 1. Self-intersection problems

The second difficulty which is a more recent issue in offset construction is the establishment of a common connectivity between the original and offset surfaces. Current literature shows that numerous incidental applications can be facilitated if the connectivity could be transported from the original surface to its offset. These areas include transport of trim curves [kumar2003] and tool path generation [park2001], and, in the present case, semi-structured mesh generation in near wall regions.

3.1 Literature Review for Front Propagation

A variety of contributions deal with the computation of offset curves and surfaces. They can be classified in two types: *direct offset methods (DOM)*, which propagate curves or surfaces directly based on a geometric construction; *indirect offset methods (IOM)*, which cast the curve or surface offset problem into a set of partial differential equations (PDE), in which, geometric information are implicitly represented.

Direct Offset Methods (DOM)

The advantage of DOM is that the entire or partial original parameterization information can be preserved, but the self-intersections problem cannot be avoided, and extra care is required for removing self-intersections. The ability to effectively eliminate these is an important criterion in the applicability of such methods in the context of an automated procedure.

The basis for constructing the offset surface is the following equation:

$$\mathbf{x}_t = F\mathbf{n} \qquad (2)$$

which relates $\mathbf{x}_t$, the time derivative of the geometric front position vector, to F the offset speed, a given function, and $\mathbf{n}$, the normal vector. The offset curves or surfaces are generated by successively solving this equation.

One representative attempt to eliminate self-intersections is the *Advancing Front Method* developed by Pirzadeh [pirzadeh1993] based on a grid-marching strategy. The solution is to simply stop the advancement of the front before self-intersections occur. Based on a similar marching idea, Sullivan [sullivan1997] presented a self-intersection detection and removal algorithm in 2D. The 3D algorithm developed by Guibault [PHD-guibault] eliminates self-intersections by first detecting tangled-loop and then re-locating the points located within this area. In the algorithm described by Glimm and al [glimm2000], a hybrid algorithm is applied to resolve self-intersections by either re-triangulating triangles after removing unphysical surfaces, or reconstructing the interface within each rectangular grid block in which crossing is detected.

Other types of techniques to eliminate self-intersections use the properties of curves and surfaces, i.e. control points, derivatives, curvature etc. Blomgren [blomgren1981], Tiller and Hanson [tiller1984], and Coquillar [coquillart1987] approached the problem by offsetting the control polygon for NURBS curves. Nachman [nachman2002] extended this idea to propagate surfaces by offsetting control points. Piegl and Tiller [piegl1999] sampled offset curves and surfaces based on bounds on the second derivatives to avoid self-intersections. In the method developed by Sun and al. [sun2004], control points are repositioned to reduce local curvature in areas where local self-intersections may occur, while the rest of the control points remain unchanged. Farouki [farouki1986] described an algorithm which first decomposes the original surfaces into parametric patches, and then uses Hermite interpolation to construct the offset surfaces.

Indirect Offset Methods (IOM)

Self-intersection problems can be avoided, but at the cost of completely losing the connectivity information stored in the original geometric front. In general, to restore a similar connectivity between the original and offset fronts is not a trivial task.

The *Level set method* developed by Sethian and Osher [osher1988], models front propagation problems as a hyperbolic differential equation.

$$\phi_t + \nabla\phi \cdot \mathbf{x}_t = 0 \qquad (3)$$

in which, $\mathbf{x}$ is the coordinate of an arbitrary point in space, ϕ is the minimum Euclidean distance from $\mathbf{x}$ to the front to be propagated. Numerical scheme and optimization method for solving Eqn. (3) are discussed in works by Sethian [sethian1996, sethian1999]. Kimmel [kimmel1993] applied the level set equation to offset NURBS curves and surfaces in 1993.

The significant improvement of this approach is that it intrinsically prevents self-intersections by constructing a weak solution to the offset problem. In this method, corners and cusps are naturally handled, and topological changes occur in a straightforward and rigorous manner. Complex motion, particularly those that require surface diffusion, sensitive dependence on normal directions to the interface, and sophisticated breaking and merging, can be straightforwardly implemented, with no user intervention ([malladi1996]).

In the level set method, the result of offsetting curves or surfaces are dependent on many other factors, such as the calculation of offset speed F and curvature k. Re-initialization of the level sets during calculation are required to maintain an accurate level set representation. Also, offset surfaces are extracted from the computed ϕ solution as iso-value surfaces, resulting in triangulated surface representation, which have to be re-parameterized, if the original surfaces are represented by two-dimensional parameterization (which may also be called topological connectivity if original surfaces are represented by triangles rather than parameterized surfaces). Transforming this parameterization or connectivity information into offset surfaces is a costly task. Several methods [sheffer2001, hormann2001] have been proposed to re-construct two-dimensional parameterizations based on triangulated surfaces but cannot be used to establish a one-to-one parametric relationship between the original and offset surfaces.

The advantages and deficiencies of the DOM and IOM methods are complementary. Among all of these reviewed propagating methods, the level set method prevails over other methods, and brings a significant improvement and elegant way in avoiding self-intersection problems. The present work proposes to combine the advantages of the two types of methods to build a new offset construction method that maintains parametric connectivity between the original and offset surfaces, and still avoids self-intersections through the use of a weak solution to the shortest distance problem.

3.2 Offset Computation Equation

The present work proposes the use of another type of PDE for surface offset construction, the *Offset Distance Equation*, which is a variation of the Eikonal equation:

$$\begin{cases} \nabla\phi \cdot \nabla\phi = 1 \\ \phi = 0 \quad if\ \mathbf{P} \in \varGamma \end{cases} \tag{4}$$

where ϕ is the minimum Euclidean distance from an arbitrary point (P) in the computational domain $(\varOmega)$ to the front $(\varGamma)$ to be propagated. Eqn. (4) expresses the condition for the shortest Euclidean distance from any space position to the boundary. It thus simplifies the level set mathematical model.

The ϕ function is very important in the present work: (1) It provides the basis for the calculation of the local normal directions to preserve the advantages of level set method. That means multi-block partition lines are propagated in the ϕ space, rather than in the geometric space; (2) It is also used to prevent global self-intersections. When ϕ decreases, the propagated point is approaching a boundary, and propagation is stopped to avoid collisions. (3) It can also be used as a direct stopping criterion when the value of ϕ at the propagated point is equal or greater than the prescribed propagated distance.

Mathematical representation

Let Γ be a continuous front to be propagated, it can be either a closed or open front; $\mathbf{P}$ be an arbitrary point in the domain of interest Ω, $\Gamma \subset \Omega$; and ϕ be the smallest Euclidean distance between $\mathbf{P}$ and Γ. To represent the ϕ function, a point $\mathbf{P_0}$ is defined which meets the following conditions: (1) $\mathbf{P_0}$ is located on Γ; (2) the Euclidean distance between $\mathbf{P}$ and $\mathbf{P_0}$ is the minimum Euclidean distance between $\mathbf{P}$ and Γ. Let $\mathbf{P} = (x, y, z)$, $\mathbf{P_0} = (x_0, y_0, z_0)$, then the ϕ function can be written as

$$\phi(x, y, z) = \sqrt{(x - x_0)^2 + (y - y_0)^2 + (z - z_0)^2} \tag{5}$$

in which, ϕ is 0 when $\mathbf{P}$ lies on Γ. After a squared summation of the first derivatives of Eqn. (5), in a 3D Cartesian frame of reference, the distance offset equation can be written as:

$$\left(\frac{\partial \phi}{\partial x}\right)^2 + \left(\frac{\partial \phi}{\partial y}\right)^2 + \left(\frac{\partial \phi}{\partial z}\right)^2 = 1 \tag{6}$$

The problem can now be formally cast as an initial value problem using Eqn. (4) for any dimension or frame of reference.

Numerical method

The focus is now on the choice of a convergent, consistent and efficient way to solve Eqn. (4). The Fast Sweeping Scheme (FSS)developed by Zhao [zhao2004] is proposed, since it improves the computational complexity from $\Theta(Nlog(N))$ in the fast marching method (FMM) developed by Sethian [sethian1999siam] to $\Theta(N)$, where N is the total number of Cartesian grid points. A brief description of the 3D numerical scheme is given to illustrate the algorithm. More details about this scheme and its validation can be found in [zhao2004, 1].

A 3D uniform Cartesian grid system is used to discretize Eqn. (6) with step sizes $\Delta x = \Delta y = \Delta z = h$. The partial derivatives $\frac{\partial \phi}{\partial x}$, $\frac{\partial \phi}{\partial y}$ and $\frac{\partial \phi}{\partial z}$ are replaced by the following Godunov upwind scheme [rouy1992],

$$\frac{\partial \phi}{\partial x} = \frac{\phi_{i,j,k}^{t+1} - a}{h},$$
$$\frac{\partial \phi}{\partial y} = \frac{\phi_{i,j,k}^{t+1} - b}{h},$$
$$\frac{\partial \phi}{\partial z} = \frac{\phi_{i,j,k}^{t+1} - c}{h}. \tag{7}$$

In the above equations, (i, j, k) are the indices for the Cartesian grid nodes in 3D, t is the sweeping time, and a, b, c are defined as

$$a = min(\phi_{i-1,j,k}^t, \phi_{i+1,j,k}^t) \quad i = 2, ..., I - 1$$
$$b = min(\phi_{i,j-1,k}^t, \phi_{i,j+1,k}^t) \quad j = 2, ..., J - 1$$

$$c = min(\phi^t_{i,j,k-1}, \phi^t_{i,j,k+1}) \quad k = 2, ..., K-1 \tag{8}$$

In Eqn. (8), I, J and K are the grid numbers in x, y and z directions, respectively. One sided-difference schemes are used for calculating a, b and c when $i = 1, or I$, $j = 1, or J$, and $k = 1, or K$.

After replacing $\frac{\partial \phi}{\partial x}$, $\frac{\partial \phi}{\partial y}$, and $\frac{\partial \phi}{\partial z}$ by the expressions in Eqn. (7), and substituting into Eqn. (6) gives,

$$\left[\left(\phi^{t+1}_{i,j,k} - a\right)^+\right]^2 + \left[\left(\phi^{t+1}_{i,j,k} - b\right)^+\right]^2 + \left[\left(\phi^{t+1}_{i,j,k} - c\right)^+\right]^2 = h^2 \tag{9}$$

where function $(m)^+$ is defined as:

$$(m)^+ = \begin{cases} m, & \text{when } m > 0 \\ 0, & \text{when } m \leq 0 \end{cases}.$$

After re-arranging a, b, and c into the order of $a < b < c$, $\phi^{t+1}_{i,j,k}$ can be calculated as:

let $\phi_{temp} = a + h$

1. if, $\phi_{temp} \leq b$, then $\phi^{t+1}_{i,j,k} = \phi_{temp}$, and it is the solution to Eqn. (6);

2. else, let $\phi_{temp} = \frac{(a+b)+\sqrt{2h^2-(a-b)^2}}{2}$
   a) if, $\phi_{temp} \leq c$, then $\phi^{t+1}_{i,j,k} = \phi_{temp}$, and it is the solution to Eqn. (6);
   b) else,

$$\phi^{t+1}_{i,j,k} = \frac{(a+b+c)+\sqrt{3h^2+2(ab+bc+ac-a^2-b^2-c^2)}}{3}, \text{ and it is the solution to Eqn. (6)}.$$

The new value $\phi^{new}_{i,j,k}$ at node (i, j, k) is chosen as: $\phi^{new}_{i,j,k} = min(\phi^t_{i,j,k}, \phi^{t+1}_{i,j,k})$. The old value $\phi^t_{i,j,k}$ is always replaced by $\phi^{new}_{i,j,k}$ in the sweeping calculation process.

4 Methodology and Application

The proposed method is presented in the context of a 3D industrial geometric model, a draft tube. The characteristics are that it is a multi-connected open domain, with sharp corners and it has both concave and convex shapes. The goal is to decompose the domain into multiple hexahedral blocks emerging from the walls or boundary surfaces. The first step of the procedure is to decompose the draft-tube surface into a set of four-sided patches (see Fig. 2). These will be propagated or swept into the domain, and the original patch and the propagated patch will constitute the top and bottom faces, respectively, of a new block.

There are five essential steps in this method: (1) define a domain Ω, and discretize it into a uniform Cartesian grid; (2) calculate ϕ for each grid node using the fast sweeping algorithm; (3) calculate normal directions for the front grid nodes; (4) propagate front points along their local normal directions according to the given distance; (5) construct blocks or the relative topological connectivities around boundary area.

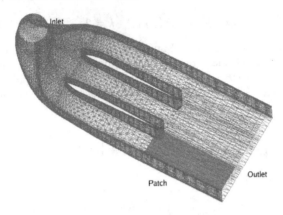

Fig. 2. Geometric model: a draft tube with 2 piers

4.1 Domain Discretization

The given geometry Γ is plunged into a domain Ω large enough to cover Γ. The domain Ω is discretized into a uniform Cartesian grid, and for each grid node, a scalar value ϕ, the minimum Euclidean distance of the Cartesian node to Γ, will be computed.

4.2 Computation of the distance field ϕ

All the grid nodes are first initialized to a large positive value. This value is chosen by the user with the only requirement is that it should be larger than the maximum possible overall distance of the computation domain. In practice, a value which is equal or greater than the size of Ω is assigned. Then two subsequent steps are applied as described below.

Initialization of Boundary-nodes

If Γ intersects an edge constructed between two grid nodes, then all the nodes belonging to the cells sharing this edge are marked as *boundary-nodes* (see Fig. 3). To enforce the boundary condition, $\phi = 0$ when $(x, y, z) \in \Gamma$, exact values are assigned to boundary-nodes. These are calculated by projecting a boundary-node onto the front segments and the minimum length of this projection line is used for initializing the ϕ value of this boundary-node. If no projection line is found for a given boundary node, then the minimum distance between this boundary-node and the node located on front segment is the exact value for this boundary-node. To accelerate boundary-node detection, the line segments in 2D (or triangles in 3D) are stored in an Alternating Digital Tree (ADT) [bonet1991].

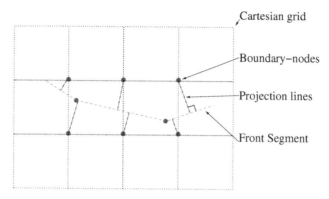

Fig. 3. Definition of Boundary-nodes

Updating the ϕ field

Gauss-Seidel iterations with alternating sweeping orderings are used to update ϕ's.
In 2D, it takes 4 sweeps to get a converged solution, each sweep is carried out
along the diagonal direction of the Cartesian grid. After each sweep, one-quadrant
data located in the destination area of the sweeping direction is updated. The same
sweeping procedure is applied for the 3D calculation. Since there are eight vertices
in a 3D Cartesian grid, it takes 8 sweeps to update the data located in 8 octants.
Algorithm 1 illustrates this ϕ updating procedure for one sweep, all the other seven
sweeps are similar.

Algorithm 1: updating ϕ (one sweep procedure)
for $k = 0, K$ do
    for $j = 0, J$ do
        for $i = 0, I$ do
            calculate a, b, and c
            calculate $\phi_{i,j,k}^{t+1}$
            calculate $\phi_{i,j,k}^{new}$
        end for
    end for
end for

4.3 Calculation of the normal directions

At each grid node, ϕ_x, ϕ_y and ϕ_z are calculated using a centered difference scheme:

$$\phi_x = \frac{\phi_{i+1,j,k} - \phi_{i-1,j,k}}{2\Delta x}, \quad i = 2, ...I - 1 \tag{10}$$

$$\phi_y = \frac{\phi_{i,j+1,k} - \phi_{i,j-1,k}}{2\Delta y}, \quad j = 2, ...J - 1 \tag{11}$$

$$\phi_z = \frac{\phi_{i,j,k+1} - \phi_{i,j,k-1}}{2\Delta z}, \quad k = 2, ...K - 1 \tag{12}$$

and a one-sided difference scheme is used for boundary nodes, when $i = 1$ or $I, j = 1$ or J and $k = 1$ or K.

At each grid node on the front, (x_P, y_P, z_P), the values of ϕ_{xP}, ϕ_{yP} and ϕ_{zP} are interpolated using the values at eight neighboring grid nodes $(\phi_{x_i}, \phi_{y_i}, \phi_{z_i})$, $i = 1, ...8$, as shown in Fig. 4. The normal directions at front points can be calculated directly using Eqn. (1).

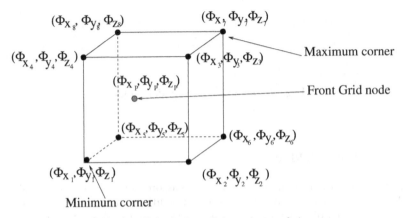

Fig. 4. Calculation of ϕ_{xP}, ϕ_{yP} and ϕ_{zP}

For an open domain, in addition to the boundary surfaces, it is necessary to specify the boundary curves which bound the boundary surface. For example, the inlet circle and outlet rectangle in Fig. 5 are such boundary curves. These lie on special boundary planes which are added for the purpose of correctly closing the domain from a topological point of view, as well as providing the support for the propagation of these boundary curves.

Within each such surface, the boundary curve propagation problem is transformed into a 2D curve propagation problem, and the algorithm described in this section is applied directly. In the present application, it is required that normal directions should lie on such special boundary planes, i.e. this algorithm cannot be used when the boundary curves are general space curves.

4.4 Point Propagation

Replacing $\mathbf{x}_t$ by a finite difference scheme, and letting $F = 1$, Eqn. (2) can be rewritten as

$$\mathbf{x}^{n+1} = \mathbf{x}^n \pm \frac{\nabla \phi}{|\nabla \phi|} dt \tag{13}$$

in which n is the propagation time, and dt is the time step. The term $\nabla \phi / |\nabla \phi|$ is calculated by a forth-order Runge-Kutta method [hoffman1992] and can be viewed as a unit propagation speed in the normal direction; positive for an outward propagation, and negative for an inward propagation.

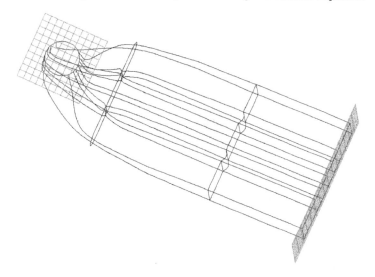

Fig. 5. *Boundary curves* and *Special boundary planes*

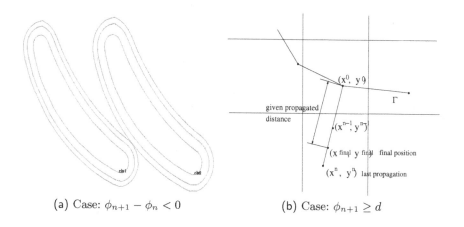

(a) Case: $\phi_{n+1} - \phi_n < 0$ (b) Case: $\phi_{n+1} \geq d$

Fig. 6. Propagation stopping criteria

For each new propagated point, there are two stopping criteria. When one of them is satisfied, the propagation is stopped (see Fig. 6): (1) If $\phi_{n+1} - \phi_n < 0$, or in other words, if ϕ starts to decrease; (2) If ϕ is equal to or greater than the specified propagation distance, d. When the value of ϕ at the final position $\mathbf{x}^{n+1}$ is greater than the given propagation distance, then linear interpolation is used to get the exact position.

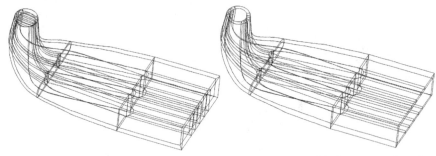

(a) The resulting blocks - closed domain (b) The resulting blocks - open domain

Fig. 7. Result comparisons (overall views)

4.5 Construction of topological connectivities

All the propagated points are sequentially connected according to their original connectivities, and corresponding points on the original and propagated geometries are used to construct blocks. Detailed descriptions about topological connectivities can be found in [PHD-guibault].

Fig. 7 illustrates the difference of the resulting blocks for a closed and open domain. In Fig. 7(a), inlet and outlet surfaces are added to Fig. 2, in this case, the original geometry can be viewed as a closed domain. Fig. 7(b) is the resulting blocks generated by directly propagating original geometry, in this case, the original domain can be viewed as an open domain, thus special boundaries are added at inlet and outlet parts. Fig. 8 is the enlarged comparisons at inlet and outlet parts. From Fig 8(a) to 8(d), we can see that inlet and outlet surfaces are propagated and blocks are generated when the original geometry is dealt as a closed domain; and boundary curves are propagated along the inlet and outlet planes when original geometry is an open domain.

Fig. 9 is the overall view of the resulting mesh, in which a structured mesh is generated within each individual block. Fig. 10 is the enlarged view for the results generated around a pier and in a sharp corner, respectively. Fig. 11 is an enlarged view of the propagating lines at a sharp corner. Theoretically, the front does not intersect itself. However, marched elements can eventually collapse causing the volume of some of the resulting elements to vanish. Elliptical smoothing of the mesh in each block could be used to attenuate this problem.

5 Conclusion

In this paper, we have presented a new approach to construct structured and semi-structured meshes near solid boundaries, based on domain decomposition. The proposed decomposition approach first proceeds by explicitly constructing an offset surface at the boundary, which is then used to topologically subdivide the domain into simple regions that are meshed independently. A key contribution of this approach

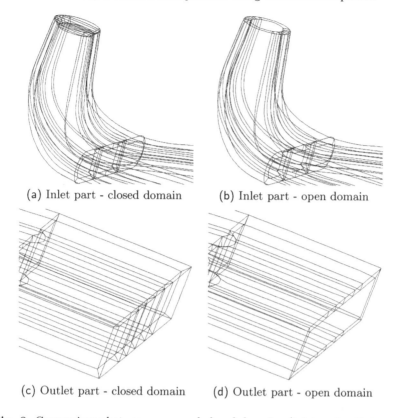

(a) Inlet part - closed domain   (b) Inlet part - open domain

(c) Outlet part - closed domain   (d) Outlet part - open domain

Fig. 8. Comparisons between open and closed domains (inlet and outlet part)

lies in the method used to construct the boundary offset surfaces and construct the structured mesh in each block. A mixed method for the solution of the offset distance equation is used, that alleviates local and global self-intersection problems during offset surface construction, while allowing to maintain a parametric relationship between the original surfaces and their corresponding offset.

This method can be applied indistinctly to continuous surfaces exported from CAD systems (e.g. NURBS) and to discrete polygonal versions of domain boundaries. The front can be an arbitrarily complex multi-connected domain. Topological connectivities are easily and naturally handled after propagation without user intervention, and the block faces exactly match each other. Local self-intersections are naturally avoided using the weak solution to the Eikonal equation, and global self-intersections can be avoided based on a simple detection criterion involving the ϕ field.

Permeable and internal surfaces can naturally be used to constrain the volume decomposition process through the introduction of special boundary planes. The current implementation only allows to treat planar fictitious boundaries, but there is no theoretical limit that prevents the extension to surfaces of arbitrary shape.

Fig. 9. Resulting mesh - overall view

6 Acknowledgments

The authors would like to express their appreciation to GE Canada (Hydro) and the National Science and Engineering Research Council of Canada (NSERC) for their support. The authors would also like to thank Ko-Foa Tchon for the valuable discussions and Mrs. Ying Zhang for supplying the 3D test cases.

[Bergman1990] M. Bergman, "Development of numerical techniques for inviscid hypersonic flows around re-entry vehicles," *PhD thesis, INP Toulouse*, 1990.

[Cordova1992] J. Cordova, "Towards a theory of automated elliptic mesh generation," *In NASA Workshop on Software Systems for Surface Modeling and Grid Generation, NASA Conference Publication, n 3143, p231*, 1992.

[Schonfeld1991] T. Schonfeld and P. Weinerfelt, "The automatic generation of quadrilateral multi-block grids by the advancing front technique," *In Numerical Grid Generatin in Computational Fluid Dynamics, A.S. Arcilla, J. Hauser, P.R. Eiseman, and J.F. Thompson (Eds.), p743, Proceedings of the 3rd International Grid Conference, North Holland, Barcelona, Spain, June*, 1991.

[Kim1995] B. Kim and S. Eberhardt, "Automatic multi-block grid generation for high-lift configuration wings," *Proceedings of the Surface Modeling, Grid Generation and Related Issues in Computational Fluid Dynamics Workshop, NASA Conference Publication 3291, p. 671, NASA Lewis Research Center, Cleveland, OH, May*, 1995.

[piperni2003] P. Piperni and R. Camarero, "A fundamental solution to the problem of domain decomposition in structued grid generation," *AIAA 2003-0951, 41st Aerospace Sciences Meeting & Exhibit, 6-9 January*, 2003.

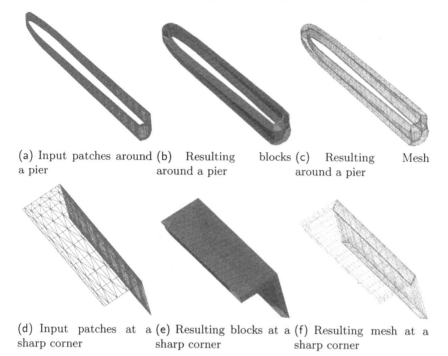

(a) Input patches around a pier

(b) Resulting blocks around a pier

(c) Resulting Mesh around a pier

(d) Input patches at a sharp corner

(e) Resulting blocks at a sharp corner

(f) Resulting mesh at a sharp corner

Fig. 10. Resulting mesh - partial views

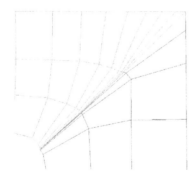

Fig. 11. Propagation behavior at a sharp corner

[ParkLee1998] S. Park and K. Lee, "Automatic multiblock decomposition using hypercube++ for grid generation," *Computers and Fluids, 27,4,May, p509-528,* 1998.

[PHD-guibault] F. Guibault, *Un mailleur hybride structuré/non structuré en trois dimensions.* PhD thesis, École Polytechnique de Montréal, 1998.

[kumar2003] R. G. V. V. Kumar, K. G. Shastry, and B. G. Prakash, "Computing offsets of trimmed nurbs surfaces," *Comp.-Aided Design*, vol. 35, no. 5, pp. 411–420, 2003.

[park2001] S. Park and B. Choi, "Uncut free pocketing tool-paths generation using pair-wise offset algorithm," *Comp.-Aided Design*, vol. 33, no. 10, pp. 739–746, 2001.

[pirzadeh1993] S. Pirzadeh, "Unstructured viscous grid generation by advancing-layers method," in *AIAA 11th Applied Aerodynamics Conference*, no. AIAA-93-3453, (Monterey, CA), pp. 420–434, Aug. August 9–11, 1993.

[sullivan1997] J. Suillivan and J. Zhang, "Adaptive mesh generation using a normal offsetting technique," *Finite Elements in Analysis and Design*, vol. 25, pp. 275–295, 1997.

[glimm2000] J. Glimm, J. Grove, X. Li, and D. Tan, "Robust computational algorithms for dynamic interface tracking in three dimensions," *SIAM J. Sci. Comp.*, vol. 21, no. 6, pp. 2240–2256, 2000.

[blomgren1981] R. Blomgren, "B-spline curves, boeing document, class notes, b-7150-bb-wp-2811d-4412," 1981.

[tiller1984] W. Tiller and E. Hanson, "Offsets of two-dimensional profiles," *IEEE Computer Graphics and Applications*, vol. 4, no. 9, pp. 36–46, 1984.

[coquillart1987] S. Coquillart, "Computing offsets of b-spline curves," *Comp.-Aided Design*, vol. 19, no. 6, pp. 305–309, 1987.

[nachman2002] A. Kulczycka and L. Nachman, "Qualitative and quantitative comparisons of b-spline offset surface approximation methods," *Comp.-Aided Design*, vol. 34, pp. 19–26, Jan. 2002.

[piegl1999] L. A. Piegl and W. Tiller, "Computing offsets of nurbs curves and surfaces," *Comp.-Aided Design*, vol. 31, pp. 147–156, Feb. 1999.

[sun2004] Y. F. Sun, A. Y. C. M. Nee, and K. S. Lee, "Modifying free-formed nurbs curves and surfaces for offsetting without local self-intersection," *Comp.-Aided Design*, vol. 36, no. 12, pp. 1161–1169, 2004.

[farouki1986] R. T. Farouki, "The approximation of non-degenerate offset surfaces," *Comp.-Aided Geom. Design*, vol. 3, no. 1, pp. 15–43, 1986.

[osher1988] S. Osher and J. A. Sethian, "Fronts propagating with curvature-dependent speed: Algorithms based on hamilton-jacobi formulations," *J. Comp. Phys.*, vol. 79, pp. 12–49, 1988.

[sethian1996] J. A. Sethian, "A fast marching level set method for monotonically advancing fronts," *Proceedings of the National Academy of Sciences of the United States of America*, vol. 93, no. 4, pp. 1591–1595, 1996.

[sethian1999] J. A. Sethian, *Level set methods and Fast Marching Methods: Evolving Interfaces in Computational Geometry, Fluid Mechanics, Computer Vision, and Materials Science*. Cambridge University Press, 1999.

[kimmel1993] R. Kimmel and A. M. Bruckstein, "Shape offsets via level sets," *Comp.-Aided Design*, vol. 25, no. 3, pp. 154–162, 1993.

[malladi1996] R. Malladi and J. A. Sethian, "Level set and fast marching methods in image processing and computer vision," *IEEE International Conference on Image Processing*, vol. 1, pp. 489–492, 1996.

[sheffer2001] A. Sheffer and E. de Sturler, "Parameterization of faceted surfaces for meshing using angle based flattening," *Engineering with Computers*, vol. 17, no. 3, pp. 326–337, 2001.

[hormann2001] K. Hormann, *Theory and Applications of Parameterizing Triangulations*. PhD thesis, Department of Computer Science, University of Erlangen, Nov. 2001.

[zhao2004] H. K. Zhao, "A fast sweeping method for eikonal equations," *Mathematics of Computation*, vol. 74, no. 250, pp. 603–627, 2005.

[sethian1999siam] J. A. Sethian, "Fast marching methods," *SIAM Review*, vol. 41, no. 2, pp. 199–235, 1999.

[1] Y.-L. Wang, F. Guibault, R. Camarero, and K.-F. Tchon, "A parametrization transporting surface offset construction method based on the eikonal equation," in *17th AIAA Computational Fluid Dynamics Conference*, 6 – 9 June 2005.

[rouy1992] E. Rouy and A. Tourin, "A viscosity solution approach to shape-from-shading," *SIAM Journal on Numerical Analysis*, vol. 29, pp. 867–884, 1992.

[bonet1991] J. Bonet and J. Peraire, "An alternating digital tree (ADT) algorithm for 3D geometric searching and intersection problems," *Int. J. Numer. Meth. Engng*, vol. 31, pp. 1–17, 1991.

[hoffman1992] J. Hoffman, *Numerical Methods for Engineers and Scientists*. McGraw-Hill Inc., 1992.

MARCHING GENERATION OF SMOOTH STRUCTURED AND HYBRID MESHES BASED ON METRIC IDENTITY

Jochen Wild[1], Peter Niederdrenk[2], Thomas Gerhold[2]

[1]DLR, Institute for Aerodynamics and Flow Technology, Lilienthalplatz 7, D-38108 Braunschweig, Germany
[2]DLR, Institute for Aerodynamics and Flow Technology, Bunsenstr.10, D-37073 Göttingen, Germany

ABSTRACT

Elliptic differential equations are derived for the generation of structured meshes and difference equations for the generation of smooth hybrid meshes from metric identity. A parabolic procedure is used to march the solution of the difference equations simultaneously for both types of meshes away from surface patches meshed by quads or triangles. An aerodynamic application for the ONERA M6 wing demonstrates how the blocking at the wing tip is simplified by using both types of meshes instead of a purely structured mesh. On the other hand it is shown that a large number of points can be saved compared to a purely hybrid grid.

Keywords: mesh generation, structured, hybrid, elliptic

1. INTRODUCTION

As long as the variations of the field variables to be resolved by the numerical solution to a set of differential equations are of the same order of magnitude in all directions the use of unstructured meshes is appropriate and furthermore avoids the manual labor of subdividing the field into blocks usually needed for structured meshes. On the other hand an adequate resolution of very thin layers with large normal gradients requires a layered structure of the mesh.

Such meshes may be generated by marching away from a fixed mesh face, say the body surface in case of a boundary layer, or away from a floating mesh face in case of a wake. When the mesh on the initial face is unstructured, the resulting three-dimensional mesh extruded from it consists of prisms of triangular cross-sections and is called hybrid. A structured mesh on the initial face, of course, leads to a fully structured three-dimensional mesh.

Structured meshes are easily clustered differently in all grid line directions. Since they need far less points for directional clustering than unstructured meshes, they add efficiency to the numerical solution procedure of the differential equations. On the other hand their generation suffers from the amount of work to be invested to achieve the blocking for complex geometries. To ease the situation we combine both kinds of surface meshes starting the three-dimensional marching generation of mixed meshes simultaneously from neighboring patches with structured and unstructured surface meshes.

Since our objective is to generate smooth meshes we will start from a set of elliptic equations. In a previous paper [1] beginning with the metric identity in its differential conservative form the well known Poisson equations for structured meshes were derived including a precise analytical definition of all 9 control functions. To solve these differential equations numerically they were parabolized with respect to the marching direction and subsequently discretized [2].

For unstructured meshes there exists no globally underlying computational space. Therefore, starting from the same basic principle means that we will begin with the metric identity in its discrete form to directly derive from it the set of algebraic equations to be solved numerically.

In the 80s marching procedures were used with the primary intention to speed up the generation process of three-dimensional structured meshes. We have added a flexible control of grid line spacing and orthogonality with no need for any initial algebraic or reference mesh [2]. With respect to the generation of hybrid meshes the majority of the publications are based on pure algebraic procedures. An exception is the paper of David Thompson [3], which makes use of a local two-dimensional expansion of the elliptic mesh generation equation originating from Patrick Knupp's work [4] on smoothing triangular surface meshes.

2. DIFFERENTIAL FORM OF METRIC IDENTITY

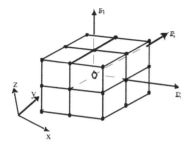

Fig. 1: Master cell

Dirichlet conditions on the boundary specify the solution of an elliptic equation. Figure 1 shows a structured master cell with 26 known mesh points on the cell boundary and just one unknown interior point. The position vector $\underline{r}$ in physical space x, y, z is a function of the computational space variables ξ, η, ζ. The normal onto a grid face $\xi = const$ is given by the cross product of the vectors along the η- and ζ-lines: $\underline{r}_\eta \times \underline{r}_\zeta$. The metric identity in its differential form simply states that the sum over the face normal vectors in $+\xi$ and $-\xi$ direction together with the sum over the remaining face normal vectors in η- and ζ-directions must vanish:

$$\left(\underline{r}_\eta \times \underline{r}_\zeta\right)_\xi + \left(\underline{r}_\zeta \times \underline{r}_\xi\right)_\eta + \left(\underline{r}_\xi \times \underline{r}_\eta\right)_\zeta = 0. \tag{1}$$

Defining normal face vectors as

$$\underline{S}_1 = \underline{r}_\eta \times \underline{r}_\zeta, \quad \underline{S}_2 = \underline{r}_\zeta \times \underline{r}_\xi, \quad \underline{S}_1 = \underline{r}_\xi \times \underline{r}_\eta \tag{2}$$

we decompose the contravariant vectors $\underline{S}_i$ into the covariant base vectors $\underline{r}_\xi, \underline{r}_\eta, \underline{r}_\zeta$ as shown below for $\underline{S}_1$

$$\left[\frac{1}{V}\left(\underline{S}_1^2 \underline{r}_\xi + \underline{S}_2 \underline{S}_1 \underline{r}_\eta + \underline{S}_3 \underline{S}_1 \underline{r}_\zeta\right)\right]_\xi + \ldots = 0 \tag{3}$$

and expand the terms in brackets in order to cast the conservative form of the identity into a non-conservative equation for the position vector $\underline{r}$:

$$\underline{S}_1^2\left\{\underline{r}_{\xi\xi} + \frac{V}{\underline{S}_1^2}\left[\left(\frac{\underline{S}_1^2}{V}\right)_\xi + \left(\frac{\underline{S}_1 \underline{S}_2}{V}\right)_\eta + \left(\frac{\underline{S}_1 \underline{S}_3}{V}\right)_\zeta\right]\underline{r}_\xi\right\} +$$

$$\underline{S}_2^2\left\{\underline{r}_{\eta\eta} + \frac{V}{\underline{S}_2^2}\left[\left(\frac{\underline{S}_2 \underline{S}_1}{V}\right)_\xi + \left(\frac{\underline{S}_2^2}{V}\right)_\eta + \left(\frac{\underline{S}_2 \underline{S}_3}{V}\right)_\zeta\right]\underline{r}_\eta\right\} +$$

$$\underline{S}_3^2\left\{\underline{r}_{\zeta\zeta} + \frac{V}{\underline{S}_3^2}\left[\left(\frac{\underline{S}_3 \underline{S}_1}{V}\right)_\xi + \left(\frac{\underline{S}_3 \underline{S}_2}{V}\right)_\mu + \left(\frac{\underline{S}_3^2}{V}\right)_\zeta\right]\underline{r}_\zeta\right\} +$$

$$2\underline{S}_1 \underline{S}_2 \underline{r}_{\xi\eta} + 2\underline{S}_2 \underline{S}_3 \underline{r}_{\eta\zeta} + 2\underline{S}_3 \underline{S}_1 \underline{r}_{\zeta\xi} = 0. \tag{4}$$

Terms in square brackets represent the so-called control functions. V is the cell volume. Only when we prescribe these control functions the metric identity becomes a mesh generating equation. The most simple prescription, setting all control functions to zero, yields the face weighted Laplace equation in computational space corresponding to the linear Laplace equation for the mesh faces ξ, η, ζ in physical space, when we interchange dependent and independent variables.

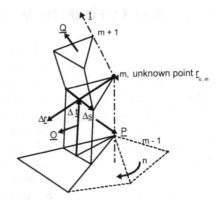

Fig. 2: Discrete triangular prism

$$\nabla^2_{x,y,z}\underline{\xi} = 0, \quad \underline{\xi} = \{\xi, \eta, \zeta\} \tag{5}$$

3. PARABOLIC STRUCTURED MESH GENERATION

In order to parabolize equation (4) we choose ζ as the independent variable in marching direction and split the line-wise second derivative of $\underline{r}$ with respect to ζ into a difference of first order up- and downwind derivatives:

$$\underline{r}_{\zeta\zeta} = \underline{r}_\zeta^d - \underline{r}_\zeta^u. \tag{6}$$

Treating $\underline{r}_\zeta^d$ as a source term yet to be specified and approximating the second line-wise derivatives by central differences we obtain

$$\underline{r}_{i,j,k} = \left\{\underline{S}_1^2 \cdot \left(\underline{r}_{i-1,j,k} + \underline{r}_{i+1,j,k}\right) + \ldots \tag{7}$$

$$+ \underline{S}_3^2 \cdot \left(\underline{r}_{i,j,k-1} + \underline{r}_\zeta^d\right) + \ldots\right\} / \left\{2\underline{S}_1^2 + 2\underline{S}_2^2 + 2\underline{S}_3^2\right\}$$

On each face $k = const$ this equation will be solved iteratively by some relaxation scheme. The source term $\underline{r}_\zeta^d$ is specified in terms of an outward spacing times a unit vector. Near body surfaces, e.g. across boundary layers, the preferred outward direction is usually the solution dependent local normal to faces $k = const$. While marching in strictly local normal direction out of concave surfaces will inevitably lead to a crossing of grid lines, it is the ellipticity of our equation in the lateral directions, which usually provides sufficient dissipation to prevent the grid from folding. Such inherent dissipation scales with $\underline{S_1}^2/\underline{S_3}^2$ and $\underline{S_2}^2/\underline{S_3}^2$ in i- and j-direction, respectively, which - under the assumption of complete orthogonality - is just the square of the scaling factors proposed by Steger and Chan [5].

For further details, especially for the controls in the lateral and marching directions, we refer the reader to [2].

4. METRIC IDENTITY FOR HYBRID MESH GENERATION

Since there is no globally underlying computational space for unstructured meshes, we start from the metric identity in its discrete form. As shown in figure 2 let us consider locally a closed collection of n triangles around a common point on three levels $m-1$, m, $m+1$.

As in the structured case we assume all points on the cell's bounding faces to be known and thus encounter a local elliptic problem for just one unknown central point. Defining normal vectors to the prism's faces, $\underline{O}$ in lateral, $\underline{P}$ in circumferential and Q in marching direction, respectively, the metric identity simply states again that the sum over all outward pointing face vectors must vanish:

$$\sum \left[\underline{O}_{n,m} + \frac{1}{4}\underline{Q}_{n,m+1} - \frac{1}{4}\underline{Q}_{n,m-1} \right] = 0 . \tag{8}$$

The face vectors $\underline{O}$, $\underline{P}$, Q are defined via cross products of the line vectors $\Delta\underline{r}$, $\Delta\underline{s}$, $\Delta\underline{t}$

$$\underline{O} = \Delta\underline{s} \times \Delta\underline{t}, \quad \underline{P} = \Delta\underline{t} \times \Delta\underline{r}, \quad \underline{Q} = \Delta\underline{r} \times \Delta\underline{s} \tag{9}$$

where $\Delta\underline{r}$ is from central point to the midpoint between neighbored corners, $\Delta\underline{s}$ points anti-clockwise in circumferential direction and $\Delta\underline{t}$ is in the outward body normal direction.

Proceeding as in the structured case we decompose the face normal vectors into the line vector directions using the box product

$$V = \langle \Delta\underline{r} \quad \Delta\underline{s} \quad \Delta\underline{t} \rangle \tag{10}$$

as shown below for the face normal $\underline{O}$

$$V \cdot \underline{O} = \langle \underline{O} \quad \Delta \underline{s} \quad \Delta \underline{t} \rangle \Delta \underline{r} + \langle \Delta \underline{r} \quad \underline{O} \quad \Delta \underline{t} \rangle \Delta \underline{s} + \langle \Delta \underline{r} \quad \Delta \underline{s} \quad \underline{O} \rangle \Delta \underline{t} \qquad (11)$$

$$= \left[\underline{O} (\Delta \underline{s} \times \Delta \underline{t}) \right] \Delta \underline{r} + \left[\underline{O} (\Delta \underline{t} \times \Delta \underline{r}) \right] \Delta \underline{s} + \left[\underline{O} (\Delta \underline{r} \times \Delta \underline{s}) \right] \Delta \underline{t}$$

$$= (\underline{O} \underline{O}) \Delta \underline{r} + (\underline{O} \underline{P}) \Delta \underline{s} + (\underline{O} \underline{Q}) \Delta \underline{t}$$

and rewrite the metric identity in terms of line vectors with coefficients consisting of scalar products of face vectors:

$$\sum_n \left\{ \left[\underline{O} \underline{O} \Delta \underline{r} + \underline{O} \underline{P} \Delta \underline{s} + \underline{O} \underline{Q} \Delta \underline{t} \right]_{n,m} \frac{1}{V_{n,m}} \right. \qquad (12)$$

$$+ \frac{1}{4} \left[\underline{Q} \underline{O} \Delta \underline{r} + \underline{Q} \underline{P} \Delta \underline{s} + \underline{Q} \underline{Q} \Delta \underline{t} \right]_{n,m+1/2} \frac{1}{V_{n,m+1/2}}$$

$$\left. - \frac{1}{4} \left[\underline{Q} \underline{O} \Delta \underline{r} + \underline{Q} \underline{P} \Delta \underline{s} + \underline{Q} \underline{Q} \Delta \underline{t} \right]_{n,m-1/2} \frac{1}{V_{n,m-1/2}} \right\} = 0.$$

Compared with the expanded differential form of the identity for structured meshes (4), which split into the Laplacian terms and the control functions, the corresponding decomposition of the discrete form of the identity (12) yields nothing similar. We are still faced with the problem to isolate the Laplacian smoothing part from the pure identity.

The problem area is, of course, confined to the unstructured mesh in the body conforming surfaces. Decomposing the discrete identity for planar surfaces

$$\sum_n \underline{O}_n = 0 \qquad (13)$$

into line vectors $\Delta \underline{r}$ and $\Delta \underline{s}$ yields

$$\sum_n \frac{1}{V_n} \left(\underline{O}_n^2 \Delta \underline{r}_n + \underline{O}_n \underline{P}_n \Delta \underline{s}_n \right) = 0 . \qquad (14)$$

Expressing also the coefficients in terms of line vectors

$$\Delta \underline{t} = const , \quad \underline{O}_n^2 = \Delta \underline{s}_n^2 , \quad \underline{O}_n \underline{P}_n = -\Delta \underline{r}_n \Delta \underline{s}_n \qquad (15)$$

we obtain in two dimensions

$$\sum_n \frac{1}{V_n} \left[\Delta \overline{\underline{s}}_n^2 \Delta \underline{r}_n - \left(\Delta \overline{\underline{s}}_n \Delta \overline{\underline{r}}_n \right) \Delta \underline{s}_n \right] = 0 . \qquad (16)$$

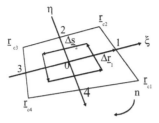

Fig. 3: planar quadrangle

In order to define the coefficients indicated by overbars as suited averages such that equation (16) will represent the discretized Laplace equation, we compare them first to the discretized form of the Laplace equation for quadrangles. Since that equation does not contain a variable volume, we assume V_n to be constant.

From structured meshes we know that all coefficients of the Laplace equation are evaluated from cell boundary points only, i.e. the central point is never involved. Thus the averaged line vectors for forming the coefficients are readily defined from opposing boundary points:

$$\Delta \bar{\underline{s}}_n = \frac{1}{2}\left(\underline{r}_{n+1} - \underline{r}_{n-1}\right), \quad \Delta \bar{\underline{r}}_n = \frac{1}{2}\left(\underline{r}_n - \underline{r}_{n+2}\right) \tag{17}$$

with n being cyclic according to figure 3.

With the so defined coefficients equation (16) becomes identical to the difference approximation of the Laplace equation in computational space. Since the above relations hold for quads only, we still have to generalize the averaging of line vectors in order to formulate the coefficients. For that purpose we take into account the angle dependence in idealized multi-cornered cells as shown in figure 4 for a hexagon.

The averaged values of $\Delta \underline{s}_n$ are defined by the formulae,

$$\Delta \bar{\underline{s}}_1 = \left[\cos\varphi_1 \Delta \underline{s}_1 + \cos\varphi_2 \Delta \underline{s}_2 + \dots + \cos\varphi_N \Delta \underline{s}_N\right]\frac{2}{N} \tag{18}$$

$$\Delta \bar{\underline{s}}_2 = \left[\cos\varphi_1 \Delta \underline{s}_2 + \cos\varphi_2 \Delta \underline{s}_3 + \dots + \cos\varphi_N \Delta \underline{s}_1\right]\frac{2}{N}$$

$$\vdots$$

$$\Delta \bar{\underline{s}}_N = \left[\cos\varphi_1 \Delta \underline{s}_N + \cos\varphi_2 \Delta \underline{s}_1 + \dots + \cos\varphi_N \Delta \underline{s}_{N-1}\right]\frac{2}{N}$$

$$\text{with} \quad \varphi_N = \frac{2\pi}{N}(n-1)$$

i.e. we measure the true $\Delta \underline{s}_n$, multiply them with the cosines of the angles of an ideally regular cell and add the terms up to obtain the average of the coefficient forming line vector $\Delta \bar{\underline{s}}_n$. Analogous formulae for $\Delta \bar{\underline{r}}_n$ only

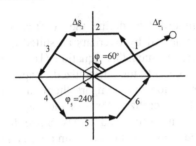

Fig. 4: Idealized hexagon

contain the midpoints of the cell's edges marked as an open circle in figure 4. The unknown central point does not enter.

Having introduced the averaged coefficients into the identity we compared our solutions with those of Winslow's discretized Laplace equation [6] for a number of example meshes consisting of quads or hexagons and found both solutions in full agreement. Since the only structured meshes made up of triangles lead to quadrangles and hexagons, our averaging procedure is not proved but only believed to hold for any collection of triangles forming an N-cornered cell.

In an earlier paper about „Winslow smoothing on two-dimensional unstructured meshes" Patrick Knupp [4] introduced pointwise a locally uniform computational space, expanded the derivatives about each mesh node in a Taylor series and solved the resulting coupled system. His procedure seems to require essentially more computational work.

Applying the averaged line vectors in three dimensions to formulate averaged face vectors according to definitions (9) the latter show some remarkable properties:

1. the face vectors $\underline{\bar{Q}}_n$ in body normal direction and

2. in consequence, also the local cell volumes $\bar{V}_n$ on the central level m both become constant for all n just as in the Laplace equation for structured meshes,

$$\underline{\bar{Q}}_n = \underline{\bar{Q}} = const \qquad \bar{V}_n = \underline{\bar{Q}} \cdot \Delta \underline{t} = const \qquad (19)$$

3. also, sums over mixed scalar products of averaged face normal vectors vanish

$$\sum \bar{\underline{O}}_n \underline{\bar{P}}_n = \sum \bar{\underline{O}}_n \underline{\bar{Q}} = \sum \underline{\bar{Q}} \underline{\bar{P}}_n = 0 . \qquad (20)$$

Expanding the differences in body normal direction about the central level m in equation (12)

$$\left(\underline{\bar{Q}}\underline{\bar{Q}}\Delta t\right)_{m+1/2} - \left(\underline{\bar{Q}}\underline{\bar{Q}}\Delta t\right)_{m-1/2} = \left(\underline{\bar{Q}}\underline{\bar{Q}}\right)_m \left(\underline{t}_{m-1/2} - 2\underline{t}_m + \underline{t}_{m+1/2}\right) \qquad (21)$$
$$+ \left(\left(\underline{\bar{Q}}\underline{\bar{Q}}\right)_{m+1/2} - \left(\underline{\bar{Q}}\underline{\bar{Q}}\right)_{m-1/2}\right) \Delta \underline{t}_m + \dots$$

we discard the difference of the coefficients between upper and lower level in accordance with the structured Laplace equation. Thus, all averaged coefficients are to be evaluated on the central level m only.

With this expansion and accounting for the above mentioned properties of the coefficients we solve equation (12) for the unknown central point $\underline{r}_0$ on level m:

$$\underline{r}_{o,m} \sum \left(\bar{\underline{O}}_n \bar{\underline{O}}_n + \frac{1}{2} \bar{\underline{Q}} \bar{\underline{Q}}_m \right) = \tag{22}$$

$$\sum \Big[\bar{\underline{O}}_{n,m} \bar{\underline{O}}_{n,m} \underline{rm}_{n,m} + \bar{\underline{O}}_{n,m} \bar{\underline{P}}_{n,m} \Delta \underline{s}_{n,m} + \frac{1}{4} \underline{\bar{Q}}_m \underline{\bar{Q}}_m \left(\underline{t}_{m-1/2} + \underline{t}_{m+1/2} \right)$$

$$+ \frac{1}{4} \bar{\underline{O}}_{n,m} \underline{\bar{Q}}_{n,m} \left(\Delta \underline{r}_{m+1/2} - \Delta \underline{r}_{m-1/2} \right) + \frac{1}{4} \underline{\bar{Q}}_{n,m} \bar{\underline{P}}_{n,m} \left(\Delta \underline{s}_{m+1/2} - \Delta \underline{s}_{m-1/2} \right) \Big],$$

where $\Delta \underline{r}_{n,m}$ is defined as the midpoint between two adjacent corner points according to the sketch in figure 5

$$\Delta \underline{r}_{n,m} = \underline{rm}_{n,m} - \underline{r}_{0,m} \quad \text{with} \quad \underline{rm}_{n,m} = 1/2 \left(\underline{rc}_{n,m} + \underline{rc}_{n+1,m} \right). \tag{23}$$

On the right hand side of equation (22) the coefficients in the first line result from the decomposition of the face vectors $\underline{Q}_{n,m}$, while the coefficients in the three following lines result from the decomposition of the face vectors $\underline{Q}_{n,m}$.

To solve equation (22) within a marching procedure we need at least three layers. Providing Dirichlet boundary conditions on the first layer initially from a known surface mesh and on the second and third layer from an algebraic forecast we iterate the solution to convergence on the intermediate layer. After each such iteration the algebraic forecast on the outer third layer will be updated. Once the solution is known on the intermediate layer, it becomes the boundary condition of the first layer for the next marching step and so forth.

The algebraic forecast needs a prescription of the spacing and the local marching direction. The latter follows from an angle weighted superposition of normal vectors to those faces around a central point, which limit the domain of visibility. Details are described in the appendix.

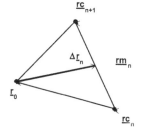

Fig. 5: Midpoint vector $\underline{rm}_n$

5. A GENERIC TEST EXAMPLE FOR HYBRID MESH GENERATION

The test example consists of a planar delta wing with extremely sharp edges and a wedge on top of it. Despite its geometric simplicity the generic configuration poses a severe test case, since we expect pure Laplacian smoothing to pull the mesh over the sharp edges and corners and also pure algebraic marching out of concavities to generate overlapped meshes.

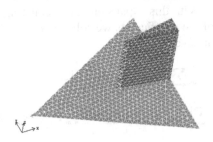

Fig. 6: Test configuration

The surface mesh is coarse (all together 1515 points, 3026 triangles), rather regular but not symmetric. Points have not been clustered towards the edges and corners - as would be preferable to better resolve these areas in fluid flow calculations - in order not to mask the cell stretching around the sharp edges expected to appear in the body conforming mesh surfaces to be generated.

Having marched the solution over 20 layers off the body surface with 20 iterations per layer the two figures below show cuts through the hybrid mesh in the symmetry plane $y = 0$ and in a section at $x = 0.8$.

As is seen from figure 7 marching out off concave corner regions does not cause any problems, even not ahead of the sharp wedge starting from a point of mixed strong convex and concave curvature, where the front edge of the wedge meets the wing in the plane $y = 0$.

But here the curvature of the cut lines changes sign when approaching the outer hull (figure 7 left) and the resulting surface mesh (figure 8) becomes unacceptable. Around those three points of mixed strong convex/concave curvature, where the wedge meets the wing, the cells are badly stretched in the direction of the convexity and are compressed in the direction of the concavity.

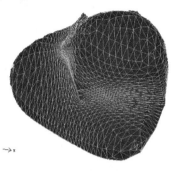

Fig. 8: Outer surface mesh (pure Laplace)

Looking for a remedy to suppress these excessively stretched cells we return to the planar case and consider just a single convex cell. For given corner points the central point (see figure 9) follows from the solution to the equation

$$\sum_n \left[\Delta \overline{\underline{s}}_n^2 \Delta \underline{r}_n - \left(\Delta \overline{\underline{s}}_n \Delta \overline{\underline{r}}_n \right) \Delta \underline{s}_n \right] = 0 \tag{24}$$

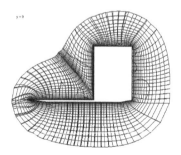

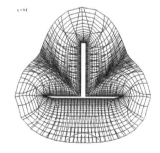

Fig. 7: Mesh cuts at $y = 0$ (left) and $x = 0.8$ (right)

as a superposition of forces $\Delta\underline{r}$ acting in the radial directions and forces $\Delta\underline{s}$ acting parallel to the cell's edges weighted by our Laplacian coefficients. The dark black arrows show the resulting force per edge acting on the central point. In figures 9 to 11 the magnitude of the arrow-heads is drawn proportional to the magnitude of the forces and the central point is marked as a circle filled black.

Switching to a cell with large differences in edge lengths putting one point to the left and five boundary points to the right hand side (see fig. 10) the solution to our Laplace equation being identical to Winslow's solution pulls over the right end of the convex hexagon. This is somewhat surprising, since

the extremum principle should hold for Laplace's equation. Obviously the extremum principle being defined for the continuous differential case does not hold for the discrete and distorted cell.

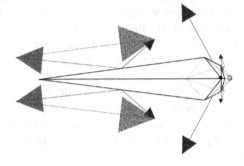

The Laplacian solution seems to deteriorate due to the large forces parallel to the edges. Reducing these

Fig. 11: Laplacian cell with reduced $\Delta\underline{s}$-forces

forces, when the cell becomes strongly stretched, i.e. when the angle between $\Delta\underline{r}$ and $\Delta\underline{s}$ becomes small, by a correction formula in an engineering fashion

$$\Delta\underline{s}_{red} = \Delta\underline{s}\cdot\left(1 - 0.25\cdot\cos^2\left(\Delta\underline{r},\Delta\underline{s}\right)\right) \tag{25}$$

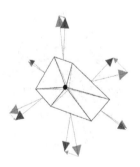

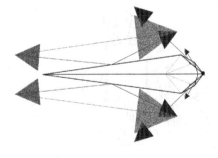

Fig. 9: Forces acting on the central point of a planar cell

Fig. 10: Folded hexagon, central point pulled over right end of cell

drives the central point from its position outside the hexagon (fig. 11, open circle) back into the convex polygon (fig. 11, black filled circle) preventing the cell from folding.

Applying the simple correction formula to the three-dimensional example the extreme stretching disappear (fig. 12) and the distribution of cells becomes rather smooth, as is also to be seen from front and rear.

While the body is symmetric, its surface triangulation is not. Therefore the full mesh has been generated. Slight deviations from symmetry are observable in the outer surface mesh.

Sure, there is still some larger stretching around the sharp edges of the wing, which could have been mitigated significantly by clustering points towards the edges on the body surface mesh to better resolve these regions. This would lead to cells being stretched in the direction of the edges on the body surface counteracting the present stretching of the cells on the outer hull perpendicular to the edges.

6. GENERATION OF MIXED MESHES

The shortcoming of purely hybrid grids is the low anisotropy of surface triangles resulting in a large number of surface grid points, which is agglomerated throughout the prismatic layers for the boundary layer resolution. Especially for the high aspect ratio wings the low anisotropy of the unstructured surface mesh leads to an unnecessary high resolution in the span-wise direction. Recalling the experience with the application of structured grids, it is known that the aspect ratios of surface quadrilaterals can be orders of magnitude higher, additionally resulting in well aligned body conforming meshes. On the other hand an increasing complexity of the configuration to be meshed limits the application of pure structured grids mainly because of intricate grid topology.

In the mixed mesh approach we use simple block structured boundary layer grids with highly stretched hexahedral cells wherever possible. Topologically difficult regions limiting the application of pure block structured grids are meshed by adjoining layers of prisms of triangular cross-section. The outer flow field is meshed using tetrahedral elements. The connection between the tetrahedral elements in the outer inviscid flow domain and the hexahedral elements in the boundary layer region is accomplished by egg-carton like pyramids. Since the structured grid approach offers direct control on point distributions by the bottom up approach, a smooth transition from the structured into the unstructured part can be achieved. It has already been shown that applying this grid technique can speed up Navier-Stokes flow computations by up to 90% without loosing solution accuracy, mainly due to the saving of points [7].

The grid generator used is the formerly purely structured DLR grid generator MegaCads [8], which allows for parametric construction of block structured grids. Besides basic CAD features like surface construction, projection and intersection, it is capable of constructing smooth grids by applying elliptic techniques as well as parabolic advancing front meshing for wall orthogonal boundary layer grids [2]. The whole grid generation process is stored in a process description file and can be re-run for modified geometries, as long as the CAD-topology is retained. Due to the parametric capabilities it is possible to guarantee grid quality in terms of smoothness, resolution and grid line angles even for larger geometrical changes. To account for the necessity to include the grid generator in an

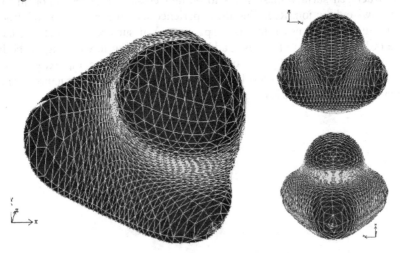

Fig. 12: Laplacian mesh with reduced $\Delta \underline{s}$-forces

optimization loop, MegaCads can be run in a non-interactive mode, repeating the grid generation process for the actual geometry.

The hybrid mesh generation procedure described in detail above in this paper has been implemented in MegaCads.

For the generation of the unstructured grid part the 3D triangulation software NETGEN of Schöberl [9] was incorporated into the framework of MegaCads as a Black-Box tool. The NETGEN software offers constrained Delaunay volume triangulation of a given triangulated surface mesh, which in the following application is the outer hull of the pyramidal interface layer and the outer prisms' faces wherever applied. The pyramidal elements connecting the structured with the unstructured mesh part are extruded from the hexahedral elements based on edge lengths in order to achieve a smooth transition of the control volumes for the reduction of numerical errors and/or instabilities.

7. APPLICATION

The ONERA M6 wing, measured by Schmidt and Valpin [11], is used for demonstration of the semi-structured grid approach using the described parabolic marching algorithms. The blunt trailing edges are closed, as this is commonly applied for CFD for this case. The rounded wing tip is replaced by a sharp cut at the end of the wing. This is done to demonstrate the prismatic capabilities. In purely structured grid generation this kind of wingtip would either lead to degenerated cells at the extension of the trailing edge or to an arbitrary closing of the wing tip within the range of one cell. Applying semi-structured grid generation the wing tip is meshed with triangular prisms (figure 13).

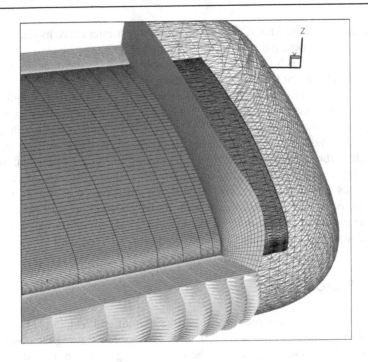

Fig. 13: Parabolic prism grid around the ONERA M6 wing tip

For comparison purposes a second purely hybrid mesh has been generated using the CENTAUR mesh generator [10]. Figure 14 shows the surface meshes for both mesh generation strategies. Both meshes have 32 cell layers normal to the wall to resolve the boundary layer and the vertical resolution is similar between both types of meshes. The mixed grid has about 215.000 points and 6.000 surface elements on the wing, while the purely hybrid grid has 1.400.000 points with about 80.000 surface elements, since for the unstructured surface mesh generation only a maximum anisotropic aspect ratio of 2 is allowed. For the semi-structured grid the maximum aspect ratio of the surface quads is around 170.

Figure 15 shows a comparison of the calculated surface pressure coefficient at two different spanwise locations for both meshes. It is observed that the solution quality is similar and the shock at the outer section is resolved better by the mixed mesh approach.

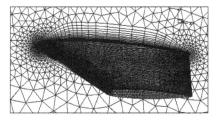

Fig. 14: Surface meshes for hybrid (left) and mixed meshes (right) around the ONERA M6 wing

8. CONCLUSIONS

In analogy to the differential equations of structured grids the face-weighted Laplace equation in discrete form has been derived from metric identity. Hybrid meshes are generated embedding the solution of the difference equation in a marching procedure with sufficient algebraic control of spacing in the structured direction. The ellipticity of the equation in the unstructured body conforming directions provides enough dissipation to prevent the mesh from folding when marching out of concave body surface areas. A simple coefficient modification avoids the formation of highly stretched distorted cells prone to overlap.

The mixed mesh strategy has been applied for the ONERA M6 wing and has been compared to a purely hybrid mesh. It has been shown that the smooth mixed mesh approach is able to capture the flow physics by using only about 20% of the points of a standard hybrid grid.

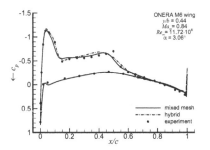
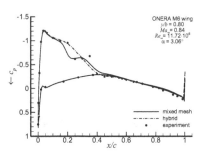

Fig. 15: Comparison of pressure distributions of the M6 wing at two different spanwise locations for a hybrid grid and a mixed mesh

REFERENCES

[1] P. Niederdrenk, "On the Control of Elliptic Grid Generation", Proc. 6th Intern. Conf. On Numerical Grid Generation in Comp. Field Simulations, London, July 1998, eds. M. Cross et al., NSF Eng. Center, Mississippi, pp. 257

[2] P. Niederdrenk, O. Brodersen, "Controlled Parabolic Marching Grid Generation", Proc. 7th Intern. Conf. On Numerical Grid Generation in Comp. Field Simulations, Whistler, British Columbia, Canada, Sept. 2000, eds. B. Soni et al., NSF Eng. Center, Mississippi, pp. 29

[3] D. Thompson, B. Soni, "Generation of Quad- and Hex-Dominant, Semi structured Meshes Using an Advancing Layer Scheme", Proc. 8th Intern. Meshing Roundtable, South Lake Tahoe, CA, USA, Oct. 1999, pp. 171

[4] P.M. Knupp, "Winslow Smoothing on Two-Dimensional Unstructured Meshes", Proc. 7th Intern. Meshing Roundtable, Sandia Natl. Labs., 1998, pp. 449

[5] W.M. Chan, J.L. Steger, "Enhancements of a Three-Dimensional Hyperbolic Grid Generation Scheme", Appl. Mathematics and Computations, 1992, Vol.51, pp. 181

[6] A. Winslow, "Numerical Solution of the Quasi-Linear Poisson Equations in a Non-Uniform Triangle Mesh", J. Comp. Phys., 1967, Vol.2, pp. 149

[7] J.Wild, "Acceleration of Aerodynamic Optimization Based on RANS-Equations by Using Semi-Structured Grids", Proc. ERCOFTAC Design Optimization: Methods & Applications, Athens (Greece), Paper ERCODO2004_221, 2004

[8] O. Brodersen, M. Hepperle, A. Ronzheimer, C.-C. Rossow, B. Schöning, "The Parametric Grid Generation System MegaCads", Proc. 5th Intern. Conf. On Numerical Grid Generation in Comp. Field Simulation, National Science Foundation (NSF), 1996, pp. 353--362

[9] J. Schöberl, "NETGEN - An advancing front 2D/3D mesh generator based on abstract rules", Computing and Visualization in Science, Vol. 1, 1997, pp. 41-52

[10]CentauerSoft, "Welcome to CentaurSoft, The solution to grid generation issues for computational engineering problems", http://www.centaursoft.com, 2004

[11]V. Schmitt, F. Charpin, „Pressure Distributions on the ONERA-M6-Wing at Transonic Mach Numbers", AGARD AR 138, 1979

APPENDIX: LOCAL MARCHING DIRECTION

The definition of the local marching direction for hybrid meshes is readily explained by an example. With respect to the configuration shown in figure 6 let us consider the point, in which the left rear edge of the wegde formed by faces 3 and 4 meets the trailing edge of the wing formed by faces 1 and 2 (see sketch below).

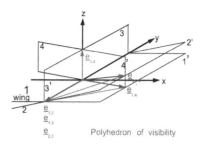

Polyhedron of visibility

To define the polyhedron of visibility for a point we only use different unit normal face vectors, i.e. we discard multiples of identical unit normals. Edge vectors $e_{j,k}$ are formed along the intersection lines of all combinations of two faces, for instance $e_{3,4}$ between faces 3 and 4. The sense of direction of the edge vector follows from the sense of the most aligned face normal, e.g. $e_{3,4}$ from n_1 or $e_{1,3}$ from n_4.

The polyhedron of visibility is given by only those edge vectors, which can see all faces, i.e. for which $e_{j,k}\, n_i$ is not negative. In the example the edge vector $e_{3,4}$ is excluded, since it cannot see face 2. The remaining edge vectors $e_{1,4}$, $e_{2,4}$ and $e_{1,2}$ form the polyhedron of visibility.

The marching direction follows from a weighted superposition of normals to those faces forming the polyhedron of visibility

$$\underline{N} = \sum w_i \underline{n}_i .$$

In our example n_3 does not contribute to the polyhedron of visibility, so it does not appear in the above formula. The weights are taken proportional to the angle between the appertaining edge vectors, for instance n_4 is weighted by the angle between $e_{1,4}$ and $e_{2,4}$.

A Hybrid Meshing Scheme Based on Terrain Feature Identification

Volker Berkhahn[1], Kai Kaapke[2], Sebastian Rath[3] and Erik Pasche[3]

[1] University of Hannover, Institute of Computer Science in Civil Engineering, Callinstraße 34, 30167 Hannover, Germany, berkhahn@bauinf.uni-hannover.de
[2] University of New South Wales, Water Research Laboratory, King Street, Sydney, 2093, Australia, k.kaapke@wrl.unsw.edu.au
[3] University of Technology Hamburg, Department of River and Coastal Engineering, Denickestraße 22, 21073 Hamburg, Germany, S.Rath@tuhh.de, Pasche@tuhh.de

1 Abstract

Hydrodynamic engineering makes profitably use of numerical simulations which rely on discrete element meshes of the topography. To cope with specific circumstances in river hydraulics, the presented hybrid meshing scheme comprises following proposals: river beds and areas of significant terrain slopes are meshed with regular elements to support user specified edge ratio and element orientation representing flow gradients appropriately; floodplains are represented as irregular triangle meshes, concatenating disconnected regular meshes while warranting high approximation quality. Automatic breakline detection approximates flow relevant changes in topographic gradients and defines borders of different mesh types. This paper presents an enhanced strategy for a terrain feature analysis based on b-spline analysis grids and on an interpolation scheme for breakline points in order to reduce the zigzag property of detected breaklines. This scheme for terrain analysis and meshing functionality is implemented in the open source software tool HybridMesh.

Keywords: Terrain feature analysis, breakline identification, hybrid element meshing scheme, regular meshes on b-spline surfaces, irregular triangle meshes, hydrodynamic simulations based on finite elements.

2 Introduction

Hydrodynamic engineering using remote sensing data sources faces growing data volumes for analysis, forecasts and assessments. Numerical simulations of water levels, flow velocities or sediment transport in river and coastal engineering require element meshes approximating the considered topography sufficiently accurate and while representing all significant terrain features. With respect to implementation

and application of these methods, element meshes are supposed to fulfill requirements regarding edge ratio, element angles and element size. In practice, these requirements also vary for different characteristic areas of a flooded domain.

The purpose of the suggested hybrid mesh generation algorithm is to generate meshes for hydrodynamic studies, providing enhanced suitability for numerical
simulations schemes. The numerical scheme under consideration describes flow characteristics based on the depth and time averaged Navier-Stokes equations, widely
known as shallow water wave equations. The hydrodynamic model for the presented
study is an ancestor of the hydrodynamic model RMA2, which is enhanced for
roughness and turbulence modeling. RMA2 uses the fully implicit implementation
of the Galerkin weighted residual technique, originally developed for the Resource
Management Associates in Lafayette. Today, RMA2 is internationally accepted as
hydraulic model for two-dimensional steady and unsteady flow and well-suited for
mapping inundation areas [FEMA-2004a]. It generally supports the use of triangular
and quadrilateral meshes as well as mixed discretisations of the terrain.

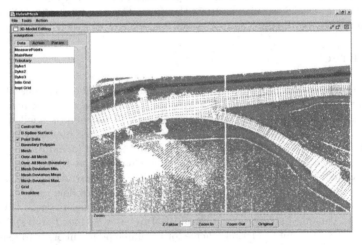

Fig. 1. Topography of the River Stoer and the tributary Bramau (case study 1)
visualized with the analysis and meshing tool HybridMesh

In this contribution the hybrid meshing scheme is exemplarily demonstrated for
two different case studies. Case study 1 represents an alluvial, partly straightened
stream: The River Stoer, located in the lowlands in Northern Germany, is a tributary
of the River Elbe. The data basis for simulations of this stream are airborne LiDAR
topographic survey data in conjunction with bathymetry measurements. Provided as
irregular point cloud with variable density, the average resolution of the LiDAR data
set offers several measures per square metre. The bathymetry is gathered from profile
measurements, compacted via linear interpolation schemes to obtain approximately
a 5 m resolution. Fig. 1 shows a plan view on the data set, revealing some lacks in
the terrain coverage due to inundation and impact of tiling.

Case study 2 represents a digital terrain model of a tributary to the River
Danube. Different characteristics of this topography are obvious: The resolution

of measurement points in the area of the river bed and the river slope is 4 m. In contrast to this, the resolution in the floodplain area is 20 m. In addition to these two different resolution areas, a fine resolution of 1 m is used to describe significant terrain features in the area of the river bed and the floodplain. These different resolution areas of the topography point out the necessity for separate considerations of different meshing areas and for a hybrid meshing scheme. Fig. 2 shows the original data set of case study 2 as well as the identified breaklines marked in red.

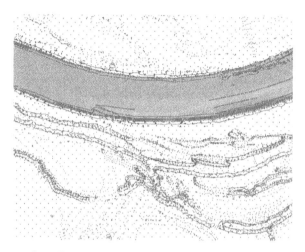

Fig. 2. Data set of a tributary to the River Danube with identified breaklines (case study 2)

The software tool HybridMesh combines the functionality of breakline identification [RathPasche-2004] and entirely automated meshing. The irregular triangle meshes are based on a Delaunay refinement [RathBajat-2004] and the regular mesh generation is based on free form surfaces [BerkhahnEtAl-2002], [BerkhahnMai-2004a] [BerkhahnMai-2004b]. This HybridMesh tool is developed by the authors of this paper and is an enhancement of the HydroMesh tool [Goebel-2005]. HybridMesh is implemented with the Java programming language and will be available as open source software.

3 Hybrid Meshing Scheme

The hybrid mesh generation scheme presented by Rath et al. [RathEtAl-2005] considers the topography of river beds and their adjacent floodplains individually. A separate consideration of river beds and their floodplains for mesh generation denotes the distinct measurement technologies for these sub domains, their accuracy demands and domination for numeric simulations. The individual consideration of different topographic sub domains is realized based on terrain feature recognition for breakline identification.

3.1 Status Quo of the Hybrid Meshing Scheme

The objective of the hybrid meshing scheme is to combine the advantages of regular mesh generation based on b-spline surfaces and irregular triangle meshes generated by a Delaunay refinement.

B-spline surfaces are beneficial for discarding blunders, while providing a topography approximation using regular element meshes, being efficiently with respect to specified edge ratio and element orientation. This approach based on free form surfaces shows conceptual limitations, if ramifications of rivers or floodplains are considered, as, in general, the domain cannot be described by a surface with 4 boundary curves. Nevertheless, b-spline surfaces allow suitable resolutions of high gradients in numerical computations of flow fields, if a user specific resolution is constituted. Consequently, b-spline surfaces are used to represent structures of the domain with dominant relevance for the flow field. In the presented case studies, structures such as river banks, levees and the river bed are approximated by b-spline surfaces.

Triangular irregular element meshes are highly adaptive to the fluvial topography. Without involving the smoothing property of b-spline surfaces triangular irregular element meshes enhance the accuracy of domain representation, whereas they are exposed to blunders in the original data set. Consequently, within the hybrid meshing scheme irregular triangle meshes are used for discretisations of the floodplains regions with inferior relevance for the flow field.

Generally, irregular triangle meshes are more suitable than b-spline surfaces to represent arbitrary shapes at a given accuracy level. The applicability of b-splines is enhanced within this hybrid scheme using the suggested breakline detection approach, which provides enhanced representations of the boundaries for these b-splines. Both, regular element meshes based on b-spline surfaces and triangular irregular element meshes are in use since decades for manual mesh generation using GIS systems. The hybrid meshing scheme joins available techniques for automated mesh generation for complex fluvial domains and enhances the representation for hydrodynamic computations based on feature detection. Its contribution facilitates the preparation of suitable meshes for hydrodynamic simulation based on vast remote sensing data sets, such as those from airborne LiDAR.

3.2 Enhancements of the Hybrid Meshing Scheme

In earlier publications [BerkhahnEtAl-2002] the regular element meshes were generated by creating the boundary curves of the b-spline surfaces manually. This means, the user had to select manually all control points defining these boundaries. The only support provided by the HydroMesh meshing tool were coloured displays of the topography. Where the colour indicates the hight of all measurement points. This is a very time consuming task to select all control points, what is not acceptable for the user.

The hybrid meshing scheme in the version presented the first time by the authors [RathEtAl-2005] involves an automatic breakline detection. These breaklines were used to define the boundary curves of the b-spline surfaces, what leads to a dramatic reduction of user interactions and, consequently, a speed up of the whole meshing process. At this stage of development the breakline detection and slope classification were performed on linear interpolation analysis grids. These linear analysis grids lead to breaklines with a significant zigzag property. Therefore, the analysis grids were enhanced by a b-spline based smoothing technique. In addition, a simple interpolation scheme is implemented in order to calculate slope values between analysis grid points. Both enhancements are explained in the following section.

4 Breakline Detection

The individual consideration of different sub domains of the topography is based on terrain feature recognition and especially on breakline identification. This terrain feature analysis and slope classification are performed on regular high resolution rectangular grids approximating the topography. Different slope determination methods are investigated by Rath and Pasche [RathPasche-2004] and are implemented within the analysis tool HybridMesh. The slope calculation for this study case is performed according to the one-over-distance method [Jones-1998]. The slope S_{ij} of a grid point $\mathbf{d}_{ij}$ is given by:

$$S_{ij} = \arctan\left(\sqrt{S_{ij\,u}^2 + S_{ij\,v}^2}\right) 180/\pi \quad \text{with} \tag{1}$$

$$S_{ij\,u} = \frac{\left(z_{i+1\,j+1} + \sqrt{2}z_{i+1\,j} + z_{i+1\,j-1}\right) - \left(z_{i-1\,j+1} + \sqrt{2}z_{i-1\,j} + z_{i-1\,j-1}\right)}{\left(4 + 2\sqrt{2}\right)|\Delta\mathbf{d}|}$$

$$S_{ij\,v} = \frac{\left(z_{i-1\,j+1} + \sqrt{2}z_{i\,j+1} + z_{i+1\,j+1}\right) - \left(z_{i-1\,j-1} + \sqrt{2}z_{i\,j-1} + z_{i+1\,j-1}\right)}{\left(4 + 2\sqrt{2}\right)|\Delta\mathbf{d}|}$$

The term $z_{i\,j}$ denotes the z-coordinate of all points of the analysis grid. The term $|\Delta\mathbf{d}|$ represents the grid edge length in the xy-plane. The application of a linear slope analysis grid is presented by the authors in earlier publications [RathPasche-2004] [RathEtAl-2005]. Fig. 3 shows a detail of the topography with measurement points and a linear rectangular slope analysis grid. Grid points exceeding the limit slope value S_{lim} of 5 degrees are indicated by red dots. Characteristic for this approach is the zigzag course of breaklines, which is handled difficultly in the presented hybrid meshing scheme.

The first key idea of the breakline identification approach presented in this contribution is the use of b-spline technology in order to generate a rectangular analysis grid. The $(N + 1) \times (M + 1)$ grid points $\mathbf{d}_{ij}$ are equidistant in the xy-plane:

$$\Delta\mathbf{d} = \mathbf{d}_{i+1\,j} - \mathbf{d}_{ij} = \mathbf{d}_{i\,j+1} - \mathbf{d}_{ij} \tag{2}$$

$$\text{for} \quad 0 \le i \le N - 1, 0 \le i \le M - 1$$

The distance $|\Delta\mathbf{d}|$ is depending on the resolution of measurement points, the dimensions of the area under consideration and finally on the computer performance.

B-spline surfaces imply beneficial properties such as local modeling or smoothing. For the generation of b-spline analysis grids, the grid points $\mathbf{d}_{ij}$ of the bilinear grid are interpreted as control points of a b-spline surface. The z-coordinates of the control points are determined by a dragging algorithm [BerkhahnEtAl-2002] which uses the z-coordinates of the bilinear grid points as starting values of the iteration process. The b-spline grid points $\mathbf{b}_{ij}$ are defined in constant parameter distances Δu and Δv according to (4):

$$\mathbf{b}_{ij} = \mathbf{b}(u_K + (i - 1)\Delta u, v_L + (j - 1)\Delta v) \tag{3}$$

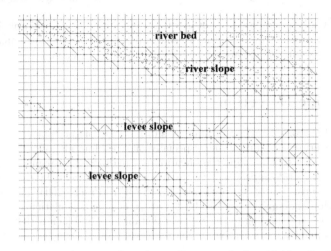

Fig. 3. Breakline identification based on a linear analysis grid (detail of case study 2: River Danube)

In order to ensure a moderate smoothing quadratic b-spline surfaces are chosen as analysis grid. In dependence of a parameter set u, v a point on the b-spline surface $\mathbf{b}(u, v)$ is defined by double sum of all control points $\mathbf{d}_{ij}$ multiplied by the corresponding b-spline functions $N_i^K(u)$ and $N_j^L(v)$:

$$\mathbf{b}(u, v) = \sum_{i=0}^{N} \sum_{j=0}^{M} \mathbf{d}_{ij} N_i^K(u)\, N_j^L(v) \tag{4}$$

$$\text{for} \quad u \in [u_K, u_{N+1}]; \; u \in [v_L, v_{M+1}]$$

In this formula the control points $\mathbf{d}_{ij}$ represent a regular grid of $(N+1) \times (M+1)$ points. The upper indices K and L indicate the degree of the b-spline functions. In order to ensure the property of local modeling the b-spline functions of degree 0 are defined as follows:

$$N_i^0(u) = \begin{cases} 1 & \text{for} \quad u \in [u_i, u_{i+1}[\\ 0 & \text{else} \end{cases} \quad \text{for} \quad i = 0, \ldots, N + K \tag{5}$$

In (5) u_i and u_{i+1} denote the lower and upper bounds of the ith parameter interval. All bounds of the parameter intervals are gathered in a knot vector $\mathbf{u}$:

$$\mathbf{u} = [u_0, \ldots, u_{N+K+1}]^T \tag{6}$$

The b-spline functions of degree r are given by the recursive formula:

$$N_i^r(u) = \frac{u - u_i}{u_{i+r} - u_i} N_i^{r-1}(u) + \frac{u_{i+r+1} - u}{u_{i+r+1} - u_{i+1}} N_{i+1}^{r-1}(u) \tag{7}$$

$$\text{for} \quad r = 1, \ldots, N + K; \quad i = 0, \ldots, N + K + 1$$

The second key idea is to eliminate the zigzag course of breaklines by an interpolation scheme for breakline points. Since the corresponding slope value is determined for every grid point, it is obvious to interpolate the points with the exact limit slope value.

A cell of the analysis grid is given by four grid points $\mathbf{b}_{ij}$, $\mathbf{b}_{i-1j}$, $\mathbf{b}_{ij-1}$ and $\mathbf{b}_{i-1j-1}$. On the edge $\mathbf{b}_{ij}\mathbf{b}_{i-1j}$ a point $\mathbf{p}_{edge1}$ is linearly interpolated by:

$$\mathbf{p}_{edge1} = \frac{S_{i-1j} - S_{lim}}{S_{i-1j} - S_{ij}}\mathbf{b}_{ij} + \frac{S_{lim} - S_{ij}}{S_{i-1j} - S_{ij}}\mathbf{b}_{i-1j} \tag{8}$$

$$\text{for}\quad \text{sign}\,(S_{i-1j} - S_{lim}) \neq (S_{ij} - S_{lim})$$

In general, this interpolation procedure for all four edges of a cell leads to two or four interpolation points $\mathbf{p}_{edge}$. The breakline point is determined by the mean value of these interpolation points. Fig. 4 shows the b-spline analysis grid, the slope points and the interpolated breakline points. These breaklines provide valuable terrain information for the river slope as well as for the floodplain. The identified terrain features are fundamental for the hybrid meshing scheme explained in the following sections.

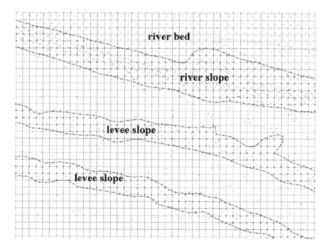

Fig. 4. Breakline identification based on a b-spline grid and breakpoint interpolation (detail of case study 2: River Danube)

5 Regular Meshes Based on B-Spline Surfaces

The topography is approximated by quadratic b-spline surfaces in order to facilitate the generation of a regular mesh with a specified edge ratio and element orientation.

5.1 B-Spline Surfaces Matching Breaklines

The identified breaklines are used for the efficient determination of control point grids defining the approximating b-spline surfaces. The user has to specify roughly some boundary points of the control point grid and the control points are generated in the specified distance and are automatically moved to the nearest breakline. Fig. 5 illustrates a detail of the definition process, where the dark blue points indicate the boundary control points of b-spline surface. The interior control points are generated via a Coons interpolation scheme [BerkhahnEtAl-2002]. Points of the regular element mesh are determined by specified parameter distances Δu and Δv and by applying (4). Figure 6 shows the regular triangle mesh approximating the river bed and matching the identified breaklines.

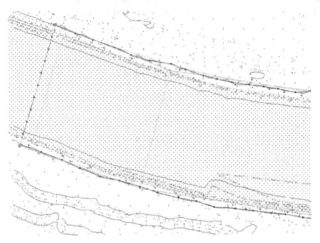

Fig. 5. Editor process with automatic identified breaklines (case study 2: River Danube)

5.2 Handling of Overlapping B-Spline Surfaces

A non ramified river bed could easily be approximated by b-spline surfaces: two opposed boundary curves of the surface approximate the shore lines of the river and the remaining two boundary curves represent the inflow and outflow cross section of the river section under consideration. This easy handling becomes more sophisticated while dealing with islands or tributaries. Berkhahn et al. [BerkhahnEtAl-2002] developed an approach to generate consistent element meshes for these cases. The key idea is to connect the different b-spline surfaces for the main stream and the tributaries consistently. This approach requires the usage of endpoint interpolating b-spline surfaces with multiple knots in the knot vectors. This manipulation of the knot vectors involves the serious disadvantage of a non uniform parameterisation of the b-spline surface and consequently the usage of constant Δu and Δv parameter distances for the generation of element nodes is impractical.

To cope with these circumstances, an approach with overlapping b-spline surfaces is developed. This approach is illustrated in figs. 7 and 8 for a small river detail of the first case study. Main river and tributary are approximated by two disconnected and overlapping b-spline surfaces. The corresponding control point grids

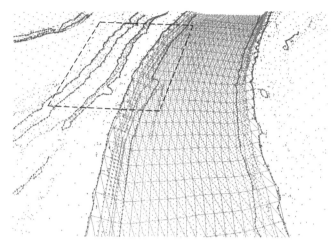

Fig. 6. Regular element mesh of the river bed matching identified breaklines (case study 2: River Danube); the dashed square indicates the detail of figs. 3 and 4

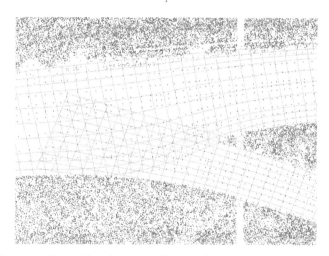

Fig. 7. Control point grids of two overlapping b-spline surfaces representing the main river and the tributary (case study 1: River Stoer)

are shown in fig. 7. In order to approximate the bank slope of the main river with higher accuracy the control points are concentrated in this area. Because both b-spline surfaces represent the same measurement points in the overlapping area the C^0 and C^1 continuity between these surfaces is ensured. Based on both surfaces, regular meshes are generated considering the specific meshing requirements. In a post process both meshes are stitched as illustrated in fig. 8. This approach of over-lapping b-spline surfaces requires no difficult and sophisticated editor functionality to generate consistent control point grids and finally, this approach is easy to handle for the user.

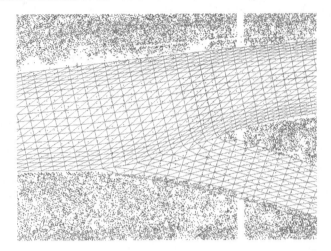

Fig. 8. Stitched element mesh of the main river and the tributary (case study 1: River Stoer)

5.3 Handling of Regular Grids with Arbitrary Boundaries

The b-spline meshing technique involves serious problems in the case of any arbitrary boundary of the considered domain. Due to the regular control point grid a b-spline surface is bounded by four boundary curves. If it is impossible to approximate the domain by a quadrilateral b-spline surface the face technology is applied. This face technology is well known in design of construction parts in free form modeling: The relevant area of a free form surface is cut out of the entire free form surface.

The same approach is used in the hybrid meshing scheme: For instance, the slope of a levee is approximated by a b-spline surface with a significant overhang. The face curve is based on the identified breaklines of the levee and represents a closed loop. All regular elements generated on the b-spline surface are deleted if at least one node of the element is located outside of the closed loop of the face curve. This approach involves the advantage to specify the edge ratio and orientation of the elements while the needless elements are neglected.

6 Irregular Triangle Meshes Adapted from Delaunay Triangulations

The hybrid meshing scheme involves irregular triangles for the floodplain. A Delaunay triangulation of point measurements efficiently provides irregular triangle meshes but involves a significant disadvantage: a one-to-one mapping of measurement points to mesh nodes is not suitable for the considered case studies, as they commonly deal with millions of points. Ruppert [Ruppert-1995] and Shewchuck [Shewchuck-2002] presented very effective refinement and coarsening schemes. In this contribution a more simple refinement and coarsening approach for the hybrid meshing scheme is used.

Aiming to generate a more or less uniformly dense, but irregular mesh on the floodplain, an initial Delaunay triangulation is performed for the original point set including all measurement points, all identified breakline points and all boundary nodes of regular partial element meshes. All points on the convex hull of the point set, on identified breaklines and on boundaries of regular partial meshes are fixed, i.e. no points are allowed to be moved or deleted.

A coarsening procedure collapses the points of a triangle or of a single edge by the corresponding center point if the edge length is below a user defined minimal edge length. The left hand side of fig. 9 shows the Delaunay triangulation of a point set (black points). This triangulation leads to two triangles each with 3 edges (red edges) below a specified minimum edge length. The corresponding points of the affected triangles are deleted and replaced by the center points (green points). The re-triangluation leads to a coarser mesh, but one single edge is still below the minimal edge length. As shown on the right hand side of fig. 9 the points corresponding to this edge (red edge) are deleted and replaced by their center point (green point). The re-triangulation leads to a coarse mesh with no edge below the minimal edge length.

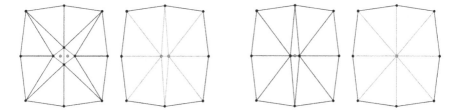

Fig. 9. Coarsening of triangles (left) and of a single edge (right)

A refinement procedure adds the corresponding center point if the edge length of a triangle or a single edge exceeds a user defined maximum edge length. The left hand side of fig. 10 shows the Delaunay triangulation of a point set (black points). This triangulation leads to two triangles each with 3 edges (red edges) exceeding a specified maximum edge length. The center points (green points) of these two affected triangles are added to the initial point set. The re-triangulation of this augmented point set leads to new edges (green edges) and to a refined triangle mesh. The right hand side of fig. 10 shows the Delaunay triangulation of a point set, where a single edge (red edge) exceeds the maximum edge length. The center point (green point) is added to the initial point set. The re-triangulation of this new point set leads to a refined triangle mesh with new edges (green edges) not exceeding the maximum edge length.

Finally, a Laplace smoothing is performed for all points not regarded to be fix. Fig. 11 shows the result of the refinement and coarsening procedure and the fixed breaklines. The triangulation and adaptations procedure concatenates regular and irregular element meshes as shown in figure 12.

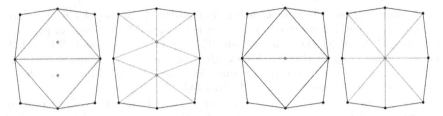

Fig. 10. Refinement of triangles (left) and of a single edge (right)

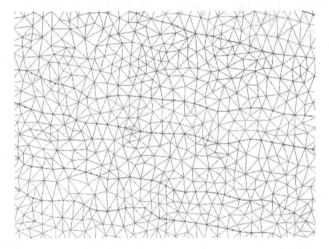

Fig. 11. Irregular triangle mesh matching identified breaklines (case study 2: River Danube)

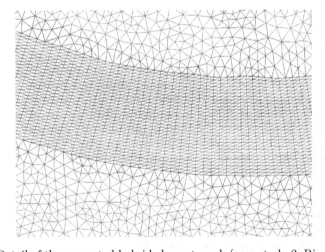

Fig. 12. Detail of the generated hybrid element mesh (case study 2: River Danube)

7 Quality Analysis for Hybrid Mesh Terrain Representation

For the case study 1 of the River Stoer the original Delaunay refinement [Ruppert-1995] [Shewchuck-2002] is applied to generate the triangle elements. The objective of this case study is the optimization of the final mesh with regard to the minimum element angle ($\alpha \geq 20$ degrees) and the maximum element area size ($A \leq 22\,\mathrm{m}^2$). The entire mesh consists of $11,060$ elements and $6,308$ nodes. This extremely dense refinement is intended to obtain a mesh, which accurately resolves the flow gradients in the river junction.

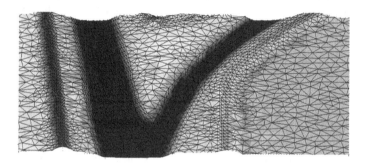

Fig. 13. Detail of the generated hybrid element mesh (case study 1: River Stoer)

The accuracy of the presented hybrid mesh is sampled for various ratios of the available measurements. A twofold classification criterion is applied, stating acceptable representations (green residuals), sufficient representations (yellow residuals) and those with significant deviation from the measured data (red residuals). With respect to the accuracy demands on LiDAR data for floodplain mapping [RathBajat-2004], the following classification for residuals is introduced.

Table 1. Classification for residuals

acceptable:	residuals $\leq \pm 0.10\,\mathrm{m}$
tolerable:	$\pm 0.10\,\mathrm{m} <$ residuals $\leq \pm 0.30\,\mathrm{m}$
significant:	$\pm 0.30\,\mathrm{m} <$ residuals

For a random sampling series 1%, 5%, 10% and 30% of the available measurements are considered. Anticipating one conclusion of this investigation, the sampling rate of 1% is sufficient to classify the mesh representation. The margin for classifications sampling more than 1% of measurements denotes less than 1.5% for each class. Further statistical measures such as the mean residual μ_R and the standard deviation σ_R of the mesh representation provide essential information about the quality of the hybrid mesh. For the case study of the River Stoer, again 1% of samples provides a suitable impression of the statistical parameters. Additional sampling reveals a margin in standard deviation below $1\,\mathrm{cm}$.

Table 2. Impact of random sampling ratios on quality assessment

Sampling Ratio	Green Residuals	Yellow Residuals	Red Residuals	Mean Residual μ_R	Standard Deviation σ_R
1%	67.6%	29.5%	2.9%	0.023 m	0.153 m
5%	69.2%	27.3%	3.5%	0.021 m	0.146 m
10%	69.2%	27.2%	3.6%	0.022 m	0.154 m
30%	68.5%	27.8%	3.7%	0.022 m	0.158 m

Accuracy standards for floodplain mapping based on LiDAR data, given by the US Federal Emergency Management Agency in accordance with the US National Standard for Spatial Data Accuracy for digital products, claim that an accurate DEM should have a maximum Root Mean Square Error of 15 cm [FEMA-2003]. Moreover, 95% of any sufficiently large sample should be less than $1.9600 \times \text{RMSE}$, holding for normally distributed differences averaging zero. A RMSE of 15 cm denotes a "30 cm accuracy at the 95% confidence level" [FEMA-2004b].

For the validation of the mesh against these requirements, the mesh is considered as ground truth estimate and the measured data is considered as corresponding ground truth. Based on table 2 the terrain representation in fig. 13 is considered suitable for floodplain mapping. With $\sigma_R \approx \text{RMSE} \approx 0.15$ m and since the magnitude of red residuals clearly ranges below 5% it is obvious, that the "30 cm accuracy at the 95% confidence level" is assured. For this case study, red residuals are partly raised by a drainage ditch, which is included in the LiDAR data set without being excluded in the hybrid mesh at the border between the b-spline mesh and its adjacent irregular triangle mesh. The relevance for a hydrodynamic simulation is negligible, though.

8 Numerical Simulation

This numerical simulation for the case study 1 shows a typical discharge scenario. Parts of the slightly elevated hook at the river junction appear as an island, surrounded by the flow of River Bramau and Stoer.

Generally, the representations of the slow velocity gradients are accurate, denoting the distinct shallow terrain. The visualisation of the water depths indicates the sensitivity of hydrodynamics with regard to the mesh topology. The drainage ditch represented in fig. 14 is represented despite a local lack of data (see fig. 1). The water depths indicate a local pit. Consequently, the velocity field in figs. 14 and 15 is dominated by turbulent structures that are likely to be representative for that spot. For coarser meshes such representations might generally lead to troublesome numerical stability demands.

9 Conclusion

This contribution presents an efficient technique to generate adaptable element meshes for an accurate representation of terrain data in numerical simulations. In-

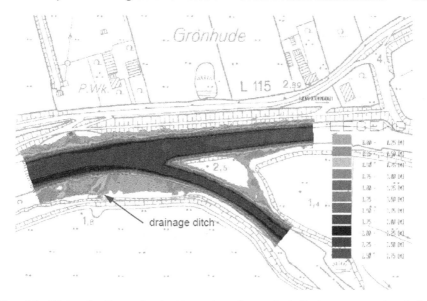

Fig. 14. Water depths and velocity vectors for typical discharge scenario of the River Stoer and Bramau (case study 1)

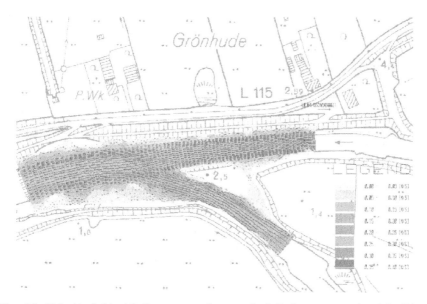

Fig. 15. Velocity field with flow vectors for a typical discharge scenario of the River Stoer and Bramau (case study 1)

terpolated breakline points determined on rectangular b-spline analysis grids are suitable to provide the essential terrain features in shallow fluvial domains with sufficient accuracy. Breakline identification is fundamental for the hybrid meshing scheme introduced by the authors. Regular element meshes use these breaklines to define the boundary for b-spline surfaces approximations for river beds, banks, groynes and any other flow relevant sloped terrain feature. In floodplain areas, represented with irregular triangle meshes, these breaklines are matched by element edges in order to represent terrain features accurately. Case studies demonstrate the usability and suitability of the presented hybrid meshing scheme based on breakline identification for hydrodynamic simulations.

10 Acknowledgements

The author Kai Kaapke wishes to acknowledge the support of the German Academic Exchange Service (DAAD) within the postgraduate research program during his year-long stay at the Water Research Laboratory in Australia.

[BerkhahnMai-2004a] Berkhahn V, Mai S (2004) Meshing bathymetries for numerical wave modeling. Proceedings of the 6th International conference on Hydroinformatics 2004. World Scientific Publishing Co. Pte. Ltd., Singapore, Vol. 1: 47-54

[BerkhahnMai-2004b] Berkhahn V, Mai S (2004) Numerical Wave Modeling based on Curvilinear Element Meshes. Proceedings of the 4th International Symposium on Environmental Hydraulics, ISEH04, Hong Kong

[BerkhahnEtAl-2002] Berkhahn V, Göbel M, Stoschek O, Matheja A (2002) Generation of Adaptive Finite Element Meshes Based on Approximation Surfaces. Proceedings of the 5th International Conference on Hydro-Science and -Engineering, ICHE 2002, Warsaw, Poland

[FEMA-2004a] FEMA, Federal Flood Emergency Management Agency (2004) Flood Hazard Mapping - Hydraulic Models Meeting the Minimum Requirement of NFIP. National Accepted Models.
http://www.fema.gov/fhm/enhydra.shtm (14.02.2005)

[FEMA-2004b] FEMA, Federal Flood Emergency Management Agency (2004) Quality Control / Quality Assurance for Flood Hazard Mapping.
http://www.fema.gov/fhm/dlcgs.shtm (14.02.2005)

[FEMA-2003] FEMA, Federal Flood Emergency Management Agency (2003) Guidelines and Specifications for Flood Hazard Mapping Partners.
http://www.fema.gov/mit/tsd/mma4b7.shtm (14.02.2005)

[Goebel-2005] Göbel M (2005) HydroMesh Meshing Tool. www.hydromesh.com (31.3.2005)

[Jones-1998] Jones K H (1998) A comparison of algorithms used to compute hill slope as a property of the DEM. Computers & Geoscience, Vol. 24: 315-323

[RathEtAl-2005] Rath S, Pasche E, Berkhahn V (2005) Modeling High Resolution Remote Sensing Data for Flood Hazard Assessment. Proceedings of ASPRS 2005, American Society for Photogrammetry & Remote Sensing, Baltimore, Maryland

[RathPasche-2004] Rath S, Pasche E (2004) Hydrodynamic Floodplain Modeling based on High-Resolution LiDAR measurements. Proceedings of 6th International Conference on Hydroinformatics 2004. World Scientific Publishing Co. Pte. Ltd., Singapore, Vol. 1: 486-493

[RathBajat-2004] Rath S, Bajat B (2004) Between Sensing, Forecasting and Risk Assessment: An Integrated Method to Model High Resolution Data for Floodplain Representations in Hydrodynamic Simulations. Proceedings of 1st Goettingen Remote Sensing Days 2004, Goettingen 07-08 October. In: Erasmi S, Cyffka B, Kappas M (eds) Remote Sensing & GIS for Environmental Studies. Goettinger Geographische Abhandlungen, Bd. 113, Verlag Erich Goltze, Goettingen, Germany

[Ruppert-1995] Ruppert J (1995) A Delaunay refinement algorithm for quality 2D-mesh generation. Journal of Algorithms, Vol. 18, No.3: 548-585

[Shewchuck-2002] Shewchuck J R (2002) Delaunay refinement algorithms for triangular mesh generation. Computational Geometry: Theory and Applications, Vol. 22 (1-3): 21-74

Meshing Piecewise Linear Complexes by Constrained Delaunay Tetrahedralizations

Hang Si and Klaus Gärtner

Weierstrass Institute for Applied Analysis and Stochastics, Berlin, Germany
{si, gaertner}@wias-berlin.de

Summary. We present a method to decompose an arbitrary 3D piecewise linear complex (PLC) into a constrained Delaunay tetrahedralization (CDT). It successfully resolves the problem of non-existence of a CDT by updating the input PLC into another PLC which is topologically and geometrically equivalent to the original one and does have a CDT. Based on a strong CDT existence condition, the redefinition is done by a segment splitting and vertex perturbation. Once the CDT exists, a practically fast cavity retetrahedralization algorithm recovers the missing facets. This method has been implemented and tested through various examples. In practice, it behaves rather robust and efficient for relatively complicated 3D domains.

1 Introduction

A fundamental problem in unstructured mesh generation is to create a mesh that represents a geometric domain bounded by piecewise linear faces and possibly with interior constraining faces. It is also referred as boundary mesh generation. Numerous applications depend on it.

This problem has been successfully solved in two dimensions. It is well known that every polygonal domain can be triangulated into triangles without adding new vertices (the Steiner points). Almost optimal algorithms (with a linear complexity in practice) have been proposed [Lee86, Chew89].

The problem is significantly more difficult for arbitrarily shaped three dimensional domains. Such domains can be described by *piecewise linear complexes* (PLCs) [Miller96], which are objects more general than polyhedra. It is known [Schoenhardt28, Bagemihl48, Chazelle84, Rambau03] even a simple polyhedron may not be tetrahedralizable without adding new vertices. The simplest example is the so-called Schönhardt polyhedron [Schoenhardt28], which is a non-convex twisted triangular prism. Moreover, the problem of deciding whether a simple polyhedron can be tetrahedralized is NP-hard [Ruppert92]. PLCs are usually much more complicated than simple polyhedra. To guarantee an arbitrary PLC can always be meshed, methods must resort to adding Steiner points. However, a number of difficult issues remain to be resolved, such as the placement of the Steiner points, the minimum bound on such points, and so on.

The recently proposed *Constrained Delaunay tetrahedralization* (CDT) (cf. [Shewchuk02]) is a Delaunay-like tetrahedralization that is constrained to respect the shape of a PLC. CDTs are obviously suitable structures for resolving the above problem. Not only they respect the boundary but also they have many nice mathematical properties inherited from Delaunay tetrahedralizations. Many applications can be envisaged after getting a CDT. For instance, it is a good initial mesh for getting a quality conforming Delaunay mesh [Shewchuk98b, Cheng04, Pav04] which is suitable for numerical methods.

A key question for constructing a CDT is to decide its existence, i.e., whether a given PLC has a CDT without adding points. So far, Shewchuk [Shewchuk98a] has proved the condition: if all segments of the PLC are *strongly Delaunay*, then the CDT exists. The hint gained from this condition is: additional points can be inserted only on the segments of the PLC.

Shewchuk [Shewchuk02] gave a segment recovery algorithm for constructing CDTs. For a PLC X, it carefully introduces additional points on some segments of X until the existence of a CDT can be guaranteed. This algorithm yields a provably good bound on edge lengths. However, it requires to compute the local feature size explicitly for protecting *sharp corners* (vertices with angles less than 90° formed by segments). In [Si04a] we proposed a new segment recovery strategy which exploits the available local geometric information to efficiently construct Steiner points on segments. It needs not to compute the local feature sizes. Moreover sharp corners are implicitly handled during the creation of the Steiner points. Both algorithms tend to use fewer additional points than other methods which do edge protect provably [Pebay98, Murphy00, Cohen-Steiner02] too.

When the existence of a CDT is known, another key issue is to recover facets of the PLC. Generally non-CDT algorithms [Pebay98, Murphy00, Cohen-Steiner02, Weatherill94, George03, Du04] will continuously insert points on the missing facets or the inside of the PLC. While CDT algorithms [Shewchuk02, Shewchuk03, Si04a] recover the missing facets without introducing additional points. This again reduces the number of Steiner points in a CDT.

By now Shewchuk has provided several facet recovery algorithms [Shewchuk00, Shewchuk02, Shewchuk03]. The incremental facet insertion algorithm [Shewchuk02] recovers facets one after one and the CDT is updated accordingly. For each facet, a gift-wrapping algorithm is used for retetrahedralizing the two cavities around it. The strategy is simple but the time complexity is poor and the performance is unstable due to the gift-wrapping algorithm. The flip-based algorithm [Shewchuk03] recovers each facet by a sequence of carefully ordered flip operations. It appears to be simple and is likely to outperform other algorithms on most inputs. In order to guarantee the correctness and termination, both algorithms require that a full perturbation has to be applied on the set of vertices to remove the degeneracies.

In this paper we present a new CDT algorithm. The main difference from other CDT algorithms is the practical exploitability of a strong CDT existence condition which requires no *local degeneracy* on the vertices of the PLC. We propose a local degeneracy removal algorithm to construct a new set of vertices out of the old one which is consistent with the constraining segments and facets of the PLC. After the strong condition is satisfied, facet recovery is done by a new cavity retetrahedralization algorithm which is fast and robust in practice.

The remainder of this paper is organized as follows. We shortly recall the definitions of PLCs and CDTs in the next section. Then the existence of CDT is discussed in section 3. Section 4 provides an overview of the proposed method. The individual algorithms are fully described in section 5, 6, and 7. Finally we present some meshing results from publicly available examples.

2 PLCs and CDTs

A *piecewise linear complexes* (PLC) proposed by Miller, Teng, Walkington, and Wang [Miller96] is a general boundary description for three-dimensional domains. Simply saying, a PLC X is a set of vertices, together with a collection of segments, and facets. Like a simplical complex, any two components of X are either disjoint or meet in a common face, e.g., two segments can only intersect at a vertex of X, and two facets can only intersect at a collection of segments and vertices of X, and so on. A PLC facet has no analogue in a simplical complex. Each facet of X is indeed a two-dimensional polygonal region embedded in three dimensions, it may not be convex and possibly contains holes, isolated segments and vertices.

PLCs are more general than polyhedra in the sense that every polyhedron is a PLC but not vice versa. For instance, a PLC containing a segment inside can't be represented by any polyhedron. Surface triangulations are one special type of PLCs - each facet is a triangle. Hence PLCs are able to approximate arbitrary complicated and curved shapes. In addition, many popular polygonal file formats (e.g., STL, OFF, PLY, and [Si04b]) can be directly used or slightly modified to describe PLCs.

Given a PLC X, Shewchuk [Shewchuk02] defined a CDT of X as follows:

Let V be the set of vertices of X. σ is any simplex (tetrahedron, triangle, edge or vertex) formed by vertices of V. σ is *Delaunay* if there exists a circumsphere of σ that encloses no vertex of V. The *Delaunay tetrahedralization* $\mathcal{D}$ of V is a tetrahedralization that all simplices of $\mathcal{D}$ are Delaunay. $\mathcal{D}$ is unique if V is general, i.e., no five or more vertices lie on a common sphere.

The visibility between two vertices p and q is *occluded* if there is a constraining facet f such that p and q lie on opposite sides of the plane that includes f, and the line segment pq intersects this facet(see Figure 1). Segments do not occlude visibility. For example, in Figure 1, c and d can see each other even if ab is a segment.

A tetrahedron t formed by vertices of X is *constrained Delaunay* if its circumsphere encloses no vertex of X which is visible from any point in the relative interior of t (see Figure 1).

A tetrahedralization $\mathcal{T}$ is a *constrained tetrahedralization* of X if $\mathcal{T}$ and X have exactly the same vertices, and the tetrahedra in $\mathcal{T}$ cover the convex hull of V. Every segment of X is an edge of $\mathcal{T}$, every facet of X is a union of triangular faces of $\mathcal{T}$.

A constrained tetrahedralization $\mathcal{T}$ of X is said to be a *constrained Delaunay tetrahedralization* (CDT) of X if each tetrahedron of $\mathcal{T}$ is constrained Delaunay.

Intuitively, the definitions of Delaunay tetrahedralization and constrained Delaunay tetrahedralization are the same except that, for the CDT, we ignore the volume of a sphere whenever the sphere passes through a facet of X.

Let $\mathcal{T}$ be a CDT of X. A facet of X is represented by a set of coplanar triangular faces of $\mathcal{T}$. Such faces are called *subfaces* for distinguishing them from other faces of $\mathcal{T}$. The set of subfaces of a facet form a two-dimensional constrained Delaunay triangulation.

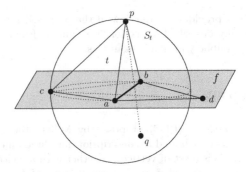

Fig. 1. Constrained Delaunay tetrahedron. The shaded region represents a facet f of X including vertices a, b, c and d. Vertices p and q lie on opposite sides of f, they can not see each other. While c and d can see each other even if ab is a segment of X. S_t is the circumsphere of tetrahedron t ($abcp$) and it encloses q but not d, t is constrained Delaunay.

3 The Existence of CDT

Given a PLC X. Generally, the CDT of X may not exist. One reason is that X may not be tetrahedralizable at all. Even the constrained tetrahedralization of X exists, it may still not have a CDT. A key question for CDT algorithms is to decide under what condition the CDT exists.

Shewchuk has proved a condition. Let σ be a simplex of X, σ is *strongly Delaunay* if there exists a circumsphere of σ that passes through and encloses no other vertices of X. Say X is *edge-protected* if every segment of X is strongly Delaunay.

Theorem 1 ([Shewchuk98a]). *If X is edge-protected, then X has a CDT.* ∎

This condition is useful in practice because it suggests that additional points can be inserted on segments only. However, this condition does not help us to construct the CDT. Suppose X has been made edge-protected (by splitting the segments into smaller segments), algorithms [Shewchuk02, Shewchuk03, Si04a] need to carefully do perturbations on the vertices of X in order to successfully recover the missing subfaces.

In this paper, we use another less general condition which guarantees the existence of CDT, too.

Definition Let $\mathcal{D}$ be the Delaunay tetrahedralization of the vertices of X, t a tetrahedron in $\mathcal{D}$ and t' an adjacent tetrahedron of t (sharing a face with t), V the set of vertices of t, t'. If all vertices of V lie on a common sphere, V is called a *local degeneracy* of X.

Remark: If X contains no local degeneracy, then $\mathcal{D}$ is unique. However, the set of vertices of X may still be degenerate (since arbitrary 5 or more vertices of X can lie on a common sphere).

Theorem 2. *If $\mathcal{D}$ contains no local degeneracy and contains all segments of X, then the CDT of X exists.*

Proof: It is easy to show: if X satisfies the above condition, then the segments of X are edge-protected. ∎

This condition is stronger than Shewchuk's condition since it implicitly satisfies it. Its advantage is: the facet recovery can be carried out efficiently. Section 7 presents such an algorithm.

4 The Algorithm

Let the initial PLC be X_0. The CDT is constructed by the following consecutive steps:

(1) Construct an initial Delaunay tetrahedralization $\mathcal{D}_0$ of the vertices of X_0.
(2) Recover the segments of X_0 in $\mathcal{D}_0$ by incrementally inserting points on missing segments, update $X_0 \rightarrow X_1$ and $\mathcal{D}_0 \rightarrow \mathcal{D}_1$ with the newly inserted points respectively.
(3) Remove the local degeneracies in X_1 by either perturbing vertices or inserting new vertices, update $X_1 \rightarrow X_2$ and $\mathcal{D}_1 \rightarrow \mathcal{D}_2$ with the newly inserted points respectively.
(4) Recover the subfaces of X_2 in $\mathcal{D}_2$ by a cavity retetrahedralization method.

In step (1), $\mathcal{D}_0$ can be efficiently constructed by any standard algorithm, such as [Edelsbrunner96]. $\mathcal{D}_0$ probably does not respect the segments and subfaces of X_0. After step (2) is done, D_1 is a Delaunay tetrahedralization of vertices of X_1 which contains all segments of X_1. However, D_1 may not respect the subfaces of X_1. Step (3) guarantees the existence of a CDT. After all local degeneracies are removed, $\mathcal{D}_2$ is the unique Delaunay tetrahedralization of vertices of X_2 and contains all segments of X_2. D_2 may still not respect the subfaces of X_2, while the existence of the CDT (of X_2) is guaranteed. Hence the step (4) can be carried out without inserting additional vertices. These processes are detailed in the following sections.

5 Segment Recovery

A segment of X_0 is *missing* if it is not in $\mathcal{D}_0$. The purpose of the segment recovery algorithm is to update $\mathcal{D}_0$ into $\mathcal{D}_1$ such that $\mathcal{D}_1$ is a Delaunay tetrahedralization and includes all segments of X_0.

Let $e_i e_j$ be a segment with endpoints e_i and e_j, $|e_i e_j|$ be its Euclidean length. A vertex is *acute* if at least two segments incident at it form an angle smaller than $90°$. We distinguish two types of segments, a segment is *type-1* if its both endpoints are not acute, it is *type-2* if only one of its endpoints is acute. If both endpoints of a segment are acute, it can be transformed into two type-2 segments by inserting a vertex.

A segment is split into *subsegments*. Subsegments inherit types from original segments. For example, let $e_i e_j$ be a subsegment of $e_1 e_2$ which is a type-2 segment and e_1 is acute, $e_i e_j$ is type-2 although none of its endpoints is acute. For any vertex v inserted on a type-2 segment (or subsegment), $O(v)$ denotes its original acute vertex. A tacit rule is used throughout this section, if $e_i e_j$ is type-2, it implies either

e_i or $O(e_i)$ is the acute vertex. In the following, unless it is explicitly mentioned, a segment can be a segment or subsegment.

A vertex *encroaches upon* a segment if it lies inside or on the diameter sphere of that segment. *Remark*: any missing segment must be encroached by at least one vertex of X.

Let $e_i e_j$ be a missing segment, it will be split by a vertex v. The *reference point* p of v, which is "responsible" for the insertion of v, is defined as follows:

- p encroaches upon $e_i e_j$;
- the circumradius of the smallest circumsphere of triangle $e_i e_j p$ is maximum over other encroaching points of $e_i e_j$.

Figure 2 illustrates how p is chosen. Notice that p may not unique (because several points can share the same circumsphere), choose an arbitrary one in this case.

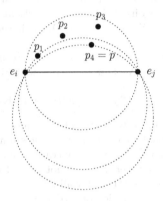

Fig. 2. The reference point p of a splitting segment $e_i e_j$. p_1, p_2, p_3 and p_4 all encroach upon $e_i e_j$. p_4 is chosen as the reference point because it forms the biggest circumsphere with $e_i e_j$.

The insertion of v to split $e_i e_j$ is governed by three rules given below. Let $S(c, r)$ denote a sphere centered at c with radius r:

1. $e_i e_j$ is type-1 (see Figure 3 (a)), then $v := e_i e_j \cap S$, where $S(c, r)$ is the sphere defined by the reference point p of v as follows:
   if $|e_i p| < \frac{1}{2} |e_i e_j|$ **then**
     $c := e_i, r := |e_i p|$,
   else if $|e_j p| < \frac{1}{2} |e_i e_j|$ **then**
     $c := e_j, r := |e_j p|$,
   else
     $c := e_i, r := \frac{1}{2} |e_i e_j|$,
   end
2. $e_i e_j$ is type-2 (see Figure 3 (b)), let $e_k := O(e_i)$, then $v := e_k e_j \cap S$, where $S(c, r)$ is the sphere defined by the reference point p of v with $c := e_k, r := |e_k p|$. However, if the subsegment $v e_j$ has length $|v e_j| < |v p|$, then we do not insert v and use rule 3 to split the segment.

3. $e_i e_j$ is type-2, let $e_k := O(e_i)$, and v' is the rejected vertex by rule 2. then
$v := e_k e_j \cap S$, where $S(c, r)$ is the sphere defined by the reference point p of v
as follows (see Figure 3 (c)):

$c := e_k$
if $|pv'| < \frac{1}{2}|e_i v'|$ **then**
   $r := |e_k e_i| + |e_i v'| - |pv'|$
else
   $r := |e_k e_i| + \frac{1}{2}|e_i v'|$
end

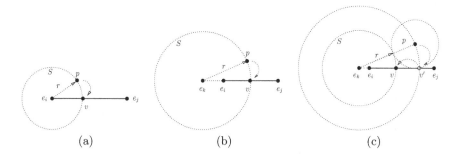

(a)   (b)   (c)

Fig. 3. Illustrations of the segment splitting rules.

For several segments sharing an acute vertex, by repeatedly using rule 2 or 3,
a protecting ball is automatically created which ensures: no other vertex can be
inserted inside the ball. The effect is shown in Figure 4. Notice, the protecting ball
is not necessarily completely created, only the missing segments will be split and
protected. Existing segments remain untouched. This reduces the number of Steiner
points.

Below is the pseudo-code of the segment recovery algorithm.

Algorithm *Delaunay Segments Recovery*
Input: $\mathcal{D}_0$, X_0.
Output: $\mathcal{D}_1$, X_1.
initialize:
    $\mathcal{D}_1 := \mathcal{D}_0$, $X_1 := X_0$;
repeat:
    form a queue Q of missing segments in $\mathcal{D}_1$;
    while $Q \neq \emptyset$ **do**
        remove a segment $e_i e_j$ from Q;
        split $e_i e_j$ using rule 1, or 2, or 3;
        update $\mathcal{D}_1$, X_1;
    end
    until no segment of X_1 is missing in $\mathcal{D}_1$

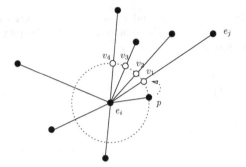

Fig. 4. The protecting ball of an acute vertex e_i. v_1, v_2, v_3, and, v_4 are points inserted on segments (by rule 2) sharing e_i. They automatically create a protecting ball of e_i.

The termination of this algorithm can be proved by showing that the length of every segment created is bounded by the local feature size divided by constant depending only on the input. For a PLC X, the *local feature size* [Ruppert95] $lfs(v)$ of any point v in X is the radius of the smallest ball centered at v that intersects two segments or vertices in X that do not intersect each other. The $lfs()$ defines a continuous map that maps every point in X into a positive value which suggests how large the ball of the empty space around this point can be. The function $lfs()$ is only defined on the input PLC X and does not change as new points are inserted.

Theorem 3 ([Si04a]). *Let e_ie_j be a finally resulting subsegment,*

- *if e_ie_j is type-1, then:*
  $|e_ie_j| \geq min\{lfs(e_i), lfs(e_j)\}$.
- *if e_ie_j is type-2, let $e_k := O(e_i)$, then:*
  $|e_ie_j| \geq \frac{1}{C}lfs(e_k)$ *when $e_i = e_k$,*
  $|e_ie_j| \geq lfs(e_k)sin(\theta)$ *when $e_i \neq e_k$.*
  where C is two times the number of segments incident at e_k and θ is the smallest angle between them.

hence the Delaunay segments recovery algorithm terminates. ∎

Practically, the algorithm terminates within a few steps creating the protection ball sector. The constant C usually is no larger than 4.

6 Removing Local Degeneracies

In order to fulfill the condition of Theorem 2, all local degeneracies have to be removed from $\mathcal{D}_1$. Techniques of perturbation [Edelsbrunner90] are effective to remove degeneracies. However, they must be carefully applied in CDT algorithms.

We say a vertex of a PLC is *perturbable* if there exists an arbitrarily small perturbation on it which does not affect the consistency of the PLC, otherwise, it is *unperturbable.* For example, a vertex on a segment or inside a facet is perturbable since it can be perturbed arbitrarily along the segment vector or inside the facet without affecting the collinearity of the segment or the coplanarity of the facet. Not every vertex of a PLC is simply perturbable. If a vertex intersected by three or more non-coplanar facets is perturbed, at least one facet becomes invalid (because it now contains a non-coplanar vertex), hence it is unperturbable.

Let Δ be a local degenerate set of vertices in X_1, i.e., Δ contains 5 vertices of X_1 which share a common sphere. If $p \in \Delta$ is perturbable, Δ can be removed by an arbitrarily small perturbation on p. We call a perturbation is *segment-safe* if after the perturbation no segment of X_1 becomes non-Delaunay. A perturbable vertex may not be segment-safe, see Figure 5 for an example.

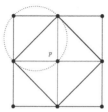

Fig. 5. A perturbable vertex which is not segment-safe. The vertices of this facet consist of a 3×3 square grid. The vertex p is perturbable but not segment-safe when the four edges opposite it are all segments. Whatever it moves in the facet, at least one segment becomes non-Delaunay.

Δ is said to be *removable* if there exits a vertex $p \in \Delta$, such that:

- p is perturbable; and
- there exists a perturbation of p which is segment-safe.

Otherwise, Δ is *unremovable.*

If Δ is unremovable. We introduce a new point v_b, called *break point*, it is chosen as follows:

- If the vertices of Δ are affinely independent. Let S_Δ be the common sphere shared by them. v_b is inside S_Δ.
- If the vertices of Δ are affinly dependent, i.e., four of them are coplanar. Let C_Δ be the common circle shared by the four vertices. v_b is coplanar with the four vertices and inside C_Δ.

v_b will break the local degeneracy (see Figure 6).

A break point may encroach upon one or more segments and subfaces of X_2. In this case, it will not be inserted. Instead, the following *boundary protection* procedure is called:

1. For each encroached segment, add its perturbed circumcenter to X_2. Updating $\mathcal{D}_2$, X_2 accordingly;

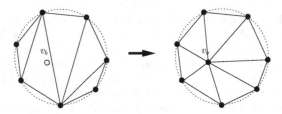

Fig. 6. The break point. On the left is a set of locally degenerate points and one of its Delaunay triangulations. v_b is a break point for removing the local degeneracy. On the right is the unique Delaunay triangulation after v_b is inserted.

2. For each encroached subface, compute its perturbed circumcenter x. If x encroaches upon any segments, it is not inserted, goto step 1 to split the encroached segments. Otherwise, add x to X_2. Updating $\mathcal{D}_2$, X_2 accordingly;

3 Call the Delaunay segment recovery algorithm to recover all missing segments of X_2 in $\mathcal{D}_2$.

The complete algorithm is listed below:

> **Algorithm** *Local Degeneracy Removal*
> Input: $\mathcal{D}_1$, X_1.
> Output: $\mathcal{D}_2$, X_2.
> **initialize:**
> $\mathcal{D}_2 := \mathcal{D}_1$, $X_2 := X_1$;
> **repeat:**
>   form a queue Q of local degeneracies of $\mathcal{D}_2$;
>   **while** $Q \neq \emptyset$ **do**
>   remove a local degeneracy Δ from Q;
>   **if** Δ is removable **then**
>   Remove Δ by a small perturbation;
>   **else**
>   Compute a v_b of Δ;
>   **if** v_b encroaches upon any segment and subface **then**
>   Push Δ into Q;
>   Call the boundary protection procedure;
>   **else**
>   Insert v_b to break Δ;
>   update $\mathcal{D}_2$, X_2;
>   **endif**
>   **endif**
>   **endwhile**
> **until** $\mathcal{D}_2$ contains no local degeneracy;

In our implementation, we choose v_b be the circumcenter of the common sphere of Δ. It is important that v_b should not create a new local degeneracy with other existing vertices. This can be checked before inserting the point. If so, we should not insert v_b but choose another location. Notice that such case only can happen when the input PLC is highly symmetric, e.g., in Figure 5. A simple strategy is to add a randomized perturbation on each v_b. After the perturbation, the probability to create a symmetric point configuration again is nearly zero. Due to that reason, the points added for boundary protecting are perturbed, too.

Some break points may locate outside X_2. They will be removed after the CDT of X_2 is constructed.

The termination of the local degeneracy removal algorithm is due to the fact that the number of unremovable local degeneracies can only be decreased and no new local degeneracies are created by break points and segment protecting points.

7 Facet Recovery

The segment recovery and local degeneracy removal algorithms have produced X_2 such that it has a CDT (guaranteed by Theorem 2). Let $\mathcal{T}$ be a CDT of X_2. Generally, $\mathcal{D}_2 \neq \mathcal{T}$ because some subfaces of $\mathcal{T}$ are non-Delaunay faces and penetrated by edges of $\mathcal{D}_2$. This section describes an algorithm which incrementally transforms $\mathcal{D}_2$ into $\mathcal{T}$. No additional points are needed in the transformation.

The general steps of our algorithm are similar to [Shewchuk02]. At initialization, let $\mathcal{T}^{(0)} := \mathcal{D}_2$; add all missing subfaces into a queue Q. The algorithm starts to recover the subfaces in Q until Q is empty. At each step i, the algorithm recovers a set of missing subfaces in $\mathcal{T}^{(i)}$, update $\mathcal{T}^{(i)}$ into $\mathcal{T}^{(i+1)}$. After m steps Q is empty, and $\mathcal{T} = \mathcal{T}^{(m)}$.

At step $i(i < m)$, several missing subfaces are recovered together. We define a *missing region* Ω to be a set of coplanar subfaces of X_2 such that

- all subfaces of Ω belong to one facet of X_2;
- the boundary edges of Ω are edges of $\mathcal{T}^{(i)}$; and
- the internal edges of Ω are missing in $\mathcal{T}^{(i)}$.

Hence Ω is a connected set of missing coplanar subfaces. It may not simply connected, i.e., Ω can contain a hole inside (see Figure 7 (a)). Each missing subface belongs to one missing region. A facet can have more than one missing regions.

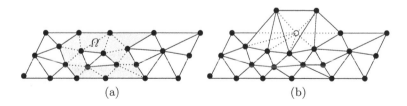

| (a) | (b) |

Fig. 7. (a) The shaded area highlights a non-simply connected missing region Ω. (b) A cavity C at (step i) is illustrated.

When a Ω is found, the *formcavity* procedure (described below) forms two cavities at each side of Ω. Each cavity is bounded by subfaces of Ω and faces of $\mathcal{T}^{(i)}$ (see Figure 7 (b)).

1. Find all tetrahedra in $\mathcal{T}^{(i)}$ that intersect the relative interior of Ω, delete them from $\mathcal{T}^{(i)}$. This creates a hole inside $\mathcal{T}^{(i)}$.
2. Insert the missing subfaces of Ω into the hole to split it into two separated cavities, one at each side of Ω.

Each cavity C is filled with a set of new tetrahedra by the following *cavity retetrahedralization* procedure. Let V be the set of vertices of C.

1. Verify C, expand C if it is necessary.
   (1) Form a queue Q containing all non-strongly Delaunay faces of C in V;
   (2) For each face $\sigma \in Q$ and σ still in C,
       let t be the tetrahedron adjacent to C and holds σ;
       remove σ from C;
       for each face Δt of t, $\Delta t \neq \sigma$ **do**
           if Δt is a face of C **then** remove Δt from C;
           else add Δt into C; **end**
       end
       Update C, V;
   (3) Repeat (1) if some faces of C are not strongly Delaunay in V.
   See Figure 8 for an example of the expansion of C.
2. Retetrahedralize C.
   (1) Construct a Delaunay tetrahedralization $\mathcal{D}_C$ of the vertices of the C.
   (2) Identify faces of C in $\mathcal{D}_C$, mark each tetrahedron of $\mathcal{D}_C$ to be "inside" or "outside".
   (3) Remove tetrahedra marked as "outside" from $\mathcal{D}_C$, and fill the remaining tetrahedra into C.
   Figure 9 illustrates a two dimensional example of the retetrahedralization procedure.

Fig. 8. The expansion of the cavity (illustrated in 2D). Left: e_1e_2 is the segment going to recover. Edges below e_1e_2 is the faces of the cavity C. p_1q_2 is not strongly Delaunay. Right: C is expanded by removing two edges p_1q_1 and p_1q_2 from C and adding one edge q_1q_2 into C.

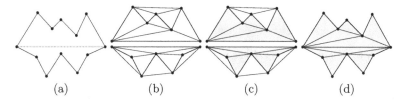

Fig. 9. Cavity triangulation (illustrated in 2D). (a) Two initial cavities separated by a constraining segment. (b) The two Delaunay triangulations constructed at each side of the segment. (c) Mark triangles as "inside" or "outside". (d) Remove "outside" triangles.

The complete facet recovery algorithm is summaried as follows:

> **Algorithm** *Facet Recovery*
> Input: $\mathcal{D}_2$, X_2.
> Output: The CDT $\mathcal{T}$ of X_2.
> **initialize:**
> $\mathcal{T} := \mathcal{D}_2$;
> **Repeat:**
>   form a queue Q of missing subfaces in $\mathcal{T}$;
>   **while** $Q \neq \emptyset$ **do**
>   remove an unrecovered subface f from Q;
>   form a missing region Ω containing f;
>   form two cavities C_1, C_2 by formcavity subroutine;
>   **for** each $C_i, i = \{1, 2\}$ **do**
>   Call cavity retetrahedralization subroutine;
>   **end**
>   **end**
> **until** no subfaces are missing in $\mathcal{T}$.

Theorem 4. *The facet recovery algorithm terminates.*

Proof: We show that the cavity retetrahedralization subroutine will always succeed, hence the missing region found at each step can be recovered without getting stuck.

The step 1 (face verification) of the cavity retetrahedralization algorithm guarantees the step 2 can be correctly executed since every strongly Delaunay simplex of V will appear in any Delaunay tetrahedralization of V. Hence the face identification in step 2 must be successful. What remains is to prove two issues in the face verification step: (1) the expansion of C terminates; and (2) the missing subfaces due to the expansion of C can be recovered later.

Let Ψ be the set of faces of $\mathcal{D}_2$ such that no face of Ψ is crossed by any subfaces of X_2. Clearly, $\Psi \neq \emptyset$ and any face $\Delta t \in \Psi$ is strongly Delaunay and exists in any $\mathcal{T}^{(i)}$. The set Ψ limits the expansion of C, i.e., C stops expanding at σ when $\sigma \in \Psi$.

Let $Vol(C)$ be the inside volume of C. During the expansion of C some subfaces are missing. Notice that the missing region Ω' formed by these missing subfaces is completely inside C, hence $Vol(C') < Vol(C)$, where C' is the cavity formed from Ω'. This relation holds if the expansion of C' still causes some other subfaces missing and results another new cavity. Such sequence will terminate since the value of

$Vol()$ can not be negative.   ∎

Theorem 5. $\mathcal{T}$ *created by the facet recovery algorithm is a CDT.*

Proof: We first show $\mathcal{T}^{(1)}$ is a CDT. $\mathcal{T}^{(1)}$ is the result of cavity retetrahedralization algorithm on $\mathcal{T}^{(0)}$. Tetrahedra of $\mathcal{T}^{(1)}$ which are outside and not adjacent to the cavities remain Delaunay. Let t be a tetrahedron created inside a cavity C, S_t is its circumsphere. Let t' be another tetrahedron in $\mathcal{T}^{(1)}$ sharing a face σ with t, v be the vertex of t' opposite to σ. We have the following cases:

(1) σ is a subface. Then t is constrained Delaunay even if S_t encloses v, i.e, the inside of t is not visible by v, otherwise at least a segment is non-Delaunay and can not exist in $\mathcal{T}^{(0)}$;

(2) σ is a face inside C, then S_t must not enclose v (guaranteed by the cavity retetrahedralization algorithm).

(3) σ is a face on C, then σ is not a subface, S_t must not enclose v. Otherwise, $\mathcal{T}^{(0)}$ is not a Delaunay tetrahedralization since the circumsphere of t' is not empty, i.e., it encloses the point of t opposite to σ.

By induction, after step i, $i > 1$, $\mathcal{T}^{(i)}$ is a CDT. Now we can show $\mathcal{T}^{(i+1)}$ is a CDT by using the similar arguments as above. The only difference is in the case (3) which is stated below:

(3) σ is a face on C, then we have two cases,

    a. σ is not a subface, S_t must not enclose v. Otherwise, $\mathcal{T}^{(i)}$ is not a CDT since the circumsphere of t' encloses the point of t opposite to σ which is also visible from the inside of t'.

    b. σ is a subface, then t is constrained Delaunay even if S_t encloses v.

Thus on the finish of the facet recovery algorithm $\mathcal{T} = \mathcal{T}^{(m)}$ is a CDT.   ∎

8 Experimental Results

This algorithm has been implemented and in our 3D quality Delaunay mesh generator - TetGen [Si04b] (http://tetgen.berlios.de). We have tested the program with a number of examples not only having simple but also relatively complicated geometries. The algorithm runs rather efficiently compares to the old version of Tet-Gen which uses gift-wrapping algorithm. For example, for a cavity having around 300 faces and 150 vertices, the gift-wrapping algorithm needs few minites to finish while the cavity retetrahedralization algorithm finishes in less than 1 second. To test the stability of the algorithm as well as our implementation, we purposely chose some models which the surface meshes are rather badly discretized. They are most likely to pull down the program if the algorithm has defect. On most of these models the program successfully produced CDTs as long as the surface meshes are PLCs.

The following two PLC models can be freely downloaded from Inria's large repository http://www-rocq1.inria.fr/gamma.

The Cup-holder (in Figure 10) has a simple geometry (918 vertices, 1848 triangles). 1261 points are added (262 break points and 999 protecting points). There are

157 missing subfaces, the biggest size of a cavity contains 40 faces. Moreover, the effect of removing local degeneracies can be visualized on the right picture.

The monster4 (in Figure 11) consists of 1392 vertices, 2784 triangles. It has complicated distribution of segments and facets as shown in the left picture. Our algorithm added 2782 Steiner points (502 break points and 2226 protecting points).

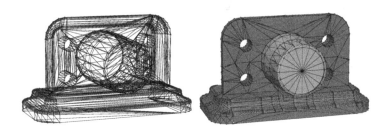

Fig. 1 . Example 1 (Cup-holder). Left: the PLC. Right: the CDT.

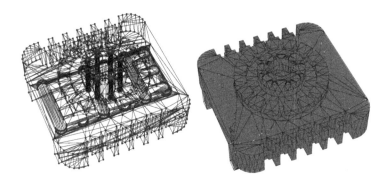

Fig. 11. Example 2 (monster4). Left: the PLC. Right: the CDT.

The geometry of Figure 12 is a human heart inside a body. The surface mesh of the heart is internal boundary separating two regions. Hence this model is made up of multiple domains. In spite of the complexities in the geometries, the point set contain no local degeneracy. Hence the CDT only require few additional points (compare to the input number of points) for protecting segments. The CDT is shown in the middle. One can see that the tetrahedra of the CDT are usually very long and skinny. The picture on the right shows the better quality tetrahedra obtained by performing a Delaunay refinement on the CDT. The number of additional points for improving the mesh quality are much bigger than that of getting a CDT.

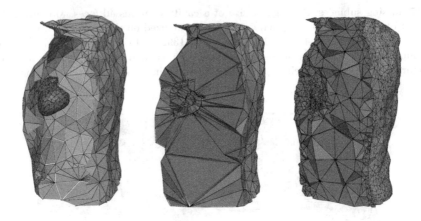

Fig. 12. Example 3 (Heart). Left: the PLC. Middle: the CDT. Right: the quality mesh. A vertical cut has been made on the meshes so that the internal boundaries can be seen.

Acknowledgements

This work is supported by the pdelib project of Weierstrass Institute for Applied Analysis and Stochastics. We are grateful to the reviewers for the valuable suggestions and comments, especially for the pointing out a potential problem in a proof of the first draft.

[Schoenhardt28] Schönhardt E (1928) Über die Zerlegung von Dreieckspolyedern in Tetraeder. Mathematische Annalen 98:309–312

[Bagemihl48] Bagemihl F (1948) On Indecomposable Polyhedra. American Mathematical Monthly 55: 411–413

[Chazelle84] Chazelle B. (1984) Convex Partition of Polyhedra: A lower Bound and Worst-case Optimal Algorithm. SIAM Journal on Computing 13(3): 488–507

[Lee86] Lee D T, Lin A K (1986) Generalized Delaunay Triangulations for Planar Graphs. Discrete Comput Geom 1:201–217

[Chew89] Chew P L(1989) Constrained Delaunay Triangulation. Algorithmica 4(1): 97–108

[Edelsbrunner90] Edelsbrunner H, Mücke M P (1990) Simulation of Simplicity: A Technique to Cope with Degenerate Cases in Geometric Algorithm. ACM Transactions on Graphics 9(1): 66–104

[Edelsbrunner96] Edelsbrunner H, Shah N R (1996) Incremental Topological Flipping Works for Regular Triangulations. Algorithmica 15: 223–241

[Ruppert92] Ruppert J (1992) On the Difficulty of Triangulating Three-dimensional Non-convex Polyhedra. Discrete Comput Geom 7: 227–253

[Ruppert95] Ruppert J (1995) A Delaunay Refinement Algorithm for Quality 2-Dimensional Mesh Generation. J Algorithms 18(3): 548–585

[Weatherill94] Weatherill N P, Hassan O (1994) Efficient Three-Dimensional Delaunay Triangulation with Automatic Point Creation and Imposed Boundary Constraints. Int. J. Numer. Meth. Engng 37: 2005–2039

[Miller96] Miller G L, Talmor D, Teng S H, Walkington N, Wang H (1996) Control Volume Meshes using Sphere Packing: Generation, Refinement and Coarsening. In: Proc. of 5th Intl. Meshing Roundtable

[Shewchuk98a] Shewchuk J R (1998) A Condition Guaranteeing the Existence of Higher-Dimensional Constrained Delaunay Triangulations. In: Proc. of the 14th Annu. Sympos. Comput. Geom., 76–85, Minneapolis, Minnesota

[Shewchuk98b] Shewchuk J R (1998) Tetrahedral Mesh Generation by Delaunay Refinement. In: Proc. of the 14th Annu. Sympos. Comput. Geom.

[Shewchuk00] Shewchuk J R (2002) Sweep Algorithms for Constructing Higher-Dimensional Constrained Delaunay Triangulations. In: Proc. of the 16th Annu. Sympos. Comput. Geom.

[Shewchuk02] Shewchuk J R (2002) Constrained Delaunay Tetrahedralizations and Provably Good Boundary Recovery. In: Proc. of 11th Intl. Meshing Roundtable

[Shewchuk03] Shewchuk J R (2003) Updating and Constructing Constrained Delaunay and Constrained Regular Triangulations by Flips. In: Proc. 19th Annu. Sympos. Comput. Geom.

[Pebay98] Pébay P (1998) A Priori Delaunay-Conformity. In: Proc. of 7th Intl Meshing Roundtable, SANDIA

[Murphy00] Murphy M, Mount D M, Gable C W (2000) A Point-Placement Strategy for Conforming Delaunay Tetrahedralization. In: Proc. of the 11th Annu. Sympos. on Discrete Algorithms

[Cohen-Steiner02] Cohen-Steiner D, Colin de Verdière E, Yvinec M (2002) Conforming Delaunay Triangulations in 3D. In: Proc. of 18th Annu. Sympos. Comput. Geom. Barcelona

[George03] George P L, Borouchaki H, Saltel E (2003) 'Ultimate' Robustness in Meshing an Arbitrary Polyhedron. Int. J. Numer. Meth. Engng 58: 1061–1089

[Rambau03] Rambau J. (2003) On a Generalization of Schönhardt's Polyhedron. MSRI Preprint 2003-13

[Du04] Du Q, Wang D (2004) Constrained Boundary Recovery for Three Dimensional Delaunay Triangulations. Int. J. Numer. Meth. Engng 61: 1471–1500

[Cheng04] Cheng S W, Dey T K, Ramos E A, Ray T (2004) Quality Meshing for Ployhedra with Small Angels. In: Proc. 20th Annu. ACM Sympos. Comput. Geom.

[Pav04] Pav S, Walkington N (2004) A Robust 3D Delaunay Refinement Algorithm. In: Proc. Intl. Meshing Roundtable.

[Si04a] Si H, Gärtner K (2004) An Algorithm for Three-Dimensional Constrained Delaunay Triangulations. In: Proc of 4th Intl. Conf. on Engineering Computational Technology, Lisbon

[Si04b] Si H (2004) TetGen, A Quality Tetrahedral Mesh Generator and Three-Dimensional Delaunay Triangulator, v1.3 User's Manual. WIAS Technical Report No. 9

Delaunay Refinement by Corner Lopping

Steven E. Pav[1] and Noel J. Walkington[2]

[1] University of California at San Diego, La Jolla, CA. spav@ucsd.edu
[2] Carnegie Mellon University, Pittsburgh, PA. noelw@andrew.cmu.edu

Summary. An algorithm for quality Delaunay meshing of 2D domains with curved boundaries is presented. The algorithm uses Ruppert's "corner lopping" heuristic [MR96b:65137]. In addition to admitting a simple termination proof, the algorithm can accept curved input without any bound on the tangent angle between adjoining curves. In the limit case, where all curves are straight line segments, the algorithm returns a mesh with a minimum angle of $\arcsin\left(1/2\sqrt{2}\right)$, except "near" input corners. Some loss of output quality is experienced with the use of curved input, but this loss is diminished for smaller input curvature.

Key words: unstructured, simplicial, planar, curved boundary, Delaunay, mesh.

1 Introduction

The Delaunay Refinement method is used for quality simplicial mesh generation in two and three dimensions. A Delaunay Refinement algorithm takes an input of points and segments (or curves) and adds Steiner Points to guarantee that the output Delaunay Triangulation conforms to the input and has high quality simplices, as measured by the circumradius to shortest edge length ratio. A Steiner Point is added to "split" an input segment into subsegments if a mesh vertex forms an obtuse angle with the segment. A Steiner Point is added at the circumcenter of a poor triangle in the mesh. Termination of the algorithm is had by proving a lower bound on the distance between Steiner Points, and applying compactness arguments [MR96b:65137, Pse2003].

Ruppert was a pioneer of the Delaunay Refinement method. Ruppert's Algorithm accepts a planar straight line graph, and outputs a Delaunay mesh where no output angle is smaller than a user-chosen parameter, which can be as large as $20.7°$. In Ruppert's analysis input segments have to meet at nonacute angles, otherwise his naïve algorithm might not terminate [MR96b:65137].

Ruppert offered two heuristic solutions to this problem. The first, "concentric shell splitting," has been adapted to a working algorithm, and allows better output quality guarantees [Sjr1997, Pse2003, MglPseWnj2005]. In this solution, segments sharing a common endpoint are split at the same distance from the endpoint, *i.e.*, on the same "shell." This simple fix gives a good lower bound, and an input-independent upper bound, on output angles. However, its analysis is involved, and does not generalize naturally to higher dimensions or curved input (due to its reliance on "power of two" arguments). Ruppert's second solution, "corner lopping," is analyzed herein, and admits a simple proof.

Delaunay Refinement for curved input was considered by Boivin and Ollivier-Gooch [BcOGc:2002]. Their analysis requires that input segments meet at an angle of at least $\pi/3$. Concentric shell splitting to deal with smaller input angles was mentioned in this context, but not shown to give a working algorithm; this fix clearly would require further modification to the output quality guarantee.

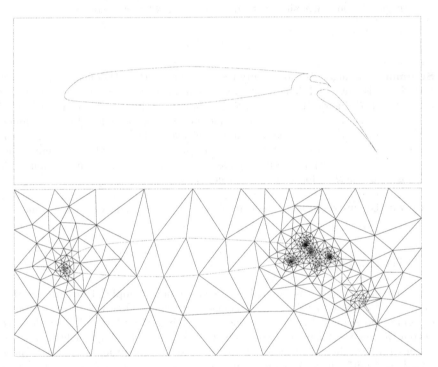

Fig. 1. The outline of a mock air foil and output from the meshing algorithm. Solving a fluid dynamics PDE would probably require further mesh refinement.

The Delaunay Refinement Algorithm has also been generalized to three dimensions. Early analysis required input segments and faces to meet at nonacute angles [MglPseWnj2002c]. As a fix, later work used protective regions around input points and segments [338236, CoVeYv01, Cswetal2004, PseWnj2004]. In that way these algorithms resemble corner lopping, which places a protective ball around acute corners in the input. Reverse from what is usually seen, these three-dimensional algorithms do not appear to be the natural generalization of any known two-dimensional algorithm.

The motivations for the present work, then, are:

1. To present an algorithm which accepts straight or curved input without a lower bound on input angle, yet admits a simple termination proof.
2. To find an algorithm which is the "projection" into the plane of recently discovered three-dimensional algorithms, and thereby to gain a better understanding of those algorithms.

2 Preliminaries

The input to the algorithm is assumed to be a PRCC.

Definition 1 (PRCC). *A set of points and a set of non-closed regular curves embedded in* $\mathbb{R}^2$, $(\mathcal{P}, \mathcal{C})$, *form a* Piecewise Regular Curve Complex *(PRCC) if*
(i) for any curve, $c \in \mathcal{C}$, *the endpoints of* c *are elements of* $\mathcal{P}$.
(ii) given two curves, $c_1, c_2 \in \mathcal{C}$, *their intersection is either empty or an endpoint (or endpoints) common to both curves.*
(iii) given $p \in \mathcal{P}$, $c \in \mathcal{C}$, *either* p *is an endpoint of* c, *or* p *is not on* c.

The goal of meshing is to produce, from input $(\mathcal{P}, \mathcal{C})$, the Delaunay Triangulation of a set of points, $\mathcal{P}'$, hereafter denoted as $\mathcal{D}(\mathcal{P}')$, such that (i) $\mathcal{P} \subseteq \mathcal{P}'$, (ii) each input curve of $\mathcal{C}$ is approximated by a piecewise linear curve which is the union of edges in $\mathcal{D}(\mathcal{P}')$, (iii) all or most of the triangles of $\mathcal{D}(\mathcal{P}')$ are "high quality." Absent of a specific interpolation problem, triangle "quality" is taken to be inversely proportional to its minimum angle. When guaranteeing a large minimum angle is not possible due to input constraints, an upper bound on the maximum angle of the mesh is often desired [BaAz76, Ci78, MglPseWnj2005].

Given two points, p, q let $\overline{pq}$ be the line segment with these points as endpoints, and let $\widetilde{pq}$ denote a curve with p and q as endpoints. Let $|p - q|$ denote the distance between p and q. For a curve c, and a point p, let

$$|p - c| = \min_{x \in c} |p - x|$$

be the distance from p to c. We use $a \vee b$ to denote the maximum of quantities a and b, and $a \wedge b$ to represent the minimum.

The local feature size, first defined by Ruppert [MR96b:65137], is used to prove termination and quality of the output mesh. We take the classical definition:

Definition 2 (Local Feature Size). *Given a PRCC,* $(\mathcal{P}, \mathcal{C})$, *define* $\mathrm{lfs}_i(x)$, *for* $i = 0, 1$, *as the distance from* x *to two mutually disjoint features of the PRCC of dimension no greater than* i. *By "feature" we mean a point in* $\mathcal{P}$ *or a curve in* $\mathcal{C}$. *Thus, for example,* $\mathrm{lfs}_0(x)$ *is the distance from* x *to the second nearest point of* $\mathcal{P}$. *Let* $\mathrm{lfs}(x) = \mathrm{lfs}_1(x)$ *be the* local feature size.

Note 1. The following facts about local feature size are immediate: (i) For any x, $\mathrm{lfs}_1(x) \leq \mathrm{lfs}_0(x)$, (ii) $\mathrm{lfs}_i(x)$ is a Lipschitz function with constant 1, *i.e.*, $\mathrm{lfs}_i(p) \leq \mathrm{lfs}_i(q) + |p - q|$, (iii) $\mathrm{lfs}_i(x)$ has a positive minimum value on $\mathbb{R}^2$.

Some authors use "local feature size" to describe a different function, the distance of a point on the input to the medial axis of the input [AnCsKrk2001]. While both definitions give Lipschitz functions, our definition yields a function defined in all of

$\mathbb{R}^2$ with a strictly positive lower bound. The latter fact is important because our local feature size describes, roughly, the size of triangles we expect to see nearby in an output mesh of the given input.

In the case of straight line input, a lower bound on the angle subtended by input segments is used to show that Steiner Points on input segments are not placed too close together. For PRCC input, we instead use the following

Definition 3 (Curve Separation). *For the sake of this definition, given curve* $c = \widetilde{xy}$, *we say that a point* z *on* c *is sufficiently far from* x *if* $|x - z| \geq \mathrm{lfs}\,(x)\,/2C_0$, *where* $C_0 = 1 + \sqrt{2}$, *is a "grading constant" (see Lemma 5).*

Given two curves c_1, c_2 *sharing a single endpoint* x, *then the separation between them is*

$$\inf_{z_1, z_2} \frac{|z_1 - z_2|}{|z_1 - x| \vee |z_2 - x|},$$

where z_i *is a point on* c_i *that is sufficiently far from* x.

If curves c_1, c_2 *share both their endpoints, say* x *and* y, *then the separation between them is*

$$\inf_{z_1, z_2} \frac{|z_1 - z_2|}{|z_1 - x| \vee |z_2 - x| \vee |z_1 - y| \vee |z_2 - y|},$$

where z_i *is a point on* c_i *that is sufficiently far from both* x *and* y.

For a given PRCC, let σ *be a lower bound on the separation between any pair of curves with at least one common endpoint. We say* σ *is the "minimum curve separation" of the PRCC.*

Given that input curves are continuous and may meet at most at their endpoints, the separation between two curves is strictly positive, and thus σ is positive as well. In the case where all input curves are straight line segments, we have $\sigma = 2\sin(\theta^*/2)$, where $\theta^* \leq \pi/3$ is a lower bound on the angle subtended by the segments.

Following Boivin and Ollivier-Gooch [BcOGc:2002], we make the following

Definition 4 (Total Variation of a Curve). *The total variation of curve* c *is* $\int_c |\mathrm{d}\theta|$, *where* θ *is the angle subtended by the tangent of the curve. Let* Δt *be an upper bound on the total variation of every curve in an input PRCC.*

For the remainder of this paper we assume $\Delta t \leq \pi/3$. Such a bound can be achieved by splitting curves in a preprocessing step.

3 The Algorithm

The Delaunay Refinement algorithm we consider maintains sets, $\mathcal{P}'$, and $\mathcal{C}'$, which are initialized, respectively, as $\mathcal{P}$ and $\mathcal{C}$. The algorithm adds *Steiner Points* to $\mathcal{P}'$, then returns $\mathcal{D}(\mathcal{P}')$ on termination. Throughout this work, "curve" or "subcurve" means a member of $\mathcal{C}'$, while "input curve" refers to a member of $\mathcal{C}$. "Input point" refers to a point of $\mathcal{P}$. We will take $\mathrm{lfs}_i(\cdot)$ to be with respect to the PRCC input to the algorithm, $(\mathcal{P}, \mathcal{C})$.

Our algorithm protects regions around the input points and (sub)curves. Following the notation of Ruppert [MR96b:65137], a (sub)curve is said to be "encroached" if there exists a vertex of the mesh inside its diametral circle, where the diametral

Algorithm 2: Algorithm for meshing via corner lopping.
Input: A PRCC, an angle bound, the splitting fraction
Output: A mesh
    $\textsc{CornerLop}((\mathcal{P}, \mathcal{C}), \kappa, \beta)$
(1) Let $\mathcal{P}' \leftarrow \mathcal{P}, \mathcal{C}' \leftarrow \mathcal{C}$.
(2) Construct $\mathcal{D}(\mathcal{P}')$.
(3) Use $\mathcal{D}(\mathcal{P}')$ to find $\mathrm{lfs}_0(\cdot)$ for points in $\mathcal{P}'$.
(4) **foreach** $p \in \mathcal{P}$
(5)   **if** p is the corner of an acute angle in $\mathcal{C}'$.
(6) $d(p) \leftarrow (\sqrt{2} - 1) \mathrm{lfs}_0(p)$
(7) $\textsc{SplitBall}(p, 1)$
(8)   **else**
(9)   Let $d(p) \leftarrow 0$.
(10) **while** any of these rules can be executed, execute one of them, with the rules listed in descending priority:
(11)   **if** there are $p \in \mathcal{P}, q \in \mathcal{P}'$, with $|p - q| < d(p)$ **then** $\textsc{SplitBall}(p, \beta)$
(12)   **if** there is $z \in \mathcal{P}', c \in \mathcal{C}'$ such that z encroaches c **then** $\textsc{SplitCurve}(c, z)$
(13)   **if** there is $\triangle xyz \in \mathcal{D}(\mathcal{P}')$ such that $\angle xyz \leq \kappa$ and the circumcenter of $\triangle xyz$ is not closer to some $q \in \mathcal{P}$ than $d(q)$ **then** $\textsc{SplitTri}(\triangle xyz)$
(14) **return** $\mathcal{D}(\mathcal{P}')$

    $\textsc{SplitBall}(p, C)$
(15) Let $d(p) \leftarrow Cd(p)$.
(16) **foreach** $\widetilde{pt} \in \mathcal{C}'$
(17)   Let x be the point on $\widetilde{pt}$ such that $|p - x| = d(p)$
(18)   Add x to $\mathcal{P}'$. Remove $\widetilde{pt}$ from $\mathcal{C}'$, replacing it with $\widetilde{px}, \widetilde{xt}$.
    $\textsc{SplitCurve}(\widetilde{xy}, [z])$
(19) **if** a z is given
(20)   Let p be *some point* on $\widetilde{xy}$, perhaps based on z.
(21) **else**
(22)   Let p be *some point* on $\widetilde{xy}$.
(23) Add p to $\mathcal{P}'$, replace $\widetilde{xy}$ in $\mathcal{C}'$ by subcurves $\widetilde{xp}$ and $\widetilde{py}$.
    $\textsc{SplitTri}(\triangle xyz)$
(24) Let p be the circumcenter of $\triangle xyz$.
(25) **if** p encroaches curve $c \in \mathcal{C}'$
(26) $\textsc{SplitCurve}(c)$
(27) **else**
(28)   Add p to $\mathcal{P}'$.

circle of curve $\widetilde{pq}$ is the circle with $\overline{pq}$ as diameter. We denote the diametral circle of curve c by $\mathcal{C}_d(c)$.

    The algorithm assigns to each input point, p, a radius $d(p) \geq 0$, and, by analogy with curves, p is said to be encroached if a vertex of the mesh is in $\mathcal{B}(p)$, which is defined to be the open ball of radius $d(p)$ centered at p.

    The algorithm, given as Algorithm 2, takes the input PRCC and two parameters: κ, the desired angle bound of the output mesh, and $\beta \in \left[\frac{1}{2}, 1\right)$, the factor by which a radius $d(p)$ is reduced when an input point p is encroached. Adjusting β may affect output mesh cardinality.

To split the curve $c = \widetilde{xy}$, an intermediate point $p \in c$ is selected and c is replaced by two subcurves $\widetilde{xp}$ and $\widetilde{py}$. In the case c is a segment, traditionally p is chosen as the midpoint of the segment. Without a clear definition of the midpoint of a curve, we assume only that the algorithm satisfies the following assumption regarding the selection of p:

Assumption 1. Assume there are constants $\eta \geq 1$ and $\mu \geq 1$ such that the algorithm is implemented so that
(1) if p is selected to split curve c by a call to SPLITCURVE(c, z), then $|p - \mathcal{C}_d(c)| \geq |p - z|/\eta$.
(2) if p is selected to split curve $c = \widetilde{xy}$ by a call to SPLITCURVE(c), then $|p - \mathcal{C}_d(c)| \geq r/\mu$, where r is the radius of the circumcircle, i.e., $|x - y|/2$.

This assumption is needed to prevent a midpoint from being added too close to the diametral circle (and any points outside this circle) of the curve it is added to split; see Figure 2.

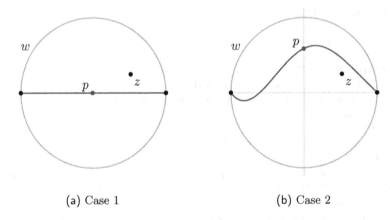

(a) Case 1   (b) Case 2

Fig. 2. For the case of segment input, as shown in (a), Assumption 1 can be satisfied with $\eta = \mu = 1$. For general curves, as in (b), a point p selected to split a curve may be near the diametral circle of the curve, i.e., near other Steiner Points which may lie outside the diametral circle.

4 Proof of Termination

We first consider some facts regarding curved input. The following lemma is a consequence of the Mean Value Theorem:

Lemma 1 (Lens Containment). *Let c be a curve with endpoints x, y, and with total variation less than Δt. Suppose z is a point on c distinct from the endpoints. Then*

$$\angle xzy \geq \pi - \Delta t.$$

This lemma claims that the worst case of a curve of total variation no more than Δt is a circular arc. That is, a curve $\widetilde{xy}$ with bounded total variation is contained in a diametral lens of segment $\overline{xy}$. The corollaries claim that by careful choice of p in step 20 and step 22, the algorithm will conform to Assumption 1, with

$$\eta = \frac{1 + \tan(\Delta t/2)}{1 - \tan(\Delta t/2)}, \qquad \mu = \frac{1}{1 - \tan(\Delta t/2)}.$$

Corollary 1. *Let $c = \widetilde{xy}$ be a curve with total variation less than Δt. Let $\mathcal{C}_d(c)$ be the diametral circumcircle of c, and let z be a point in this circle. Then there is a point p on c such that*

$$|p - \mathcal{C}_d(c)| \geq |p - z| \frac{1 - \tan(\Delta t/2)}{1 + \tan(\Delta t/2)}.$$

Corollary 2. *Let $c = \widetilde{xy}$ be a curve with total variation less than Δt. Let $\mathcal{C}_d(c)$ be the diametral circumcircle of c. Then there is a point p on c such that*

$$|p - \mathcal{C}_d(c)| \geq r\,(1 - \tan(\Delta t/2)),$$

where r is the radius of the diametral circle.

Both the corollaries are proved by taking p to be the intersection of the curve with the perpendicular bisector of segment $\overline{xy}$, as shown in Figure 2(b).

The following lemma, which is proved by basic geometry and use of the Mean Value Theorem, guarantees that a point encroaching a subcurve cannot be a midpoint on the same input curve.

Lemma 2. *Let $c' = \widetilde{xy}$ be a subcurve of c, which is a curve with total variation less than $\Delta t \leq \pi/2$. Let z be a point on c which is not on c'. Then $\angle xzy \leq \pi/2$.*

The next lemma is needed since we protect curves with the diametral circle of their secant segments, but add "midpoints" on the curve. Thus, in general, the diametral circle of a curve may not wholly contain the diametral circles of its subcurves.

Lemma 3 (Diametral Circle Protection). *Let $c = \widetilde{xy}$ be a curve with total variation less than Δt. Let p be a point which does not encroach the diametral circle of c. Letting $|p - c|$ be the distance from p to the curve c, then*

$$\frac{|p - c|}{|p - x| \wedge |p - y|} \geq \frac{1}{\zeta},$$

where $\zeta = \sqrt{2}\sin(\Delta t/2)/\left(\sqrt{1 + \sin \Delta t} - 1\right)$.

Proof. We assume that $|p - x| \leq |p - y|$. Let ℓ_1, ℓ_2 be the two arcs of points subtending angle $\pi - \Delta t$ with x, y. By Lemma 1, c is between these two arcs. Without loss of generality, assume p is "above" ℓ_1. Let w be the center of arc ℓ_1.

We consider two cases:

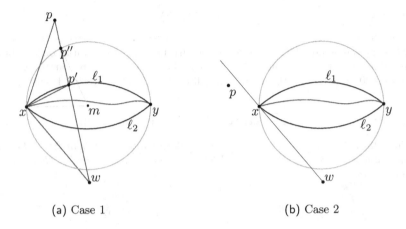

(a) Case 1   (b) Case 2

Fig. 3. Two cases in the proof of Lemma 3. In (a), $\overline{wp}$ intersects the arc ℓ_1, while there is no such intersection in (b).

1. The first case is if $\overline{pw}$ and ℓ_1 intersect. Let p' be their point of intersection. Let p'' be the intersection of $\overline{pw}$ with the diametral circle of segment $\overline{xy}$, as in Figure 3(a).

   Then, using the sine rule,

   $$\frac{|p-c|}{|p-x| \wedge |p-y|} = \frac{|p-c|}{|p-x|} \geq \frac{|p-p'|}{|p-x|} = \frac{\sin \angle p'xp}{\sin \angle pp'x} \geq \frac{\sin \angle p'xp''}{\sin \angle p''p'x} = \frac{|p''-p'|}{|p''-x|}.$$

   Note that since $\angle pp'x$ is obtuse, then $\angle p'xp$ is acute and we can indeed conclude that $\sin \angle p'xp \geq \sin \angle p'xp''$. Let $R = |w-x|$, and let $r = |m-x|$, where m is the midpoint of x and y. Because $\angle xwy = \Delta t$, then $r = R \sin(\Delta t/2)$. Let $\phi = \angle p''mx$. Then

   $$|p''-p'| = |w-p''| - R = \sqrt{(R \cos(\Delta t/2) + r \sin \phi)^2 + (r \cos \phi)^2} - R$$
   $$= R \left(\sqrt{1 + \sin \Delta t \sin \phi} - 1 \right).$$

   Then

   $$\frac{|p''-p'|}{|p''-x|} = \frac{R\sqrt{1 + \sin \Delta t \sin \phi} - R}{2r \sin(\phi/2)} = \frac{\sqrt{1 + \sin \Delta t \sin \phi} - 1}{2 \sin(\Delta t/2) \sin(\phi/2)}.$$

   This quantity decreases as ϕ increases, thus it takes minimum value when $\phi = \pi/2$. Thus

   $$\frac{|p''-p'|}{|p''-x|} \geq \frac{\sqrt{1 + \sin \Delta t} - 1}{\sqrt{2} \sin(\Delta t/2)}.$$

2. The other case is that $\overline{pw}$ and ℓ_1 do *not* intersect, as shown in Figure 3(b). That is, $\angle mxp > \pi/2 + \Delta t/2$. Looking at the circle centered at w through ℓ_1, then $|p-c| \geq |p-\ell_1| = |p-x|$. Thus

   $$\frac{|p-c|}{|p-x| \wedge |p-y|} \geq 1 > \frac{1}{\sqrt{2}} \geq \frac{\sqrt{1 + \sin \Delta t} - 1}{\sqrt{2} \sin(\Delta t/2)}.$$

As expected, this lower bound is "worse," *i.e.*, smaller, for larger Δt. However, we can make the rough uniform bound, $\zeta < 2$ for $0 \leq \Delta t \leq \pi/3$.

For $p \in \mathcal{P}$, let $\mathcal{B}(p)$ be the open ball of radius $d(p)$ centered at p during any step of the execution of colop. Let $\partial \mathcal{B}(p)$ be the boundary of $\mathcal{B}(p)$, and let $\overline{\mathcal{B}}(p)$ be the closed ball of radius $d(p)$ centered at p. If $d(p) = 0$, let $\mathcal{B}(p)$ be the empty set and let $\overline{\mathcal{B}}(p)$ be the point p.

After $d(p)$ is set for all $p \in \mathcal{P}$, then for $p, q \in \mathcal{P}$,

$$d(p) + d(q) \leq \left(\sqrt{2} - 1\right)\left(\text{lfs}_0(p) + \text{lfs}_0(q)\right) < \frac{\text{lfs}_0(p) + \text{lfs}_0(q)}{2} \leq |p - q|$$

Since $d(p)$ can only shrink for a given $p \in \mathcal{P}$ during the lifetime of the algorithm, we make the following claim.

Claim. For distinct $p, q \in \mathcal{P}$, at all times $\overline{\mathcal{B}}(p) \cap \overline{\mathcal{B}}(q) = \emptyset$.

In the following analysis, we will refer to points of $\mathcal{P}$ as "input;" points which are added to split curves, that is at step 7 or step 23, as "midpoints;" triangle circumcenters added at step 28 are called "circumcenters."

The following two lemmata assert that a ball $\mathcal{B}(p)$ is only split by a midpoint on an input edge disjoint from p, and thus $d(p)$ is always bounded below by a constant times lfs (p).

Lemma 4. *If a point t is added to $\mathcal{P}'$ such that $t \in \mathcal{B}(p)$ for some $p \in \mathcal{P}$ then t must be a midpoint on a curve of $\mathcal{C}$ disjoint from p.*

Proof. Consider the possible identity of such a t:
- It cannot be that $t \in \mathcal{P}$, as $\overline{\mathcal{B}}(p) \cap \overline{\mathcal{B}}(t) = \emptyset$.
- It cannot be that t was added to $\mathcal{P}'$ in step 7, as if t is on $\partial \mathcal{B}(p)$ then it is not in $\mathcal{B}(p)$, and if $t \in \partial \mathcal{B}(q)$ for some $q \in \mathcal{P}$, then since $\overline{\mathcal{B}}(p)$ does not meet $\overline{\mathcal{B}}(q)$, q is not in $\mathcal{B}(p)$.
- Suppose t is the midpoint of some curve, c, which was added in step 23 or step 26. Then c cannot have been encroached by some $q \in \mathcal{B}(p)$ with $q \neq p$ as then the rule SPLITBALL would have been preferred to SPLITCURVE or SPLITTRI. Thus c cannot have endpoint p, as otherwise its diametral circle would be contained in $\overline{\mathcal{B}}(p)$, and thus it could not be encroached by such a q. That is, c must be disjoint from p.
- It cannot be that t was the circumcenter of a triangle, as this kind of circumcenter is explicitly prohibited from being added to $\mathcal{P}'$.

Lemma 5. *During execution of colop, if $p \in \mathcal{P}$, and $d(p) > 0$ then*

$$\text{lfs}(p) \leq C_0 d(p),$$

where $C_0 = \left(1 + \sqrt{2}\right) \approx 2.41$.

Proof. A nonzero $d(p)$ is set at two places in the algorithm, step 6 and step 15:

(step 6) In this case, because lfs $(p) \leq \text{lfs}_0(p)$,

$$\text{lfs}(p) \leq \text{lfs}_0(p) = \frac{1}{\sqrt{2} - 1} d(p) = C_0 d(p).$$

(step 15) In this case, by Lemma 4, the input point is encroached by a point t on a curve disjoint from p. Then before $d(p)$ is reduced to $\beta d(p)$ it is the case that lfs $(p) \leq |p - t| \leq d(p)$. Considering the new value of $d(p)$, this is lfs $(p) \leq d(p)/\beta \leq 2d(p) < C_0 d(p)$.

Thus the radius $d(p)$ associated with an input point is never too much smaller than the local feature size, which is determined by the input. This implies that for a given input, SPLITBALL(p, C) is called a finite number of times. Next we show that this lower bound on $d(p)$ gives a bound on the distance between midpoints on nondisjoint input curves.

Lemma 6. *Suppose p, q are midpoints in $\mathcal{P}'$. Furthermore assume the midpoints are on distinct nondisjoint curves of $\mathcal{C}$. Then*

$$\text{lfs}\,(p) \leq \frac{1 + C_0}{\sigma} |p - q|,$$

where $C_0 = \left(1 + \sqrt{2}\right)$ is the constant from Lemma 5.

Proof. Let the two curves share input point x. Since p and q are on curves sharing this point, both $|p - x|$ and $|q - x|$ are at least $d(x)$, which is bounded from below by lfs $(x)/C_0$, by Lemma 5. Applying Definition 3,

$$|p - q| \geq \sigma\left(|p - x| \vee |q - x|\right) \geq \sigma |p - x|.$$

By the Lipschitz property, then using Lemma 5, then bounding $|p - x|$ and using the above,

$$\text{lfs}\,(p) \leq |p - x| + \text{lfs}\,(x) \leq |p - x| + C_0 d(x) \leq (1 + C_0)|p - x| \leq \frac{1 + C_0}{\sigma} |p - q|$$

The following theorem asserts that the spacing between points in $\mathcal{P}'$ is bounded by the local feature size of the input, *i.e.*, the output of the algorithm is "well graded." Moreover, this theorem proves termination of the algorithm, by a ball-packing argument [MR96b:65137, Pse2003].

Theorem 1. *Suppose p is to be added to $\mathcal{P}'$ during the execution of colop. Suppose the algorithm is implemented such that Assumption 1 is satisfied. Let q be a point in $\mathcal{P}'$ at the time p is to be added. Then there are constants C_1, C_2 such that*

$$\text{lfs}\,(p) \leq \begin{cases} C_1 |p - q| & \text{if } p \text{ is a curve midpoint,} \\ C_2 |p - q| & \text{if } p \text{ is a triangle circumcenter.} \end{cases}$$

Proof. We consider the cases:

1. Suppose p is a midpoint on input curve c. If q is an input point disjoint from c, or a midpoint on a curve disjoint from c, then lfs $(p) \leq |p - q|$, and it suffices to take $1 \leq C_1$. If q is an input endpoint of c, then since $d(q) \leq |p - q|$, by Lemma 5, lfs $(p) \leq |p - q| + \text{lfs}\,(q) \leq |p - q| + C_0 d(q) = (1 + C_0)|p - q|$. Thus $1 + C_0 \leq C_1$ suffices.

   If q is a midpoint on a curve nondisjoint from c, then by Lemma 6, it suffices to take

$$\boxed{\frac{1 + C_0}{\sigma} \leq C_1}$$

   Thus we can assume q is a circumcenter, or another midpoint on c. Now consider the subcases for the identity of p:

a) Suppose p is a point on input curve $c = \widetilde{xy}$, added to $\mathcal{P}'$ in step 7. Since no circumcenters have been added to $\mathcal{P}'$ when step 7 is executed, q must be a midpoint on c. Since at most two midpoints are added to c during the grooming phase, p, q are associated with the endpoints, x, y respectively. That is $|p - x| = d(x)$, and $|q - y| = d(y)$. Then

$$|p - q| = |x - y| - d(x) - d(y) \geq \frac{1}{\sqrt{2} - 1} \left(d(x) \vee d(y) \right) - 2 \left(d(x) \vee d(y) \right)$$

$$= \frac{3 - 2\sqrt{2}}{\sqrt{2} - 1} \left(d(x) \vee d(y) \right).$$

Because x, y are input points

$$\text{lfs} (p) \leq |p - x| \vee |p - y| = d(x) \vee (|p - q| + d(y))$$

$$\leq \left(1 + \frac{\sqrt{2} - 1}{3 - 2\sqrt{2}} \right) |p - q| = (1 + C_0) |p - q|$$

Thus $1 + C_0 \leq C_1$ suffices.

b) Suppose p is added in step 18 when the ball around input point x is reduced. Let $\mathcal{B}_- (x)$ be ball around x before $d(x)$ is reduced, i.e., the ball of radius $d_-(x) = d(x)/\beta$, where $d(x)$ reflects the value after step 15 has been executed. Then we have $|p - x| = d(x)$, and the distance from x to the boundary of $\mathcal{B}_- (x)$ is $(1 - \beta)d(x)/\beta$.

By Lemma 4, $d(x)$ is only reduced if there is some t on a curve disjoint from x with $t \in \mathcal{B}_- (x)$. In this case lfs $(x) \leq |x - t| \leq d_-(x) = d(x)/\beta$.

We entertain the two possible locations of q. The first is that q is outside $\overline{\mathcal{B}}_- (x)$, and thus $|p - q| \geq (1 - \beta)d(x)/\beta$. Then

$$\text{lfs} (p) \leq |p - x| + \text{lfs} (x) \leq (1 + 1/\beta) \, d(x) = \left(\frac{1 + \beta}{1 - \beta} \right) \left(\frac{1 - \beta}{\beta} \right) d(x)$$

$$\leq \frac{1 + \beta}{1 - \beta} |p - q|.$$

Thus we must take

$$\boxed{\frac{1 + \beta}{1 - \beta} \leq C_1}$$

The second possibility is that q is inside $\overline{\mathcal{B}}_- (x)$. We assumed q is a circumcenter or on the input curve c also containing p. However, by design of the algorithm, q cannot be a circumcenter in $\mathcal{P}'$ and be inside $\overline{\mathcal{B}}_- (x)$. The remaining possibility is that q be x itself. In this case lfs $(p) \leq |p - x| + \text{lfs} (x) \leq |p - x| + C_0 d(x) = (1 + C_0) |p - x|$, and so the requirement $1 + C_0 \leq C_1$ suffices.

c) Suppose p is added to $\mathcal{P}'$ by a call to SPLITCURVE(c', z). If z is an input point or midpoint on an input curve distinct from c, then by arguments above

$$\text{lfs} (p) \leq \frac{1 + C_0}{\sigma} |p - z|.$$

By Lemma 2, z cannot be a midpoint on c. If z is a circumcenter, then at the time it was added it did not encroach a supercurve of c', as otherwise it would have been rejected. By Lemma 3, since p is a point on that supercurve, $|z - p| \geq r_z / \zeta$, where r_z is the radius of the triangle that z killed. Using this result inductively when z was added, lfs $(p) \leq |p - z| + $ lfs $(z) \leq |p - z| + C_2 r_z \leq (1 + \zeta C_2) |p - z|$. Since Assumption 1 is satisfied, $|p - z| \leq \eta |p - \mathcal{C}_d (c')|$, and thus

$$\text{lfs} (p) \leq \max \left\{ \frac{1 + C_0}{\sigma}, 1 + \zeta C_2 \right\} \eta |p - \mathcal{C}_d (c')|.$$

Similarly, if q is a point of $\mathcal{P}'$ which encroaches c', then by the preceding arguments,

$$\text{lfs} (p) \leq \max \left\{ \frac{1 + C_0}{\sigma}, 1 + \zeta C_2 \right\} |p - q|.$$

If q does not encroach c', then $|p - \mathcal{C}_d (c')| \leq |p - q|$, and so it suffices to take

$$\boxed{\eta \frac{1 + C_0}{\sigma} \leq C_1} \quad \text{and} \quad \boxed{\eta (1 + \zeta C_2) \leq C_1}$$

d) Suppose p is added to $\mathcal{P}'$ by a call to SPLITCURVE(c') at step 26. Then there was some circumcenter x which encroached c', but was rejected. Inductively lfs $(x) \leq C_2 |x - t|$, where t is either endpoint of c'. Letting r be the radius of c', i.e., half the length of its secant, this gives lfs $(x) \leq \sqrt{2} C_2 r$, by the classical argument [MR96b:65137]. Then lfs $(p) \leq |p - x| + $ lfs $(x) \leq (1 + \sqrt{2} C_2) r$.

Since circumcenter x was being inserted, SPLITTRI was called, thus c' was not encroached. Thus q does not encroach c', so $|p - \mathcal{C}_d (c')| \leq |p - q|$. Using Assumption 1, then lfs $(p) \leq (1 + \sqrt{2} C_2) \mu |p - \mathcal{C}_d (c')|$, and so it suffices to take

$$\boxed{\mu \left(1 + \sqrt{2} C_2 \right) \leq C_1}$$

2. Suppose p is the circumcenter of a triangle $\triangle xyz \in \mathcal{D} (\mathcal{P}')$, with $\angle xyz \leq \kappa$. Without loss of generality, assume z was added to $\mathcal{P}'$ after x. If x, z are points of $\mathcal{P}$, then lfs $(z) \leq |x - z|$. Otherwise, if z is not an input point then, inductively, lfs $(z) \leq (C_1 \vee C_2) |x - z|$. By the sine rule, $|x - z| = 2 |p - z| \sin \theta \leq 2 |p - z| \sin \kappa$. Thus

$$\text{lfs} (p) \leq |p - z| + \text{lfs} (z) \leq (1 + 2 \sin \kappa [C_1 \vee C_2]) |p - z|.$$

Because the triangle is Delaunay, it must be that $|p - z| \leq |p - q|$. Thus it suffices to take

$$\boxed{1 + 2 \sin \kappa [C_1 \vee C_2] \leq C_2}$$

Then the following choices of constants work:

$$C_1 = \max \left\{ \frac{1 + \beta}{1 - \beta}, \frac{\eta (1 + C_0)}{\sigma}, \frac{\eta (1 + \zeta)}{1 - 2 \zeta \eta \sin \kappa}, \frac{\mu (1 + \sqrt{2})}{1 - 2 \mu \sqrt{2} \sin \kappa} \right\},$$

$$C_2 = 1 + 2 C_1 \sin \kappa,$$

as long as $\kappa < \arcsin (1 / 2 \mu \sqrt{2}) \wedge \arcsin (1 / 2 \zeta \eta)$.

Corollary 3. *If* $\kappa < \arcsin (1 / 2 \mu \sqrt{2}) \wedge \arcsin (1 / 2 \zeta \eta)$, *the algorithm will terminate.*

The requirement $\kappa < \arcsin\left(1/2\zeta\eta\right)$ is probably more restrictive than necessary, as it is built on an unlikely worst case scenario from Lemma 2.

Lemma 7. *Let $(\mathcal{P}', \mathcal{C}')$ be the point and curve sets at the termination of colop. Then*
1. *For every $c \in \mathcal{C}$ there are $c_0, c_1, \ldots, c_n \in \mathcal{C}'$ such that c is the union of the c_i, and each of the c_i has empty diametral circle with respect to $\mathcal{P}'$, and thus each c_i corresponds to an edge in $\mathcal{D}\left(\mathcal{P}'\right)$.*
2. *If $\triangle xyz \in \mathcal{D}\left(\mathcal{P}'\right)$, then either the minimum angle of the triangle is no less than κ, or there is a point $p \in \mathcal{P}$ with*

$$\max\left(|x - p|, |y - p|, |z - p|\right) \leq 2d(p).$$

Note that, if midpoints are chosen properly (*cf.* Corollary 1 and Corollary 2), then as $\Delta t \to 0$, that $\zeta \to \sqrt{2}$, and η and μ go to 1, and thus the bound on κ goes to $\arcsin\left(1/2\sqrt{2}\right)$. This is the classical output bound of the segment input case [Pse2003, MR96b:65137].

Thus the user can select κ arbitrarily close to $\arcsin\left(1/2\sqrt{2}\right)$ by subdividing his/her input to make Δt sufficiently small. It is straightforward to automate this process for commonly used curves [BcOGc:2002]. However, this preprocessing step is likely unnecessary; rather if κ is set to the desired level, curves with too large variation will automatically be split. An argument of this type was used by the authors in the analysis of a three-dimensional algorithm [PseWnj2004].

From Theorem 1, and the definition of σ, it should be clear that the algorithm can accept input with curves that meet at zero tangent angle. See Figure 4 for an example.

5 Results

A prototype of the algorithm was implemented using CGAL [Fa2001]. The code accepts straight segments, circular arcs, and quadratic and cubic curves. Curves are automatically split to meet a user-specified curvature bound, Δt.

The input in Figure 4 contains several curves which meet at a zero tangent angle. Where input curves meet with small curve separation, the output mesh contains a large number of large number of triangles, as predicted by the reliance of the grading constants on $1/\sigma$. This is shown in detail in Figure 5, which shows the region just above where the two circles meet a line segment and two cubic curves, all at zero tangent angle. There is small curve separation near the tangency point, which results in a number of small angle triangles.

6 Discussion and Improvements

The presence of large angles in the output meshes is an obvious annoyance, for which a fix has been discovered: The fix involves protecting the ball $\mathcal{B}(p)$ with "pseudo-input" arcs, and enforcing the empty diametral circle property for these arcs. The arcs need to be split to ensure a bound of $\pi/3$ on total curvature, which introduces pseudo-midpoints. When a ball $\mathcal{B}(p)$ is split, the pseudo-input arcs are removed from consideration, while the pseudo-midpoints remain in the mesh.

(a) Input

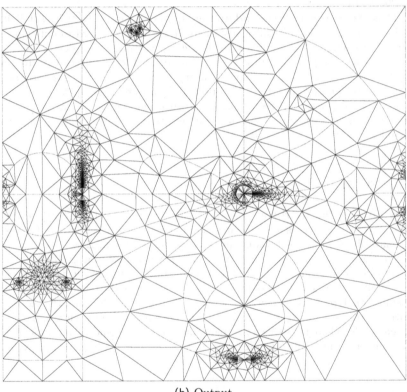

(b) Output

Fig. 4. Results of the prototype code applied to an example PRCC. The input was first subdivided with Steiner Points to insure $\Delta t = 0.5$. The code was executed with $\beta = 0.95$ and $\kappa = \arcsin\left(1/2\sqrt{2}\right)$. The minimum output angle, however, was only about 0.29 , due to the presence of small angles in the input. The mesh contains 2503 points.

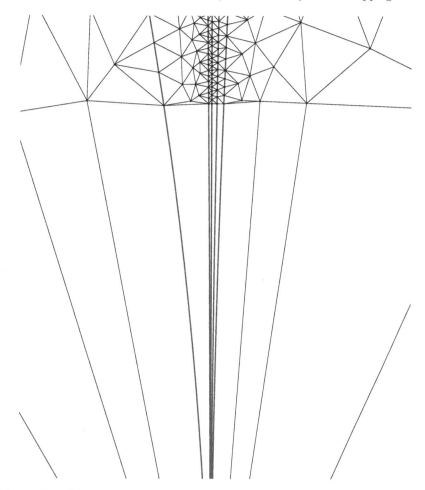

Fig. 5. Detail from Figure 4, showing the region near where the two larger circles touch a line segment and two cubic curves. All five curves meet at a zero tangent angle. The input is shown in bold red lines. Note the presence of triangles with large angles facing small angle triangles emanating from the point of tangency (which is not shown).

Then when a ball $B(p)$ is split, check the pseudo-midpoints it would generate. If one of these has nearest neighbor closer than $d(p)/2$, split the ball again. If one of the pseudo-midpoints would encroach a segment, split the segment if its diameter is less than $d(p)$, otherwise split the ball again. If the ball need not be split again, add the pseudo-input points to $\mathcal{P}'$.

Then protect the pseudo-input segments as real segments. In particular, if a circumcenter encroaches one, split the pseudo-input segment instead of adding the circumcenter. The only other modification is to, not attempt to kill a poor quality triangle which is wholly contained in a ball $B(p)$.

The changes in the proof are minimal: C_0 needs to be increased to 3, the proof of Lemma 5 has to include the new reasons a ball $\mathcal{B}(p)$ might be split, and minor changes need to be made in Theorem 1. This modification is not considered in the main of the paper because of the added complication of the description of the algorithm and its proof, and because it has not yet been implemented.

The use of pseudo-input arcs would make the algorithm similar to the initialization part of the three-dimensional Delaunay Refinement Algorithm of these authors, wherein the boundaries of input faces are protected by portions of arcs [PseWnj2004]. The use of an implicit boundary to protect $\mathcal{B}(p)$ in colop, as opposed to an explicit circle of pseudo-segments, is more akin to the three-dimensional algorithm of Cheng et al. [Cswetal2004]. The presence of large angles in the meshes produced by this two-dimensional algorithm raises doubts about the three-dimensional algorithms which appear to generalize it: do they not also produce simplices with large (solid) angles, even in the absence of small angles, dihedral and otherwise, in the input? It is possible this problem is avoided by the more aggressive approach of Cheng and Poon, who place explicit spherical patches around input facet boundaries to protect small dihedral angles [MR1974932]. Among these three algorithms, only that of Cheng et al. appears to have been implemented, so they cannot be compared directly. It seems that an aggressive protection strategy may be necessary in three dimensions, but this creates meshes with large cardinalities, and papers with impenetrable technicalities.

The ambiguity in the choice of a midpoint by SPLITCURVE, i.e., that it follows Assumption 1, allows room for heuristic modifications to the algorithm. In particular, it may not always be desirable to split a curve $\widetilde{xy}$ by the point where it intersects the bisector of x and y. In the case where point z encroaches $\widetilde{xy}$, one may instead choose the intersection of the angle bisector of $\angle xzy$ with the curve as the splitting point. In some situations this heuristic can slightly improve the maximum output angle of the mesh.

The choice of β appears to affect mesh cardinality for some input. For the input of Figure 4, the use of $\beta = 0.5$ results in a mesh with approximately 32% Note that the term $(1 + \beta)/(1 - \beta)$ which bounds C_1 from below is modest for $\beta \approx \frac{1}{2}$. That is, it should be smaller than the other terms which bound C_1, so there appears to be little lost by increasing β, as was seen in this case. By no means, however, is there a clear relationship between β and mesh cardinality.

[MR96b:65137] Ruppert J. "A Delaunay refinement algorithm for quality 2-dimensional mesh generation." *J. Algorithms*, vol. 18, no. 3, 548–585, 1995. Fourth Annual ACM-SIAM Symposium on Discrete Algorithms (SODA) (Austin, TX, 1993)

[Pse2003] Pav S.E. *Delaunay Refinement Algorithms*. Ph.D. thesis, Department of Mathematics, Carnegie Mellon University, Pittsburgh, Pennsylvania, May 2003. URL http://www.math.cmu.edu/\~{}nw0z/publications/phdtheses/pav/p%avabs

[Sjr1997] Shewchuk J.R. *Delaunay Refinement Mesh Generation*. Ph.D. thesis, School of Computer Science, Carnegie Mellon University, Pittsburgh, Pennsylvania, May 1997. Available as Technical Report CMU-CS-97-137

[MglPseWnj2005] Miller G.L., Pav S.E., Walkington N.J. "When and Why Delaunay Refinement Algorithms Work." *Internat. J. Comput. Geom. Appl.*, vol. 15, no. 1, 25–54, 2005. Special Issue: Selected Papers from the 12th International Meshing Roundtable, 2003

[BcOGc:2002] Boivin C., Ollivier-Gooch C.F. "Guaranteed-Quality Triangular Mesh Generation For Domains with Curved Boundaries." *Internat. J. Numer. Methods Eng.*, vol. 55, no. 10, 1185–1213, 2002

[MglPseWnj2002c] Miller G.L., Pav S.E., Walkington N.J. "Fully Incremental 3D Delaunay Mesh Generation." *Proceedings of the 11th International Meshing Roundtable*, pp. 75–86. Sandia National Laboratory, September 2002

[338236] Murphy M., Mount D.M., Gable C.W. "A point-placement strategy for conforming Delaunay tetrahedralization." *Proceedings of the eleventh annual ACM-SIAM symposium on Discrete algorithms*, pp. 67–74. Society for Industrial and Applied Mathematics, 2000

[CoVeYv01] Cohen-Steiner D., de Verdiere E.C., Yvinec M. "Conforming Delaunay Triangulations in 3D." CS 4345, INRIA, 2001

[Cswetal2004] Cheng S.W., Dey T.K., Ramos E.A., Ray T. "Quality meshing for polyhedra with small angles." *SCG '04: Proceedings of the Twentieth Annual Symposium on Computational Geometry*, pp. 290–299. ACM Press, 2004

[PseWnj2004] Pav S.E., Walkington N.J. "Robust Three Dimensional Delaunay Refinement." *Proceedings of the 13th International Meshing Roundtable*, pp. 145–156. Sandia National Laboratory, September 2004

[BaAz76] Babuška I., Aziz A.K. "On the angle condition in the finite element method." *SIAM J. Numer. Anal.*, vol. 13, no. 2, 214–226, 1976

[Ci78] Ciarlet P.G. *The Finite Element Method for Elliptic Problems*. North–Holland, 1978

[AnCsKrk2001] Amenta N., Choi S., Kolluri R.K. "The power crust." *SMA '01: Proceedings of the sixth ACM symposium on Solid modeling and applications*, pp. 249–266. ACM Press, New York, NY, USA, 2001

[Fa2001] Fabri A. "CGAL - The Computational Geometry Algorithm Library." *Proceedings of the 10th International Meshing Roundtable*, pp. 137–142. Sandia National Laboratory, October 2001

[MR1974932] Cheng S.W., Poon S.H. "Graded conforming Delaunay tetrahedralization with bounded radius-edge ratio." *Proceedings of the Fourteenth Annual ACM-SIAM Symposium on Discrete Algorithms (Baltimore, MD, 2003)*, pp. 295–304. ACM, New York, 2003

Robust Construction of 3-D Conforming Delaunay Meshes Using Arbitrary-Precision Arithmetic

Konstantin Bogomolov

Institute of Mathematical Modeling, Moscow kbogomolov@mail.ru

An algorithm for the construction of 3-D conforming Delaunay tetrahedralizations is presented. The boundary of the meshed domain is contained within Voronoï cells of the boundary vertices of the resulting mesh. The algorithm is explained heuristically. It has been implemented. The problem of numerical precision is shown to be a major obstacle to robust implementation of the algorithm. The Automatic Arbitrary-Precision Arithmetic Library is introduced to solve this problem. The resulting program is intended to be applicable to any mathematically correct input. It has performed successfully on a number of test cases, including a known difficult case for tetrahedral meshing. It is available on the Internet. The Arithmetic Library may be useful for resolving numerical precision problems in any application, and as a base for experimenting with new meshing strategies.

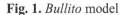

Fig. 1. *Bullito* model   **Fig. 2.** Delaunay mesh of *Bullito*

1 Introduction

The Delaunay tetrahedralization and its dual, the Voronoï diagram, are attractive approaches to three-dimensional mesh generation due to several useful properties of these objects. The Voronoï diagram, in particular, seems well suited for application to 3-D numerical simulation problems because it divides the space into domains naturally "belonging" to each of the points of the input set.

The difficulty of Delaunay / Voronoï methods lies in obtaining a mesh that fits into the specified boundary. The basic Delaunay tetrahedralization covers the convex hull of the set of input points. If the domain to be meshed is not convex, there is no guarantee that all of its faces and edges will appear in the tetrahedralization.

An overview of the known approaches to this problem is contained in the paper by Jonathan Richard Shewchuk [1]. These include the conforming Delaunay, the "almost Delaunay" and the constrained Delaunay approaches. The conforming Delaunay approach is justifiably criticized for adding more points than the other approaches and for its difficulty in controlling the quality of the mesh. However, of the three approaches, only this one generates a truly Delaunay mesh.

The conforming Delaunay tetrahedralization is a normal Delaunay tetrahedralization of the set of input points, together with some additional points that are introduced, typically, onto the domain boundary. It must contain all of the domain's edges and faces as unions of its own edges and faces. The task of a conforming Delaunay tetrahedralization algorithm is to determine the positions of additional points needed to recover the given domain boundary.

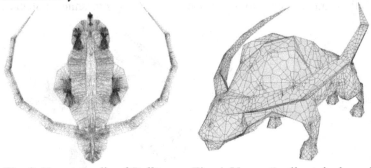

Fig. 3. Voronoï cells of *Bullito*   **Fig. 4.** Voronoï cells at the boundary

It seems that the Delaunay property would be important for the user of a Delaunay algorithm, and not merely due to the quality of the Delaunay tetrahedra (especially considering that quality Delaunay meshing can be actually difficult)! The other good property of Delaunay meshes is their duality to the Voronoï diagrams. Perhaps the user of a Delaunay meshing program will be more interested in the Voronoï cells than in the tetrahedral mesh, - and in that case, the conforming Delaunay approach should be the most straightforward one, as it immediately provides the covering of the domain with the Voronoï cells in the usual metric.

Fig. 5. A domain with sharp angles   **Fig. 6.** Delaunay mesh

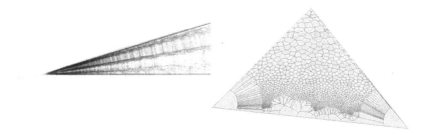

Fig. 7. Voronoï cells   **Fig. 8.** Voronoï cells on the boundary

My algorithm and program strive to achieve a stricter conformation. I want the Voronoï cells of the points belonging to the domain boundary to contain within them this entire boundary. They will constitute the outer layer of the domain's Voronoï decomposition, and it will be possible to trim them with the boundary, thus achieving a boundary-fitting Voronoï mesh (figures 3, 4). The cells of the interior points will all be contained strictly inside the domain, without any of them touching the boundary or sticking outside. The internal two-sided boundaries (which can also be present in the domain) will also be fully covered by their Voronoï cells. This construction (also known as the empty smallest circumsphere property for the boundary triangles) is a known way to make Delaunay meshing simpler, at the price of adding more points. However, it may perhaps be also useful to the potential user of the Voronoï mesh: for example, it could simplify the specification of boundary conditions (or of the special processing near the boundary) in 3-D numerical simulation problems. Besides, it helps to make the pictures look nice!

Despite the conceptual simplicity of the conforming Delaunay tetrahe-dralizations, it seems that the first proof of their existence for arbitrary domains was published only in 2000, by Michael Murphy, David M. Mount and Carl W. Gable [2]. Algorithms for constructing them have also been published by Cohen-Steiner, de Verdière and Yvinec [3], Cheng and Poon [4], Pav and Walkington [5], and Cheng, Dey, Ramos and Ray [6]. These papers all present proofs of the algorithms' correctness. Cohen-Steiner, de Verdière and Yvinec also report an implementation of their algorithm that requires knowledge of the domain's local feature size.

In this article, a simple but unproven conforming Delaunay algorithm that does not require additional information about the input domain is presented. It has been successfully implemented. I tried to develop a black-box code that works correctly with arbitrary, mathematically correct input. The solution to numerical precision problems I have implemented may be of interest by itself. The code, including the Library, is available for download at http://kbogomolov.informatics.ru.

I would like to thank Professor Vladimir Tishkin for many useful discussions, and the reviewers of this article, whose comments were extremely helpful for improving this publication.

The work was supported by ISTC grant 1820. The *Bullito* model is courtesy of Primal Software, Moscow.

2 The Problem

The input for the algorithm is a Piecewise-Linear Complex (defined, for example, in the article [3]). There is an additional requirement that some of the faces of the PLC are marked as "boundary": these faces should divide the space into the interior ("the domain") and the exterior. All other faces should belong to the interior; they play the role of 2-sided internal boundaries. If we need to mesh an arbitrary PLC, it is possible convert it to the required form by enclosing it inside a polyhedron and marking all of the polyhedron's faces as boundary. This discrimination between the interior and the boundary faces is useful, in practice, for avoiding the unnecessary refinement caused by the interactions of the tetrahedra outside the domain.

All of the points added into the Delaunay tetrahedralization (the vertices of the domain, the interior points and the points added during the conformation process) will be called throughout this paper "points" or "vertices". The vertices of the original input domain will also sometimes be called "corners". The PLC will sometimes be called "domain" or "boundary".

It is additionally required of the algorithm that the faces of the PLC must be fully contained inside the Voronoï cells of their points. To illustrate this requirement, a conformed Delaunay triangle on a boundary face may be considered. Its presence does mean that the cells of its three vertices share a common Voronoï edge somewhere along the line that is orthogonal to the face and passes through the triangle's circumcenter. It does not guarantee, however, that this common edge intersects the boundary; thus, some part of the boundary may be outside the outer layer of cells, and some interior cells may be sticking outside the domain. So, in order for this condition to be fulfilled, it is required that all of the tetrahedra adjacent to the boundary have their circumcenters on the same side of the boundary as themselves. In 2-D, this requirement would translate into triangles near border having acute angles opposite the border. In 3-D, I will sometimes call such tets *sufficiently acute*. Since the tetrahedra outside the domain are removed once the mesh is constructed, they are not required to be sufficiently acute. Boundary Voronoï cells are trimmed by the boundary for output, without requiring access to the deleted exterior tets.

So, the entire problem the algorithm and program must solve is this: build a conforming Delaunay tetrahedralization of the given PLC such that all of the interior tetrahedra adjacent to the input faces have their circumcenters on the same side of the face as themselves.

3 The Algorithm

3.1 Delaunay Tetrahedralization

The first step in solving the stated problem is to construct an unconstrained Delaunay tetrahedralization of the set of input points. I use the simple incremental point addition (Bowyer-Watson) algorithm, described, for example, in the book [7]. The search for the first hit tetrahedron is performed using a spatial binary tree, with each node subdividing the parent's region in half, and the axis of subdivision changing with each new level.

The most important numerical operation in this algorithm is the check whether the added point is inside a tet's circumsphere. For the program to be robust, it is essential for this check to always give the correct result. The computations performed are as follows:

Let $\mathbf{p}$ be the added point and let the tet's vertices be $\mathbf{0}, \mathbf{b}, \mathbf{c}, \mathbf{d}$ (we translate the origin into the first vertex). The vertices should be ordered so that $\mathbf{b} \cdot \mathbf{c} \times \mathbf{d} > 0$. Then, the point shall be strictly inside the tet's circumsphere if and only if

$$\begin{vmatrix} |\mathbf{b}|^2 & \mathbf{b}_x & \mathbf{b}_y & \mathbf{b}_z \\ |\mathbf{c}|^2 & \mathbf{c}_x & \mathbf{c}_y & \mathbf{c}_z \\ |\mathbf{d}|^2 & \mathbf{d}_x & \mathbf{d}_y & \mathbf{d}_z \\ |\mathbf{p}|^2 & \mathbf{p}_x & \mathbf{p}_y & \mathbf{p}_z \end{vmatrix} > 0. \tag{1}$$

So, in order to implement this check robustly, we must be able to determine the sign of an expression that depends on the coordinates of the input vertices and contains the operations of addition, subtraction and multiplication. A method for doing this is described in Section 4.

3.2 Conforming Tetrahedralization

Once the unconstrained Delaunay mesh is constructed, we must add points on the boundary in order to recover it and to make the adjacent tets in the interior sufficiently acute.

The following ideas provide the base from which the algorithm for this operation has been constructed. However, I do not have a formal proof of the algorithm's termination. The algorithm is listed in the pseudocode below.

The condition of the boundary Voronoï cells containing entire boundary means that the vertices owning these cells must be packed closely enough on the boundary, so that the cells of other vertices could not "pierce" their layer. In the terms of Delaunay mesh, for any triangle on the boundary to be present in the tetrahedralization, it is sufficient that its smallest circumscribed sphere does not contain any vertex inside it. This condition (*empty smallest circumsphere*) is also a necessary and sufficient one for the adjacent tets to be sufficiently acute. Therefore, if we add points on a boundary face so densely that its 2-D Delaunay triangles have empty smallest circumspheres, we will achieve the desired conforming tetrahedralization. But the radius of the smallest circumsphere is the circumradius of the Delaunay triangle; therefore, we can use a 2-D Delaunay refinement algorithm, such as Ruppert's [8], to minimize it!

Of course, the difficulty lies in refining the faces so that the refinement of one face does not interfere with the refinement of others. If all faces were disjoint, there would be no problem here: we would simply refine the triangulation on each face until the circumcircles of all triangles were small enough, e.g. smaller than the minimal distance between the faces. However, input faces usually have common edges and vertices, and interference between them may happen, especially between the faces at an acute dihedral angle to each other.

In 2-D, there is a similar problem. If we recover domain boundary, for example, by splitting each non-recovered segment at its midpoint, we may get infinite looping near domain corners with sharp angles. A solution is to split the segments adjacent to corners by putting points at some fixed distances from the corner – e.g. powers of 2. This way, points on the edges near corners will arrange themselves into "wheels" of some radius, and there will be no interference between edges near each corner even if they are at a sharp angle to each other. This notion of powers of two distances appears in a number of places: Ruppert's concentric shell splitting [8], Shewchuk's work [9], Pav's thesis [10], etc.

The solution in 3-D, it seems, could be similar; however, we need to construct such protecting structures both at the corners and the edges of the boundary, since the adjacent faces may meet at both.

For each corner, we can simply put points at a fixed distance from it on all incident edges and faces. It is possible to show that if these points are added on the faces densely enough, fans of triangles will appear near each corner with each triangle having empty smallest circumsphere, and, therefore, sufficiently acute adjacent tets. In order to avoid interference between faces through a common edge, the first and the last face point of each fan is situated so that it makes a fixed angle with the edge, for example, one the sine of which is a power of 2.

It remains to protect the common edges of the faces, away from the corners. Suppose we have already recovered one face, and are now processing a face that is adjacent to it via a small dihedral angle. Then let us copy some of the points from the already recovered face on the current face by *rotating them around the common edge*. If we do that at least with the first layer of points neighboring the edge, we will get identical triangulations on both faces near this common edge, and will, therefore, avoid interference and achieve conformation.

This description is, of course, incomplete, because I did not specify exactly which points are transferred between the faces and under which conditions. The precise formulation is contained in the algorithm listed below, but I am not sure it is a correct one, i.e. that it guarantees the algorithm to always terminate. That is the obstacle to proving the algorithm.

However, it seems that a provably correct algorithm can be constructed from the same basic ideas: 1) each face can be refined independently of all others until the circumcircles of its 2-D Delaunay triangles are small enough (while obeying the radii of the corners by refining the arcs around them); 2) interference between the faces through the common edges can be neutralized by transferring the points between faces through rotations around edges. Note that when we add points to a face after transferring, the circumradii of the face's triangles *do not increase*.

Here is the algorithm I have implemented. **Bold** letters indicate a definition of a "routine", *italics* indicate a call to a routine. The entry point is *recovery of the domain*. The unconstrained mesh is supposed to be constructed at this point.

Recovery of the domain:
> *compute radii of the corners*;
> *recover edges*;
> *recover faces*.

Computation of the corner radii:
> for each corner of the domain:
>> find its nearest neighbor point in the tetrahedralization;
>> determine d = distance to it;
>> determine the radius $R = 2^q$ such that $0.2d \leq 2^q \leq 0.4d$ (q is an integer)

Recovery of the edges:
> while any of the segments into which the domain edges are already subdivided *requires subdivision*:
>> *add a point* on such *segment*;
>> *recalculate corner radii after adding a new point*.

Check whether a segment of an edge requires subdivision:
> if it is absent from the tetrahedralization, then
>> requires.
>
> Otherwise:
>> for each of the PLC faces incident to the edge:
>>> find the triangle of the 2-D Delaunay triangulation on this face that is incident to the segment;
>>> if its angle opposite the segment is not acute, then the subdivision is required.
>
> Otherwise (if the segment is present and all adjacent 2-D Delaunay triangles have acute angles), the subdivision is not required.

Addition of a point on a segment:
> if both ends of the segment are domain corners, or if both aren't:
>> put a point approximately in the middle;
>
> otherwise (if only one end is a corner):
>> if the segment's length is larger than 3 radii of the corner,
>>> put a point approximately in the middle;
>>
>> otherwise
>>> put a point at the radius distance from the corner.
>
> (See Section 4.3 for details on adding a point approximately in the middle.)

Recalculation of corner radii after adding a new point:
>for each of the neighbors of the newly added point:
>>if the neighbor is a corner of the PLC:
>>>check if its distance to the new point is less than or equal to its radius;
>>>if it is, reduce the radius by halving it repeatedly until it becomes strictly less than the distance to the added point.

Recovery of the faces:
>repeat for all faces of the PLC until all are fully recovered:
>>Let F be the current face that needs recovery.
>>Repeat while F is not recovered:
>>>*recover all edges of the face*;
>>>*recover triangles of the face (adding one point)*;
>>mark F as "once recovered"

Recovery of the face's edges:
>while any of the segments into which the face's edges are already subdivided *requires subdivision*:
>>*add a point on* such *segment*;
>>*recalculate corner radii after adding a new point.*

(Note that this is similar to the *Recovery of the edges* function, but works only for one face.)

Recovery of triangles of the face (adding one point):
>Find one of the following on the PLC face F:
>>a triangle that is present in the 2-D Delaunay triangulation of F but not in the current tetrahedralization,

or

>>a triangle on F that is present in the tetrahedralization but has adjacent tets the circumcenters of which are on a wrong side of F (check two tets for interior faces and one tet for boundary ones).
>*Add point on the found triangle.*

Addition of point on triangle:
>temporarily (so that it can be removed later) add a point into the tetrahedralization at the circumcenter of the triangle. Let this point be P.
>For each point Q to which P becomes a neighbor:
>>if Q is a corner of the current face,
>>>if P is closer to Q than the radius of Q,
>>>>cancel the addition of P;
>>>>*add point P' on the arc around corner Q;*

recalculate corner radii after adding a new point
return from function.
else, if Q belongs to an edge of the current face:
gather candidates for transferring by point Q.
Look among all candidates for transferring for the suitable one. If found, transfer and add it instead of P, otherwise confirm the addition of P.
Recalculate corner radii after adding a new point.

Addition of point on the arc around domain corner:
Let P be the point we were trying to add on the face F that has violated the radius of the F's corner Q.
Find the sector between the neighbors of Q belonging to F which was hit by P. Add P' on the arc of Q's radius in this sector. If this sector is an extreme one near F's border, put the point at a fixed angle from the border (the sine of which is a power of 2); if the sector is strictly inside the face, add P' approximately in the middle of it (see Section 4.3 about adding point approximately in the middle).

Gathering candidates for transferring:
Let Q be a point lying on the domain edge E that is incident to the face F.
For each neighbor vertex N of Q:
    if it belongs to a PLC face $G \neq F$:
        if G is incident to F via the edge E:
            if G is marked as "once recovered":
                select N as a candidate.

Looking for the suitable candidate for transferring:
Let P be the point we were trying to add on the face F and C be the total set of candidates for transferring gathered after adding P.
For each element of C, calculate the coordinates of the result of its transfer onto F and the distance of the result from P.
Choose the candidate the result of which gets nearest to P while obeying the following requirements: 1) it is strictly inside the circumcircle of the triangle for the refining of which we were adding P; 2) it does not violate the radii of the corners of F; 3) it is strictly inside F.
If none of the candidates obey all of these rules, do not select any.

In this algorithm, we initially subdivide the domain edges until 1) they all become connected, and 2) all triangles near edges in 2-D triangulations of the domain faces have acute angles opposite the edge. This guarantees that any point added on a face at the circumcenter of a 2-D Delaunay triangle will be strictly inside this face.

The refinement of faces is performed by adding points at the circumcenters of those triangles that are not present in the tetrahedralization, or the adjacent tets of which are not sufficiently acute. If the added point is close to a corner or an edge of the face, we may, instead of adding this point, refine the fan of the corner, or transfer a point around the edge from an already recovered face. We add points on faces one by one; before adding each point we ensure that all triangles near the border of the face are acute.

Currently, the algorithm tries to arrange the points so that the first layer of a face's vertices near an edge is identical for all faces sharing this edge. The transferring of points around the edge, and the setting of fan vertices at fixed angles from the edge, both serve this purpose.

After the boundary is recovered, the program marks the tets outside the domain by passing over them with an advancing front, starting from domain boundary. This allows to check easily whether a point with given coordinates belongs to the domain's interior: simply locate the tet to which it belongs (using the spatial tree) and see if it is an interior one. It is possible to perform addition of points at the circumcenters of the interior tets. Because Voronoï cells of the boundary points contain the entire boundary, we may be sure that all of the interior tets have their circumcenters strictly inside the domain.

4 Numerical Precision

4.1 Numerical Precision and Robustness

The presented algorithm is very sensitive to numerical precision errors in several places: the main point-in-tet-circumsphere check, distance checks when updating corner radii, point-inside-polygon check, etc. For the code to be robust, these checks must be implemented so as to give the correct result in the maximum number of cases, desirably for all acceptable inputs. There exist methods for performing such checks robustly and efficiently when the input is integer or machine-precision floating-point [9][11][12]. Unfortunately, in my algorithm, the points added during the conformation stage may require more precision than a floating-point variable may contain, and may even be irrational. For example, the computation of a triangle's circumcenter coordinates involves division, and may, therefore, result in an infinite binary fraction. Some vertices that are introduced during the conformation process may have such coordinates, and the geometry checks, ideally, should be able to cope with them. Even worse, rotating a point around an edge, or putting a point on a line at a fixed distance from a given point, may introduce square roots, and, therefore, irrational coordinates for the geometric predicates to handle.

I do not see any simple way to construct a robust conforming Delaunay algorithm that avoids placing points with such bad coordinates. At least, no limited-precision arithmetic seems to be sufficient as an easy solution to this problem. Such a limited-precision arithmetic (no matter fixed- or floating-point) will constrain the input points onto some grid. It is always possible to imagine a face or an edge the vertices of which belong to this grid, but none of the interior points do. If this face or edge needs to be recovered, it will not be possible to place a point exactly on it.

If we use a limited-precision arithmetic, the point will be placed a little away from the plane of the face. In this case, there is a possibility that a Delaunay tet will be constructed with one triangle on the face and the 4th vertex at the new point (for example, if the face is on external boundary and the point a little way inside the domain; in this case, even though the tet will have big circumradius, the circumscribing sphere may be almost entirely outside the domain, where there may be no points to block it).

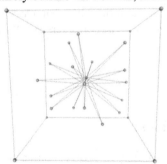

Fig. 9. PLC similar to hard case in [1]

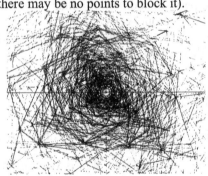

Fig. 10. Delaunay mesh

Fig. 11. Voronoï cells

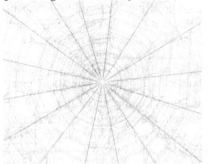

Fig. 12. Voronoï cells, another view

It is, perhaps, possible to handle such situations, for example, with some sliver removal technique; however, that would complicate things as the sliver removal might interfere with the conformation. Most importantly, placing the new points not exactly where they should go would make the code non-robust (on extreme inputs) by design. It would be difficult to track whether a given anomaly during the execution of the program was caused by a fault in the algorithm, an implementation bug or by the rounding of coordinates.

Because of all these difficulties with limited-precision arithmetic, I have implemented the presented algorithm using arbitrary-precision arithmetic. I believe the benefits of this approach to be these:

1. It allows to empirically verify the unproven algorithm. The arithmetic library may serve as a test base for experimenting with new strategies in meshing or other fields.

2. The code is almost totally robust by design (almost, because there remains one adjustable parameter for choosing speed or robustness in one extreme case).

3. The library may also be useful in other applications where a high level of robustness is sought for; it would probably help to save much work (compared to algorithm-specific modification) to achieve the required level of robustness. Also, the library may help reduce constraints on the input data (e.g., no limit on how different the size of the domain and the size of small features are), and possibly reduce the number of tweakable "epsilon" parameters.

A method for robust computation of *staged geometric predicates* has been recently presented by Shewchuk [11] and Nanevski, Blelloch and Harper [12]. Unlike the approach presented here, this method makes use of the hardware-supported floating-point numbers, by exploiting the properties of their roundoff errors. This allows to achieve a significantly better speed of computation than with the approach presented here; however, the following presents justifications for developing my method.

First, the staged geometric predicates operate on points with floating-point coordinates. This means that the expression the sign of which is evaluated by the predicate is fixed: the predicate is either hand-coded or generated from a given expression by a compiler. My approach, however, allows the coordinates of input points to be expressions themselves; it puts no limit on the level of nesting of operations. The expression the sign of which is being determined can be constructed during the execution of the program.

Second, the staged geometric predicates that operate with floating-point variables require that the exceptional conditions of overflow and underflow do not occur during the computation [12]. If they do occur, it is suggested that the computation should be rerun in another, slower form of exact arithmetic. The presented library should be suitable exactly for that!

4.2 The Automatic Arbitrary-Precision Arithmetic Library

The following notes describe the implemented arbitrary-precision strategy for avoiding numerical error problems.
1. Use arbitrary-precision arithmetic for all computations.
2. Keep all results that can be reused (e.g., vertex coordinates) in memory, in the form of a graph of expressions. Each expression in this graph is linked to its arguments, which may be other expressions or constants. Each expression also stores its value in the numerical form, exactly or with some precision.
3. Make the arithmetic library able to resume computations, increasing the precision of any expression starting from the previously reached precision.
4. Make the library automatically choose the precision needed for the intermediate results, in order to get the requested precision for the final result.

The strategy has been implemented in the Automatic Arbitrary-Precision Arithmetic Library. It can compute, in accordance with the formulated rules, arbitrary expressions containing the 4 arithmetic operations, square roots, and finite-bit-length constants.

The input vertices of the mesh may be usual floating-point constants. They are converted and stored as the Library's finite-bit-length constants. The coordinates of vertices added during the conformation are expressions that depend on these input constants, and, possibly, on some previously constructed expressions. Many different expressions may point to the same expression as an argument, and, if the value of this argument expression has been computed to some precision, all of the depending expressions will be able to use it. The entire graph of expressions for vertex coordinates remains in memory while the mesh is being constructed. This is necessary since the addition of new vertices may require to increase the precision of already existing vertices. The expressions for performing geometry checks, such as the point-in-circumsphere check, are usually not reused, and, therefore, exist in memory only for the duration of the check. This applies only to the "top" part of the expression graph for a given check, e.g. to the multiplications and additions that comprise the determinant (1). The argument expressions (vertex coordinates) are those stored in the main, permanent, graph.

To perform a geometric check, the sign of some expression needs to be evaluated, for example, the sign of the determinant (1) for the point-in-circumsphere check. The program constructs the graph for this determinant and calls the Refine function for the top expression in the graph. The Refine function increases the precision of an expression by the given amount of bits. It first determines what precision the arguments must have in order to achieve the requested precision in the result. If the arguments are not constant, and their current precision is not sufficient, Refine is called for them, recursively, in order to achieve the required precision.

So, in order to know the sign of the determinant, we call Refine for its expression, repeatedly, increasing its precision by some amount of bits each time, until it becomes sufficient, i.e., until we get to a non-zero bit.

A special case is when the determinant is zero, and, because of divisions or roots, its numerical evaluation results in an infinite sequence of zero bits. In this case, we cannot reliably determine whether it is really zero, or just a very small positive or negative number, the leading "1" bit of which we have not yet reached. Currently, when the program reaches the 300th bit after the decimal point without encountering any 1s, it accepts the number as zero; this is the only situation when the program's logic is not fully robust. This constant (-300) can be changed if necessary. It may also be possible to implement an analytical transformation of any expression showing such behavior that eliminates all divisions and roots and checks if it is in fact zero. My current implementation of this operation, however, has complexity proportional to 2^q, where q is the number of roots in the expression, and was disabled for performance reasons.

The Library is implemented with C++ classes. The basic class, `CPrecise`, stores a number (as an arbitrary-length array of bits). The finite-bit-length constants are stored as objects of this class. Expressions are implemented as classes inherited from `CPrecise`. In addition to the bit array (that can be expanded, if necessary, to accommodate the increasing precision), they store the pointers to their arguments, as well as the information about current precision. This precision information has the form of integer w such that

$$l_{cur} \leq l_{ex} \leq l_{cur} + 2^w, \qquad (2)$$

where l_{cur} is the approximate absolute value currently stored in the bit array, and l_{ex} is absolute value of the exact number this expression represents. I call this w value the *error bit*.

When Refine function is called, for example, for an addition object, it checks the error bits of the arguments to see what precision we can reach with the current state of the arguments. E.g., if we are adding two positive numbers, one having error bit w_1, and another w_2, we might truncate both arguments to the bit $w_{max} = \max(w_1, w_2)$, add them, and set the error bit of the result to $w_{max} + 2$, which can be shown to be sufficient for the result to conform to the inequality (2). From this, we can determine how much the arguments should be refined for the result to have the precision requested by the caller. Similar considerations are used with all other operations. The logic is contained in the Refine functions for all operations and is completely transparent to the user. The user only specifies by how many bits the end result should be refined; the Library refines all intermediate values as necessary, choosing their precision automatically.

The Library also handles a special case of numbers that are so close to zero that their signs are not yet known. For such an "around zero" number, the error bit is specified so that the number is inside the $[-2^w, 2^w]$ segment. Refine functions for all operations perform special handling if any of the arguments is an "around zero" number. When evaluating the sign of determinant, we actually call the Refine function repeatedly until the determinant's expression becomes a regular, non-"around zero" number.

The implementation of all 5 operations and the handling of all special cases resulted in a quite big and complex code. The Library has built-in self-verification (for debug builds only), that, when turned on, verifies all results and error bounds with simple bitwise checks. At the time of writing there are no known bugs; however, *the Library should not be used in any application the reliability of which has critical effect in the real world*.

4.3 Performance and Memory Use

If only a fixed-precision (or the common floating-point) arithmetic was used, each vertex of the mesh would store its coordinates in a few bytes, e.g. 8 per coordinate. With the Library, the coordinates of vertices are the nodes in the expression graph, which also contains all of the intermediate values. Each vertex, however, adds only a fixed number of expressions to the graph (exactly how many, depends on what kind of vertex it is: one added at a circumcenter, or on an edge, or on a circle around corner, etc.). Therefore, the number of nodes in the graph grows linearly with the number of points added during the conformation.

Each node, however, stores an arbitrary-length array of bits. In order to understand how the presented approach will scale with increasing problem size, we need to consider the impact of this increase on the precision with which the intermediate results shall have to be computed. It is very important that this precision does not grow too fast, as the algorithms for arbitrary-precision arithmetic may have bad asymptotic properties in relation to requested precision (e.g. $O(N^2)$) for the current implementation of multiplication).

When the sign of an expression needs to be determined, this expression is refined until either its leading "1" bit, or the 300th bit after the decimal point, is reached. This means that we'll need to perform only a fixed amount of work for the top expression in the graph. The total amount of work, however, will depend on how many bits will need to be computed in the intermediate expressions.

For addition and subtraction, in order to refine the result down to a given bit, each of the arguments must be refined down to that bit and by 2 bits more. For multiplication, the arguments must each have the number of significant bits roughly equal to the required amount of significant bits in the result. For the reciprocal ($1/x$), the argument must have, roughly, the same amount of significant bits as the result. The argument of the square root must have about 2 times more significant bits than the result.

This means that the amount of bits needed to be calculated for intermediate expressions may grow with the increasing depth of dependence in the graph. However, the fastest, really prohibitive, growth happens only with square roots; luckily, the algorithm does not require any nesting of square roots (they appear only in unit edge vectors and normals of the input PLC, and when putting point at an angle with a power of 2 sine).

Note that in the algorithm pseudocode it is said that a point is sometimes added approximately in the middle of an edge or an arc. This is done in order to reduce the nested dependence between points. Suppose, for example, that an edge of the PLC has the points $\mathbf{a}$ and $\mathbf{b}$ at the ends, and the points $\mathbf{p}_1$ and $\mathbf{p}_2$ already added on it. Suppose also that we need to subdivide the edge further by adding a point between $\mathbf{p}_1$ and $\mathbf{p}_2$. If we put it exactly at midpoint, i.e. at 0.5 ($\mathbf{p}_1 + \mathbf{p}_2$), we will introduce a dependence of the new point on $\mathbf{p}_1$ and $\mathbf{p}_2$ (that, in their own turn, probably depend already on the edge ends $\mathbf{a}$ and $\mathbf{b}$). If the point can be placed not precisely in the middle between $\mathbf{p}_1$ and $\mathbf{p}_2$, we can compute its coordinates as $\mathbf{a} + t$ ($\mathbf{b} - \mathbf{a}$) with some finite-precision constant t, and thus avoid one level of dependence. Similar considerations are used when subdividing protecting arcs around PLC vertices (instead of making the new point dependent on its neighbors on the arc, we add the point at an angle with finite-precision sine or cosine from edge).

The level of dependence for other operations can grow with increasing problem size. This happens in 2 cases: (a) the circumcenter of a triangle depends on its 3 corner vertices; (b) a vertex transferred from another face by rotation around edge depends on the original vertex. It seems that the growth in the level of nesting of operations shouldn't be too fast in practice. Note that, if the level of nesting of circumcenters is increased by 1, the amount of points that can be placed on the face is, roughly, tripled. Also, perhaps, the (a) dependence can be totally avoided by placing points not exactly at the circumcenters (but exactly on the face!), making them dependent only on the PLC corners. The (b) dependence, it seems, cannot be completely avoided, but it is unlikely to be common that a vertex jumps consecutively around many different edges.

These considerations show that the presented approach is likely to scale well with increasing problem size in practice. That, however, has not been verified, as the current implementation suffers from an extremely inefficient ($O(N^2)$-like) search for unconformed triangles and edges (avoiding it poses no theoretical problems, but has not yet been implemented).

References

1. Jonathan Richard Shewchuk (2002) Constrained Delaunay Tetrahedralizations and Provably Good Boundary Recovery. In: Proceedings of the 11th International Meshing Roundtable 193-204
2. Michael Murphy, David M. Mount, Carl W. Gable (2000) A Point-Placement Strategy for Conforming Delaunay Tetrahedralization. In: Proceedings of the 11th Annual ACM-SIAM Symposium on Discrete Algorithms 67-74
3. David Cohen-Steiner, Éric Colin de Verdière, Mariette Yvinec (2002) Conforming Delaunay Triangulations in 3D. In: Proceedings of the 18th ACM Symposium on Computational Geometry 199-208
4. Siu-Wing Cheng, Sheung-Hung Poon (2003) Graded Conforming Delaunay Tetrahedralization with Bounded Radius-Edge Ratio. In: Proceedings of the 14th Annual ACM-SIAM Symposium on Discrete Algorithms 295-304
5. Steven E. Pav, Noel J. Walkington (2004) Robust Three Dimensional Delaunay Refinement. In: Proceedings of the 13th International Meshing Roundtable
6. Siu-Wing Cheng, Tamal K. Dey, Edgar A. Ramos, Tathagata Ray (2004) Quality Meshing for Polyhedra with Small Angles. In: Proceedings of the 20th Annual ACM Symposium on Computational Geometry 290-299
7. Timothy J. Baker (1999) Delaunay-Voronoï Methods. In: Joe F. Thompson, Bharat K. Soni, Nigel P. Weatherill (eds) Handbook of Grid Generation 16. CRC Press, Boca Raton London New York Washington, D.C.
8   Jim Ruppert (1995) A Delaunay Refinement Algorithm for Quality 2-Dimensional Mesh Generation. In: J. Algorithms 18,3:548-585
9   Jonathan Richard Shewchuk (1997) Delaunay Refinement Mesh Generation. PhD Thesis, Carnegie Mellon University

10 Steven Elliot Pav (2003) Delaunay Refinement Algorithms. PhD Thesis, Carnegie Mellon University

11 Jonathan Richard Shewchuk (1996) Robust Adaptive Floating-Point Geometric Predicates. In: Proceedings of the 12th Annual Symposium on Computational Geometry

12 Aleksandar Nanevski, Guy Blelloch, Robert Harper (2003) Automatic Generation of Staged Geometric Predicates. In: Higher-Order and Symbolic Computation, vol. 16, issue 4, 379-400

Meshing Volumes Bounded by Smooth Surfaces*

Steve Oudot[1], Laurent Rineau[2], and Mariette Yvinec[1]

[1] INRIA, BP 93 06902 Sophia Antipolis, France
[2] ENS, 45 rue d'Ulm, 75005 Paris, France

Summary. This paper introduces a three-dimensional mesh generation algorithm for domains bounded by smooth surfaces. The algorithm combines a Delaunay-based surface mesher with a Ruppert-like volume mesher, to get a greedy algorithm that samples the interior and the boundary of the domain at once. The algorithm constructs provably-good meshes, it gives control on the size of the mesh elements through a user-defined sizing field, and it guarantees the accuracy of the approximation of the domain boundary. A noticeable feature is that the domain boundary has to be known only through an oracle that can tell whether a given point lies inside the object and whether a given line segment intersects the boundary. This makes the algorithm generic enough to be applied to a wide variety of objects, ranging from domains defined by implicit surfaces to domains defined by level-sets in 3D grey-scaled images or by point-set surfaces.

1 Introduction

Simplicial meshes are one of the most popular representations for surfaces, volumes, scalar fields and vector fields, in applications such as Geographic Information Systems (GIS), computer graphics, virtual reality, medical imaging and finite element analysis. However, constructing discrete representations of continuous objects can be time-consuming, especially when the geometry of the object is complex. In this case, mesh generation becomes the pacing phase in the computational simulation cycle. Roughly speaking, the more the user is involved in the mesh generation process, the longer the latter is. An appealing example is given in [Mav00], where the mesh generation time is shown to be 45 times that required to compute the solution. This motivates the search for fully-automated mesh-generation methods, which inherently require the use of guaranteed-quality meshing algorithms.

---

*Work partially supported by the IST Programme of the EU as a Shared-cost RTD (FET Open) Project under Contract No IST-006413 (ACS - Algorithms for Complex Shapes) and by the European Network of Excellence AIM@shape (FP6 IST NoE 506766).

Delaunay refinement is recognized as one of the most powerful techniques for generating meshes with guaranteed quality. It allows the user to get an easy control on the sizes of the mesh elements, for instance through a (possibly non-uniform) sizing field. Moreover, it constructs meshes with a good grading, able to conform to quickly varying sizing fields. The pioneer work on Delaunay refinement is due to Ruppert [Rup95], who proposed a two-dimensional mesh generator for domains with piecewise linear boundaries and constraints. Provided that the boundaries and constraints do not form angles smaller than $\frac{\pi}{3}$, Ruppert's algorithm guarantees a lower bound on the smallest angle in the mesh. Furthermore, this bound is achieved by adding an asymptotically optimal number of Steiner vertices. Later on, Shewchuk improved the handling of small angles in two dimensions [She02] and generalized the method to the meshing of three-dimensional domains with piecewise linear boundaries [She98]. The handling of small angles is more puzzling in three dimensions, where dihedral angles and facet angles come into play. Using the idea of protecting spheres around sharp edges, first proposed by Cohen-Steiner et al. [CCY04], Cheng and Poon [CP03] provided a thoroughful handling of small input angles formed by boundaries and constraints. Cheng et al. [CDRR04] turned the same idea into a simpler and practical meshing algorithm.

In three-dimensional space, Delaunay refinement is able to produce tetrahedral meshes with an upper bound on the radius-edge ratios of the tetrahedra, where the radius-edge ratio of a tetrahedron is the ratio between its circumradius and the length of its shortest edge. This eliminates from the mesh all kinds of degenerate tetrahedra, except the ones called *slivers*. A sliver can be described as a tetrahedron formed by four vertices close to the equatorial circle of a sphere and roughly equally spaced on this circle. Cheng et al. [CDE+00], and later on Cheng and Dey [CD02], proposed to exude slivers from the mesh by turning the Delaunay triangulation into a weighted Delaunay triangulation with carefully-chosen small weights applied to the vertices. Li and Teng [LT01] proposed to avoid slivers by relaxing the choice of refinement vertices inside small areas around the circumcenters of the elements to be refined.

The main drawback of the above techniques is that they deal exclusively with domains with piecewise linear boundaries, whereas in many applications, objects have curved boundaries. In such applications, time is spent discretizing the boundary B of the object into a polyhedron P, before the interior of the object can be sampled. Then, the original boundary B is dropped away and replaced by its discretized version, P. On one hand, mesh generation algorithms based on advancing front methods [FBG96], as well as some Delaunay refinement techniques, like the unit edge mesher of [GHS90, GHS91], construct meshes that conform strictly to the discretized boundary P. On the other hand, Ruppert-like methods [She98] refine the boundary mesh: whenever a point should be inserted on B, it is in fact inserted on P. However, in both cases, the quality of the resulting mesh and the accuracy of the boundary approximation depend highly on the initial surface mesh P.

Several methods have been proposed for meshing two-dimensional or three-dimensional domains with curved boundaries. Most of them deal only with specific types of boundaries (parametric, implicit etc.) [SU01], or they simply come with no guarantee regarding the topology of the ouput mesh, or the quality of its elements, or even the termination of the process [DOS99, LBG96, ACSYD05]. One noticeable exception is [BOG02], where the algorithm is able to handle any two-dimensional domain bounded by piecewise smooth curves, of any type, provided that a small number of geometric quantities can be estimated, such as the curvature of a given curve at a given point or the total variation of the unit tangent vector between two points on a given curve. The problem with this method is that it is designed exclusively for the two-dimensional case. Moreover, estimating the required geometric quantities can be time-consuming on certain types of curves.

In this paper, we take advantage of recent results on the front of smooth surface meshing and approximation using Delaunay refinement [BO05], to build a fully-automated algorithm that can mesh three-dimensional domains bounded by smooth surfaces. Specifically, we combine the surface mesher of [BO05] with a Ruppert-like volume mesher, to get a greedy Delaunay-based algorithm that samples the interior and the boundary of the domain at the same time. A noticeable feature of this algorithm is that the boundary of the object has to be known only through an oracle that can answer two simple geometric questions: whether a given point lies inside the object, and whether a given line segment intersects the boundary. This makes the algorithm generic enough to be applied to objects with a wide variety of boundary types, such as implicit surfaces, level-sets in 3D grey-scaled images, point-set surfaces, etc. Concerning guarantees, our algorithm terminates and constructs good-quality meshes for domains whose boundaries are (not necessarily connected) smooth surfaces. The sizes of the mesh elements are controlled through a user-defined sizing field. Moreover, the accuracy of the approximation of the original boundary is guaranteed, and the size of the output mesh is bounded.

The paper is organized as follows. Section 2 recalls a few known facts about restricted Delaunay triangulations and surface approximation. Section 3 describes the main algorithm. Section 4 deals with the accuracy of the approximation of the object by the output mesh. In Section 5, we prove that the meshing algorithm terminates, and we bound the number of vertices of the output mesh. Section 6 addresses the practicality of the algorithm: it gives some details about the choice of the sizing field and it explains how to remove slivers. Finally, Section 7 provides a few examples and experimental results.

2 Preliminary definitions

In the sequel, $\mathcal{O}$ denotes a bounded open subset of the Euclidean space $\mathbb{R}^3$, and $\bar{\mathcal{O}}$ denotes the topological closure of $\mathcal{O}$. We call $\partial\mathcal{O}$ the boundary of $\mathcal{O}$, and we assume that $\partial\mathcal{O}$ is $C^{1,1}$, *i.e.* that its normal vector field is well-defined and 1-Lipschitz.

Definition 1.
- *The* medial axis *M of $\partial\mathcal{O}$ is the topological closure of the set of points of $\mathbb{R}^3$ that have at least two nearest neighbors on $\partial\mathcal{O}$. Every point of M is the center of an open ball that is maximal w.r.t. inclusion among the set of open balls included in $\mathbb{R}^3 \setminus \partial\mathcal{O}$. Such a ball is called a* medial ball.
- *Given a point $x \in \mathbb{R}^3$, we call* distance to the medial axis at x, *or $d_M(x)$, the Euclidean distance from x to M.*

It is well-known [Fed70] that, since $\partial\mathcal{O}$ is $C^{1,1}$, the infimum of d_M over $\partial\mathcal{O}$ is positive. This infimum is called the *reach* of $\partial\mathcal{O}$. The class of surfaces with positive reach has been intensively studied in the recent years, and the distance to the medial axis was used to define a notion of *good* sample, called ε-sample [AB99]:

Definition 2. *Let $\mathcal{P}$ be a finite set of points. $\mathcal{P}$ is a ε-sample of $\partial\mathcal{O}$ if it is included in $\partial\mathcal{O}$ and if $\forall x \in \partial\mathcal{O}$, $d(x, \mathcal{P}) \leq \varepsilon\, d_M(x)$.*

Definition 3. *Let $\mathcal{P}$ be a finite set of points.*
- *The* Voronoi cell *of $p \in \mathcal{P}$ is the set of all points of $\mathbb{R}^3$ that are closer to p than to any other $p' \in \mathcal{P}$.*
- *The* Voronoi diagram *of $\mathcal{P}$, $\mathcal{V}(\mathcal{P})$, is the cellular complex formed by the Voronoi cells of the points of $\mathcal{P}$.*

It is well known that, if the points of $\mathcal{P}$ are in general position, then the dual complex of $\mathcal{V}(\mathcal{P})$ is a tetrahedrization of the convex hull of $\mathcal{P}$, called the *Delaunay triangulation* ($\mathcal{D}(\mathcal{P})$ for short). The meshing strategy described in this paper relies on a subcomplex of $\mathcal{D}(\mathcal{P})$, defined below.

Definition 4. *Let $\mathcal{P}$ be a finite point set.*
- *The Delaunay triangulation of $\mathcal{P}$ restricted to $\mathcal{O}$, or $\mathcal{D}_{|\mathcal{O}}(\mathcal{P})$ for short, is the subcomplex of $\mathcal{D}(\mathcal{P})$ formed by the tetrahedra whose dual Voronoi vertices lie in $\mathcal{O}$.*
- *The Delaunay triangulation of $\mathcal{P}$ restricted to $\partial\mathcal{O}$, or $\mathcal{D}_{|\partial\mathcal{O}}(\mathcal{P})$ for short, is the subcomplex of $\mathcal{D}(\mathcal{P})$ formed by the triangles whose dual Voronoi edges intersect $\partial\mathcal{O}$.*

Given a facet f of $\mathcal{D}_{|\partial\mathcal{O}}(\mathcal{P})$ and its dual Voronoi edge e, every point of $e \cap \partial\mathcal{O}$ is the center of an open ball containing no point of $\mathcal{P}$, and whose bounding sphere passes through the vertices of f. This ball is called a *surface Delaunay ball of $\mathcal{P}$*.

The main idea of our algorithm is to sample $\mathcal{O}$ and $\partial\mathcal{O}$ greedily and simultaneously, using $\mathcal{D}_{|\mathcal{O}}(\mathcal{P})$ and $\mathcal{D}_{|\partial\mathcal{O}}(\mathcal{P})$ to drive the choice of the next point to insert. The output is a point set whose restriction to $\partial\mathcal{O}$ is a *loose ε-sample* of $\partial\mathcal{O}$ [BO05]:

Definition 5. *Let $\mathcal{P}$ be a finite point set, and ε be a positive value. $\mathcal{P}$ is a loose ε-sample of $\partial\mathcal{O}$ if the following conditions hold:*

L1 $\mathcal{P} \subset \partial\mathcal{O}$;
L2 $\mathcal{D}_{|\partial\mathcal{O}}(\mathcal{P})$ *has vertices on every connected component of $\partial\mathcal{O}$;*
L3 *the center c of any surface Delaunay ball of $\mathcal{P}$ is closer to $\mathcal{P}$ than $\varepsilon\, d_M(c)$.*

Notice that ε-samples verify Assertions L1 and L3. Moreover, if $\varepsilon < 0.1$, then L2 is verified as well, by Theorem 2 of [AB99]. It follows that any ε-sample is a loose ε-sample, for $\varepsilon < 0.1$. Loose ε-samples enjoy many properties [BO05], which we summarize below:

Theorem 1. *If $\mathcal{P}$ is a loose ε-sample of $\partial\mathcal{O}$, with $\varepsilon \leq 0.09$, then $\mathcal{D}_{|\partial\mathcal{O}}(\mathcal{P})$ is a closed 2-manifold ambient isotopic to $\partial\mathcal{O}$, at Hausdorff distance $O(\varepsilon^2)$ from $\partial\mathcal{O}$, and its normals approximate the normals of $\partial\mathcal{O}$ within an error of $O(\varepsilon)$. Moreover, $\partial\mathcal{O}$ is covered by the surface Delaunay balls of $\mathcal{P}$, and $\mathcal{P}$ is a $\varepsilon(1 + 8.5\varepsilon)$-sample of $\partial\mathcal{O}$.*

Ambient isotopy and Hausdorff approximation are most interesting for our problem. As for normal and curvature approximations, they are useful in all applications where the user wants to estimate differential quantities on surfaces.

3 Main algorithm

The algorithm takes as input the domain $\mathcal{O}$ to be meshed, a sizing field σ, and two parameter values α and B. The domain is known through an oracle that can tell whether a given point lies inside $\mathcal{O}$ or outside. The oracle can also detect whether a given segment intersects $\partial\mathcal{O}$ and, in the affirmative, return all the points of intersection (which are finitely many, generically). The sizing field is a positive function $\sigma : \bar{\mathcal{O}} \to \mathbb{R}^{+}$ defined over $\bar{\mathcal{O}}$ and assumed to be 1-Lipschitz.

The algorithm first constructs an initial point set $\mathcal{P}_i \subset \partial\mathcal{O}$ that is a $\frac{1}{3}$-sparse 0.09-sample of $\partial\mathcal{O}$, that is:

- $\forall x \in \partial\mathcal{O}$, $\mathrm{d}(x, \mathcal{P}_i) \leq 0.09 \, \mathrm{d}_M(x)$;
- $\forall p \in \mathcal{P}_i$, $\mathrm{d}(p, \mathcal{P}_i \setminus \{p\}) \geq 0.03 \, \mathrm{d}_M(p)$.

The construction of such a point set is described extensively in [BO05], thus we skip it here. Once $\mathcal{P}_i$ is built, the algorithm constructs $\mathcal{P}$ iteratively, starting with $\mathcal{P} = \mathcal{P}_i$ and inserting one point in $\mathcal{P}$ per iteration. In the meantime, the restricted Delaunay triangulations $\mathcal{D}_{|\mathcal{O}}(\mathcal{P})$ and $\mathcal{D}_{|\partial\mathcal{O}}(\mathcal{P})$ are maintained, using the oracle.

At each iteration, one element of the mesh (a facet of $\mathcal{D}_{|\partial\mathcal{O}}(\mathcal{P})$ or a tetrahedron of $\mathcal{D}_{|\mathcal{O}}(\mathcal{P})$) is refined. To refine a tetrahedron, the algorithm inserts its circumcenter in $\mathcal{P}$. A facet f of $\mathcal{D}_{|\partial\mathcal{O}}(\mathcal{P})$ may be circumscribed by several surface Delaunay balls. Thus, to refine f, the algorithm inserts in $\mathcal{P}$ the center of the surface Delaunay ball $B(c, r)$ circumscribing f with largest ratio $r/\sigma(c)$. The choice of the next element to be refined is driven by the following rules, considered in this order:

R1 if a facet f of $\mathcal{D}_{|\partial\mathcal{O}}(\mathcal{P})$ does not have its three vertices on $\partial\mathcal{O}$, then refine f;

R2 if a facet f of $\mathcal{D}_{|\partial\mathcal{O}}(\mathcal{P})$ has a surface Delaunay ball $B(c, r)$ with ratio $r/\sigma(c) > \alpha$, then refine f;

R3 if a tetrahedron t of $\mathcal{D}_{|\mathcal{O}}(\mathcal{P})$ has a circumradius greater than $\sigma(c)$, where c is the circumcenter of t, or if t has a radius-edge ratio greater than B, then consider the circumcenter c of t:

    R3.1 if c is not included in any surface Delaunay ball, then insert c in $\mathcal{P}$;

    R3.2 else, insert in $\mathcal{P}$ the center of one surface Delaunay ball containing c.

The algorithm terminates when the triggering conditions of Rules R1, R2 and R3 are no longer met. Upon termination, every facet of $\mathcal{D}_{|\partial\mathcal{O}}(\mathcal{P})$ has its three vertices on $\partial\mathcal{O}$ (Rule R1) and every surface Delaunay ball $B(c, r)$ has a radius $r \leq \alpha \, \sigma(c)$ (Rule R2). Moreover (Rule R3), every tetrahedron t of $\mathcal{D}_{|\mathcal{O}}(\mathcal{P})$ has a circumradius $r \leq \min\{\sigma(c), B \, l_{min}\}$, where c is the circumcenter of t and l_{min} is the length of the shortest edge of t.

4 Approximation accuracy

In this section, we assume that the algorithm terminates. Termination is discussed in Section 5, which uses several results stated here. From now on, $\mathcal{P}_i$ denotes the initial point set and $\mathcal{P}$ the output point set. Let $\mathcal{P}_{|\partial\mathcal{O}} = \mathcal{P} \cap \partial\mathcal{O}$.

Since $\mathcal{P}_i$ is a 0.09-sample of $\partial\mathcal{O}$, $\mathcal{P}_{|\partial\mathcal{O}}$ is also a 0.09-sample of $\partial\mathcal{O}$, since no point is deleted during the course of the algorithm. Thus, $\mathcal{D}_{|\partial\mathcal{O}}(\mathcal{P}_{|\partial\mathcal{O}})$ is a closed 2-manifold with the same topology type as $\partial\mathcal{O}$, by Theorem 1. Therefore, to have topological guarantees on the output of the algorithm, it suffices to prove that the boundary of $\mathcal{D}_{|\mathcal{O}}(\mathcal{P})$ is equal to $\mathcal{D}_{|\partial\mathcal{O}}(\mathcal{P}_{|\partial\mathcal{O}})$.

There exists a strong relationship between the boundary of $\mathcal{D}_{|\mathcal{O}}(\mathcal{P})$ and $\mathcal{D}_{|\partial\mathcal{O}}(\mathcal{P})$:

Lemma 1. *The boundary of $\mathcal{D}_{|\mathcal{O}}(\mathcal{P})$ is a subcomplex of $\mathcal{D}_{|\partial\mathcal{O}}(\mathcal{P})$. Moreover, if every edge of the Voronoi diagram $\mathcal{V}(\mathcal{P})$ intersects $\partial\mathcal{O}$ at most once, and transversally, then the boundary of $\mathcal{D}_{|\mathcal{O}}(\mathcal{P})$ is equal to $\mathcal{D}_{|\partial\mathcal{O}}(\mathcal{P})$.*

Proof. Since $\mathcal{D}_{|\mathcal{O}}(\mathcal{P})$ is a union of Delaunay tetrahedra, its boundary is a union of Delaunay facets. Let f be a facet of the boundary of $\mathcal{D}_{|\mathcal{O}}(\mathcal{P})$. By definition, it belongs to two Delaunay tetrahedra, one of which has its dual Voronoi vertex inside $\mathcal{O}$, whereas the other one has its dual Voronoi vertex outside $\mathcal{O}$. It follows that the Voronoi edge dual to f intersects $\partial\mathcal{O}$, which means that $f \in \mathcal{D}_{|\partial\mathcal{O}}(\mathcal{P})$.

Let us now assume that every edge of $\mathcal{V}(\mathcal{P})$ intersects $\partial\mathcal{O}$ at most once. Let f be a facet of $\mathcal{D}_{|\partial\mathcal{O}}(\mathcal{P})$. By definition, the Voronoi edge dual to f intersects $\partial\mathcal{O}$. Since this edge intersects $\partial\mathcal{O}$ only once, one of its vertices lies inside $\mathcal{O}$ whereas the other one (which may be at infinity) lies outside $\mathcal{O}$. It follows, by definition of $\mathcal{D}_{|\mathcal{O}}(\mathcal{P})$, that one of the Delaunay tetrahedra incident to f belongs to $\mathcal{D}_{|\mathcal{O}}(\mathcal{P})$, while the other one does not. Hence, f belongs to the boundary of $\mathcal{D}_{|\mathcal{O}}(\mathcal{P})$. $\square$

In our case, $\mathcal{D}_{|\partial\mathcal{O}}(\mathcal{P})$ is precisely the boundary of $\mathcal{D}_{|\mathcal{O}}(\mathcal{P})$, due to the following result:

Lemma 2. *Every edge of $\mathcal{V}(\mathcal{P})$ intersects $\partial\mathcal{O}$ at most once, and transversally.*

Proof. Among the edges of $\mathcal{V}(\mathcal{P})$, only those whose dual Delaunay facets have their three vertices on $\partial\mathcal{O}$ can intersect $\partial\mathcal{O}$, thanks to Rule R1. Let e be such an edge. It is included in an edge e' of $\mathcal{V}(\mathcal{P}_{|\partial\mathcal{O}})$. Since $\mathcal{P}_{|\partial\mathcal{O}}$ is a 0.09-sample of $\partial\mathcal{O}$, Lemma 3.6 of [BO05] tells that e' intersects $\partial\mathcal{O}$ at most once, and transversally, which yields the lemma. $\square$

Corollary 1. *The boundary of $\mathcal{D}_{|\mathcal{O}}(\mathcal{P})$ is $\mathcal{D}_{|\partial\mathcal{O}}(\mathcal{P})$.*

It follows from Corollary 1 that, if we can prove that $\mathcal{D}_{|\partial\mathcal{O}}(\mathcal{P}) = \mathcal{D}_{|\partial\mathcal{O}}(\mathcal{P}_{|\partial\mathcal{O}})$, then the boundary of $\mathcal{D}_{|\mathcal{O}}(\mathcal{P})$ will be equal to $\mathcal{D}_{|\partial\mathcal{O}}(\mathcal{P}_{|\partial\mathcal{O}})$, which approximates $\partial\mathcal{O}$ topologically. We need an intermediate result.

Lemma 3. $\mathcal{D}_{|\partial\mathcal{O}}(\mathcal{P})$ *has vertices on all the connected components of $\partial\mathcal{O}$.*

Proof. By Rule R1, every edge e of $\mathcal{V}(\mathcal{P})$ that intersects $\partial\mathcal{O}$ has a dual Delaunay facet f whose three vertices are in $\mathcal{P}_{|\partial\mathcal{O}}$. Since $\mathcal{P}_{|\partial\mathcal{O}}$ is a 0.09-sample of $\partial\mathcal{O}$, the point $c = e \cap \partial\mathcal{O}$ lies at distance at most $0.09\, d_M(c)$ from the vertices of f. It follows, by Lemma 8 of [AB99], that c and the vertices of f lie on the same connected component of $\partial\mathcal{O}$. As a consequence, to prove the lemma, it suffices to show that every connected component of $\partial\mathcal{O}$ is intersected by at least one Voronoi edge.

Notice that every connected component $\mathcal{C}$ of $\partial\mathcal{O}$ is the fronteer between two connected components Ω_1 and Ω_2 of $\mathbb{R}^3 \setminus \partial\mathcal{O}$, so that every connected path from Ω_1 to Ω_2 crosses $\mathcal{C}$. Therefore, to prove that $\mathcal{C}$ is intersected by a Voronoi edge, it suffices to prove that Ω_1 and Ω_2 both contain Voronoi vertices, since the graph made of the Voronoi vertices and edges is connected.

Let us assume for a contradiction that some component Ω of $\mathbb{R}^3 \setminus \partial\mathcal{O}$ contains no Voronoi vertex. Since the Delaunay balls centered at the Voronoi vertices (including the ones at infinity) cover $\mathbb{R}^3$, at least one such ball (say $B(c,r)$) contains a point x of $M \cap \Omega$. Since c lies outside Ω while x lies inside, the line segment $[c,x]$ intersects the boundary of Ω (which is part of $\partial\mathcal{O}$). Let y be a point of intersection. The ball centered at y, of radius $\mathrm{d}(x,y)$, is contained in the interior of $B(c,r)$. Therefore, it contains no point of $\mathcal{P}$. Now, its radius is $\mathrm{d}(x,y)$, which is at least the distance from y to M since $x \in M \cap \Omega$. It follows that y is farther from $\mathcal{P}$ than $\mathrm{d}_M(y)$, which contradicts the fact that $\mathcal{P}_{|\partial\mathcal{O}}$ is a 0.09-sample of $\partial\mathcal{O}$. It follows that Ω contains at least one Voronoi vertex, which ends the proof of Lemma 3.   □

We can now prove that $\mathcal{D}_{|\partial\mathcal{O}}(\mathcal{P}) = \mathcal{D}_{|\partial\mathcal{O}}(\mathcal{P}_{|\partial\mathcal{O}})$, by using the fact that $\mathcal{D}_{|\partial\mathcal{O}}(\mathcal{P})$ is the boundary of a three-dimensional object, namely $\mathcal{D}_{|\mathcal{O}}(\mathcal{P})$ (Corollary 1).

Lemma 4. $\mathcal{D}_{|\partial\mathcal{O}}(\mathcal{P}) = \mathcal{D}_{|\partial\mathcal{O}}(\mathcal{P}_{|\partial\mathcal{O}})$.

Proof. Thanks to Rule R1, all the facets of $\mathcal{D}_{|\partial\mathcal{O}}(\mathcal{P})$ have their three vertices in $\mathcal{P}_{|\partial\mathcal{O}}$, hence their dual Voronoi edges are included in edges of $\mathcal{V}(\mathcal{P}_{|\partial\mathcal{O}})$. It follows that $\mathcal{D}_{|\partial\mathcal{O}}(\mathcal{P})$ is a subcomplex of $\mathcal{D}_{|\partial\mathcal{O}}(\mathcal{P}_{|\partial\mathcal{O}})$.

To prove the lemma, it suffices then to show that every facet of $\mathcal{D}_{|\partial\mathcal{O}}(\mathcal{P}_{|\partial\mathcal{O}})$ is also a facet of $\mathcal{D}_{|\partial\mathcal{O}}(\mathcal{P})$. Let us assume for a contradiction that there exists a facet f of $\mathcal{D}_{|\partial\mathcal{O}}(\mathcal{P}_{|\partial\mathcal{O}})$ that is not a facet of $\mathcal{D}_{|\partial\mathcal{O}}(\mathcal{P})$. Let C be the connected component of $\mathcal{D}_{|\partial\mathcal{O}}(\mathcal{P}_{|\partial\mathcal{O}})$ to which f belongs. By Lemma 8 of [AB99], the vertices of C belong to a single component $\mathcal{C}$ of $\partial\mathcal{O}$. By Lemma 3, at least one vertex v of $\mathcal{D}_{|\partial\mathcal{O}}(\mathcal{P})$ lies on $\mathcal{C}$. Let C' be the connected component of $\mathcal{D}_{|\partial\mathcal{O}}(\mathcal{P})$ that contains v. Since $\mathcal{D}_{|\partial\mathcal{O}}(\mathcal{P})$ is a subset of $\mathcal{D}_{|\partial\mathcal{O}}(\mathcal{P}_{|\partial\mathcal{O}})$, C' is included in C (which is a connected 2-manifold without boundary). Moreover, since f is not included in C' while v is, C' has a boundary. Now, by Corollary 1, C' is a connected component of the boundary of $\mathcal{D}_{|\mathcal{O}}(\mathcal{P})$. Thus, C' cannot have a boundary, which raises a contradiction.   □

It follows from the previous results that the boundary of $\mathcal{D}_{|\mathcal{O}}(\mathcal{P})$ is equal to $\mathcal{D}_{|\partial\mathcal{O}}(\mathcal{P}_{|\partial\mathcal{O}})$, which is ambient isotopic to $\partial\mathcal{O}$, by Theorem 1. In addition to this topological result, we would like to give a bound on the Hausdorff distance between $\partial\mathcal{O}$ and the boundary of $\mathcal{D}_{|\mathcal{O}}(\mathcal{P})$, depending on the input sizing field σ. Let $\varepsilon = \min\{0.09,\ \sup_{x\in\partial\mathcal{O}} \frac{\alpha\ \sigma(x)}{\mathrm{d}_M(x)}\}$. Our bound will depend on ε. So far, we know that $\mathcal{P}_{|\partial\mathcal{O}}$ is a 0.09-sample of $\partial\mathcal{O}$.

Lemma 5. *The surface Delaunay balls of $\mathcal{P}$ and those of $\mathcal{P}_{|\partial\mathcal{O}}$ are the same.*

Proof. Since every edge of $\mathcal{V}(\mathcal{P})$ that intersects $\partial\mathcal{O}$ is included in an edge of $\mathcal{V}(\mathcal{P}_{|\partial\mathcal{O}})$, the surface Delaunay balls of $\mathcal{P}$ are also surface Delaunay balls of $\mathcal{P}_{|\partial\mathcal{O}}$. Let us show that the converse is true. Let e be an edge of $\mathcal{V}(\mathcal{P}_{|\partial\mathcal{O}})$. If $e \cap \partial\mathcal{O} \neq \emptyset$, then $|e \cap \partial\mathcal{O}| = 1$, by Lemma 3.6 of [BO05]. Moreover, the Delaunay facet dual to e belongs to $\mathcal{D}_{|\partial\mathcal{O}}(\mathcal{P})$, by Lemma 4. This means that e contains an edge e' of $\mathcal{V}(\mathcal{P})$, such that $|e' \cap \partial\mathcal{O}| \geq 1$. Hence $e \cap \partial\mathcal{O} = e' \cap \partial\mathcal{O}$.   □

Thanks to Lemma 5, Rule R2 controls the radii of all the surface Delaunay balls of $\mathcal{D}_{|\partial\mathcal{O}}(\mathcal{P}_{|\partial\mathcal{O}})$, which implies that, upon termination of the algorithm, $\mathcal{P}_{|\partial\mathcal{O}}$ is a loose ε-sample of $\partial\mathcal{O}$. Hence, $\mathcal{D}_{|\partial\mathcal{O}}(\mathcal{P}_{|\partial\mathcal{O}})$ (and therefore the boundary of $\mathcal{D}_{|\mathcal{O}}(\mathcal{P})$) approximates $\partial\mathcal{O}$ both topologically and geometrically.

Theorem 2. $\mathcal{D}_{|\mathcal{O}}(\mathcal{P})$ *is a 3-manifold ambient isotopic to $\bar{\mathcal{O}}$, at Hausdorff distance $O(\varepsilon^2)$ from $\bar{\mathcal{O}}$, where $\varepsilon = \min\{0.09,\ \sup_{x \in \partial\mathcal{O}} \frac{\alpha\,\sigma(x)}{d_M(x)}\}$. Moreover, the surface Delaunay balls of $\mathcal{P}$ cover $\partial\mathcal{O}$.*

Proof. By Corollary 1 and Lemma 4, the boundary of $\mathcal{D}_{|\mathcal{O}}(\mathcal{P})$ is equal to $\mathcal{D}_{|\partial\mathcal{O}}(\mathcal{P}_{|\partial\mathcal{O}})$. Since $\mathcal{P}_{|\partial\mathcal{O}}$ is a loose ε-sample of $\partial\mathcal{O}$, we know by Theorem 1 that there exists an ambient isotopy $h : [0,1] \times \mathbb{R}^3 \to \mathbb{R}^3$ that maps $\partial\mathcal{O}$ to $\mathcal{D}_{|\partial\mathcal{O}}(\mathcal{P}_{|\partial\mathcal{O}})$. The map $h(1,.) : \mathbb{R}^3 \to \mathbb{R}^3$ is an ambient homeomorphism that maps the compact 3-manifold $\bar{\mathcal{O}}$ to a compact 3-manifold bounded by $\mathcal{D}_{|\partial\mathcal{O}}(\mathcal{P}_{|\partial\mathcal{O}})$. Now, the only compact 3-manifold bounded by $\mathcal{D}_{|\partial\mathcal{O}}(\mathcal{P}_{|\partial\mathcal{O}})$ is $\mathcal{D}_{|\mathcal{O}}(\mathcal{P})$ itself[3]. Thus, we have $h(1, \bar{\mathcal{O}}) = \mathcal{D}_{|\mathcal{O}}(\mathcal{P})$, which means that $\mathcal{D}_{|\mathcal{O}}(\mathcal{P})$ is ambient isotopic to $\bar{\mathcal{O}}$.

Since $\mathcal{D}_{|\mathcal{O}}(\mathcal{P})$ and $\bar{\mathcal{O}}$ are both compact, their Hausdorff distance is achieved by a pair of points lying on their boundaries. Hence, we have $d_H(\mathcal{D}_{|\mathcal{O}}(\mathcal{P}),\ \bar{\mathcal{O}}) = d_H(\mathcal{D}_{|\partial\mathcal{O}}(\mathcal{P}_{|\partial\mathcal{O}}),\ \partial\mathcal{O})$, which is $O(\varepsilon^2)$ since $\mathcal{P}_{|\partial\mathcal{O}}$ is a loose ε-sample of $\partial\mathcal{O}$. As for the fact that the surface Delaunay balls of $\mathcal{P}$ cover $\partial\mathcal{O}$, it is a direct consequence of Theorem 1 and Lemma 5. □

Observe that the results of this section do not rely on Rule R3. Hence, they hold not only upon termination, but also during the course of the algorithm, each time neither Rule R1 nor Rule R2 can be applied. In particular, Theorem 2 holds every time Rule R3 is triggered. This observation will be instrumental in proving Lemma 6 of Section 5.

5 Termination and size of the output

In this section, we provide conditions on parameters α and B to ensure that the algorithm terminates. We assume that the sizing field σ is 1-Lipschitz over $\bar{\mathcal{O}}$.

Our strategy is to prove an upper bound on the size of the point sample constructed by the algorithm. The termination of the algorithm results from this bound.

Definition 6. *Given a point p inserted in $\mathcal{P}$ by the algorithm, the* insertion radius *of p, or $r(p)$ for short, is the Euclidean distance from p to $\mathcal{P}$ right before its insertion[4]. The insertion radius of a point p of the initial point set $\mathcal{P}_i$ is the Euclidean distance from p to $\mathcal{P}_i \setminus \{p\}$.*

Our first task is to provide a lower bound on the insertion radius of every point of $\mathcal{P}$. In fact, we will prove a stronger result, stated as Lemma 6. We define a sizing field σ_0 which can be considered as an extension of d_M over $\bar{\mathcal{O}}$:

$$\forall x \in \bar{\mathcal{O}},\ \sigma_0(x) = \inf\ \{d(x, x') + d_M(x') \mid x' \in \partial\mathcal{O}\}$$

As proved in [MTT99, TW00, ACSYD05], σ_0 is a 1-Lipschitz function, equal to $d_M(x)$ on $\partial\mathcal{O}$. In fact, σ_0 is the pointwise maximal 1-Lipschitz function which is at most d_M on $\partial\mathcal{O}$. Let $\sigma'(p) = \min\{\alpha\,\sigma(p),\ 0.03\,\sigma_0(p)\}$, $\forall p \in \bar{\mathcal{O}}$. Notice that, since σ and σ_0 are 1-Lipschitz, σ' is γ-Lipschitz, where $\gamma = \max\{\alpha,\ 0.03\}$.

---

[3] $\mathcal{D}_{|\mathcal{O}}(\mathcal{P})$ is compact because it is a finite union of finite tetrahedra.

[4] Notice that it is also the length of the smallest Delaunay edge created when p is inserted.

Lemma 6. *If $\alpha < \frac{1}{5}$ and $B \geq \frac{4}{1-5\gamma}$, then the following conditions are verified:*

$C1$ $\forall p \in \mathcal{P}$, $r(p) \geq \sigma'(p)$;

$C2$ $\forall p \in \mathcal{P} \setminus \mathcal{P}_{|\partial\mathcal{O}}$, $\delta(p) \geq \frac{1}{1-\gamma} \sigma'(p)$, *where $\delta(p)$ is the Euclidean distance from p to $\partial\mathcal{O}$.*

Proof. We prove the lemma by induction. Initially, we have $\mathcal{P} = \mathcal{P}_i$, and every point of $\mathcal{P}_i$ verifies C1, since $\mathcal{P}_i$ is a $\frac{1}{3}$-sparse 0.09-sample. Moreover, the points of $\mathcal{P}_i$ belong to $\partial\mathcal{O}$, thus C2 is also verified. Let us now assume that C1 and C2 are verified by every point of $\mathcal{P}$, up to a certain step where point c is inserted in $\mathcal{P}$. We will prove that c also verifies C1 and C2.

• If Rule R1 is being applied, then c is the center of a surface Delaunay ball of $\mathcal{P}$ whose bounding sphere passes through a point $p \in \mathcal{P} \setminus \mathcal{P}_{|\partial\mathcal{O}}$. The insertion radius of c is the radius of the surface Delaunay ball, *i.e.* $r(c) = \mathrm{d}(c,p)$. Moreover, $\mathrm{d}(c,p)$ is at least the distance $\delta(p)$ from p to $\partial\mathcal{O}$, which is at is at least $\frac{1}{1-\gamma} \sigma'(p)$ by C2. Since σ' is γ-Lipschitz, we have $\sigma'(p) \geq (\sigma'(c) - \gamma \, \mathrm{d}(c,p))$, hence:

$$\mathrm{d}(c,p) \geq \frac{1}{1-\gamma} \left(\sigma'(c) - \gamma \, \mathrm{d}(c,p) \right) \ \Rightarrow \ r(c) = \mathrm{d}(c,p) \geq \sigma'(c)$$

It follows that C1 is verified for c. Moreover, C2 is also verified, since c belongs to $\partial\mathcal{O}$.

• If Rule R2 is applied, then c is the center of a surface Delaunay ball of radius greater than $\alpha \, \sigma(c) \geq \sigma'(c)$, thus the insertion radius of c is at least $\sigma'(c)$, which satisfies C1. Moreover, C2 is satisfied since c belongs to $\partial\mathcal{O}$.

• If Rule R3.1 is applied, then c is the center of a tetrahedron t, and the insertion radius $r(c)$ is the circumradius r of t. According to Rule R3.1, r is either greater than $\sigma(c)$ or greater than $B \, l_{min}$, where l_{min} is the length of the shortest edge of t. In the first case, we have $r > \sigma(c) > \alpha \, \sigma(c) \geq \sigma'(c)$, since $\alpha < 1$. In the second case, we have $r > B \, l_{min}$. Among the vertices of the shortest edge of t, let p be the one inserted last. We have $r(p) \leq l_{min}$, thus $r > B \, r(p)$. Moreover, by C1, we have $r(p) \geq \sigma'(p)$. Hence, $r \geq B \, \sigma'(p)$. Since σ' is γ-Lipschitz, $B \, \sigma'(p)$ is at least $B \left(\sigma'(c) - \gamma \, \mathrm{d}(c,p) \right) \geq B \left(\sigma'(c) - \gamma r \right)$. It follows that $r \geq \frac{B}{1+B\gamma} \sigma'(c)$, which means that C1 is verified for c if B satisfies:

$$B \geq \frac{1}{1-\gamma} \tag{1}$$

To check C2, we notice that, in both cases ($r > \sigma(c)$ and $r > B \, l_{min}$), $r(c)$ is bounded from below by $\frac{B}{1+B\gamma} \sigma'(c)$. Let q be the point of $\partial\mathcal{O}$ that is closest[5] to c. We have $\delta(c) = \mathrm{d}(c,q) \geq \mathrm{d}(c, \mathcal{P}_{|\partial\mathcal{O}}) - \mathrm{d}(q, \mathcal{P}_{|\partial\mathcal{O}})$, where $\mathrm{d}(c, \mathcal{P}_{|\partial\mathcal{O}}) \geq r(c) \geq \frac{B}{1+B\gamma} \sigma'(c)$.

Since Rule R3 is applied only when R1 and R2 are fulfilled, Theorem 2 holds right before c is inserted. Hence, the surface Delaunay balls of $\mathcal{P}$ cover $\partial\mathcal{O}$, and q belongs to a surface Delaunay ball B'', of center c'' and radius r''. Let $p \in \mathcal{P}$ be a vertex of the facet of $\mathcal{D}_{|\partial\mathcal{O}}(\mathcal{P})$ circumscribed by B''. $\mathrm{d}(q,p)$ is at most $2r''$ because p and q both belong to B''. Due to Rule R2, r'' is at most $\alpha \, \sigma(c'')$, which is at most $\alpha(\sigma(q) + r'')$ since σ is 1-Lipschitz. It follows that $\mathrm{d}(q, \mathcal{P}_{|\partial\mathcal{O}}) \leq \mathrm{d}(q,p) \leq \frac{2\alpha}{1-\alpha} \sigma(q)$,

---

[5]if there are more than one such points, then choose any of them

which is less than $3\alpha\,\sigma(q)$ because $\alpha < \frac{1}{3}$. Moreover, since $\mathcal{P}_i \subseteq \mathcal{P}_{|\partial\mathcal{O}}$ is a 0.09-sample of $\partial\mathcal{O}$, $d(q, \mathcal{P}_{|\partial\mathcal{O}})$ is also bounded by $0.09\ d_M(q) = 3\,(0.03\,\sigma_0(q))$. Thus, $d(q, \mathcal{P}_{|\partial\mathcal{O}}) \le 3\,\sigma'(q)$.

Hence, $\delta(c) = d(c, q) \ge \frac{B}{1+B\gamma}\,\sigma'(c) - 3\,\sigma'(q)$. Since σ' is γ-Lipschitz, we have $\sigma'(q) \le \sigma'(c) + \gamma\,d(c, q)$, thus $\delta(c) = d(c, q) \ge \frac{1}{1+3\gamma}\left(\frac{B}{1+B\gamma} - 3\right)\sigma'(c)$. It follows that C2 is verified for c if B satisfies:

$$\frac{1}{1+3\gamma}\left(\frac{B}{1+B\gamma} - 3\right) \ge \frac{1}{1-\gamma}\ , \quad i.e.\ \ B \ge \frac{4}{1-5\gamma} \qquad (2)$$

- If Rule R3.2 is applied, then c is the center of a surface Delaunay ball B, of radius $r = r(c)$, containing the circumcenter c' of a tetrahedron t' of circumradius $r' \ge \frac{B}{1+B\gamma}\,\sigma'(c')$ (see case R3.1). Since σ' is γ-Lipschitz, we have $\sigma'(c') \ge \sigma'(c) - \gamma\,r(c)$. Moreover, the circumsphere of t' is empty, thus $r' \le d(c', p)$, for any point p of $\mathcal{P}$ lying on the bounding sphere of B. Since B contains both p and c', $d(c', p)$ is at most $2r(c)$. Hence,

$$2r(c) \ge d(c', p) \ge r' \ge \frac{B}{1+B\gamma}\,(\sigma'(c) - \gamma\,r(c)), \quad i.e.\ \ r(c) \ge \frac{B}{2+3B\gamma}\,\sigma'(c)$$

Therefore, C1 is verified for c if B satisfies:

$$B \ge \frac{2}{1-3\gamma} \qquad (3)$$

Moreover, C2 is verified because $c \in \partial\mathcal{O}$.

To conclude, Conditions C1 and C2 are verified if B and γ satisfy Eqs. (1), (2) and (3), which is granted if we choose $\gamma < \frac{1}{5}$ (and hence $\alpha < \frac{1}{5}$) and $B \ge \frac{4}{1-5\gamma}$. $\quad\square$

From now on, we assume that $\alpha < \frac{1}{5}$ and that $B \ge \frac{4}{1-5\gamma}$, where $\gamma = \max\{\alpha,\ 0.03\}$. Given $p \in \mathcal{P}$, we define $B(p)$ as the open ball centered at p, of radius $\varrho(p) = \frac{1}{2(1+\gamma)}\,\sigma'(p)$.

Lemma 7. *The balls $(B(p))_{p\in\mathcal{P}}$ are pairwise disjoint.*

Proof. Given two points p, q of $\mathcal{P}$, we assume without loss of generality that q was inserted in $\mathcal{P}$ before p. The distance from p to q is then at least $r(p)$. By Lemma 6 (Condition C1), we have $r(p) \ge \sigma'(p)$, which is at least $\frac{1}{1+\gamma}\,\sigma'(q)$ since σ' is γ-Lipschitz. Thus, $d(p, q) \ge \frac{1}{1+\gamma}\max\{\sigma'(p), \sigma'(q)\}$. It follows that $\frac{1}{2(1+\gamma)}\,\sigma'(p) + \frac{1}{2(1+\gamma)}\,\sigma'(q) \le d(p, q)$, which means that $B(p)$ and $B(q)$ are disjoint. $\quad\square$

To compute an upper bound on the size of the output point sample, we need another result, which states that every ball $B(p)$ lies partly inside $\mathcal{O}$, and that the volume of the part included in $\mathcal{O}$ can be lower-bounded.

Lemma 8. *For any $p \in \mathcal{P}$, $B(p) \cap \mathcal{O}$ contains a ball of radius $\frac{1}{2}\,\varrho(p)$.*

Proof. We distinguish between two cases:

• If p lies inside $\mathcal{O}$, then, according to Lemma 6 (Condition C2), the distance $\delta(p)$ from p to $\partial\mathcal{O}$ is at least $\frac{1}{1-\gamma}\sigma'(p)$, which is greater than $\frac{1}{2}\varrho(p)$. Hence, the ball centered at p, of radius $\frac{1}{2}\varrho(p)$, is included in $\mathcal{O}$.

• Otherwise, p lies on $\partial\mathcal{O}$. There are two medial balls B_i and B_o tangent to $\partial\mathcal{O}$ at p. One of them (say B_i) is included in $\mathcal{O}$, whereas the other one is included in $\mathbb{R}^3 \setminus \mathcal{O}$. Since B_i is a medial ball, its radius is at least $\mathrm{d}_M(p) > \sigma'(p)$. Moreover, the radius of $B(p)$ is $\varrho(p) < \sigma'(p)$. It follows that the intersection of $B(p)$ with B_i contains a ball of radius $\frac{1}{2}\varrho(p)$. □

Theorem 3. *If $\alpha < \frac{1}{5}$ and $B \geq \frac{4}{1-5\gamma}$ (where $\gamma = \min\{\alpha,\ 0.03\}$), then the output point sample $\mathcal{P}$ verifies:*

$$|\mathcal{P}| = O\left(\iiint_{\mathcal{O}} \frac{dx}{\sigma_0^3(x)} + \frac{1}{\alpha^3}\iiint_{\mathcal{O}} \frac{dx}{\sigma^3(x)}\right)$$

where σ_0 depends only on $\mathcal{O}$ (not on σ).

Proof. We use a standard scheme [BO05]. We will bound the integral of $1/\sigma'^3$ over $\mathcal{O}$, where σ' is the minimum of $\alpha\,\sigma$ and of $0.03\,\sigma_0$. Since $B(p)\cap\mathcal{O} \subseteq \mathcal{O}$ for any $p \in \mathcal{P}$, we have $\iiint_{\mathcal{O}} \frac{dx}{\sigma'^3(x)} \geq \iiint_{\bigcup_{p\in\mathcal{P}}(B(p)\cap\mathcal{O})} \frac{dx}{\sigma'^3(x)}$. Moreover, the balls $B(p)$ are pairwise disjoint, by Lemma 7, thus $\iiint_{\bigcup_{p\in\mathcal{P}}(B(p)\cap\mathcal{O})} \frac{dx}{\sigma'^3(x)} = \sum_{p\in\mathcal{P}}\iiint_{(B(p)\cap\mathcal{O})} \frac{dx}{\sigma'^3(x)}$. In addition, since σ' is γ-Lipschitz, we have:

$$\forall x \in B(p),\ \sigma'(x) \leq \sigma'(p) + \gamma\,\mathrm{d}(x,p) \leq \sigma'(p) + \gamma\,\varrho(p) = \left(1 + \frac{\gamma}{2(1+\gamma)}\right)\sigma'(p)$$

It follows that $\sum_{p\in\mathcal{P}}\iiint_{(B(p)\cap\mathcal{O})} \frac{dx}{\sigma'^3(x)} \geq \sum_{p\in\mathcal{P}} \frac{\mathrm{Vol}(B(p)\cap\mathcal{O})}{\left(1+\frac{\gamma}{2(1+\gamma)}\right)^3\sigma'^3(p)}$. Now, by Lemma 8, the volume of $B(p) \cap \mathcal{O}$ is at least $\frac{4}{3}\pi\frac{1}{64(1+\gamma)^3}\sigma'^3(p)$, which yields:

$$\sum_{p\in\mathcal{P}} \frac{\mathrm{Vol}\left(B(p)\cap\mathcal{O}\right)}{\left(1+\frac{\gamma}{2(1+\gamma)}\right)^3\sigma'^3(p)} \geq \sum_{p\in\mathcal{P}} \frac{\frac{\pi}{48(1+\gamma)^3}\sigma'^3(p)}{\left(1+\frac{\gamma}{2(1+\gamma)}\right)^3\sigma'^3(p)} = \frac{\pi}{6\left(2+3\gamma\right)^3}|\mathcal{P}|$$

which is at least $\frac{1}{34}|\mathcal{P}|$ since $\gamma < \frac{1}{5}$. Hence, $|\mathcal{P}|$ is at most $34 \iiint_{\mathcal{O}} \frac{dx}{\sigma'^3(x)}$. Now, $\sigma'(x)$ is defined as the minimum of $\alpha\,\sigma(p)$ and of $0.03\,\sigma_0(p)$, which are positive functions. It follows that $\iiint_{\mathcal{O}} \frac{dx}{\sigma'^3(x)}$ is at most $\iiint_{\mathcal{O}} \frac{dx}{\alpha^3\,\sigma^3(x)} + \iiint_{\mathcal{O}} \frac{dx}{0.03^3\,\sigma_0^3(x)}$, which ends the proof of the theorem. □

Since σ_0 and σ are both positive and continuous over $\bar{\mathcal{O}}$, which is compact, the bound given in Theorem 3 is finite. The next result follows, since the algorithm inserts one point in $\mathcal{P}$ per iteration and never removes points from $\mathcal{P}$.

Corollary 2. *The algorithm terminates.*

6 Practicality of the algorithm

6.1 Sizing field

The meshing algorithm presented in the previous sections takes as input a sizing field $\sigma : \bar{\mathcal{O}} \to \mathbb{R}^+$ which, for the purpose of the analysis in Section 5, is assumed to be 1-Lipschitz. This section explains how to deal with user-defined sizing fields that are not Lipschitz or not defined everywhere in $\bar{\mathcal{O}}$.

Let us assume that the user wants a mesh whose grading conforms to a sizing field σ_u that is not 1-Lipschitz. Then we can use the technique of Miller, Talmor and Teng [MTT99] to derive from σ_u a new sizing field σ'_u that is 1-Lipschitz:

$$\forall x \in \bar{\mathcal{O}}, \ \sigma'_u(x) = \inf \ \{\mathrm{d}(x, x') + \sigma_u(x') \mid x' \in \bar{\mathcal{O}}\}$$

Notice that $\sigma'_u(p) \leq \sigma_u(p)$, $\forall p \in \bar{\mathcal{O}}$. The field σ'_u is the best 1-Lipschitz approximation of σ_u [ACSYD05], because any 1-Lipschitz function that is pointwise at most σ_u is also pointwise at most σ'_u.

The meshing algorithm can be run using the sizing field σ'_u, however it is not necessary to compute σ'_u inside $\mathcal{O}$. Indeed, the algorithm requires an evaluation of the sizing field at internal points only in Rule R3, in order to trigger the refinement of a tetrahedron. The refinement of a tetrahedron t is triggered either for a size reason (*i.e.* $r \geq \sigma(c)$, where c and r are respectively the circumcenter and the circumradius of t) or for a shape reason (*i.e.* the radius-edge ratio of t is greater than B). A careful look at the proof of termination shows that a 1-Lipschitz lower bound on the circumradius of tetrahedra refined for size reason is sufficient for the proof. Then, since $\sigma'_u(p) \leq \sigma_u(p)$ for any $p \in \mathcal{P}$, the proofs still hold if rule R3 is triggered by the condition $r \geq \sigma_u(p)$. Besides saving some sizing field evaluations, this variant of the algorithm constructs a sparser mesh whose density conforms to the user-defined sizing field, with a grading bounded only by the bound B on the radius-edge ratio.

In the case where the user has no particular sizing requirements, the 1-Lipschitz sizing field used in the analysis is the field σ_0 introduced at the beginning of Section 5. Here again, the algorithm does not need to evaluate σ_0 inside $\mathcal{O}$. It may simply skip the size test for tetrahedra and consider for refinement only the badly-shaped tetrahedra (*i.e.* those with a radius-edge ratio greater than B). Since the occasions of refining tetrahedra are fewer in this variant than in the original version of the algorithm, it is clear that this variant also terminates. Its output is a mesh whose sizing is $\varepsilon \, \mathrm{d}_M$ on $\partial\mathcal{O}$ and grows as fast as possible (regarding the bound on the radius-edge ratio) when moving towards the medial axis.

In any case, the algorithm needs to compute $\mathrm{d}_M(x)$ at some points on $\partial\mathcal{O}$ in order to test whether the precondition of Rule R2 is met. This point has been adressed in [BO05], where the authors propose an approximation of $\mathrm{d}_M(x)$ based on the notion of λ-medial axis introduced by [CL04].

6.2 Sliver removal

Optimizing radius-edge ratios prevents our meshes from containing any bad tetrahedra, except possibly slivers. Following the definition of Edelsbrunner and Guoy [EG02], we say that a tetrahedron t is bad if the ratio between the radius of its inscribed

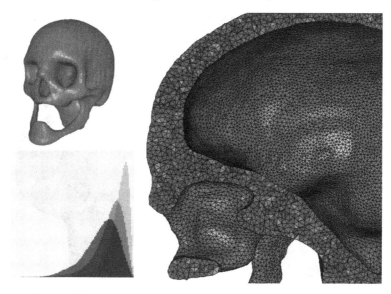

Fig. 1. Skull model: $89,245$ vertices and $442,542$ tetrahedra.

sphere and its circumradius is less than 0.15, which corresponds to a minimal dihedral angle less than 5 degrees. This ratio is called the *radius-radius ratio* of t. In order to remove slivers from our meshes, we use the pumping algorithm of [CDE+00] as a post-process. This algorithm consists in assigning carefully chosen weights to the vertices of the mesh, so that their weighted Delaunay triangulation contains as few slivers as possible. Although the guaranteed theoretical bound on radius-radius ratios is known to be miserably low [CDE+00], the method is efficient in practice and generates almost sliver-free meshes [EG02].

7 Implementation and results

The algorithm has been implemented in C++, using the geometric library CGAL [CGAL] which provided us with an efficient and flexible implementation of the three-dimensional Delaunay triangulation.

Figures 1 and 2 show two meshes generated by our algorithm coupled with the post-processing step described in Section 6.2. Each figure is composed of two views of the output mesh: one shows the boundary (top left), the other shows a zoom on the interior, cut by a plane[6] (right). The bottom-left corner of each figure shows the distribution of the radius-radius ratios of the tetrahedra, represented on a linear scale ranging from 0 to $\frac{1}{3}$ (which corresponds to the radius-radius ratio of a regular tetrahedron). The histograms are normalized with respect to area, so that we can make fair comparisons between meshes of different sizes.

In Figure 1, the boundary of the domain is a level set in a 3D grey-scaled image. Its diameter is about 280 millimeters, and its reach approximately 1 millimeter.

---

[6]The screenshots were obtained using Medit [Med].

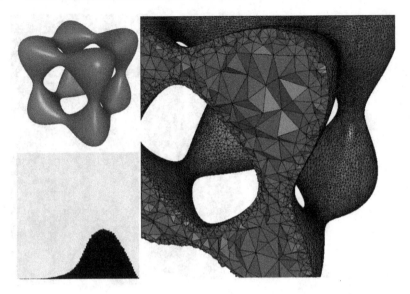

Fig. 2. Tanglecube model: $57,293$ vertices and $226,010$ tetrahedra.

Although our theoretical results require strict conditions on σ, α and B, in practice the algorithm works well under weaker conditions. For instance, in this example we used a uniform sizing field of 2 millimeters, with $\alpha = 1$ and $B = 2$, which is far beyond the theoretical limits. Note that the topology of the domain has been captured, and that the boundary has been accurately approximated.

The radius-radius ratios distribution of our algorithm (in medium grey) has been superimposed with those obtained by two other algorithms: the unit edge mesher of [GHS90, GHS91] (in dark grey), which priviledges the running time (approximately ten seconds for the skull model on a Pentium IV at 1.7 GHz), and the variational mesher of [ACSYD05] (in light grey), which priviledges the quality of the output. These two programs, run with our initial surface mesh $\mathcal{D}_{|\partial\mathcal{O}}(\mathcal{P}_i)$ as input, generated approximately the same number of vertices as our mesher. It turns out that, in practice, our algorithm carries out a good compromise between running time and quality of the output.

In Figure 2, the boundary of the domain is an algebraic surface of degree four and genus five, called *tanglecube*. We used no sizing field inside the domain and $\sigma_0 = 0.09 \, d_M$ on its boundary, as described in Section 6.1. The bound B on the radius-edge ratios was set to 2, which enforced the grading of the output mesh. Although the overall appearance of the radius-radius ratios distribution is deteriorated due to the non-uniformity of the sizing field, the quality of the output mesh remains quite acceptable.

8 Conclusion

We have introduced a new method for meshing three-dimensional domains bounded by smooth surfaces. This method is a combination of existing work on smooth surface meshing on the one hand, on piecewise linear volume meshing on the other hand. We have given theoretical guarantees on the output of the algorithm, regarding its size, the quality of its elements, and the accuracy of the approximation of the original object. We have also provided experimental evidence that the algorithm works well in practice.

The main advantage of our method is that it samples the object $\mathcal{O}$ and its boundary $\partial\mathcal{O}$ at the same time, which lets the user free to decide which density he wants inside $\mathcal{O}$ but also on $\partial\mathcal{O}$. Moreover, the algorithm takes as input the object itself, which makes it independent from any original discretization of the object's boundary and allows to approximate $\partial\mathcal{O}$ within any desired accuracy. In addition, the required a-priori knowledge of $\mathcal{O}$ is minimal, since the algorithm needs only to know the object through an oracle capable of answering two basic geometric questions.

Notice that our algorithm is also able to mesh domains with smooth constraints. The difference between a constraint and a boundary is that both sides of the constraint have to be meshed, whereas only one side of the boundary has to. It turns out that our proofs hold for constraints as well.

Several possible extensions of the work presented in this paper should be addressed in a near future:

- The bound in Theorem 3 depends highly on α, whereas the latter influences the density of the mesh only in the vicinity of $\partial\mathcal{O}$. It would be nice to decompose the bound into two terms: one depending on α and on the integral of $1/\sigma^2$ over $\partial\mathcal{O}$, the other depending on the integral of $1/\sigma^3$ over $\mathcal{O}$.
- Our theoretical results assume that the input domain has a smooth boundary. However, the method has been tested with some success on domains whith singularities. Further work for a systematic handling of singularities is in progress. Another related direction of research is to mesh domains with non-manifold constraints.

[AB99]   N. Amenta and M. Bern. Surface reconstruction by Voronoi filtering. *Discrete Comput. Geom.*, 22(4):481–504, 1999.

[AB02]   D. Attali and J.-D. Boissonnat. Approximation of the medial axis. Technical Report ECG-TR-124103-01, INRIA Sophia-Antipolis, 2002.

[ACK01]   N. Amenta, S. Choi, and R. K. Kolluri. The power crust, unions of balls, and the medial axis transform. *Comput. Geom. Theory Appl.*, 19:127–153, 2001.

[ACSYD05] P. Alliez, D. Cohen-Steiner, M. Yvinec, and M. Desbrun. Variational tetrahedral meshing. In *Proceedings SIGGRAPH*, 2005.

[BGH+97]   H. Borouchaki, P.-L. George, F. Hecht, P. Laug, and E. Saltel. Delaunay mesh generation governed by metric specifications. Part I: algorithms. *Finite Elem. Anal. Des.*, 25(1-2):61–83, 1997.

[BGM97]   H. Borouchaki, P.-L. George, and B. Mohammadi. Delaunay mesh generation governed by metric specifications. Part II: applications. *Finite Elem. Anal. Des.*, 25(1-2):85–109, 1997.

[BO05]   J.-D. Boissonnat and S. Oudot. Provably good sampling and meshing of surfaces. *Graphical Models, special issue of Solid Modeling '04*, 2005. Article in press.

[BOG02]   C. Boivin and C. Ollivier-Gooch. Guaranteed-quality triangular mesh generation for domains with curved boundaries. *International Journal for Numerical Methods in Engineering*, 55(10):1185–1213, 2002.

[CCY04]   D. Cohen-Steiner, E. Colin de Verdière, and M. Yvinec. Conforming delaunay triangulations in 3d. *Computational Geometry: Theory and Applications*, pages 217–233, 2004.

[CD02]   S.-W. Cheng and T. K. Dey. Quality meshing with weighted delaunay refinement. In *SODA '02: Proceedings of the thirteenth annual ACM-SIAM symposium on discrete algorithms*, pages 137–146, 2002.

[CDE+00]   S.-W. Cheng, T. K. Dey, H. Edelsbrunner, M. A. Facello, and S.-H. Teng. Silver exudation. *J. ACM*, 47(5):883–904, 2000.

[CDRR04]   S.-W. Cheng, T. K. Dey, E. A. Ramos, and T. Ray. Quality meshing for polyhedra with small angles. In *SCG '04: Proceedings of the twentieth annual symposium on Computational geometry*, pages 290–299. ACM Press, 2004.

[CGAL]   The CGAL Library. Release 3.1 (http://www.cgal.org).

[Che93]   L. P. Chew. Guaranteed-quality mesh generation for curved surfaces. In *SCG '93: Proceedings of the ninth annual symposium on Computational geometry*, pages 274–280. ACM Press, 1993.

[CL04]   F. Chazal and A. Lieutier. Stability and homotopy of a subset of the medial axis. In *SM '04: Proceedings of the 9th ACM symposium on Solid modeling and applications*, 2004.

[CP03]   S.-W. Cheng and S.-H. Poon. Graded conforming delaunay tetrahedralization with bounded radius-edge ratio. In *SODA '03: Proceedings of the fourteenth annual ACM-SIAM symposium on Discrete algorithms*, pages 295–304. Society for Industrial and Applied Mathematics, 2003.

[DG03]   T. K. Dey and S. Goswami. Tight cocone: a water-tight surface reconstructor. In *SM '03: Proceedings of the eighth ACM symposium on Solid modeling and applications*, pages 127–134. ACM Press, 2003.

[DOS99]   S. Dey, R. O'Bara, and M. S. Shepard. Curvilinear mesh generation in 3d. In *Proc. 8th Internat. Meshing Roundtable*, pages 407–417, 1999.

[EG02]   H. Edelsbrunner and D. Guoy. An experimental study of sliver exudation. *Engineering With Computers, Special Issue on 'Mesh Generation' (10th IMR 2001)*, 18(3):229–240, 2002.

[FBG96]   P. J. Frey, H. Borouchaki, and P.-L. George. Delaunay tetrahedrization using an advancing-front approach. In *Proc. 5th International Meshing Roundtable*, pages 31–43, 1996.

[Fed70]   H. Federer. *Geometric Measure Theory*. Springer-Verlag, 1970.

[GHS90]   P.-L. George, F. Hecht, and E. Saltel. Fully automatic mesh generator for 3d domains of any shape. *Impact of Comuting in Science and Engineering*, 2:187–218, 1990.

[GHS91]   P.-L. George, F. Hecht, and E. Saltel. Automatic mesh generator with specified boundary. *Computer Methods in Applied Mechanics and Engineering*, 92:269–288, 1991.

[LBG96]   P. Laug, H. Bourouchaki, and P.-L. George. Maillage de courbes gouverné par une carte de métriques. Technical Report RR-2818, INRIA Rocquencourt, 1996.

[LT01]   X.-Y. Li and S.-H. Teng. Generating well-shaped delaunay meshed in 3d. In *SODA '01: Proceedings of the twelfth annual ACM-SIAM symposium on Discrete algorithms*, pages 28–37. Society for Industrial and Applied Mathematics, 2001.

[Mav00]   F. Mavriplis. Cfd in aerospace in the new millenium. *Canadian Aeronautics and Space Journal*, 46(4):167–176, 2000.

[Med]   Medit, a scientific visualization tool (`http://www-rocq.inria.fr/gamma/medit/medit.html`).

[MTT99]   G. L. Miller, D. Talmor, and S.-H. Teng. Data generation for geometric algorithms on non uniform distributions. *International Journal of Computational Geomety and Applications*, 9(6):577–599, 1999.

[Rup95]   J. Ruppert. A Delaunay refinement algorithm for quality 2-dimensional mesh generation. *J. Algorithms*, 18:548–585, 1995.

[She98]   J. R. Shewchuk. Tetrahedral mesh generation by Delaunay refinement. In *Proc. 14th Annu. ACM Sympos. Comput. Geom.*, pages 86–95, 1998.

[She02]   J. R. Shewchuk. Delaunay refinement algorithms for triangular mesh generation. *Computational Geometry: Theory and Applications*, 22:21–74, 2002.

[SU01]   A. Scheffer and A. Ungor. Efficient adaptive meshing of parametric models. In *Proc. 6th ACM Sympos. Solid Modeling and Applications*, pages 59–70, 2001.

[TW00]   S.-H. Teng and C. W. Wong. Unstructured mesh generation: Theory, practice, and perspectives. *Int. J. Comput. Geometry Appl.*, 10(3):227–266, 2000.

An Approach for Delaunay Tetrahedralization of Bodies with Curved Boundaries

Sergey N. Borovikov[1], Igor A. Kryukov[2], and Igor E. Ivanov[1]

[1] Moscow Aviation Institute (State Technical University), Moscow, Russia
s_borovikov@mtu-net.ru
[2] Institute for Problems in Mechanics, Russian Academy of Sciences, Moscow,
Russia

Summary. Problem of tetrahedral meshing of three-dimensional domains whose boundaries are curved surfaces is wide open. Traditional approach consists in an approximation of curved boundaries by piecewise linear boundaries before mesh generation. As the result mesh quality may deteriorate. This paper presents a technique for Delaunay-based tetrahedralization in which a set of constrained facets is formed dynamically during face recovery and mechanisms for mutual retriangulation of the curved faces and the tetrahedralization are suggested. The proposed algorithm is constructed in such a way that a facet that was once added in the set of constrained facets is never split into small triangles. It allows retaining the high quality of surface mesh in the tetrahedralization, because during boundary recovery the surface mesh on the curved faces and the tetrahedralization are refined conjointly.

Key words: Delaunay tetrahedralization, curved boundary, boundary recovery, tetrahedral meshes, meshing of parametric models

List of Abbreviations and Symbols

t_{ps}^{Γ} Triangulation in the parametric space of some curved face Γ of input body for which the Delaunay tetrahedralization is being constructed.

t_{3d}^{Γ} An image of t_{ps}^{Γ} in three-dimensional space.

T_{Γ} A set of triangular faces of tetrahedra that are representation of curved face Γ in the tetrahedralization (usually does not coincide with t_{3d}^{Γ}).

1 Introduction

The Delaunay tetrahedralization is widely used for the mesh generation. But the Delaunay tetrahedralization is defined for bodies with piecewise linear boundaries. And when it is needed to tetrahedralize a body with curved boundary usually the input body is approximated with facets and piecewise linear complex (PLC) is created. Then the tetrahedralization is build from the linear model using either the advancing front method or the Delaunay triangulation method [GE91, DU04]. This paper concerns the Delaunay tetrahedralization method in which tetrahedralization is build from the point set and then boundary recovery is performed, while the advancing front method is not considered.

Direct use of the approach when a body with curved boundaries is approximated with piecewise linear elements has two disadvantages. First disadvantage is connected with the fact that the piecewise linear approximation of the input body might have geometrical artifacts (e.g. degenerate and disconnected facets). It might happen, especially if the linear model is taken from a CAD system. However this disadvantage is not essential since 1) there are methods that allow generating surface mesh without geometrical artifacts and 2) CAD data repairing methods are developed.

Second and more serious disadvantage is the fact that quality of resulting tetrahedral mesh is very sensitive to the quality of the initial surface mesh. It occurs because curved face is approximated with linear elements and all linear elements both facets and edges must present in the final tetrahedralization. And if the piecewise linear approximation contains a facet with an acute angle, then this angle remains in tetrahedralization. Thus surface mesh of tetrahedralization cannot be better then the input surface mesh because during tetrahedralization remeshing is possible only within a facet. However, it should be pointed out that there is a common approach when small coplanar facets are unified into one facet. And only this new facet is used during mesh generation instead of the set of the small facets. In some cases it allows significantly improving the quality of the final tetrahedralization. But facets from the approximation of curved face can rarely be unified into one.

Also it should be mentioned that there are methods that allow reconstructing of surface mesh and improving its quality at the same time (e.g. [BE02, FR00]). But all improvements are made before mesh generation and during the tetrahedralization the reconstruction of surface mesh is not executed and each facet is considered individually what can lead to poor quality of the final mesh. Quality deteriorates when Steiner points are inserted into the facets.

Nevertheless, the approach when a body with curved boundaries is approximated with piecewise linear elements is widely used. Usually for general bodies if the initial surface mesh is small enough, then after the tetrahedralization of an initial point set most of facets is already recovered and others can be recovered by adding a few Steiner points. So the quality does not become much worse. Problems can arise 1) if the size of the initial surface mesh is larger than required size of the tetrahedral mesh or 2) in the place where two faces subtend a sharp dihedral angle. In these cases many Steiner are generally added.

This paper considers issues of the Delaunay constrained tetrahedralization for bodies with curved boundaries given analytically, i.e. for each edge a corresponding curve is specified and for each face corresponding parametrization of a surface is specified. Major part of the paper is focused on the problem of boundary recovery.

In comparison with the described above approach when a set of constrained facets F is specified before mesh generation and boundary recovery is performed for the set F, this paper suggests an approach in which the set F is formed dynamically during face recovery and facets (triangles) are added to F when it is known that they have already been recovered. Furthermore, the algorithm is constructed in such a way that a triangle that was once added in F is never split into small triangles. Hence the surface mesh is transferred to the tetrahedralization without changes.

Face recovery is based on comparison of the face triangulation in the parametric space with the tetrahedral mesh. However, it is not required that each triangle of the two-dimensional triangulation in the parametric space presents as a face of some tetrahedron in the tetrahedralization because it is not always possible to get such conformity. An algorithm, which is described in section 3, seeks for the triangles that will be an approximation of the curved face in three-dimensional space. The search is carried out among faces of tetrahedrons of the constructed tetrahedralization. Requirements for such faces are discussed in section 2. A set of the appropriate faces of tetrahedrons may not only consist of images of the triangles of the triangulation in the parametric space. If all appropriate faces are found (it means that a curved face is recovered), then they are added to the set of the constrained facets F, otherwise the curved face is additionally refined. A crucial difference of the suggested approach is the fact that the facets from the approximation of a curved face are added to F if and only if all of them present in the tetrahedralization what means that the curved face is recovered. It allows retaining the high quality of surface mesh in the tetrahedralization, because during boundary recovery the surface mesh on the curved faces and the tetrahedralization are refined conjointly.

As it was noted before, small coplanar facets can be unified into one facet. And only new facet is used during mesh generation instead of the set of the small facets. Actually, the paper proposes an extension of this approach and the facets are unified if they belong to the same curved face and mechanisms for mutual retriangulation of the curved faces and the tetrahedralization are suggested. We compare face triangulation in the parametric space with the tetrahedral mesh as if we tried to reconstruct the triangulation in the parametric space in accordance with the configuration really occurred in the three-dimensional space.

2 Initial mesh generation

The generation of the constrained Delaunay tetrahedralization begins from the Delaunay tetrahedralization of the point set lying on boundary of body. So an initial triangular surface mesh must be constructed prior tetrahedral mesh generation. Since the surface mesh will be compared with the tetrahedralization it is important to choose appropriate surface meshing method, which must satisfy two criteria. Firstly, the generated surface mesh must be as closer as possible to the Delaunay triangulation on a curved surface [CH93]. Secondly, method for surface meshing must allow inserting a new vertex into the existed surface triangulation. The authors use two-dimensional anisotropic mesh generation in the parametric space of surface [CH97, BO96].

As the result of the surface mesh generation a set of 3D vertices is obtained. Next step (see Fig. 1) is the Delaunay tetrahedralization for the set of vertices. Specifically, a three-dimensional algorithm based on flips was used [JO91]. The acquired tetrahedralization will be an initial mesh. As the Delaunay triangulation is the strictly defined triangulation, which does not allow vertices to be connected arbitrary, next step is the boundary recovery in the three-dimensional space.

In the spatial case the boundary recovery problem consists of two tasks: edge recovery and face recovery. In our algorithm the recovery of missing edges must be done prior to the face recovery. Edge recovery is not the complex task for the current state of mesh generation and it has been discussed in many papers. In the particular case a variant of Rupert's algorithm [SC02] was used with modifications for handling curved boundaries [BO02] and with some generalizations to three dimensions (e.g. diametrical spheres were used instead of diametrical circles).

In our algorithm constructing Delaunay tetrahedralization requires the maintenance of tetrahedralization and a separate triangulation in the parametric space of each curved face.

Generate surface mesh of input body;
Build the Delaunay tetrahedralization of the vertices lying on surface of body;
Perform recovery of missing curved edges;
Perform first stage of curved faces recovery;
For each curved face Γ_k do
      Perform second stage of curved face recovery for face Γ_k; enddo
Insert Steiner points inside of body;

Fig. 1. An overview of the algorithm for constructing Delaunay tetrahedralization of bodies with curved boundaries.

3 Recovery of missing curved faces

Consider the recovery of missing curved faces, which is the most complex problem of the boundary recovery. First of all, it should be discussed what means that a curved face is recovered. A curved edge (face) will be considered to be recovered in the tetrahedralization if there is a piecewise linear approximation of the curved edge (face) and each linear element of it is present in the tetrahedralization. For a curved edge linear elements are segments, which present as edges of tetrahedra. For a curved face Γ linear elements are triangles and they must present as faces of tetrahedra in the tetrahedralization. A set T_Γ of such triangular faces of tetrahedra must satisfy the requirements:

   I. Vertices of the triangular faces lie on the curved face Γ.
  II. Each 3D vertex that are an image of some vertex from the triangulation t_{ps}^Γ in the parametric space belongs to at least one triangle from the set T_Γ.
 III. Interiors of the triangular faces do not intersect each other.
 IV. Each edge of triangular face from the set T_Γ that is a segment of the approximation of some external curved edge of Γ belongs to only one triangular face from the set T_Γ. Such edges are called external.
  V. Each edge of triangular face from the set T_Γ that is not external belongs to only two triangular faces from the set T_Γ. Such edges are called internal.
 VI. Consider triangular faces from T_Γ that share the same vertex v lying on Γ and their edges. These triangular faces form a subset S. For each edge e that has the vertex v sum how many triangular faces from the subset S have the edge e. Internal edge must belong to only two triangular faces from the subset S, whereas external edge must belong to only one triangular face. This requirement is implied by conditions IV and V, nevertheless, it has been formulated here since it will be used in the section 3.2.

These six requirements are substitute for conditions of absence of gaps and self-intersections in comparison with the definition of two-dimensional triangulation.

Thus for the curved face Γ three different objects are defined: triangulation t_{ps}^{Γ} in the parametric space, its image in three-dimensional space t_{3d}^{Γ} and the set T_{Γ}, which is the representation (or approximation) of the curved face Γ in the tetrahedralization. Generally, the sets T_{Γ} and t_{3d}^{Γ} do not coincide.

During face recovery our goal is to find the set T_{Γ} (if it exists) or reconstruct the mesh in such a way that the approximation T_{Γ} is appeared in the tetrahedralization. Faces are recovered in two stages. During the first stage all curved faces are handled together while during the second stage curved faces are handled one by one.

3.1 First stage of curved faces recovery

In some algorithms for constructing Delaunay tetrahedralization equatorial spheres are used for the boundary recovery [SC98]. The equatorial sphere of a triangle is the smallest sphere passes through the three vertices of the triangle. A triangle is encroached if a vertex lies inside or on its equatorial sphere. If a triangle is not encroached it appears in tetrahedralization as a triangular face of tetrahedron. This idea with some modifications can be used for curved faces recovery.

Definition 1. *A vertex v and a triangle t is said to be adjacent if at least one vertex of t and v lie on the same curved face (see Fig. 2).*

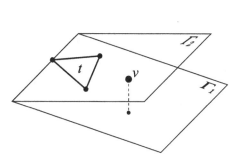

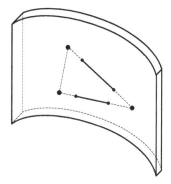

Fig. 2. A triangle t lying on a face Γ_1 and a vertex v lying on a face Γ_2 are adjacent.

Fig. 3. If two curved faces are close to each other then it may be situation when a triangle of t_{3d}^{Γ} of one face intersects other face. The first stage allows significantly reducing a number of such triangles or eliminating them at all.

During first stage triangles of triangulations t_{3d}^{Γ} and their equatorial spheres are examined. The equatorial sphere of each triangle can contain only allowed vertices. All adjacent vertices of a triangle are allowed to lie inside its equatorial sphere. If the equatorial sphere of a triangle t^{3d} contains a prohibited vertex v^{3d} then a corresponding triangle t^{ps} in the parametric space is split by inserting a new vertex w^{ps}

at the center of circumscribed ellipse of t^{ps} [CH97, BO96]. And a three-dimensional vertex w^{3d} that is an image of w^{ps} is also inserted into the tetrahedralization. Insertion a new vertex both into t^{Γ}_{ps} and into the tetrahedralization implies performing all necessary actions (e.g. flips) to maintain the Delaunay property.

If the new vertex w^{ps} falls within a triangle Δt^{ps}; and if Δt^{3d} and v^{3d} are adjacent, the vertex w^{ps} is not inserted; instead v^{3d} is allowed to lie inside of the equatorial sphere of t^{3d}.

The first stage is executed until all triangles of all curved faces have the equatorial spheres containing only allowed vertices. Since we do not require that the equatorial spheres must strictly be empty the endless cycle of infinite encroachment does not occur and the first stage always terminates. (All modifications have been introduced only to prevent the endless cycle.)

In case of curved boundaries the first stage has very important meaning. If two curved faces are close to each other then a situation is possible when a triangle of t^{Γ}_{3d} of one curved face intersects other curved face (see Fig. 3). The first stage allows significantly reducing a number of such triangles or eliminating them at all (see Fig. 4). Actually, if two curved faces Γ_1 and Γ_2 are close to each other then it is not generally allowed equatorial spheres of triangles of Γ_1 contain vertices of Γ_2. So two triangulations of Γ_1 and Γ_2 "feel" each other and they are conjointly refined during the first stage.

Fig. 4. An example of how the algorithm works with thin-walled bodies. At left, a coarse mesh that consists of triangulations t^{Γ}_{3d} of input body before the first stage of recovery is depicted. The mesh has self-intersections. At right, a tetrahedralization after face recovery (including the second stage) is depicted. During the first stage the surface mesh had been refined in such a way that it became without self-intersections. Although the final tetrahedralization is still coarse but it is valid.

A direct check on self-intersections is not performed because of computational complexity of this task. Thus there is no theoretical guarantee that self-intersections will be removed during the first stage in all cases. Nevertheless, our experiments with thin-walled bodies demonstrated that the algorithm perfectly works with such bodies. Moreover, some self-intersections can be removed during the second stage of face recovery.

It should be mentioned that after the first stage the tetrahedralization is strictly Delaunay.

3.2 Second stage of curved face recovery

Curved faces are recovered one by one and recovery procedure does not switch to the next curved face until previous face is fully recovered. The recovery procedure does not spoil once recovered faces. Each curved face Γ is recovered as follows. First, a list that contains triangular faces of tetrahedra being images of triangles of the triangulation t_{ps}^{Γ} in the parametric space is created. These triangular faces are marked as triangles belonging to the approximation T_{Γ} of the curved face Γ. If each triangle in the parametric space is present in the tetrahedralization as a triangular face and an obtained set of triangular faces satisfies conditions of topological integrity of mesh (see below), then the curved face is already recovered in the tetrahedralization. If the list is empty, i.e. no triangles of t_{ps}^{Γ} are present in the tetrahedralization, then it means that the triangulation in the parametric space t_{ps}^{Γ} differs sharply from the Delaunay triangulation. So it is why so important to choose an appropriate method for surface meshing. In this case the additional refinement of the mesh t_{ps}^{Γ} is performed and the refinement is stopped if at least one triangle of the triangulation t_{ps}^{Γ} is present in the tetrahedralization as a triangular face of some tetrahedron. Usually a few additional points need to be added. It has to be mentioned that face triangulation t_{ps}^{Γ} is refined together with the tetrahedralization.

Then a main loop of recovery procedure is started. While the list is non-empty the following actions take place. A triangular face t is removed from the beginning of the list. For each internal edge e of t other triangular face that will be a neighbor for the triangular face t through the edge e has to be found. First, the search is performed among already marked triangles. And if such triangular face exists (it can be only one), then corresponding neighbor is found and next internal edge of t is considered. Certainly, many triangular faces of tetrahedralization can share the same edge e but only two marked triangular faces can share the edge e (see the requirement V) including the triangular face t. So for a marked triangular face through an internal edge only one neighbor is possible. The requirement V is always fulfilled for marked triangles if surface of face does not have singularities and seam edges (see details in section 3.3).

If the neighbor has not been found among marked triangles, then triangular faces, satisfying conditions of topological integrity of mesh, are searched among all triangular faces that have the edge e. These triangular faces will be candidates to be the neighbor for the triangular face t through the edge e.

Elucidate aforesaid in figure 6. At left a fragment of triangulation t_{ps}^{Γ} of curved face in the parametric space is shown, while at right a corresponding configuration in the three-dimensional space. For a triangle AMN from the parametric space a corresponding triangular face $A'M'N'$, which is a face of some tetrahedron of tetrahedralization, exists. So the triangular face $A'M'N'$ is marked as belonging to the approximation T_{Γ} of the curved face and pushed into the list. Let the triangular face $A'M'N'$ be now extracted from the list. Adjoining triangular faces that also will belong to the approximation T_{Γ} must be found for the triangular face $A'M'N'$. For this purpose internal edges of $A'M'N'$ are examined. Consider one such edge

Find triangular faces of tetrahedra being images of triangles of t^{Γ}_{ps};
Mark them as belonging to T_{Γ};
Add them to the end of the list L;
While the list L is nonempty do //It is a main loop
    Take first element (triangular face t) of the list L and remove it from the list L;
    For each internal edge e_i of t, $i \in \{0,1,2\}$
        If (other triangular face that shares the same edge e_i exists)
            {go to the next edge e_{i+1};}
        Else Switch (how many possible triangular faces to be the neighbor for t through e_i have been found)
        1: {marked this triangular face as belonging to T_{Γ} and add it to the end of the list L;}
        0 or > 1: {neighbor is not found}
        end Switch
        end If
    end For
    If (at least one neighbor is not found)
        {Add t to the end of the list L; } end If
    If (a pass over the list does not give any new marked triangular faces)
        {curved face is not recovered, break;} end If
end do
If (face is not recovered) {Remesh curved face and tetrahedralization; repeat the second stage again;} end If
Curved face has been successfully recovered

Fig. 5. Second stage of curved face recovery.

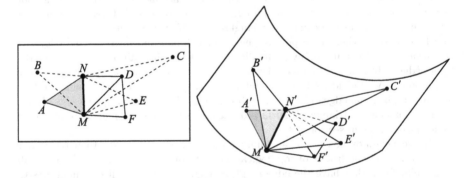

Fig. 6. Left: Fragment of two-dimensional triangulation of curved face in the parametric space. Right: the same configuration in the three-dimensional space.

$M'N'$. First, it is checked if there is other marked triangular face with the edge $M'N'$. Such triangle face can be only one by construction. If the marked triangle face has been found, then it means that the neighbor for the triangular face $A'M'N'$ through the edge $M'N'$ is found.

Suppose the marked triangle face has not been found. Return to the figure 6. The edge MN also belongs to the triangle MND but there is no triangular face $M'N'D'$ in the three-dimensional space. Instead of it the triangular face $M'N'F'$ presents in the tetrahedralization but a corresponding triangle MNF is absent from the parametric space. Besides the triangular face $A'M'N'$ and $M'N'F'$, the edge $M'N'$ also belongs to following triangular faces of tetrahedra: $M'N'B'$, $M'N'C'$ and $M'N'E'$. All of them ($M'N'F'$, $M'N'B'$, $M'N'C'$ and $M'N'E'$) are candidates to be the neighbor for the triangular face $A'M'N'$ through the edge $M'N'$. But only one of them can really become such neighbor. Among all possible candidates find those that meet conditions of topological integrity of mesh. These conditions are similar to requirements described at the beginning of the section 3.

Condition 1. Vertices of the triangular face lie on curved face that is being recovered.

Condition 2. A vertex (C' see Figure 7) of a triangular face-candidate ($M'N'C'$) that is opposite of considering edge ($M'N'$) is not fully recovered. A vertex is said to be fully recovered if it satisfies with the requirement VI provided that the set T_Γ is constituted only from marked triangular faces adjoining the vertex. Thus the vertex C' depicted in figure 7 is fully recovered and the triangular face-candidate $M'N'C'$ will be rejected as not meeting the condition 2.

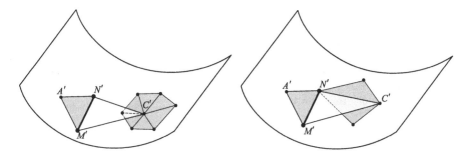

Fig. 7. A vertex C' is fully recovered. Therefore, a triangular face $M'N'C'$ will be rejected as not meeting the condition 2.

Fig. 8. An edge of tetrahedron $N'C'$ is fully recovered. Therefore, a triangular face $M'N'C'$ will be rejected as not meeting the condition 3.

Condition 3. Two edges of a triangular face-candidate ($M'C'$, $N'C'$ see Figure 8) connecting vertices (M' and N') to opposite vertex (C') of the considering edge ($M'N'$) are not fully recovered. An internal edge is said to be fully recovered if it belongs to two marked triangular faces. An external edge is said to be fully recovered if it belongs to one marked triangular face. This condition is equivalent to the requirements III and IV. Illustrated in figure 8 the edge $N'C'$ is fully recovered and the triangular face-candidate $M'N'C'$ will be rejected as not meeting the condition 3.

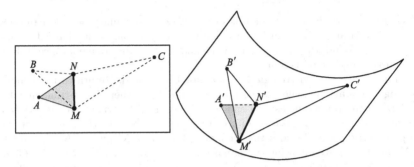

Fig. 9. Marked faces of tetrahedra must not produce mesh in the parametric space with self-intersections.

Condition 4. A vertex (A' see Figure 9) of extracted from the list triangular face ($A'M'N'$) and a vertex of a triangular face-candidate (B' or C'), which are opposite of the considering edge ($M'N'$), lie in the parametric space in different half-planes separated by line (MN) that connects the vertices of the considering edge in the parametric space. Thus the triangular face-candidate $B'M'N'$ does not meet condition 4, whereas the triangular face-candidate $M'N'C'$ meets this condition.

It is easy to notice that conditions 2 and 3 do not allow an internal edge to belong to more than two marked triangular faces and an external edge to belong to more than one marked triangular face.

The condition 4 differs from other conditions because it is not a part of the requirements described at the beginning of the section 3. The first three conditions are needed to guarantee that obtained approximation T_Γ is correct, whereas the condition 4 allows finding the approximation T_Γ that is different from t_{3d}^Γ. Thus if the condition 4 were not required, face recovery would not be possible in many cases, because it rarely occurs when each triangle of the triangulation t_{ps}^Γ is present in the tetrahedralization as a triangular face.

3.3 Handling seams and singularities

There are some difficulties with checking of the condition 4. These difficulties are connected with the fact that mapping from the parametric space to the three-dimensional space may not be biunique. Usually non-uniqueness arises at the boundaries of parametrization domain and can be two types. First type is seam edges. For example (see Figure 10), when a cylinder with a hole is unfolded, edges A_1B_1, A_2B_2, C_1D_1 and C_2D_2 are arisen in the parametric space what leads to creation of artificial seam edges $A'B'$ and $C'D'$ in the three-dimensional space. It takes place because parametrization of cylinder is periodic in one parameter. Seam edges in the parametric space must be handled together in such a way that it will be possible to join surface mesh in the three-dimensional space.

Second type of non-uniqueness is degeneracy (or singularity) on a surface. Singularity is a place where the parametrization collapses so that a range of parameter values corresponds to a single three-dimensional point. For example, sphere has two singularities corresponding with its poles. Cone has one singularity (its vertex).

Fig. 10. When a cylinder with a hole is unfolded, edges A_1B_1, A_2B_2, C_1D_1 and C_2D_2 are arisen in the parametric space what leads to creation of artificial seam edges $A'B'$ and $C'D'$ in the three-dimensional space.

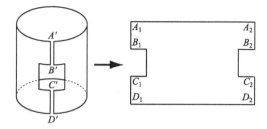

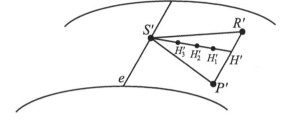

Fig. 11. Determination of parametric coordinates (u, v) for a point S' lying on a seam edge e.

In case of seam edge usually it is possible to perform backward transformation and find parametric coordinates (u, v) for a point lying on a seam edge e. Let u be the periodic coordinate. Let S' be a vertex of triangular face $P'R'S'$ and S' lies on seam edge (fig. 11). Let H' be any point lying in the interior of segment $P'R'$. Consider three points H'_1, H'_2 and H'_3 lying in the interior of $S'H'$ so that $d(S', H'_3) < d(S', H'_2) < d(S', H'_1)$, where d — distance between points, and for these points find the closest points $(H_1(u_1, v_1), H_2(u_2, v_2), H_3(u_3, v_3))$ lying on the curved face. If $u_1 < u_2 < u_3$, then $S(u_S, v_S)$ lies in the parametric space on a seam edge with the largest value of periodic coordinate, if $u_1 > u_2 > u_3$, then $S(u_S, v_S)$ lies on a seam edge with the lowest value of periodic coordinate; otherwise the periodic coordinate cannot be determined. The value of parametric coordinate depends on triangular face in which a point lies. For example, if we consider other triangular face to which the vertex S' belongs, then the value of the periodic coordinate might be other.

In case of singularities it is not possible to find parametric coordinates for a point lying on degeneracy. If the condition 4 cannot be checked, assume that it is fulfilled. A main problem of curved face recovery is that usually more than one neighbor is possible. Our task is to find only one appropriate neighbor using some criteria (four conditions). If condition 4 is assumed to be fulfilled, it makes the task of face recovery be more difficult. If condition 4 is not checked, it does not lead to incorrect mesh, since conditions 1-3 guarantee that that obtained approximation T_Γ is correct.

Presence of seam edges and singularities is the cause why it is impossible to introduce a definition determining that a curved face Γ is recovered if every non-degenerate triangle of the triangulation t_{ps}^Γ in the parametric space is present as a triangular face of tetrahedron in the tetrahedralization. A non-degenerate triangle is a triangle with three different 3D vertices (some vertices of t_{ps}^Γ may coincide in the three-dimensional space). Examine a triangulation t_{ps}^Γ in the parametric space depicted in figure 12 and suppose that every triangle has corresponding triangular face in the three-dimensional space. Let e_1 and e_2 be seam edges, which coincide in

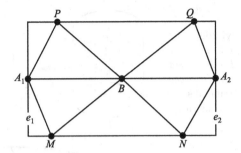

Fig. 12. Two-dimensional triangulation of a curved face in the parametric space t_{ps}^{Γ}. Edges e_1 and e_2 are seam edges. If each of the triangles A_1PB, A_1MB, BQA_2 and BNA_2 is present as a triangular face of some tetrahedron in the tetrahedralization, then four triangular faces of the set T_{Γ} share the same edge $A'B'$.

the three-dimensional space. In this case an image of the vertices A_1 and A_2 is the same point A' in the three-dimensional space. Respectively, edges A_1B and BA_2 are the same edge $A'B'$ of some tetrahedron. As the result four triangular faces that are a part of the approximation T_{Γ} share the same edge $A'B'$ but it is inadmissible. Thus even if every triangle has corresponding triangular face in three-dimensional space it does not exempt from necessity to check conditions of topological integrity of mesh. So before starting the main loop of recovery, already marked triangles must be checked if they meet requirements IV and V. If not, additional refinement of the face mesh is performed.

3.4 Remeshing curved face and tetrahedralization

Return to the face recovery method. If only one triangular face is remained after all four conditions have been checked then this triangular face is marked as belonging to the approximation T_{Γ} and added to the end of the list, then the next internal edge of t is considered. If more then one or no triangular faces fulfill all four conditions, the extracted triangular face t, for which neighbors have been searched, is added to the end of the list again. The triangular face t is added to the end of the list (not to the beginning) since others triangular faces of the list should be examined before t is considered again. They can produce new marked triangular faces and when t is extracted again, neighbors for t might be found in spite of the fact that previous attempt to do it ended in failure.

The main loop continues until one more pass over the list gives any new marked triangular faces (the recursion is stopped to prevent endless cycle) or the list becomes empty. In the former case the curved face is not recovered and a new vertex should be inserted into the triangulation t_{ps}^{Γ} and the tetrahedralization to locally change triangulation in a non-recovered area.

The list contains triangular faces for which neighbors have not been found yet, so when the list is empty, it means that separate marked triangles are joined to each other and a curved face is recovered. Triangular faces that represent a curved face and are in the tetrahedralization (they are called "marked" triangles and have abbreviation T_{Γ}) should be stored somewhere. These triangular faces are added to the set of constrained facets F. Also they are locked to prevent their deletion during maintaining the Delaunay property. Since the triangular faces are locked, the recovery of a next face never makes a previously recovered face disappear. Usually this approach leads to the constrained Delaunay tetrahedralization. But it has been shown [JO89] that in some cases "locking" faces does not always give the constrained Delaunay tetrahedralization. For example, during a 3-2 flip, when two triangular faces are eliminated and a new is created, if the first face is locked whereas the second is not, then the flip cannot be performed and the second triangular face might not be locally Delaunay. Nevertheless, this fact is not essential. First, if an input body does not have an edge that belongs to only one face (e.g. no internal boundaries) then the second triangular face will also be locked after a corresponding curved face is recovered. Second, usually the Delaunay tetrahedralization itself is rarely required since it is not optimal in three-dimensional space. After tetrahedral mesh improvement (smoothing and local transformations) it looses all properties of Delaunay triangulation anyway.

Additional refinement of curved face and tetrahedralization is performed in the following cases.

1. If none of triangles of the triangulation t_{ps}^{Γ} is present in the tetrahedralization. A few points must be inserted into the triangulation t_{ps}^{Γ} and their images into the tetrahedralization. Then it is necessary to check if this requirement is satisfied.

2. To resolve the configuration depicted in figure 12. In that case a bad triangle (e.g. $\triangle A_1 P B$) is eliminated by inserting a new vertex at the center of circumscribed ellipse of the bad triangle [CH97, BO96]. A three-dimensional image of the inserted vertex must also be inserted into the tetrahedralization. Insertion a new vertex both into t_{ps}^{Γ} and into the tetrahedralization implies performing all necessary actions (e.g. flips) to maintain the Delaunay property.

3. When an attempt to find the neighbor for the triangle t through the edge e ended in failure. There are two subcases here. 3a) The triangular face t has a corresponding triangle t_Γ in the triangulation t_{ps}^{Γ}. In this subcase the edge e has a corresponding edge e_Γ. The edge e_Γ belongs to two triangles: t_Γ and Δt_Γ. A new vertex is inserted at the center of circumscribed ellipse of the triangle Δt_Γ. 3b) This subcase occurs when the subcase 3a is not applicable. Take the edge e and find its midpoint. For the midpoint find the closest point $p(u,v)$ lying on the face. For the point $p(u,v)$ find a triangle Δt_Γ of the triangulation t_{ps}^{Γ} in which the point lies. A new vertex is inserted at the center of circumscribed ellipse of the triangle Δt_Γ. A three-dimensional image of the inserted vertex must also be inserted into the tetrahedralization. After the new vertex is inserted the second stage must be repeated.

4. Aforesaid method for the face recovery does not guarantee that each three-dimensional vertex, being an image of some vertex of the triangulation t_{ps}^{Γ} in the parametric space, belongs to at least one marked triangular face (requirement II). A roundabout path not including some vertices lying on the curved face may be found. To resolve this situation for vertices that are missed in T_Γ additional refinement of face mesh around them must be done. Then the second stage must be repeated.

When mesh is being refined, curved edges must remain to be recovered.

So the face recovery procedure is the heuristic iterative process. During one iteration the triangular faces that belong to face three-dimensional triangulation T_Γ are looked for. The triangulation in the parametric space t_{ps}^{Γ} is compared with the tetrahedralization and corresponding triangular faces are marked. It is similar as if we tried to reconstruct triangulation t_{ps}^Γ in accordance with that actually takes place in the three-dimensional space.

3.5 Termination

In this section some possible termination issues are discussed.

In order to start the main loop of the recovery procedure it is required that at least one triangle of the triangulation t_{ps}^Γ is present in the tetrahedralization as a triangular face of some tetrahedron. While this usually works out in practice; there is no guarantee, in theory, no matter how finely face mesh is refined.

The main loop of recovery is terminated when all neighbors for all marked triangles are found. It might get stuck in an attempt to find a neighbor for some triangular face.

To investigate these issues, this algorithm has been tested for real-world bodies with curved geometry. Bodies contained up to 2000 curved faces with different geometries. Surfaces of curved faces varied form the simplest (planes, cylinders, cones and spheres) to fairly complicated (tori, B-surfaces, swept and spun surfaces). Our attempts to find a body for which the algorithm doesn't work out ended in failure. It always terminated.

Nevertheless, in order to guarantee termination in all cases this algorithm is deliberately stopped to prevent endless loop in the following cases.

1) When quantity of point of the triangulation t_{ps}^Γ has been doubled but no triangles of the triangulation t_{ps}^Γ are present in the tetrahedralization.

2) When quantity of the marked triangular faces for which neighbors are not found is slightly changed from iteration to iteration. After algorithm is stopped curved face is replaced with piecewise linear approximation and each facet is recovered separately. It should be emphasize that there was no need to do it as these two cases never occurred and the algorithm terminated for all bodies on which it was tested.

So described in the paper algorithm can be regarded as a preprocessor step before boundary recovery using piecewise linear complex. Most likely, boundary recovery will be done during the preprocessor step; if not, piecewise linear complex can be considered.

Carried out tests demonstrated good performance. For example, for bodies depicted in an appendix mesh generation time does not exceed some minutes on a computer with Pentium IV processor. And this time includes all stages of mesh generation: surface mesh generation, initial mesh generation, boundary recovery and refinement of tetrahedral mesh.

[GE91] George P.L., Hecht F., Saltel. E. Automatic mesh generator with specified boundary. *Computer Methods in Applied Mechanics and Engineering*, 92(3):269–288, 1991.

[DU04] Du Q., Wang D. Boundary recovery for three dimensional conforming Delaunay triangulation. In *Comp. Meth. Appl. Mech. Engr*, 193(23–26), pages 2547–2563, 2004.

[BE02] Béchet E., Cuilliere J. C., Trochu. F Generation of a finite element MESH from stereolithography (STL) files. *Computer Aided Design*, 34(1):1–17, 2002.

[FR00] Frey P. J. About surface remeshing. In *Proc. of the 9th Int. Meshing Roundtable*, New Orleans, Louisiana, October 2000.

[CH93] Chew L. P. Guaranteed-quality mesh generation for curved surfaces. In *Proc. of the 9th Annual ACM Symp. on Computational Geometry*, pages 274–280, May 1993.

[CH97] Chen H., Bishop J. Delaunay triangulation for curved surfaces. In *Proc. of the 6th Int. Meshing Roundtable*, Park City, Utah, USA, October 1997.

[BO96] Bossen F. J., Heckbert P. S. A pliant method for anisotropic mesh generation. In *Proc, of the 5th Int, Meshing Roundtable*, Pittsburgh, Pennsylvania, USA, October 1996.

[JO91] Joe B. Construction of three-dimensional Delaunay triangulations using local transformations. *Computer Aided Geometric Design*, 8:123–142, 1991.

[SC02] Shewchuk. J. R. Delaunay refinement algorithms for triangular mesh generation. *Computational Geometry: Theory and Applications*, 22(1–3):21–74, May 2002.

[BO02] Boivin C., Ollivier-Gooch C. Guaranteed–quality triangular mesh generation for domains with curved boundaries. *Int. J. for Numerical Methods in Engineering*, 55(10):1185–1213, August 2002.

[SC98] Shewchuk. J. R. Tetrahedral mesh generation by Delaunay refinement. In *Proc. of the 14th Annual Symp. on Computational Geometry*, pages 86–95, Minneapolis, Minnesota, June 1998.

[JO89] Joe B. Threedimensional triangulations from local transformations. *SIAM Journal on Scientific and Statistical Computing*, 10:718–741, 1989.

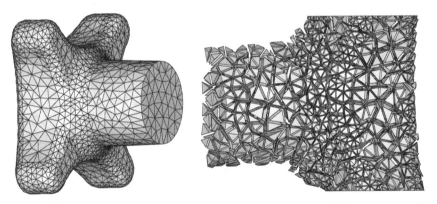

Fig. 13. Example 1. Boundary of the body consists of 51 curved faces separated by 120 curved edges. Mesh contains 2920 vertices and 10760 tetrahedra.

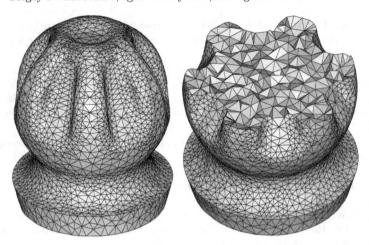

Fig. 14. Example 2. Boundary of the body consists of 87 curved faces separated by 204 curved edges. Mesh contains 7489 vertices and 34082 tetrahedra.

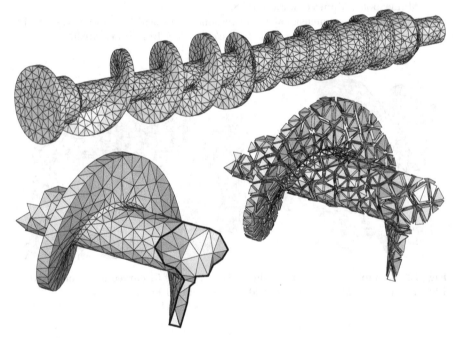

Fig. 15. Example 3. Boundary of the body consists of 75 curved faces separated by 186 curved edges. Mesh contains 4989 vertices and 18856 tetrahedra.

Fig. 16. Example 4. Boundary of the body consists of 753 curved faces separated by 1982 curved edges. Mesh contains 18571 vertices and 53444 tetrahedra.

Stitching and Filling: Creating Conformal Faceted Geometry

Paresh S. Patel[1], David L. Marcum[2], and Michael G. Remotigue[3]

[1] Computational Simulation and Design Center, ERC, Mississippi State University, Mississippi State, MS 39762, U.S.A. patel@erc.msstate.edu
[2] Computational Simulation and Design Center, ERC, Mississippi State University, Mississippi State, MS 39762, U.S.A. marcum@erc.msstate.edu
[3] Computational Simulation and Design Center, ERC, Mississippi State University, Mississippi State, MS 39762, U.S.A. remo@erc.msstate.edu

Summary. Consistent and accurate representation of geometry is required by a number of applications such as mesh generation, rapid prototyping, manufacturing, and computer graphics. Unfortunately, faceted Computer Aided Design (CAD) models received by downstream applications have many issues that pose problems for their successful usability. Automatic or semi-automatic tools are needed to process the geometry to make it suitable for these downstream applications. An algorithm is presented to detect commonly found geometrical and topological issues in the faceted geometry and process them with minimum user interaction. The present algorithm is based on the iterative vertex pair contraction and expansion operations called *stitching* and *filling* respectively. The combination of generality, accuracy, and efficiency of this algorithm seems to be a significant improvement over existing techniques. Results are presented showing the effectiveness of the algorithm to process two- and three-dimensional configurations.

1 Introduction

Computational design, analysis, optimization, and manufacturing have become an integral part of the product development process in automotive, aerospace, electronics, and many other industries. The simulation-based design process begins with creating a detailed geometry model in a Computer Aided Design (CAD) system. This CAD model is the starting point for many downstream applications such as mesh generation, structural/fluid/thermal analysis, rapid prototyping, numerical controlled machining, casting, computer graphics, real time rendering. Each of these downstream applications has specific requirements for the geometry definition and representation. Hence, success of the downstream application strongly depends on accuracy and consistency of the input geometry.

The Computational Fluid Dynamics (CFD) simulation process has several stages including pre-processing, flow solution, and post-processing of the results. Typically, pre-processing involves geometry cleanup and mesh generation to discretize the computational domain. In recent years, many automatic structured and unstructured mesh generation methods emerged. Most of these methods require a suitable (i.e. a clean well connected water-tight) geometry to start the grid generation process. Unfortunately, CAD data translated through neutral file formats like IGES [IGE88] and STL [STL89] have many geometrical and topological issues that prevent automatic creation of a water-tight geometry. They have many gaps, cracks, holes, overlaps, T-connections, invalid topology, inconsistent orientations. As a result of these issues or errors, true automation of the grid generation process is still elusive. The analyst has to manually clean the geometry to make it suitable for grid generation. This cleanup (pre-meshing) process is very time consuming, expensive, and tedious task for a design/analysis engineer. For realistic simulations, this is the single most labor-intensive task in the process, preventing true auto-meshing.

An algorithm is developed to detect the commonly found geometrical and topological issues and process them automatically to build topology information. The present algorithm is based on the iterative vertex pair contraction and expansion operations called *stitching* and *filling* respectively. The algorithm closes small gaps/overlaps via the *stitching* operation and fills big gaps by adding new faces through the *filling* operation to process the model accurately. This algorithm is general and can process manifold as well as non-manifold geometry models. Moreover, the present algorithm uses a spatial data structure, octree, for searching and neighbor finding to process large models efficiently.

2 Related Work

Adaptive Cartesian grid generation method [Aft97] based approaches [HLBZ02, WS02] can be used for CAD cleanup and surface triangulation. Hu et al. [HLBZ02] has utilized an overlay grid, obtained through Cartesian grid generation, to cleanup and reconstruct the geometry. Intersecting points of the overlay grid and geometry (water-tight volume) is reconstructed using a point cloud. In their work, it has been reported that this approach does not work for a complex configuration. Wang et al. [WS02] have also demonstrated the use of an adaptive Cartesian grid generation method for 'dirty' geometry clean up to get a surface triangulation of a complex configuration. Geometry is used to get the intersection/projection points to reconstruct the surface from these points. Further improvement of these approaches [HLBZ02, WS02] can be achieved by targeting only the bad areas with gaps and overlaps. Cartesian mesh-based approaches reconstruct the geometry approximately using the intersecting/projection points information; hence, output geometry is not accurate. Even if there is a small error in some part of the geometry, these approaches rebuild the entire model approximately, and accuracy depends on the refined Cartesian mesh cell size. Moreover, it may not be efficient to find many intersection/projection points if the Cartesian mesh is very fine, which is needed to achieve accuracy.

Many computer graphics and real time rendering applications also require an error free input geometry model. Baum et al. [BMSW91] developed a series of algorithms to preprocess the input geometry to meet the requirements of mesh based radiosity computation algorithms. Murali et al. [MF97] described an algorithm based on space subdivision to construct a consistent solid and boundary representation from polygons. This technique is simple and works well in the absence of degeneracies and narrow angles among neighboring polygons. It also requires significant amount of time to process even small models with few hundred polygons. Gueziec et al. [GTLH01] developed greedy strategies to convert a set of non-manifold polygonal surfaces to a manifold. The aim of their work is to modify the topology of surfaces, not to correct geometrical errors. Stereo Lithography (STL) is a widely used data exchange format in the Rapid Prototyping industry. In order to manufacture models correctly, input geometry must be geometrically and topologically correct. However, real world geometries translated through the .STL files generally have many geometrical and topological errors like gaps, overlaps, intersections, inconsistent orientations. Rock and Wozny [RW92] have used an AVL tree data structure to locate neighbor vertices efficiently to build model topology from a given set of unordered triangular facets. Bohn and Wozny [BW92] described a solution to achieve shell-closure of polyhedral CAD-models by extracting and triangulating the directed Jordan curves in three-dimensions to fill gaps. Makela and Dolenc [MD93] developed methods to handle overlapping and intersecting triangles efficiently. Sheng and Meier [SM95] have used a technique based on incremental matching and merging of boundaries of surface models to repair gaps. Their technique merges small as well as large gaps to process the models. Hence, it is not accurate to process large gaps due to missing geometry or polygons. Morvan and Fadel [MF96b] developed a virtual environment to correct the errors in a given model interactively, which is very time consuming and expensive to process large models. Barequet et al. [BK97, BS95] used a computer vision technique called geometric hashing [KSSS86, SS87] to repair geometrical and topological errors in the boundary representation (b-rep) of two-manifold geometry models. Recently, Patel et al. [PMR05] developed a technique based on a modified iterative vertex pair contraction operation called *stitching* to build topology information for manifold and non-manifold models. This work is an extension of the topology generation algorithm [PMR05] to process geometry models more accurately.

It seems previous mesh-based efforts [BK97, BS95, BW92, MD93, MF96b, RW92, SM95] assume that geometry models to be processed are two-manifold or use some special procedure to be able to handle non-manifold model like a two-manifold model. This assumption poses many restrictions not only the input model topology type but also on the processing algorithm design. The geometry model processing algorithm alters the topology of the input models. Hence, it is possible that even if the input geometry model is manifold, the processed model can be non-manifold. The present algorithm explicity supports manifold and non-manifold geometry models. Capability of handling manifold and non-manifold topologies during the geometry processing makes the procedure more general and flexible. Some of the previous approaches [BK97, BS95, BW92] collected the Jordan curves that are non-intersecting three-dimensional closed polygonals. These polygons are then triangulated to fill the gaps and holes. These approaches assume that boundary edges form a closed polygonal loop. However, it is possible that a set of boundary edges may not form Jordon curves and produce a valid triangulation of holes or require some user interaction in these situations. The *filling* process of the present algorithm uses a different approach to handle such situations. It does not require to form Jordan curves to fill gaps with new triangles. Moreover, the present algorithm stitches small gaps/overlaps and fills big gaps by adding new triangles to process the model accurately. In addition, it uses the octree data structure for searching and neighbor finding to process large models efficiently. In this way, the present procedure offers combined benefits of generality, accuracy and efficiency for automatic processing of faceted geometry.

3 Geometry and Topology Representation

CAD models are often represented as a set of triangulated surfaces in three-dimensional Euclidean space R^3 [J. 98]. Let us define a geometry model $M = (V, F)$ as a set of vertices V and a set of triangular faces F. The vertex list $V = (v_1, v_2, ..., v_m)$ is an ordered sequence of vertices. Each vertex v_i is defined by three coordinates (x_i, y_i, z_i) and a unique index. The face list $F = (f_1, f_2, ..., f_n)$ is also an ordered sequence of faces. Every face or triangle f_i is defined by an ordered list of three vertex indices (j, k, l) and a unique index. A face made of vertices v_j, v_k and v_l can be denoted as $\triangle v_j v_k v_l$. An edge e_i is defined by an ordered sequence of two vertex indices (j, k). It can be denoted as $\overline{v_j v_k}$. An edge with one incident face is called a boundary edge and its end points are called boundary vertices. An edge with two and more than two incident faces is called a manifold edge and non-manifold edge respectively. Note that the geometry model does not have topology or adjacency information. Topology information tells how geometric objects are connected. Many downstream applications need such information for further use of geometry models. The goal is to develop an algorithm to process geometrical and topological issues and build the topology information (neighbor maps) with minimum user interaction.

4 The Topology Generation Algorithm

The present algorithm is based on the iterative vertex pair contraction and expansion operations to process the geometrical and topological issues. An edge-split operation is introduced to make vertex pair contraction [GH97, PH97] more reliable and accurate. Mainly, it consists of boundary detection, boundary vertex pair generation, iterative vertex pair contraction and expansion with the following specific steps:

(i) Read and pre-process the input geometry model represented by vertices and indexed faces.
(ii) Detect and mark boundary edges and vertices.
(iii) Build the Octree data structure [H. 90a, H. 90b] for efficient searching.
(iv) Generate a list of boundary vertex pairs. For each of the boundary vertices search for other boundary vertices or edges within a user specified resolution tolerance, ϵ_r, to pair with and insert into the list.
(v) Sort the list of boundary vertex pairs using a cost function that is dependent on the distance between the paired vertices.
(vi) Iteratively remove a boundary vertex pair from the sorted list with minimum cost. Perform the vertex pair contraction operation, if the cost of the vertex pair is less then the user specified glue tolerance, ϵ_g, otherwise perform the vertex pair expansion operation. Update the connectivity information during the vertex pair contraction and expansion operations.
(vii) Output the processed geometry model with adjacency information.

Detailed description of these steps is presented in the following sections.

4.1 Pre-processing

First, build and pre-process the input geometry model represented by vertices and indexed faces. In this step, initialize the data structure and generate the list of vertices and faces and classify them. At this point the connectivity among the faces is not known. The goal is to find the matching boundary edges and merge them to build the topology information and correct the geometrical issues for the entire model. Now, detect the boundary edges and vertices by finding the number of faces attached to each edge. If an edge has one incident face then it is a boundary edge and incident vertices of a boundary edge are boundary vertices. Geometric entities are created, classified and marked with flags during this step for further processing.

4.2 Spatial Data Structure

To build the list of the boundary vertex pairs, for a given boundary vertex, other boundary vertices or edges within the resolution tolerance, ϵ_r, need to be searched. Hence, efficiency of the algorithm strongly depends on the choice of the data structure used for answering such queries. There are many spatial data structures [H. 90a, H. 90b] that can be used for this type of range queries. However, the octree, a simple yet powerful spatial data structure, is used in the present algorithm. A bounding box covering the entire geometry model is the root or parent cell of the octree. This root cell is recursively subdivided into eight children until each of the children contains few geometric objects. The aim is to search for the boundary objects in nearby region. Hence, only boundary objects are inserted into the tree to reduce the amount of data associated with the spatial search. Once the tree data structure is built, finding the geometric objects lying in a given search range is very fast. For a detailed description of the octree data structure a bit of old but classic references [H. 90a, H. 90b] can be reviewed.

4.3 Boundary Detection and Vertex Pairs

At this point, all the boundary objects are marked and the octree data structure is built for efficient searching. For each boundary vertex, find other boundary vertices/edges within the resolution tolerance using the octree search. As shown in figure 1(a), vertex v_i and v_j are paired without splitting the boundary edge for contraction, if $|v_i - v_j| \leq \epsilon_r$. The boundary vertex pair generation procedure strongly depends on the relative position of the boundary vertices to be paired. For example, there is no clear correspondence between boundary vertices v_i and v_j in figure 1(b). A large ϵ_r is required to pair them up. But, it is important to note that an appropriate choice of the ϵ_r is very important. Too small ϵ_r can leave many potential boundary vertex candidates un-paired. On the other hand, a large ϵ_r may pair inappropriate boundary vertices and makes the procedure less reliable. Moreover, the vertex pair contraction without edge-split can not handle the T-joints (end point of one edge lies within another edge) situations. This usually occurs when a big surface is in the neighborhood of two small surfaces and forms a T-like shape near the junction of surfaces or when two neighboring curves are discretized using different point distribution functions.

To make the procedure more reliable and handle T-joints, boundary vertex pair contraction with edge-split is introduced. Consider the situation shown in figure 1(b), there is no other boundary vertex within a smaller ϵ_r to pair vertex v_i. However, boundary edge $\overline{v_j v_k}$ split operation can create a vertex v_p and boundary vertices v_i and v_p can be paired without increasing ϵ_r. To find a nearby boundary edge, check if orthogonal projection of vertex v_i on a nearby boundary edge $\overline{v_j v_k}$ is possible. To perform this check, lets define $a = \overrightarrow{v_j v_i}$ and $b = \overrightarrow{v_j v_k}$. In theory, orthogonal projection is possible as long as the following inequality is satisfied.

$$0 \leq \frac{a \cdot b}{|b|} \leq |b| \tag{1}$$

In practice, there may be a problem due to numerical inaccuracies to check the inequality given in the equation (1). To make it numerically more stable and robust, a threshold parameter α can be used and the resultant inequality relation is,

$$\alpha \leq \frac{a \cdot b}{|b|} \leq |b| - \alpha \tag{2}$$

It can be checked using equation (2) that orthogonal projection of the boundary vertex v_i on the boundary edge $\overline{v_j v_k}$ is possible. To compute the location of projection point v_p the following equation can be used,

$$v_p = \frac{b}{|b|} d \tag{3}$$

where $d = |v_j - v_p| = |a| \cos \theta = a \cdot b / |b| =$ projection of a on b, and $\theta = \angle v_i v_j v_k$. Now, split the edge $\overline{v_j v_k}$ with respect to projection point location v_p and pair the boundary vertices v_i and v_p. Edge split operation adds a new vertex v_p and face $\triangle v_p v_j v_k$. The connectivity information for corresponding vertices and faces is also updated that modifies the topology. It is possible that the projection point v_p lies very close to vertex v_j when $\theta \approx 90^0$ and creates a new vertex and face due to edge-split operation. It can be avoided by checking the distance d and reject the edge-split operation, if d is small. The same check can be performed without computing d, the parameter α in equation (2) can be chosen in such a way that the inequality is not satisfied if the projection vertex v_p and v_j happen to be very close.

In this way, there are two advantages on using the parameter α in equation (2). First, it takes care of the instabilities due to the numerical inaccuracies and makes the check more stable. Second, it avoids the edge-split operation if the distance d is small without any extra computations. The value of the parameter α can be $0 \leq \alpha \leq 0.5 |b|$. As mentioned earlier, $\alpha = 0$ makes the check numerically unstable and may create projection vertex v_p close to the boundary vertex v_j. While $\alpha = 0.5 |b|$ does not allow the edge-split operation in most cases but it may make boundary vertex pair generation procedure less reliable. It is essentially the vertex pair generation without edge-split operation. The present algorithm uses $\alpha = \epsilon_g$. A boundary vertex pair of vertices v_i and v_j can be denoted as pair (v_i, v_j). All the boundary vertex pairs generated with and without the edge-split operation are inserted in a heap keyed on cost with the minimum cost pair at the top. The cost function of a boundary vertex pair (v_i, v_j) is $C(v_i, v_j) = |v_i - v_j|$.

4.4 Iterative Vertex Pair Contraction

The topology generation algorithm iteratively removes a vertex pair (v_i, v_j) with minimum cost from the heap and performs the contraction operation, if $C(v_i, v_j) \leq$

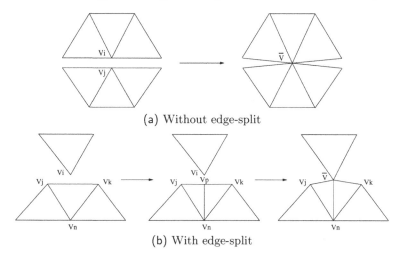

(a) Without edge-split

(b) With edge-split

Fig. 1. Vertex pair contraction operation

ϵ_g. The boundary vertex pairs are merged to process the geometrical and topological issues. A boundary vertex pair (v_i, v_j) contraction moves boundary vertices v_i and v_j to a new location v_i, v_j or $\bar{v} = (v_i + v_j)/2$. Merging boundary vertices actually modifies the geometry and processes the geometrical issues such as gaps, intersections, overlaps, etc. The vertex pair contraction operation also replaces faces and edges incident to the v_j with v_i. This step modifies the topology of the model to process the topological issues. The boundary vertex pair contraction operation generally does not collapse faces. However, in case of geometry with long skinny surfaces, the vertex pair contraction operation may produce degenerate faces that are removed from the processed model. Hence, a boundary vertex pair contraction operation merges two vertices and updates the connectivity information. It deletes a boundary vertex and may delete one or more faces. Figure 1(a) and 1(b) show boundary vertex pair contraction operations with and without edge-split respectively. The processing algorithm iteratively removes the pair with minimum cost from the heap and performs the vertex pair contraction operation, if the cost of the vertex pair is less than the glue tolerance. This process is called *stitching*. Note that pair vertex contraction moves boundary vertices and modifies the geometry model. An error is introduced in the processed geometry model that is bounded by the resolution tolerance. Consider the effect of the edge-split operation on the error introduced in the processed model. As mentioned earlier, vertex pair contraction without edge-split requires to use a larger resolution tolerance than with edge-split to pair boundary vertices and produces more error. Hence, vertex pair contraction with edge-split operation is not only more reliable and robust but also more accurate.

4.5 Iterative Vertex Pair Expansion

The vertex pair contraction operation with edge-split introduces less error in the processed geometry model than that of without edge-split. It moves the vertices physically in order to process the errors smaller than the glue tolerance. Hence, if the errors are large in the model to be processed, then the vertex pair contraction operation would require to choose a larger glue tolerance and produce more error in the processed model. A new operation, vertex pair expansion, is developed to fill the larger gaps with new triangles without moving the vertices as opposed to the vertex pair contraction operation. It fills the gaps larger than the glue tolerance and smaller than the resolution tolerance with new triangles without introducing any error in the processed model. The boundary vertex pairs near the juction of surfaces are penalized by setting their cost function to be zero. In this way, the boundary vertex pairs are forced to be contracted to join disjoint surfaces.

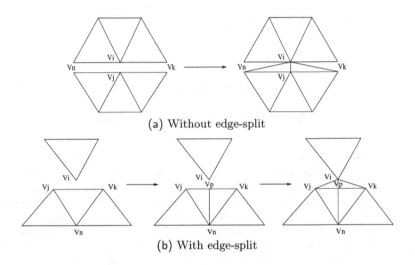

(a) Without edge-split

(b) With edge-split

Fig. 2. Vertex pair expansion operation

The topology generation algorithm removes a vertex pair (v_i, v_j) with minimum cost from the heap and performs the expansion operation, if $\epsilon_g < C(v_i, v_j) \leq \epsilon_r$. The boundary vertex pair (v_i, v_j) expansion adds one or more new triangles to fill the gap. Addition of new triangles actually modifies the geometry and processes the geometrical issues like gaps. The pair expansion also updates the list of incident faces to the vertices. This step modifies the topology of the model to process the topological issues. Hence, the boundary vertex pair expansion operation adds new faces and updates the topology information. Figure 2(a) shows the vertex pair (v_i, v_j) expansion operation without edge-split. It adds two new triangles $\triangle v_i v_j v_k$ and $\triangle v_i v_j v_n$. Figure 2(b) shows the vertex pair expansion with edge-split. It also adds two new triangles $\triangle v_i v_p v_k$ and $\triangle v_i v_p v_j$. As mentioned earlier, if the value of the resolution tolerance is big-enough, then the vertex pair expansion would add a new triangle $\triangle v_i v_j v_k$ without splitting the edge $\overline{v_j v_k}$. The vertex pair expansion may also add one triangle near already stitched or juction of surfaces that will be disccused later in details. This iterative vertex pair expansion process is called *filling*. Note that *filling* does not move the vertices like *stitching* operation to process the model. Therefore, it does not introduce any error in the processed model.

In summary, the output of the algorithm is a well-suited discrete CAD model along with the necessary topology information that can be an input for many downstream applications. In this mode the output mesh can be used as a background mesh to subsequently generate a high-quality unstructured mesh using Advancing Front Local Reconnection (AFLR) [MW95] algorithm. The current procedure is based on an assumption of the locality of geometrical and topological issues. It is assumed that the neighboring surface patches have small gaps/overlaps due to the translation errors, numerical inaccuracies, tolerance settings in different systems, etc. In practice, gaps/overlaps between the surfaces are generally very small so the key assumption of locality is reasonable. However, there are many other issues that may cause large gaps/overlaps. For example untrimmed or missing surfaces may create large gaps and overlaps. These issues may violate the assumption of locality and can be processed up to some extent.

5 Results and Discussion

The topology generation algorithm following the outline given in the previous section is developed and implemented using object oriented technology in C++ on a unix workstation. CAD models obtained from various systems are processed to build correct topology information. Consider a simple plate as shown in Figure 3(a). There is a small gap and overlap between two triangulated surfaces that needs to be processed. Figure 3(b) shows the extracted boundary edges and boundary vertex pairs for further processing. The resultant geometry after the *stitching* operation is shown in Figure 3(c). Note that the *stitching* operation moves vertices physically by changing their coordinates and introduces an error of the order of the resolution tolerance. Figure 3(d) shows that there is no gap and overlap between two surfaces in the processed geometry. Moreover, the topology (connectivity map) information is available and can be used for downstream applications.

The *stitching* operation modifies the original model to process the geometrical and topological issues. The same plate model can also be processed via *filling* operation. Figure 4 shows the processed geometry model. Unlike the *stitching* operation, the *filling* does not move vertices and modify the original model. The original model can be processed without introducing any error but it adds many self-intersecting and skinny triangles in the boundary region of the processed model.

The *stitching* operation merges boundary vertices to process the geometry and produces more error than the *filling* operation. However, if only *filling* operation is used then it may introduce many new triangles in the boundary regions for large models. To minimize the error in the processed model and reduce adding too many triangles, the *stitching* and *filling* operations are used together as shown in figure 5. Notice that small gaps/overlaps are merged together and large gaps are filled with new triangles. In this way, combined use of the *stitching* and *filling* operations makes the geometry processing more accurate and practical.

Figure 6(a) shows two cubes sharing a common curve with cracks and overlaps between surfaces. The common curve is shared by four faces and it is required to provide an explicit support for non-manifold topology to process this model. As mentioned earlier the topology generation algorithm can handle non-manifold situations that provides more flexibility and generality. Figure 6(b) shows the geometrically

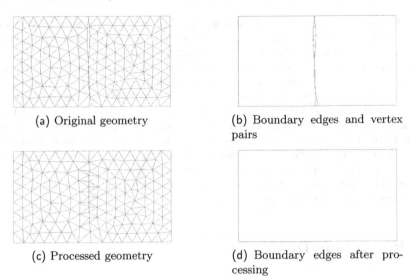

(a) Original geometry

(b) Boundary edges and vertex pairs

(c) Processed geometry

(d) Boundary edges after processing

Fig. 3. Plate showing the stitching operation

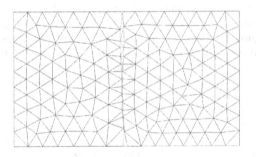

Fig. 4. Plate showing the filling operation

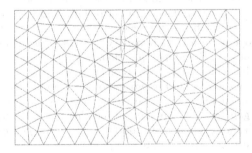

Fig. 5. Plate showing the stitching and filling operations

and topologically well-defined model obtained after processing. Note that the topology generation algorithm locally modifies the topology of the input model. Hence, it is possible that even if the input geometry model is manifold the output or intermediate model may be non-manifold. The data structure used in the implementation of present algorithm can support manifold and non-manifold topology.

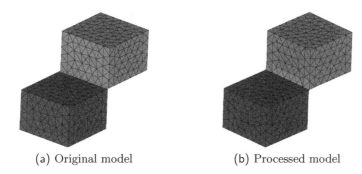

      (a) Original model   (b) Processed model

Fig. 6. Two cubes sharing a common edge (non-manifold topology)

Figure 7(a) shows the triangulated model of a flying minnow. Figure 7(b) shows the detected boundary edges of the same model. Figure 8(b) shows the remaining gaps after the *stitching* operation. These gaps are then triangulated via the *filling* operation. The flying minnow model consists of 5566 vertices and 9062 faces. The resolution tolerance, ϵ_r, and the glue tolerance, ϵ_g, used are 0.1 and 0.01 mm respectively. The boundary vertex pair generation process has formed 1162 vertex pairs, of which about 1127 pairs are contracted and only 35 pairs are expanded during the *stitching* and *filling* operations respectively. The *filling* operation may introduce self-intersecting triangles in the boundary regions. Boundary edge detection on the processed flying minnow model found no edge because it forms a closed volume and each of the edges in the model is two-manifold.

Figure 9(a) shows the body shell of Infiniti G35 car model. Figure 9(b) shows the boundary edges before processing. This model has 4283 vertices and 7448 faces. Figure 10(a) is the processed vehicle model after the *stitching* and *filling* process. To verify the repairing process, boundary edges of the modified geometry are extracted as shown in figure 10(b). It shows the boundary curves that are shared by only one surface. Moreover, the topology information is also built during the processing. The total number of boundary vertex pairs generated was 420 for the value of ϵ_r and ϵ_g to be 1.0 and 0.01 mm respectively.

The discrete geometry of Infiniti G35 doors is shown in Figure 11(a). The model consists of 6443 vertices and 10814 faces. It has a large gap near the upper-right corner due to a missing surface. Figure 11(b) shows the enlarged view of the gap that has to be processed. It is important to note that the size of the gaps and some of the surfaces in the model are of the same order. Figure 12(a) and figure 12(b) show the processed model and enlarged view of the rectangular region. The large gap is filled with new triangles via the *filling* operation. The topology generation algorithm performed about 827 vertex pair contractions and 92 vertex pair expansions in order to process the model with ϵ_r and ϵ_g of 2.25 and 0.1 mm respectively. This test

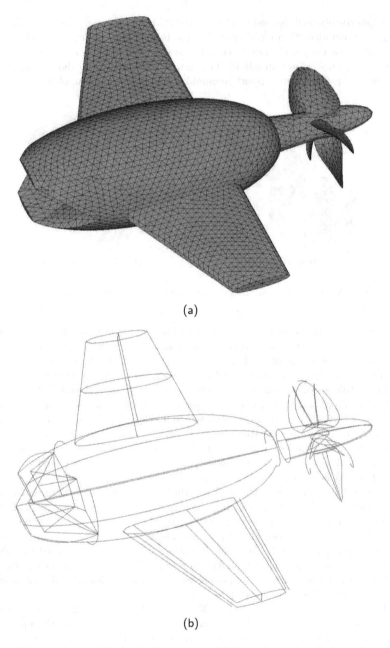

(a)

(b)

Fig. 7. Flying minnow (a) Original geometry. (b) Boundary edges before processing.

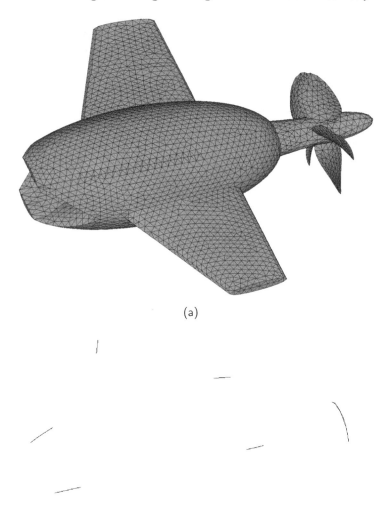

(a)

(b)

Fig. 8. Flying minnow (a) Processed geometry. (b) Gaps filled with new triangles.

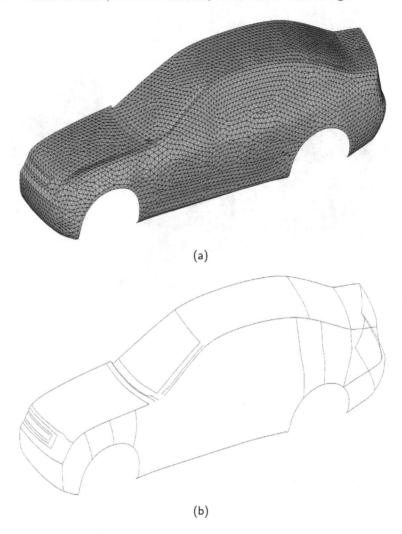

(a)

(b)

Fig. 9. Body shell of Infiniti G35 (a) Original geometry. (b) Boundary edges before processing.

case shows that the topology generation algorithm can even create small missing geometry entities up to some extent. For example, it did add new faces in place of a small missing surface near the upper-right corner of the door.

6 Conclusion

Automatic detection and processing of commonly found geometrical and topological issues such as gaps, overlaps, intersections, T-connections, invalid or no topology,

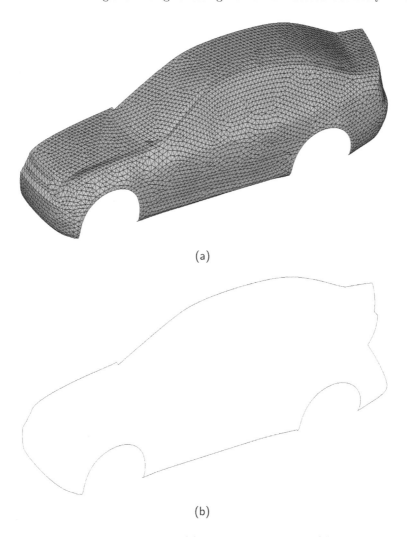

(a)

(b)

Fig. 10. Body shell of Infiniti G35 (a) Processed geometry. (b) Boundary edges after processing.

etc. is achieved for two- and three-dimensional configurations. Unlike a CAD model repair procedure that requires significant user interaction, the proposed methodology is highly automated. Results demostrate that the generality, accuracy and efficiency of the topology generation algorithm appears to be a significant improvement over existing methodologies. In addition, the processed models are guaranteed to be free of self-intersecting boundary edges. This work is a step towards automatic geometry processing for mesh generation applications. There are many issues that need to be addressed to automate the pre-processing of geometry for the same application. For example, the user has to provide an appropriate value of distance threshold. It may

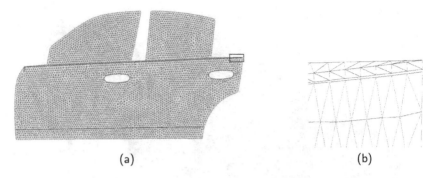

(a)   (b)

Fig. 11. Doors and windows of Infiniti G35 (a) Original model. (b) Enlarged view of the rectangular region showing the gap due to a missing surface.

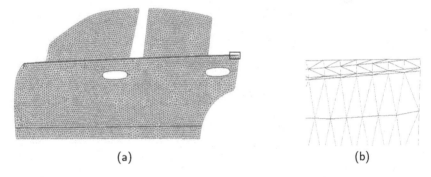

(a)   (b)

Fig. 12. Doors and windows of Infiniti G35 (a) Processed model. (b) Enlarged view of the rectangular region shows that the gap is filled with triangles in the processed model.

require the user to try different threshold values until a reasonable value is found. It is found that small variation in distance threshold does not affect the results much in most cases. However, selection of the distance threshold should be automatic and adaptive. Currently, work is in progress to address these issues.

[Aft97]   M. J. Aftosmis. Solution adaptive cartesian grid methods for aerody-
           namics flows with complex geometries. von Karman Institute of Fluid
           Dynamics, Lecture series 1997-02, 1997.
[BK97]   G. Barequet and S. Kumar. Repairing cad models. In *Proceedings:
           IEEE Visualization*, pages 363–370, Phoenix, AZ, 1997.
[BMSW91]   D. R. Baum, S. Mann, K. P. Smith, and J. M. Widget. Making ra-
           diosity usable: Automatic preprocessing and meshing techniques for the
           generation of accurate radiosity solutions. In *Proceedings: SIGGRAPH*,
           volume 25, pages 51–60, Las Vegas, Nevada, 1991. ACM.
[BS95]   G. Barequet and M. Sharir. Filling gaps in the boundary of a polyhe-
           dron. *Computer Aided Design*, 12(2):207–229, 1995.

[BW92]   J. H. Bohn and M. J. Wozny. Automatic cad-model repair: Shell-closure.
            In *In H. L. Marcus et al., eds., Proc. Solid Freeform Fabrication Symp.*,
            pages 86–94, Austin, TX, 1992. The Univ. of Texas.
[GH97]   M. Garland and P. S. Heckbert. Surface simplification using quadratic
            error metrics. In *Proceedings: SIGGRAPH Computer Graphics*, pages
            209–216. ACM, 1997.
[GTLH01]   A. Gueziec, G. Taubin, F. Lazarus, and B. Horn. Cutting and stitching:
            converting sets of polygons to manifold surfaces. *IEEE Transaction on
            Computer Graphics*, 7(2):136–151, 2001.
[H. 90a]   H. Samet. *Applications of Spatial Data Structures: Computer Graphics,
            Image Processing, and GIS*. Addison-Wesley, Reading, MA, 1990. ISBN
            0-201-50300.
[H. 90b]   H. Samet. *The Design and Analysis of Spatial Data Structures*. Addison-
            Wesley, Reading, MA, 1990. ISBN 0-201-50255-0.
[HLBZ02]   J. Hu, Y. K. Lee, T. Blacker, and J. Zhu. Overlay grid based geometry
            cleanup. In *Proceedings: 11th International Meshing Roundtable*, pages
            313–324. Sandia National Laboratories, 2002.
[IGE88]   Initial graphics exchange specification (iges) version 4.0. NBSIR 88-
            3813, 1988.
[J. 98]   J. F. Thompson and B. K. Soni and N. P. Weatherill. *Handbook of Grid
            Generation*. CRC press, Boca Raton, Florida, 1998.
[KSSS86]   A. Kalvin, E. Schonberg, J. T. Schwartz, and M. Sharir.   Two-
            dimensional model-based, boundary matching using footprints. *Int J.
            of Robotics Research*, 5(4):38–55, 1986.
[MD93]   I. Makela and A. Dolenc. Some efficient procedures for correcting tri-
            angulated models. In *In H. L. Marcus et al., eds., Proc. Solid Freeform
            Fabrication Symp.*, pages 126–134, Austin, TX, 1993. The Univ. of
            Texas.
[MF96a]   S. M. Morvan and G. M. Fadel. Ivecs: An interactive virtual environment
            for the correction of .stl files. In Conference on Virtual Design, Univ. of
            California, Irvine, CA, 1996.
[MF96b]   S. M. Morvan and G. M. Fadel. Ivecs, interactively correcting .stl files
            in a virtual environment. In *In H. L. Marcus et al., eds., Proc. Solid
            Freeform Fabrication Symp.*, Austin, TX, 1996. The Univ. of Texas.
[MF96c]   S. M. Morvan and G. M. Fadel. Virtual prototyping using .stl files.
            International Body Engineering Conference (IBEC), Detroit, MI, 1996.
[MF97]   T. M. Murali and T. A. Funkhouser. Consistent solid and boundary
            representation from arbitrary polygonal data. In *In Proc. Symp. On
            Interactive 3D Graphics*, pages 155–162, Providence, RI, 1997.
[MW95]   D. L. Marcum and N. P. Weatherill. Unstructured grid generation us-
            ing iterative point insertion and local reconnection. *AIAA Journal*,
            33(9):1619–1625, 1995.
[PH97]   J. Popovic and H. Hoppe. Progressive simplicial complexes. In *Proceed-
            ings: SIGGRAPH Computer Graphics*, pages 217–224. ACM, 1997.
[PMR]   P. S. Patel, D. L. Marcum, and M. G. Remotigue. Automatic cad
            model topology generation. *International Journal of Numerical Method
            in Fluids, under review*.
[PMR05]   P. S. Patel, D. L. Marcum, and M. G. Remotigue. Building topological
            information for triangulated models. SIAM Computational Science and
            Engineering Conference, February 12-15, Orlando, Florida, 2005.

[RW92]   S. J. Rock and M. J. Wozny. Generating topological information from a bucket of facets. In *In H. L. Marcus et al., eds., Proc. Solid Freeform Fabrication Symp.*, pages 251–259, Austin, TX, 1992. The Univ. of Texas.

[SM95]   X. Sheng and I. R. Meier. Generating topological structures for surface models. *IEEE Computer Graphics and Applications*, 15(6):35–41, 1995.

[SS87]   J. T. Schwartz and M. Sharir. Identification of partially obscured objects in two and three dimensions by matching noisy characteristics curves. *Int J. of Robotics Research*, 6(2):29–44, 1987.

[STL89]   Stereolithography interface specification (stl). valencia, ca. 3D Systems Publications, 1989.

[WS02]   Z. J. Wang and K. Srinivasan. An adaptive cartesian grid generation method for "dirty" geometries. *International Journal of Numerical Method in Fluids*, 39:703–717, 2002.

Polygon Crawling: Feature-Edge Extraction from a General Polygonal Surface for Mesh Generation

Soji Yamakawa[1] and Kenji Shimada[2]

The Department of Mechanical Engineering, Carnegie Mellon University
[1]soji@andrew.cmu.edu, [2]shimada@cmu.edu

Abstract.

This paper describes a method for extracting feature edges of a polygonal surface for mesh generation. This method can extract feature edges from a polygonal surface typically created by a CAD facet generator in which typical feature edge extraction methods fail due to severe non-uniformity and anisotropy. The method is based on the technique called "polygon crawling," which samples a sequence of points on the polygonal surface by moving a point along the polygonal surface. Extracting appropriate feature edges is important for creating a coarse mesh without yielding self-intersections. Extensive tests have been performed with various CAD-generated facet models, and this technique has shown good performance in extracting feature edges.

1. Introduction

This paper describes a method for extracting feature edges of a polygonal surface for mesh generation; the input to the proposed method is not limited to a dense uniform triangular mesh. Feature edge extraction methods have been developed for a dense triangular mesh as shown in Fig. 1 (a), which is typically created by a laser range scanner and a uniform mesh as shown in Fig. 1 (b). The proposed method targets a general polygonal surface that includes concave and high aspect-ratio polygons. This method is particularly advantageous when feature edges must be extracted from a triangular facet generated by a CAD facet generator, which includes triangles with high aspect ratio. Existing feature edge extraction schemes do not perform well on such triangular facets. This method also tolerates certain discretization noise created by a CAD facet generator. A CAD facet generator often yields concavity where it is convex or convexity where it is concave due to a bug in the facet generator, a bad set of parameters given to the facet generator, or a deficiency of the original CAD model.

In this paper, feature edges are defined as a set of edges that must be preserved, or constrained, to obtain a valid mesh. A mesh is valid if no element intersects with another, and if any point on the mesh is two-manifold -- except along the boundary of the open polygonal surface. The proposed method focuses on extracting the following three types of edges: (1) edges along sharp corners, (2) edges in which surface curvature, measured in the direction perpendicular to the edge, changes significantly across the edge, and (3) edges on the cylindrical surfaces.

| (a) A dense triangular mesh typically created by a laser range scanner | (b) A good quality uniform mesh | (c) A general polygonal surface including concave and high aspect-ratio polygons, which the proposed method deals with |

Fig. 1. Typical polygonal surfaces

Fig. 2 demonstrates the effects of feature edges on mesh generation. Fig. 2 (a) shows the input model. Fig. 2 (b) shows a meshed model without constraining edges at all, which clearly does not describe the input model well. Figures 2 (c) and 2 (d) show edges that pertain to type one and a meshed model with type one edges constrained. The mesh does not describe the fillets and the cylindrical surfaces. Figures 2 (e) and 2 (f) show edges of all three types extracted by the proposed method and a meshed model with constraining edges of all three types; of the three meshed examples it best describes the original geometry.

A simple method can detect only type one edges; it is more difficult to detect edges that pertain to types two and three. The goal of the proposed method is to detect all three types of feature edges. The proposed method can extract edges along the boundaries of any surface as type two edges -- a plane, a cylindrical surface, a spherical surface, a saddle surface, etc. -- because along such edges a sudden change of curvature occurs frequently. However, a closed (or nearly closed) cylindrical surface does not include such sudden change of curvature on the surface. The proposed method seeks out such cylindrical surfaces and extracts edges on these surfaces as type three edges.

Extracting feature edges from a polygonal surface is a much more challenging problem than extracting feature edges from a CAD model with smooth surfaces from which exact curvature at any point on the surface can be calculated. However, a polygonal surface itself does not have such information, and curvature on the surface can only be estimated. Feature edges thus can be more accurately extracted from a CAD model than from a polygonal surface. However, we often have only access to a polygonal surface for variety of reasons including incompatibility of a CAD data file between systems (we may not be able to open a received CAD file,) confidentiality requirements of the contract, or the loss of the original CAD model caused by a malfunction of a computer or operator error. In such cases it is necessary to develop a method for estimating curvature and extracting feature edges from a polygonal surface.

The proposed method uses a procedure called polygon crawling for estimating curvature and finding feature edges. The polygon crawling method samples sequence of points, called a "crawling path," on the polygonal surface. The curvature of the polygonal surface is estimated by analyzing the crawling path. The method locates the feature edges where the estimated curvature changes significantly from one side of the edge to the other. The cylindrical surface in the polygonal surface can be found by (1) finding a principal curvature direction at a point on the polygonal surface, and (2) fitting a circle to the crawling path starting from the point toward the principal curvature direction.

The organization of the paper is as follows. Section 2 reviews previous work of the feature edge extraction. Section 3 explains the polygon crawling method. Section 4 explains the parameters that control the results of the proposed method. Section 5 describes how the curvature is estimated, and Section 6 presents the algorithm of extracting feature edges based on the curvature estimation. Section 7 shows some results, and Section 8 concludes the paper.

2. Related Work

Mangan and Whitaker present a method for partitioning a dense 3D mesh [1]. Their method calculates a curvature (or some other height function) at each vertex, and merges regions based on the curvature (or the height). Feature edges can be extracted by taking edges between partitions. Their method, however, is not designed for a mesh that includes high aspect ratio triangles.

Nomura and Hamada present a method for extracting feature edges from a triangular mesh [2]. Their method first calculates a measurement called "*concavity-convexity*," and extracts edges between a convex triangle and a concave triangle. The method also assumes a dense uniform mesh as input.

Jiao and Heath present a method for extracting feature edges [3, 4]. Their method finds a set of edges that the dihedral angle exceeds the given threshold called θ-strong edges, and takes as a feature edge a θ-strong edge that is likely to form a smooth curve when it is connected to a neighboring θ-strong edge. Their method is designed for a uniform mesh, and more work needs to be done to extend this method to a general polygonal surface.

Sun et al. present a method for detecting feature edges [5]. This method computes the likelihood of an edge to be a feature edge -- called the "edge strength," by taking vote of tensors calculated from the normal vectors of the vertices weighted by the geodesic distance from the edge. Their method is designed for segmenting a dense uniform mesh, such as a mesh created by a laser range scanner, and more work must be done to extend this method to a polygonal surface created by a CAD facet generator.

Baker presents a method for identifying feature edges from a triangular mesh [6]. His method uses two angle thresholds: one identifies dominant edges, and the other identifies sub-dominant edges. Dominant edges are automatically extracted as feature edges, and some sub-dominant edges are extracted based on the connection to the dominant edges. The method assumes a sufficient degree of regularity, and cannot be applied to a general polygonal surface unless some modification to the method is made.

There has also been some researches on extracting feature edges from point clouds [7, 8]. These methods extract feature edges from dense sample points. These methods are not designed for taking a general polygonal surface; therefore the goal of the proposed method differs from the goals of these methods.

In summary, the existing feature edge extraction techniques for a polygonal surface can roughly be classified into two groups: (1) the techniques for a dense polygonal surface such as scanned data created by a laser range scanner, and (2) the techniques for a uniform triangular mesh. Neither type can deal effectively with a general polygonal surface that includes polygons with more than four vertices, concave polygons, and high aspect-ratio polygons.

The goal of the proposed method is to extract feature edges from a general polygonal surface, mainly a facet created by a CAD facet generator. The method tolerates a certain amount of noise imposed by a CAD facet generator. It also finds cylindrical surfaces and extracts feature edges on those cylindrical surfaces.

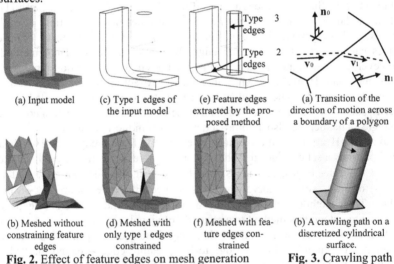

(a) Input model

(c) Type 1 edges of the input model

(e) Feature edges extracted by the proposed method

(a) Transition of the direction of motion across a boundary of a polygon

(b) Meshed without constraining feature edges

(d) Meshed with only type 1 edges constrained

(f) Meshed with feature edges constrained

(b) A crawling path on a discretized cylindrical surface.

Fig. 2. Effect of feature edges on mesh generation

Fig. 3. Crawling path

3. Polygon Crawling

The proposed feature edge detection technique relies on the technique called polygon crawling, which samples sequence of points along a path on the polygonal surface called a crawling path. A crawling path is analogue to a geodesic path of a smooth surface. However, a crawling path is a set of line segments as opposed to a smooth geodesic path of a smooth surface.

A crawling path is defined by an initial point on a polygonal surface, an initial direction, and termination criteria. An initial direction must be perpendicular to the normal of the polygonal surface at the initial point. Termination criteria can be anything; typically it is based on the length and/or the total angular change of the direction.

Assume that a sample point is located at the given initial point and moving toward the initial direction. The sample point maintains the same direction until it reaches the boundary of the polygon – an edge or a vertex. When the sample point moving toward $\mathbf{v}_0$ direction hits the boundary and moves out of the polygon whose unit normal is $\mathbf{n}_0$, and it moves into a polygon whose unit normal is $\mathbf{n}_1$, the moving direction changes to $\mathbf{v}_1 = \mathbf{R}(\mathbf{n}_0 \rightarrow \mathbf{n}_1)\mathbf{v}_0$ where $\mathbf{R}(\mathbf{n}_0 \rightarrow \mathbf{n}_1)$ is a 3×3 affine transformation matrix that rotates $\mathbf{n}_0$ into $\mathbf{n}_1$ about $\mathbf{n}_0 \times \mathbf{n}_1$ as shown in Fig. 3 (a). The sample point keeps moving until the termination criteria is met. A crawling path is then written as $C = \{\mathbf{p}_0, \mathbf{p}_1, ..., \mathbf{p}_N\}$ where $\mathbf{p}_i$ is the point at which the sample point crosses an edge or a vertex for $(i+1)$ th time. Fig. 3 (b) shows an example of a crawling path on a discretized cylindrical surface.

We also define a "crawling plane" as $(\mathbf{x} - \mathbf{o}_{cp}) \cdot (\mathbf{v}_0 \times \mathbf{n}_0) = 0$, where $\mathbf{x}$ is a point on the crawling plane, $\mathbf{o}_{cp}$ is the initial point of the crawling path, $\mathbf{v}_0$ is a unit vector parallel to the initial direction of the crawling path, and $\mathbf{n}_0$ is the normal of the polygon on which the initial point is lying. A crawling plane is shown in Fig. 3 (b) by the semi-transparent rectangle.

As can be seen from Fig. 3 (b), a crawling path can move away from the crawling plane. We define a "twist angle" at each point on a crawling path for measuring a directional deviation of the crawling direction from the crawling plane. If the direction of the crawling path is $\mathbf{v}_n$ ($\mathbf{v}_n$ is a unit vector) at point $\mathbf{x}$ on the crawling path, the twist angle is defined as $\theta_w = \mathrm{asin}(|\mathbf{v}_n \cdot \mathbf{n}_{cp}|)$, where $\mathbf{n}_{cp}$ is the normal of the crawling plane written as $\mathbf{n}_{cp} = (\mathbf{v}_0 \times \mathbf{n}_0)$.

4. User-Specified Parameters

The result of the proposed method can be controlled by eight parameters as shown in Table 1. For obtaining the best result, these parameters must be adjusted for the meshing requirements and the behavior of the CAD facet generator. However, because it is difficult to quantify the behavior of the CAD facet generator, a good combination of these parameters can only be found empirically. The experiments performed in this research suggest that the same set of parameters works well for the polygonal surfaces created by the same CAD facet generator.

Smaller $\omega_{shallow}$, ω_{sharp}, ω_{twist}, γ_1, and γ_2 capture more feature edges, and increases the sensitivity to the noise. Parameters ω_{cyl1} and ω_{cyl2} also control sensitivity to noise, and smaller ω_{cyl1} and ω_{cyl2} capture more cylindrical surfaces, and increases sensitivity to noise. Hence, if the CAD facet generator creates a polygonal surface with little unevenness, these parameters can be smaller, and the method will give satisfactory results. Or if the CAD facet generator creates an uneven (noisy) polygonal surface, these parameters should be increased. N_{cyl} depends on how smooth a cylindrical surface needs to be when it is meshed. Larger N_{cyl} will yield a smoother mesh on a cylindrical surface. However, it may excessively increase the number of elements. A set of parameters that gave a good performance for majority of the test models is presented in Section 7.

Table 1. Input parameters

$\omega_{shallow}$	If two vectors make an angle less than $\omega_{shallow}$, two vectors are considered to be equivalent.
ω_{sharp}	An angle threshold for finding sharp corners. It is also used for termination condition of the crawling method detailed in Section 5.2.
ω_{twist}	An angle threshold used for a termination condition of the crawling method detailed in Section 5.2.
γ_1, γ_2	Two thresholds for finding feature edges in which the curvature changes significantly, detailed in Section 6.2.
ω_{cyl1}, ω_{cyl2}	Two angle thresholds for finding cylindrical surfaces, detailed in Section 6.3.
N_{cyl}	The required number of subdivision around a complete cylinder, detailed in Section 6.3.

5. Estimating Curvature by Polygon Crawling

The proposed method estimates curvature on the polygonal surface by analyzing a crawling path. The goal is to estimate a curvature of the polygonal surface at sample point x measured in direction v_0. In the following, a curvature estimation scheme of two-dimensional line segments is briefly reviewed, and it is then extended for estimating curvature of a crawling path, which is essentially the curvature estimation of the polygonal surface measured in the initial crawl direction.

5.1. Curvature Estimation of 2D Line Segments

The curvature of a curve is defined as $\kappa = 1/r_c$ where r_c is the curvature radius. Hence, if two tangential vectors at the two end points of a partial circular arc of length l_{arc} makes angle θ, the curvature of the arc is calculated as $\kappa = 1/r = \theta / l_{arc}$, because $l_{arc} = r\theta$ as shown in Fig. 4 (a). Therefore, the curvature of a curve can be approximated by the angular change of a tangential vector divided by the arc length required for yielding the change.

Similarly, the curvature of line segments can be estimated by the angular change of the average tangential vector divided by the length required to yield the change. If $\mathbf{p}_0$, $\mathbf{p}_1$, ..., $\mathbf{p}_N$ are the vertices of line segments, average tangential vectors are calculated as: $\mathbf{t}_0 = (\mathbf{p}_1 - \mathbf{p}_0)/|\mathbf{p}_1 - \mathbf{p}_0|$ at $\mathbf{p}_0$, $\mathbf{t}_N = (\mathbf{p}_N - \mathbf{p}_{N-1})/|\mathbf{p}_N - \mathbf{p}_{N-1}|$ at $\mathbf{p}_N$, and $\mathbf{t}_i = (\mathbf{p}_{i+1} - \mathbf{p}_{i-1})/|\mathbf{p}_{i+1} - \mathbf{p}_{i-1}|$ at $\mathbf{p}_i$ ($0 < i < N$).

Curvature on line segments between $\mathbf{p}_i$ and $\mathbf{p}_j$ ($i < j$) can be estimated as $\kappa(\mathbf{p}_i \rightarrow \mathbf{p}_j) = \theta_{seg}(\mathbf{p}_i \rightarrow \mathbf{p}_j)/l_{seg}(\mathbf{p}_i \rightarrow \mathbf{p}_j)$, where $\theta_{seg}(\mathbf{p}_i \rightarrow \mathbf{p}_j)$ is the angular change of the average tangential vector between $\mathbf{p}_i$ and $\mathbf{p}_j$, and $l_{seg}(\mathbf{p}_i \rightarrow \mathbf{p}_j) = \sum_{K=i}^{j-1} |\mathbf{p}_{K+1} - \mathbf{p}_K|$.

For example, when line segments are defined by six points, $\mathbf{p}_0$, $\mathbf{p}_1$, ..., and $\mathbf{p}_5$, curvature of the line segments between $\mathbf{p}_1$ and $\mathbf{p}_4$ is estimated as: $\kappa(\mathbf{p}_1 \rightarrow \mathbf{p}_4) = \theta_{seg}(\mathbf{p}_1 \rightarrow \mathbf{p}_4)/(\sum_{K=1}^{3} |\mathbf{p}_{K+1} - \mathbf{p}_K|)$. As can be seen from Fig. 4 (b), $\sum_{K=1}^{3} |\mathbf{p}_{K+1} - \mathbf{p}_K|$ is analog to the arc length of Fig. 4 (a).

5.2. Curvature Estimation of a Crawling Path

To extend the 2D curvature estimation scheme to a crawling path, the sign of the angular change of the average tangential vector must be taken into account because the curvature must be positive if the polygonal surface is convex along the crawling path, and negative if it is concave. Let $\mathbf{p}_0$, $\mathbf{p}_1$, ..., and $\mathbf{p}_N$ be points of a crawling path, $\overline{\mathbf{p}}_i$ ($0 \leq i \leq N$) is the projection of $\mathbf{p}_i$ on the crawling plane, and $\overline{\mathbf{t}}_0, \overline{\mathbf{t}}_1, ..., \overline{\mathbf{t}}_N$ are average tangential vectors calculated from $\overline{\mathbf{p}}_0, \overline{\mathbf{p}}_1, ..., \overline{\mathbf{p}}_N$. The signed angular change of the tangential vector from $\overline{\mathbf{t}}_i$ to $\overline{\mathbf{t}}_{i+1}$ is calculated as: $\theta_{seg}(\overline{\mathbf{p}}_i \rightarrow \overline{\mathbf{p}}_{i+1}) = \mathrm{acos}(\overline{\mathbf{t}}_i \cdot \overline{\mathbf{t}}_{i+1})$ if $(\mathbf{n}_0 \times \mathbf{v}_0) \cdot (\overline{\mathbf{t}}_i \times \overline{\mathbf{t}}_{i+1}) \geq 0$, $\theta_{seg}(\overline{\mathbf{p}}_i \rightarrow \overline{\mathbf{p}}_{i+1}) = -\mathrm{acos}(\overline{\mathbf{t}}_i \cdot \overline{\mathbf{t}}_{i+1})$ otherwise, where $\mathbf{n}_0$ is the normal vector of the polygonal surface at the initial point of the crawling path. The angular change of the tangential vector from $\overline{\mathbf{t}}_i$ to $\overline{\mathbf{t}}_j$ ($i < j$) is calculated as: $\theta_{seg}(\overline{\mathbf{p}}_i \rightarrow \overline{\mathbf{p}}_j) = \sum_{K=i}^{j-1} \theta_{seg}(\overline{\mathbf{p}}_K \rightarrow \overline{\mathbf{p}}_{K+1})$. The curvature of the projected crawling path between $\overline{\mathbf{p}}_i$ and $\overline{\mathbf{p}}_j$ can be estimated as: $\kappa_{seg}(\overline{\mathbf{p}}_i \rightarrow \overline{\mathbf{p}}_j) = \theta_{seg}(\overline{\mathbf{p}}_i \rightarrow \overline{\mathbf{p}}_j)/l_{seg}(\overline{\mathbf{p}}_i \rightarrow \overline{\mathbf{p}}_j)$.

For estimating the curvature of the polygonal surface at $\mathbf{x}$ in $\mathbf{v}_0$ direction, the following two issues still need to be considered: (1) termination criteria for sampling a crawling path, and (2) choice of i and j.

The proposed curvature estimation method uses three user-specified parameters, sharp angle threshold $-\ \omega_{sharp}$, twist angle threshold $-\ \omega_{twist}$, and shallow angle threshold $-\ \omega_{shallow}$ for defining the termination criteria. The proposed method samples two crawling paths $C_0 = \{a_0, a_1, ..., a_N\}$, and $C_1 = \{b_0, b_1, ..., b_M\}$. Projection of C_0 on the crawling plane is $\overline{C}_0 = \{\overline{a}_0, \overline{a}_1, ..., \overline{a}_N\}$, and C_1 $\overline{C}_1 = \{\overline{b}_0, \overline{b}_1, ..., \overline{b}_M\}$. The initial point of both C_0 and C_1 are x, and the initial crawling direction of C_0 is v_0 and of C_1 is $-v_0$. For both directions, the points are sampled until: (1) total angular change of the average tangential vector on the crawling plane exceeds ω_{sharp}, i.e., both $\theta_{seg}(\overline{a}_0 \rightarrow \overline{a}_N)$ and $\theta_{seg}(\overline{b}_0 \rightarrow \overline{b}_M)$ do not exceed ω_{sharp}, (2) angular change of the normal vectors of polygons in one step of crawling exceeds ω_{sharp}, or (3) twist angle exceeds ω_{twist}.

The two crawling paths are then merged into a crawling path: $C_{all} = \{p_{-(M+1)}, p_{-M}, ..., p_N\}$ $= \{b_M, b_{M-1}, ..., b_0, a_0, ..., a_N\}$. Projection of C_{all} on the crawling plane is $\overline{C}_{all} = \{\overline{p}_{-(M+1)}, \overline{p}_{-M}, ..., \overline{p}_N\}$. Fig. 5 (a) shows an example of a merged crawling path.

From the merged crawling path, $\overline{C}_{all}$, $\kappa_{seg}(\overline{p}_i \rightarrow \overline{p}_j)$ can be computed for any i and j that satisfy $i < j$, $-(M+1) \leq i$, $i \leq N-1$, $-M \leq j$, and $j \leq N$, and the curvature of the polygonal surface at x in v_0 direction is found by choosing appropriate i and j. However, it is very difficult to automatically choose appropriate i and j. Instead of choosing only one set of i and j, the proposed method estimates the curvature from some pairs of $l_{seg}(\overline{p}_i \rightarrow \overline{p}_j)$ and $\theta_{seg}(\overline{p}_i \rightarrow \overline{p}_j)$, or length-angle pairs, written as $h_k = (l_k, \theta_k)$, by the following algorithm.

Let $k=0$

Let i be the largest integer that satisfies $\theta_{seg}(\overline{p}_i \rightarrow \overline{p}_0) \geq \omega_{shallow}$ and $i < 0$.

(If none satisfies, $i = -(M+1)$)

Let j be the smallest integer that satisfies $\theta_{seg}(\overline{p}_{-1} \rightarrow \overline{p}_j) \geq \omega_{shallow}$ and $0 \leq j$.

(If none satisfies, $j = N$)

Do:

Set $h_k = (l_k, \theta_k) = (l_{seg}(\overline{p}_i \rightarrow \overline{p}_j), \theta_{seg}(\overline{p}_i \rightarrow \overline{p}_j))$

Find the next value of j, j', which satisfies:

   $\theta_{seg}(\overline{p}_j \rightarrow \overline{p}_{j'}) \geq \omega_{shallow}$, $j \leq N$, and $j < j'$. (If none satisfies, $j' = N$)

Find the next value of i, i', which satisfies:

   $\theta_{seg}(\overline{p}_{i'} \rightarrow \overline{p}_i) \geq \omega_{shallow}$, $-(M+1) < i$, and $i' < i$. (If none satisfies, $i' = -(M+1)$)

If $l_{seg}(\overline{p}_i \rightarrow \overline{p}_0) < l_{seg}(\overline{p}_{-1} \rightarrow \overline{p}_j)$ replace i with i', otherwise replace j with j'.

Increment k.

Until one of $\theta_{seg}(\overline{p}_i \rightarrow \overline{p}_j) > \omega_{sharp}$ or ($i = -(M+1)$ and $j = N$) is satisfied.

This algorithm updates i and j so that the growth of $l_{seg}(\overline{p}_i \rightarrow \overline{p}_0)$ and $l_{seg}(\overline{p}_{-1} \rightarrow \overline{p}_j)$ are balanced. Fig. 5 (b), (c), and (d) show examples of length-angle pairs.

Finally, the curvature is estimated from the length-angle plot. On length-angle plot with length as the lateral axis and angle as the vertical axis, the slope of a function $\theta = f(l)$ describes the curvature. Hence, the curvature can be estimated by the least-square fitting of function $\theta = \kappa_{est}l$ to $\mathbf{h}_k$ as shown in Fig. 6; and κ_{est} is written as: $\kappa_{est}(\mathbf{x},\mathbf{v}_0)=\sum_k l_k\theta_k/\sum_k l_k^2$, which gives an estimated curvature at $\mathbf{x}$ measured in $\mathbf{v}_0$ direction.

5.3. Reliability of the Curvature Estimation for a Crawling Path

A reliability measure of the estimated curvature must be taken into account to avoid excessive extraction of feature edges. If the estimated curvature is subject to considerable error, it makes a sudden change of the estimated curvature across an edge, and the proposed method may falsely consider some edges as feature edges.

The curvature estimation scheme explained in Sections 5.1 and 5.2 becomes inaccurate where the curvature changes from negative to positive. For example, for the crawling path shown in Fig. 7 (a), the exact curvature could be zero because the sample point is on a straight line connecting two curves as shown in Fig. 7 (b), or it could be non-zero because the sample point is on one of the two curves represented by the sampled crawling path.

The proposed method tests the reliability of the curvature estimation as follows. Let I and J be smallest i and largest j used for sampling length-angle pairs. The curvature estimation is considered to be unreliable if all three conditions of the following are satisfied: (1) $\theta_{seg}(\mathbf{p}_I\rightarrow\mathbf{p}_0)\theta_{seg}(\mathbf{p}_{-1}\rightarrow\mathbf{p}_J)<0$, (2) $|\theta_{seg}(\mathbf{p}_0\rightarrow\mathbf{p}_J)|>\omega_{shallow}$, and (3) $|\theta_{seg}(\mathbf{p}_I\rightarrow\mathbf{p}_0)|>\omega_{shallow}$. These conditions check whether the crawling path bends considerably to both positive and negative curvature directions. If at least one of the three conditions is not satisfied, the curvature estimation is considered to be reliable.

5.4. Finding Principal Curvature

The principal curvature at a point on the polygonal surface can be found by numerically maximizing the absolute value of the curvature estimated by the scheme explained in Section 5.2. Let $\mathbf{x}$ be a point on the polygonal surface, and $\mathbf{n}$ be the normal of the polygonal surface at $\mathbf{x}$. $\mathbf{i}$ is an arbitrary unit vector that satisfies $\mathbf{n}\cdot\mathbf{i}=0$, and $\mathbf{j}=\mathbf{n}\times\mathbf{i}$. Any unit vector parallel to the polygon can be described as $\mathbf{v}(\phi)=\mathbf{i}\cos\phi+\mathbf{j}\sin\phi$. The problem of finding the principal curvature at $\mathbf{x}$ thus becomes a one-dimensional optimization problem: finding ϕ that maximizes $|\kappa_{est}(\mathbf{x},\mathbf{v}(\phi))|$. A conventional numerical optimization scheme such as bi-section search, Newton's method, or Powell's method, can solve this problem.

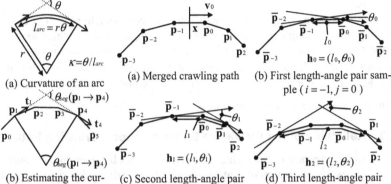

(a) Curvature of an arc

(a) Merged crawling path

(b) First length-angle pair sample ($i=-1, j=0$)

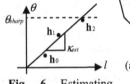

(b) Estimating the curvature of line segments

(c) Second length-angle pair sample ($i=-1, j=1$)

(d) Third length-angle pair sample ($i=-2, j=1$)

Fig. 4. Estimating curvature in 2D

Fig. 5. Crawling path C_{all} ($M=N=2$) and length-angle pairs.

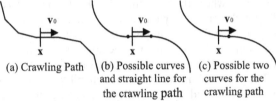

Fig. 6. Estimating the curvature by fitting a linear function to length-angle pairs

(a) Crawling Path

(b) Possible curves and straight line for the crawling path

(c) Possible two curves for the crawling path

Fig. 7. Ambiguity of possible curves and lines represented by the crawling path

6. Extracting Feature Edges

The proposed method extracts three different types of feature edges in the following order: (1) Feature edges on sharp corners, (2) Feature edges where the curvature changes suddenly across an edge, (3) Feature edges on the cylindrical surface.

For finding types (1) and (2) feature edges, the proposed method tests each edge. For type (3) feature edges the proposed method takes one sample point per polygon and tests whether the polygon is a part of a cylindrical surface. The total computational time for finding feature edges is thus roughly proportional to the number of edges and the number of polygons. The following section explains the methods for extracting feature edges of the three types.

6.1. Extracting Feature Edges on Sharp Corners

The proposed method first extracts feature edges on sharp corners. The proposed method tests each edge of the polygonal surface shared by two polygons; if the normal vectors of the two polygons sharing the edge make a larger angle than a given angle threshold ω_{sharp}, the edge is extracted as a feature edge.

6.2. Extracting Feature Edges Based on Sudden Curvature Change

The proposed method extracts feature edges by finding a sudden change of estimated curvature across an edge. For this purpose, the curvature is measured in the direction perpendicular to the edge at two points – one point each on a polygon sharing the edge.

The proposed method visits each edge shared by two polygons and estimates the curvature for each polygon sharing the edge as follows: Let e_1 and e_2 be the vertices of the edge being tested, and n is the normal of the polygon. The proposed method shoots two rays from $(e_1+e_2)/2$ to one of the directions $(e_2-e_1) \times n$ or $(e_1-e_2) \times n$ whichever points inside the polygon and finds the first intersection e_x. A point inside the polygon x_{mid} is then found as: $x_{mid}=(e_x+(e_1+e_2)/2)/2$, as shown in Fig. 8. Since the edge is used by two polygons, two of such points, x_{mid1} and x_{mid2}, are found. The proposed method then estimates curvatures at x_{mid1} and x_{mid2} as: $\kappa_{est1} = \kappa_{est}(x_{mid1}, v_{mid1})$, and $\kappa_{est2} = \kappa_{est}(x_{mid2}, v_{mid2})$, where v_{midK} is a unit vector parallel to $(x_{midK}-(e_1+e_2)/2)$, $\{K=1,2\}$. An example of the curvature estimation at two points on the both sides of an edge is shown in Fig. 9.

Finally, if one of the following two conditions is met, the edge is extracted as a feature edge.

(1) κ_{est1} and κ_{est2} are reliable with respect to the conditions described in Section 5.3, and the ratio of $|\kappa_{est1}|$ and $|\kappa_{est2}|$ is greater than a given threshold γ_1 (i.e., $|\kappa_{est1}|>\gamma_1|\kappa_{est2}|$ or $|\kappa_{est2}|>\gamma_1|\kappa_{est1}|$)

(2) Only one of κ_{est1} and κ_{est2} is reliable with respect to the conditions explained in Section 5.3, and the ratio of $|\kappa_{est1}|$ and $|\kappa_{est2}|$ is greater than a given threshold γ_2 (i.e., $|\kappa_{est1}|>\gamma_2|\kappa_{est2}|$ or $|\kappa_{est2}|>\gamma_2|\kappa_{est1}|$)

The threshold γ_1 describes how much change of curvature across an edge is required for the edge to be extracted as a feature edge. The edge is extracted as a feature edge if the curvature changes γ_1 times across the edge. Hence, γ_1 must be chosen based on the user's requirement.

The choice of γ_2 must depend on the percentage of the error caused by the unreliability described in Section 5.3. For example, if the estimated curvature κ_{est} is subject to an error ratio of E ($E>1$), κ_{est} can vary in the range of $\kappa/E \leq \kappa_{est} \leq \kappa E$, where κ is the exact curvature. To take the error into account, γ_2 must be $E\gamma_1$. However, because the percentage of error E cannot be obtained, γ_2 can only be found empirically.

6.3. Extracting Feature Edges on the Cylindrical Surfaces

The third procedure in the proposed method extracts feature edges on the cylindrical surfaces. It is desirable to constrain four to six edges on the cylindrical surface that are parallel to the axis of the cylinder for the meshing as shown in Fig. 2.

It is also important to constrain edges on the co-axial cylindrical surfaces consistently so that meshing does not produce self-intersections. For example, if two co-axial cylindrical surfaces have constrained edges as shown by thick lines in Fig. 10, meshing produces self-intersections when large elements need to be created for reducing the number of elements. Hence, the proposed method needs to find cylindrical surfaces from the polygonal surface, and extract feature edges on them consistently so that the chance of yielding such self-intersections is minimized.

If some polygons represent a complete cylinder (a cylinder whose cross section is a complete circle,) the cylinder can be found by fitting a circle to a crawling path in the principal curvature direction and checking the fitness of the circle to the crawling path. However, it is difficult to identify some polygons representing a partial cylinder (a cylinder whose cross section makes a partial circle) because such polygons can also be interpreted as a part of a non-cylindrical surface.

For this reason, a heuristic approach is used for finding partial cylinders. Often a complete cylinder is co-axial to a partial cylinder; when the proposed method finds some polygons that potentially represent a partial cylinder, those polygons are considered to be a part of a cylindrical surface if a complete cylinder co-axial to the partial cylinder is located close to the partial cylinder.

The proposed method uses two angle thresholds – ω_{cyl1} and ω_{cyl2} ($\omega_{cyl1} < \omega_{cyl2}$) – to find cylindrical surfaces. When the proposed method locates a set of polygons that may represent a cylindrical surface, it calculates the angle of arc of the cross section of the potential cylindrical surface, denoted as θ_{cyl}. The surface represented by the polygons is considered to be a complete cylinder if $\omega_{cyl2} \le \theta_{cyl}$. If $\omega_{cyl1} \le \theta_{cyl} < \omega_{cyl2}$, the surface represented by the polygons is considered to be a partial cylinder if a co-axial complete cylinder is found. The surface is not considered to be a part of a cylindrical surface if $\theta_{cyl} < \omega_{cyl1}$. Since θ_{cyl} is calculated by taking an angle of revolution made by a crawling path around the surface, we call θ_{cyl} a "revolution angle" of the cylinder.

After locating cylindrical surfaces represented by sets of polygons the proposed method makes groups of cylindrical surfaces based on the proximity and axial directions of the cylinders. A coordinate frame is assigned to each group, and feature edges are extracted on the cylindrical surfaces based on the coordinate frame. Details of the algorithm are described below.

The proposed method calculates a principal curvature κ_1 and the principal curvature direction $v_{\kappa 1}$ at $\mathbf{x}$ by the method described in Section 5.4 where $\mathbf{x}$ is a point on the polygonal surface. The method then finds two crawling paths starting from $\mathbf{x}$ to the directions $\pm v_{\kappa 1}$ until: (1) total angular change of the average tangential vector exceeds 180 degrees, (2) angular change of the normal vectors of polygons in one step of crawling exceeds ω_{sharp}, or (3) twist angle exceeds ω_{twist}.

The two crawling paths described as $\{a_0, a_1, ..., a_N\}$ and $\{b_0, b_1, ..., b_M\}$ are merged into a crawling path: $C_{cyl} = \{\mathbf{p}_{-M-1}, \mathbf{p}_{-M}, ..., \mathbf{p}_N\} = \{\mathbf{b}_M, \mathbf{b}_{M-1}, ..., \mathbf{b}_0, \mathbf{a}_0, \mathbf{a}_1, ..., \mathbf{a}_N\}$.

The revolution angle θ_{cyl} is calculated from C_{cyl} as: $\theta_{cyl} = \theta_{seg}(\mathbf{p}_{-M-1} \rightarrow \mathbf{p}_N)$, which is the total angular change of the tangential vector of the crawling path from $\mathbf{p}_{-M-1}$ to $\mathbf{p}_N$. If the crawling path is on a complete cylindrical surface and makes a complete revolution around the cylinder, θ_{cyl} is 360 degrees. If the crawling path is on a partial cylindrical surface, θ_{cyl} is less than 360 degrees.

If the crawling path is on a cylindrical surface, the plot of normal vectors lies on a plane, and the normal of the plane is parallel to the axial direction as shown in Fig. 11. The proposed method fits a plane to the normal vectors of the polygons that the crawling path passes by the least-square method. The plane is denoted as P_n, and the axial direction is denoted as $\mathbf{a}$.

The points of the crawling path are then projected on P_n, and a circle that fits to the projected points is calculated by the least-square method [9, 10]. The center of the circle is $\mathbf{o}$, and the diameter is R. To test the fitness of the circle to the projected points, the proposed method calculates the maximum and the minimum distances from a projected point to the center of the circle, R_{max} and R_{min}.

Finally, the proposed method considers that the polygons passed by the crawling path are part of a cylindrical surface if the following three conditions are satisfied: (1) $\omega_{cyl1} \leq \theta_{cyl}$, (2) $0.95 \leq R_{max}/R \leq 1.05$, and (3) $0.95 \leq R_{min}/R \leq 1.05$.

The proposed method tests the above conditions at a sample point per polygon and makes a list of detected cylindrical surfaces. If the polygon is convex, the sample point for the polygon is the center of the polygon. If the polygon is concave, the proposed method picks randomly a point inside the polygon.

Groups of cylindrical surfaces are then made by the following algorithm. Let $C_{cyl,1}, C_{cyl,2}, ..., C_{cyl,N}$ be the detected cylinders sorted in the descending order of the revolution angle, and $\theta_{cyl,k}$ be the revolution angle of $C_{cyl,k}$ ($1 \leq k \leq N$); $\theta_{cyl,i} \geq \theta_{cyl,j}$ ($i < j$). Assume that $C_{cyl,I}$ is the first cylinder that is not yet in any group and satisfies $\omega_{cyl2} \leq \theta_{cyl,I}$. The proposed method makes a new group of cylindrical surfaces, $G_{cyl,K}$ (K th group), that initially includes only $C_{cyl,I}$. It then compares $C_{cyl,I}$ with $C_{cyl,J}$ where $I+1 \leq J \leq N$, and $C_{cyl,J}$ is added to $G_{cyl,K}$ if

(1) the axial directions of $C_{cyl,I}$ and $C_{cyl,J}$ are not different by more than $\omega_{shallow}$, and (2) $C_{cyl,I}$ and $C_{cyl,J}$ intersect each other, or one encloses the other. The two conditions checks if $C_{cyl,I}$ and $C_{cyl,J}$ are co-axial. The two conditions, however, can be satisfied if the axis of one cylinder is off center of the other. The two conditions are intentionally made loose for tolerating discretization errors. The proposed method repeats this process until all the cylinders that satisfy $\omega_{cyl2} \le \theta_{cyl,I}$ are included in a group. Some cylinders with less than ω_{cyl2} revolution angle may not belong to any group, and feature edge will not be extracted on such cylinders.

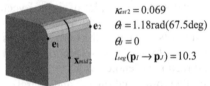

$\kappa_{est1} = 0.49$
$\theta_I = 0.18\text{rad}(10.3\deg)$
$\theta_J = 0.98\text{rad}(56.2\deg)$
$l_{seg}(\mathbf{p}_I \rightarrow \mathbf{p}_J) = 2.34$

(a) Crawling path originated from x_{mid1}

$\kappa_{est2} = 0.069$
$\theta_I = 1.18\text{rad}(67.5\deg)$
$\theta_J = 0$
$l_{seg}(\mathbf{p}_I \rightarrow \mathbf{p}_J) = 10.3$

(b) Crawling path originated from x_{mid2}

Fig. 9. Example of curvature estimation on both sides of an edge

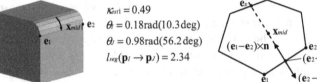

Fig. 8. Finding a point Inside a polygon

Yield self-intersection

Fig. 10. Inappropriately constrained edges

Finally, the proposed method calculates a coordinate frame for each group and makes feature edges on the surfaces of the cylinders included in the group. $\mathbf{a}_K$ is a unit vector that is parallel to the axis of the first cylinder of the group $G_{cyl,K}$, $\mathbf{i}$ is an arbitrary unit vector that is perpendicular to $\mathbf{a}_K$, and $\mathbf{j}$ is $\mathbf{i} \times \mathbf{a}_K$. The coordinate frame is defined by the three vectors $\mathbf{i}$, $\mathbf{j}$, and $\mathbf{a}_K$. The proposed method assigns to each polygon included in the cylinders a clockwise rotation angle measured from $\mathbf{i}$ to the normal of the polygon $\mathbf{n}$, ($\theta_h = \text{acos}(\mathbf{i} \cdot \mathbf{n})$ if $\mathbf{j} \cdot \mathbf{n} \ge 0$, $\theta_h = 2\pi - \text{acos}(\mathbf{i} \cdot \mathbf{n})$ otherwise.) The proposed method extracts as a feature edge an edge shared by two polygons included in a cylinder of the group that satisfies $\lfloor \theta_{h1}N_{cyl}/2\pi \rfloor \ne \lfloor \theta_{h2}N_{cyl}/2\pi \rfloor$ where θ_{h1} and θ_{h2} are the angles assigned to the two polygons sharing the edge, N_{cyl} is a user-specified parameter that defines the number of feature edges needed around a complete cylinder, and $\lfloor x \rfloor$ is the largest integer that is less than x. For example, when N_{cyl} is four, $\lfloor \theta_h N_{cyl}/2\pi \rfloor$ varies zero to three, and a feature edge is extracted where $\lfloor \theta_h N_{cyl}/2\pi \rfloor$ changes as shown in Fig. 12.

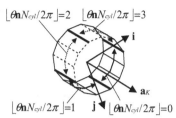

(a) A discretized cylindrical surface and normal vectors

(b) A plot of normal vectors lying on a plane

Fig. 11. Normal vector plot of a discretized cylindrical surface

Fig. 12. Feature edges (Thick lines) around a cylinder

7. Results

The proposed method has been tested with many polygonal surfaces including more than fifty facet models used for plastic injection molding simulation, and has shown good performance. This section shows some results from the experiments. The parameters used for creating these models are shown in Table 2.

Figures 13 and 14 demonstrate the ability of the proposed method in capturing sharp corners, fillet boundaries, and cylindrical surfaces. Both models include concave polygons with more than four nodes. The proposed method successfully captures the cylindrical surfaces and extracted six feature edges on each cylindrical surface. Fig. 13 (b) shows that the proposed method extracts feature edges on the two co-axial cylinders so that a feature edges on one of the two cylinders is located close to a feature edge on the other cylinder in order to reduce the possibility of producing self-intersections when large elements must be created on the cylinders. Both models are successfully meshed into a very coarse mesh without producing self-intersections. Creating a very coarse mesh without producing self-intersections is very difficult and is only possible when appropriate edges are constrained.

Fig. 15 shows a more complex example. The input model includes concave polygons with more than four vertices. The proposed method successfully extracts feature edges from the input model, and the model is meshed into a coarse triangular mesh.

Fig. 16 shows a PDA cover. This model is created by a CAD facet generator and includes many high aspect-ratio triangles. The model has some small cylindrical holes as shown in Fig. 16 (b); such holes create self-intersections when a coarse mesh is needed unless feature edges are appropriately constrained. The proposed method successfully extracts feature edges on the holes, and a valid triangular mesh is shown in Figures 16 (e) and 16 (f).

Table 2. Parameters used for creating examples

$\omega_{shallow}$	5 degrees	ω_{twist}	45 degrees	γ_2	10	ω_{cyl1}	45 degrees
ω_{sharp}	45 degrees	γ_1	2	N_{cyl}	6	ω_{cyl2}	270 degrees

8. Discussions and Conclusions

The proposed method extracts feature edges from a general polygonal surface. The method extracts (1) edges on sharp corners, (2) edges where the curvature changes significantly across the edge, and (3) edges on the cylindrical surfaces. Such feature edges must be constrained for meshing, or some geometric features will be lost unless sufficiently small elements are used. The method has been tested with many polygonal surfaces and has shown a good performance on the test models.

The method helps create a coarse mesh without producing self-intersections. However, the method does not necessarily help create good quality elements. For creating a valid and good quality mesh, the proposed method must be used together with an appropriate element sizing function.

The method captures cylindrical surfaces and extracts edges on such cylindrical surfaces. However, the method does not capture when the cross-section is an ellipse, and/or the surface is significantly tapered. Capturing such elliptic cylinders and conic surfaces is a topic for future research.

References

[1] A. P. Mangan and R. T. Whitaker, "Partitioning 3D Surface Meshes Using Watershed Segmentation", *IEEE Transactions on Visualization and Computer Graphics*, vol. 5, pp. 308-321, 1999.

[2] M. Nomura and N. Hamada, "Feature Edge Extraction from 3D Triangular Meshes Using a Thinning Algorithm," Proceedings of SPIE - Vision Geometry X, pp. 34-41, 2001.

[3] X. Jiao and M. T. Heath, "Feature Detection for Surface Meshes," Proceedings of 8th International Conference on Numerical Grid Generation in Computational Field Simulations, pp. 705-714, 2002.

[4] X. Jiao and M. T. Heath, "Overlaying Surface Meshes, Part II: Topology Preservation and Feature Detection", *International Journal on Computational Geometry and Applications*, vol. 14, pp. 403-419, 2004.

[5] Y. Sun, D. L. Page, J. K. Paik, A. Koschan, and M. A. Abidi, "Triangle Mesh-Based Edge Detection and Its Application to Surface Segmentation and Adaptive Surface Smoothing," Proceedings of International Conference on Image Processing, pp. 825-828, 2002.

[6] T. J. Baker, "Identification and Preservation of Surface Features," Proceedings of 13th International Meshing Roundtable, pp. 299-309, 2004.

[7] S. Gumhold, X. Wang, and R. MacLeod, "Feature Extraction from Point Clouds," Proceedings of 10th International Meshing Roundtable, pp. 293-305, 2001.

[8] M. Pauly, R. Keiser, and M. Gross, "Multi-Scale Feature Extraction on Point-Sampled Surfaces," Proceedings of Eurographics, pp., 2003.

[9] W. Gander, G. H. Golub, and R. Strebel, "Least Square Fitting of Circles and Ellipses", *BIT*, vol. 34, pp. 558-578, 1994.

[10] I. D. Coope, "Circle Fitting by Linear and Nonlinear Least Squares", *Journal of Optimization Theory and Applications*, vol. 76, pp. 381-388, 1993.

 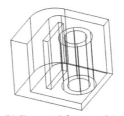 

(a) Input model (235 vertices and 200 polygons) (b) Extracted feature edges (Less than 0.1 second [*]) (c) A triangular mesh (165 vertices and 326 triangles)

Fig. 13. A Mechanical Part

(a) Input model (356 vertices and 348 polygons) (b) Extracted feature edges (Less than 0.1 second [*]) (c) A triangular mesh (204 vertices and 428 triangles)

Fig. 14. A Bent Beam with Holes

(a) Input model (1,661 vertices and 1,082 polygons) (b) Extracted feature edges (1.3 seconds [*]) (c) A triangular mesh(444 vertices and 884 triangles)

Fig. 15. B767

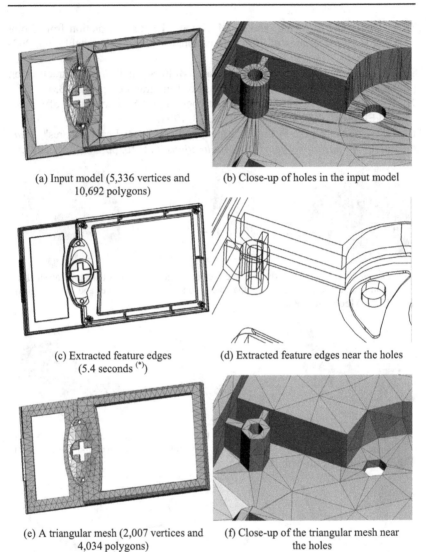

(a) Input model (5,336 vertices and 10,692 polygons)

(b) Close-up of holes in the input model

(c) Extracted feature edges (5.4 seconds [*])

(d) Extracted feature edges near the holes

(e) A triangular mesh (2,007 vertices and 4,034 polygons)

(f) Close-up of the triangular mesh near the holes

Fig. 16. A front cover of PDA

[*] Computational time is measured with a PC with AMD Athlon 64 2.2GHz.

Generation of Mesh Variants via Volumetrical Representation and Subsequent Mesh Optimisation

Katrin Bidmon and Thomas Ertl

Visualization and Interactive Systems Group, University of Stuttgart
Universitaetsstrasse 38, 70569 Stuttgart, Germany
(bidmon|ertl)@vis.uni-stuttgart.de
http://www.vis.uni-stuttgart.de/

Summary. Having reliable finite element (FE) meshes is one of the basics of reliable FE simulations. As development times i.e. in the car industry are expected to decrease, engineers need to edit and optimise FE meshes without access to the underlying CAD geometry. If meshes are not only locally effected by the editing operation, simple mesh optimisations such as mesh relaxation or local remeshing are not sufficient to make the mesh suitable for numerical simulation again and global remeshing is needed. To avoid the traditionally used time-consuming remeshing strategy, we developed a tool to remesh an FE surface model – taking into account the needs for good FE meshes – via volumes. We first voxelise the surface and then generate a new quad mesh via isosurface extraction and subsequent mesh optimisation. This method provides the opportunity to directly couple editing operations on the volumetrical representation with the remeshing procedure.

Key words: FE mesh, remeshing, warping, optimization, voxelization, isosurface extraction

1 Introduction

Virtual prototyping more and more replaces real mock-ups and experiments in industrial product development such as in automotive industry. The fast increasing processing power of modern computers together with more and more efficient algorithms allows to calculate complex non-linear and highly dynamic processes such as crash worthiness simulations within few days. So, expensive car prototypes are increasingly replaced by virtual simulations based on Finite Element Analysis (FEA).

Usually the car parts are designed as analytical surfaces using computer aided design (CAD) and have to be transformed into an FE mesh for simulations. For this

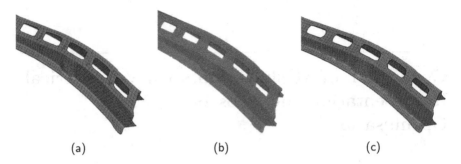

(a)   (b)   (c)

Fig. 1. Surface in different representations: (**a**) original FE mesh representation, (**b**) volumetric representation and (**c**) reconstructed FE mesh representation

purpose several meshing algorithms have been developed (e.g. [BS91, ZZHW91]), but most of them are tailored to fit a specific kind of simulation and to preserve a special surface property. So, a lot of expert knowledge and manual intervention is still needed to adapt those meshes to fit the prerequisites of a reliable numerical simulation. A recent improvement in simulation algorithms made it possible to mesh each car component individually, which was a huge step forward in giving the engineer more flexible tools and to speed up the development cycle. Now single car parts can be exchanged or varied without the necessity of remeshing the whole car model. This fact induced the desire for having tools to directly manipulate and edit the surfaces in the FE mesh representation(e.g. [BRE04]) instead of going back to the CAD department for each change during the development. If the changes in the surface only affect a small region in the mesh, the mesh properties required for simulation can be regained by local optimisation as relaxation or – in more serious cases – local remeshing. But if the mesh gets deformed too much during editing or manipulation, these repairing mechanisms will be insufficient and remeshing is needed. To avoid the frequent and thus time-consuming way back to CAD, the engineers need a fast method to generate new meshes suitable for numerical simulation.

Due to the potential possibility to directly combine this necessary remeshing with coarse-scale modifications of the surface in an intermediate step, we decided for a volume-based approach to remeshing (see Fig. 1): the surface is voxelised as described in Sect. 3, then reconstructed by isosurface extraction taking into account the desired properties of FE meshes (see Sect. 4). As isosurface extraction usually leads to triangle meshes, but quad meshes are needed for our structural mechanics simulations, the triangle mesh is converted to a quad dominated mesh in a subsequent processing step described in Sect. 4.2.

FE models are not very tolerant concerning mesh deformations, whereas generating quad meshes on curved surfaces always introduces warping: the problem that the vertices of a quad are not bound to lie on a common plane as vertices of a triangle inevitably do. But warping, which also can arise in the original mesh resulting from CAD, leads to huge problems during the numerical simulation and must be eliminated from the mesh before the simulation. This process is described in Sect. 5.1 in more detail. Additionally some quads in the mesh might be oriented diagonally to the other quads (see Fig. 9(a)) which may also lead to numerical problems during simulation. Thus, these quads need to be detected and removed by either splitting them into triangles or merging and re-splitting them together with neighbouring elements. This approach is explained in Sect. 5.2.

The paper presents some previous work in the following section and results and some conclusions in Sect. 6, as well as an outlook to enhancements planned for future work.

2 Related Work

There is a lot of work and publications about voxelisation, often bound to special needs of the application they were developed for. An algorithm fitting all interests at the same time does not exist. Especially for closed surfaces and real volumes many sometimes simple but powerful algorithms have been invented – such as binary volumes (e.g. [Kau87, HYFK98]) or distance volumes (e.g. [Gib98, VKK+03]) – some also taking into account special requirements like accuracy of surface details [Sra01, HLC+01]. A nice overview of the voxelisation literature is given in [COK95].

In our case we do not have closed surfaces but bounded ones, which leads to new problems: how to handle the boundary, how to treat distances measured to the boundary. Using binary volumes leads to problems with thin surfaces and the surface's thickness would have to be increased to get a hole-free voxelisation. Additionally it would lead to double surfaces during reconstruction, as the surface between "inside" and "outside" would be extracted – which is of no use for our purpose. So we decided for signed distance volumes, to be able to extract the zero isosurface as a single surface. Another approach, presented in [KBSS01], is to directly save the points where voxel and surface intersect each other instead of storing a scalar distance value per voxel.

Consequently, there is also a lot of related work in the field of isosurface extraction from volume data. Starting from the fundamental *Marching Cubes* algorithm [LC87], many variants of this method exist: the algorithm has been simplified to *Discretized Marching Cubes* [MSS94] where each intersection of the marching cube with the isosurface is set to the middle of the respective cube's edge. In addition, many extensions and improvements have been applied to the Marching Cubes algorithm in order to solve ambiguous cases and maintain topological properties (e.g. [Che95] and [LLVT03]), geometrical properties like sharp edges (e.g. [KBSS01]), or to get smoother surfaces (e.g. [LB03]).

Marching Cubes and its derivatives usually generate triangle meshes. In FE simulations we need quad meshes or at least quad dominated meshes with only few triangles since triangle meshes lead to unstable numerical simulations in FEA. So we have to extract a quad mesh from the volume, similar as in the *Dual Marching Cubes* [Nie04] where the dual grid is used to enhance the triangle mesh. Generating a mesh dual to the one generated by a Marching Cubes variant leads to new problems on bounded surfaces as data might be lost at the boundary (see Sect. 4.2).

In the following sections we describe how we combined the existing body of knowledge to make it available in the context of Finite Elements. By implementing our algorithms into the commercially available preprocessing tool *scFEMod* [sci04] we made this functionality available for productive use in the CAE departments of major German car manufacturers.

3 Voxelisation of the FE Mesh

Generating a volume representation of the FE mesh is done by distance calculation. The voxel size (uniform in each direction respectively) depends on the elements' globally minimal edge length to take fine surface features into account. When the volume dimensions are calculated, the shortest Euclidean distance to the FE mesh is computed for each voxel. If the distance to the surface is larger than a specified threshold (e.g. more than the doubled diagonal length of the voxels), the voxel value is set to 255, representing an "invalid" value that should not be taken into account for volume rendering or later isosurface extraction.

The point P on the surface having shortest distance to the voxel might have different properties depending on its location in respect to the bounded mesh (see Fig. 2): in the ideal case P lies inside an element (Fig. 2(a)) and the distance vector $\mathbf{v}$ is perpendicular to the element. Another possibility is, that the point P lies on an element edge or vertex being part of the surface boundary (see Fig. 2(b)). In this case the distance stored in the voxel would lead to wrong assumptions during volume rendering or surface reconstruction.

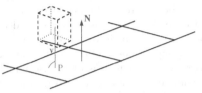
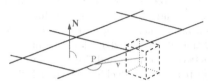

(a) Closest point on surface lies inside an element

(b) Closest point on surface lies on mesh boundary

Fig. 2. Two possible locations of the closest point on the surface in respect to the voxel's position

In order to avoid these problems voxels are skipped and their value is set to invalid if the angle enclosed by the corresponding distance vector $\mathbf{v}$ and the normal of the element the edge belongs to is larger than a specified threshold ($\mathbf{v} \cdot \mathbf{N} > \varepsilon$). In the other cases P lies on an inner edge or an inner vertex and the distance vector is compared to the average of the neighbouring elements' normals in the same way.

If a shortest distance is measured from the same voxel to an edge as well as to an element, the element gets favoured (see example in Fig. 3).

In Fig. 4 an example of a voxelised FE mesh is given, showing that this method preserves surface features such as small holes.

4 Surface Reconstruction

In order to retrieve a meshed surface, an isosurface has to be extracted from the volume. The Marching Cubes algorithm computes for each cube the zero points along the edges by linear interpolation between the values at the cell's vertices. If one vertex value is negative and one positive, there has to be a zero point on this edge

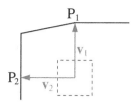

Fig. 3. In case a shortest distance from one voxel is measured to an edge (P_1) and to an element (P_2), the element gets favoured.

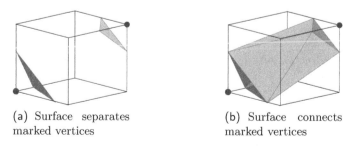

Fig. 4. Volumetric representation of the surface – preserving small surface features

which means an intersection of the cube with the surface. As we use signed distance fields to generate the volume, the zero isosurface is the surface we are looking for.

4.1 Isosurface Extraction

The classical Marching Cubes look-up table contains some ambiguous configurations that may lead to topological problems like holes in the mesh. If there are e.g. only two negative values on the vertices connected by an inner diagonal of the cube, the surface inside the cube may separate or connect these two vertices as depicted in Fig. 5.

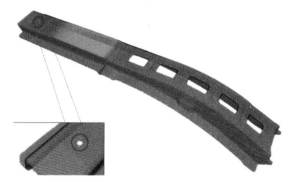

(a) Surface separates
marked vertices

(b) Surface connects
marked vertices

Fig. 5. Two possible triangulations with the same configuration of values on the vertices

Choosing the wrong configuration would lead to problems when connecting the generated triangles to the ones of the neighbouring cubes, possibly generating cracks in the triangle mesh. The development of enhanced look-up tables taking into account and solving these ambiguities, e.g. in [Che95], allowed algorithms – such as [LLVT03] – preserving the original topology and getting rid of unintentional holes in the mesh. This algorithm, an enhanced version of the *Marching Cubes 33* [Che95], not only takes into account the vertices' values, but also interpolates values at inner points or on surfaces to distinguish between configurations. This is important to ensure topologically correctness. To provide consistent triangulations neighbouring configurations have to be additionally considered.

As avoiding cracks in the mesh is among the most important prerequisites for FEA, we decided to base our meshing method on this algorithm: Using the look-up table of [LLVT03] we determine the connectivity and so the triangulation within each cube. But as a triangle mesh is not suitable for FE applications, we have to convert the obtained mesh into a quad mesh.

4.2 Quad Mesh Generation

In [Nie04] a triangle mesh is smoothed by applying the dual operator to the mesh two times. Applying this duality only once, we automatically obtain a quad mesh: the triangles are always constructed within a cube and connected to the triangles inside neighbouring cubes.

As a first step we consider only cubes of the inner surface, thus not containing triangles of the surface's boundary. Imagine four cubes connected in one conjoint edge. Connecting the centres of these four cubes leads to a square. This method can now be applied to the triangulation obtained as described before. Considering

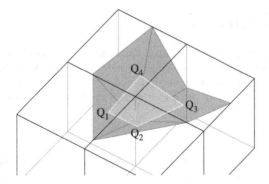

Fig. 6. Construction of a quad element dual to the triangle mesh generated by Marching Cubes

connected triangles within one cube as a single polygon, we can connect the centre of the polygon (Q_i) with the ones of the neighbouring cubes, as depicted in Fig. 6, and get quads instead of triangles. Even if there is more than one polygon within a cube, the statement holds for each of these surfaces respectively. The only problems arise on the surface boundary as the quad mesh is slightly smaller than the triangle mesh, due to the duality construction. To avoid that and to retain the original surface boundary, polygons are added and subsequently split or merged to quads. This method is a trade-off between keeping the shape of boundary lines and exchanging triangles in boundary corners by quads (see Fig. 7).

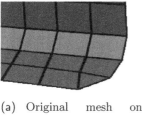

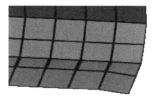

(a) Original   mesh   on boundary

(b) Reconstructed mesh on boundary

Fig. 7. Boundary reconstruction

4.3 Feature Line Preservation

One of the main problems with isosurface extraction is the fact that sharp feature lines are being wiped out. To avoid this we explicitly treat these feature lines. When generating the volume representation, feature lines are detected depending on the angle between two neighbouring elements and the voxels crossed by the lines are marked. Since we use the dual grid described in Sec. 4.2 our new vertices are located in the centre of the polygon extracted by Marching Cubes within each voxel. So, if the voxel is one of those originally crossed by a feature line, we move the new voxel to the closest point on the extracted feature line. If there is a feature point within the voxel, e.g. a corner in the surface boundary, the new vertex is moved to this point. As the vertices are being moved only slightly within the voxel the shape of the elements is only little affected but the feature lines are well preserved (see Fig. 8).

With these methods applied the resulting mesh already looks pretty appropriate for FE simulation. Nonetheless the quad mesh still lacks some enhancements – described in the following section – to fulfil the demands of Finite Element Analysis.

5 Quad Mesh Optimisation

Elements in FE meshes have to satisfy special properties to be able to be used for simulation. The mesh should consist of quads shaped as close to squares as possible, oriented all in the same direction (see Fig. 9). If needed few triangles can be included. Extreme angles as well as warping influence the simulation speed and result. Since the mesh obtained by the algorithm described above might lack these prerequisites in some elements, these properties have to be checked.

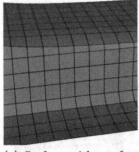

(a) Surface without feature preservation

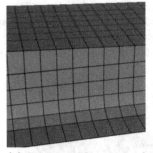

(b) Surface with preserved features

Fig. 8. Preservation of surface feature lines

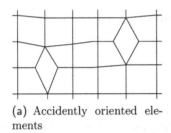

(a) Accidently oriented elements

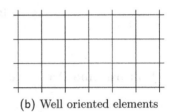

(b) Well oriented elements

Fig. 9. Orientation of quads in the FE mesh

5.1 Warping Removal

Compared to triangle meshes quad meshes bear the risk of warping. This might occur already during construction or as a result of manipulations of the elements.

Due to the quad construction described in Sect. 4.2 our mesh usually is warped in some elements. To make these elements suitable for simulation again they have to be detected and unwarped. Our meshing tool allows to mark warped elements by colour as seen in Fig. 10(a). To detect warping each quad is divided into two triangles the normals N_1 and N_2 of which get compared. If the angle is above a specified threshold, the diagonal is marked in dark/red, if it is below the threshold but the element still warped this is marked in light/yellow.

Both possible triangulations of a quad are examined and the direction corresponding to the wider angle between the two normals is the one considered further on during unwarping. To remove warping, all elements are analysed and if the angle exceeds the threshold the element is marked to be unwarped. Using the average of the two normals the linear smoothing plane through the middle point P of the four vertices $\mathbf{v}_i$ is calculated:

$$n_1 x + n_2 y + n_3 z - d = 0 \tag{1}$$

with $\mathbf{N} = (n_1, n_2, n_3)^T = \mathbf{N}_1 + \mathbf{N}_2$, $d = \sum n_i p_i$ and $P = (p_1, p_2, p_3) = \sum \mathbf{v}_i$. Then the vertices' new coordinates are calculated projecting them vertices onto this plane

$$\mathbf{v}'_i = \mathbf{v}_i + t\,\mathbf{N} \tag{2}$$

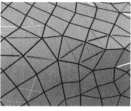

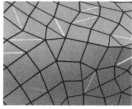

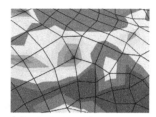

(a) Original mesh with warping

(b) Mesh after unwarping severely warped elements

(c) Euclidean distance between warped and unwarped surface

Fig. 10. (a) Highlighting and (b) unwarping of warped elements: slight warping coloured light/yellow, severe warping marked dark/red. (c) Comparison between warped mesh and unwarped mesh using distance mapping. Euclidean distance is colour-coded: yellow/light $< 0.01\,\mathrm{mm}$ to red/dark up to $1\,\mathrm{mm}$

with $t = (d - \mathbf{N} \cdot \mathbf{v}_i)/\|\mathbf{N}\|$.

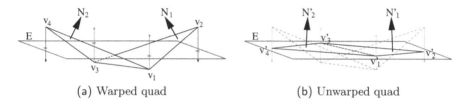

(a) Warped quad

(b) Unwarped quad

Fig. 11. Warping removal: Vertices of warped quads are projected on the linear smoothing plane

As neighbouring elements get affected by this treatment, the new coordinates are set no earlier than all elements are checked and the new coordinates are the average of the calculated coordinates for each vertex respectively. This calculation is repeated iteratively until no element is warped more than allowed by the specified angle threshold.

This algorithm is also useful to remove warping in the original mesh as well as after local editing operations (see [BRE04]) as the shape of the surface is only very little affected by unwarping the elements (see Fig. 10(c)). Fixing warped elements may slightly change the feature line's run but they are not wiped out since they do not run through the elements but along element edges.

5.2 Fixing Quad Orientation

The mesh constructed using the dual grid approach described in Sect. 4.2 usually contains rhombic shaped quads or quads oriented diagonally to its neighbours as depicted in Fig. 12. These artefacts lead to problems during numerical simulation and therefore should be removed.

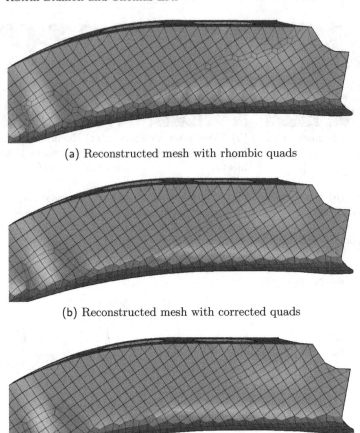

(a) Reconstructed mesh with rhombic quads

(b) Reconstructed mesh with corrected quads

(c) Reconstructed mesh with corrected quads and mesh relaxation

Fig. 12. Removing of rhombic quads

As can be seen in the picture, the vertices of a rhombic quad have a different degree than the ones around it: the vertices within the regular grid have even degree, the ones part of the rhombus have odd degree. Taking that fact as a reference to detect these quads, they can be removed by contracting the diagonal shared by the vertices with lower degree to one single vertex (see Fig. 12(b)). This leads to four equally oriented quads. To further enhance the quads' shape and equalise the elements' size and inner angles, relaxation as described in [BRE04] can be applied to the mesh additionally (see Fig. 12(c)).

6 Results and Conclusion

Modifying and editing FE meshes is part of daily engineering work. As the quality of Finite Element Analysis simulations is very sensitive to the FE mesh's quality, one important aim is to provide tools that fit the specific needs of Finite Elements. If the FE mesh is edited locally, it is usually only little deformed and minimal interfering mechanisms like mesh relaxation or local remeshing are sufficient to retain the prerequisites for the simulation. On the other hand, especially in the early design stage sometimes larger scaled modifications of the surfaces are desired and the mechanisms mentioned above will not be sufficient to make the mesh suitable for numerical simulation again. In this case new approaches are needed. In this paper we presented a remeshing method via volumetric representations. We first convert the surface defined by an FE mesh to a signed distance volume. From this volume we reconstruct the surface by isosurface extraction. To retrieve the required quad mesh we apply a duality algorithm to extract a quad mesh out of the modified Marching Cubes algorithm. This resulting mesh is enhanced by removing badly oriented quads, by fixing warped elements and by mesh relaxation (see Fig. 12).

Considering a mesh consisting of some 2300 nodes the calculation of the distance volume as well as the reconstruction takes only a few seconds on a P4 system with 2.8 GHz. The additional mesh enhancement methods work instantly.

The presented method implies the capability to combine the intermediate volumetric representation with the editing operation itself, which is planned for future work. The described procedures are ongoing work. Therefore, the resulting surfaces still show problems we are about to solve, e.g. the reconstruction of small intended holes in the surface, sometimes only of the size of one element in the original mesh.

Acknowledgements

We would like to thank the *BMB+F* project *AutoOpt* [1] for founding our research as well as the engineers at *BMW AG* for their cooperation and fruitful discussions within this project and for giving insight into today's engineering problems. Additionally we would like to thank Ove Sommer at *science+computing ag* and Alexander Kramer for providing many lines of code being part of this work.

[BRE04] K. Bidmon, D. Rose, and T. Ertl. Intuitive, Interactive, and Robust Modification and Optimization of Finite Element Models. In *Procceedings 13th International Meshing Roundtable*, pages 59–69, 2004.

[BS91] T. D. Blacker and M. B. Stephenson. Paving: A new approach to automated quadrilateral mesh generation. *International Journal for Numerical Methods in Engineering*, 32:811–847, 1991.

[Che95] E. V. Chernyaev. Marching cubes 33: Construction of topologically correct iso-surfaces. Technical report, CERN CN 95–17, 1995.

[COK95] D. Cohen-Or and A. Kaufman. Fundamentals of surface voxelization. *Graph. Models Image Process.*, 57(6):453–461, 1995.

---

[1]`www.azto-opt.de` (only in German)

[Gib98]   S. F. F. Gibson. Using distance maps for accurate surface representation in sampled volumes. In *VVS '98: Proceedings of the 1998 IEEE symposium on Volume visualization*, pages 23–30, New York, NY, USA, 1998. ACM Press.

[HLC⁺01]   J. Huang, Y. Li, R. Crawfis, S. C. Lu, and S. Y. Liou. A complete distance field representation. In *VIS '01: Proceedings of the conference on Visualization '01*, pages 247–254. IEEE Computer Society, 2001.

[HYFK98]   J. Huang, R. Yagel, V. Filippov, and Y. Kurzion. An accurate method for voxelizing polygon meshes. In *VVS '98: Proceedings of the 1998 IEEE symposium on Volume visualization*, pages 119–126. ACM Press, 1998.

[Kau87]   A. Kaufman. Efficient algorithms for 3d scan-conversion of parametric curves, surfaces, and volumes. In *SIGGRAPH '87: Proceedings of the 14th annual conference on Computer graphics and interactive techniques*, pages 171–179, New York, NY, USA, 1987. ACM Press.

[KBSS01]   L. P. Kobbelt, M. Botsch, U. Schwanecke, and H.-P. Seidel. Feature sensitive surface extraction from volume data. In *SIGGRAPH '01: Proceedings of the 28th annual conference on Computer graphics and interactive techniques*, pages 57–66. ACM Press, 2001.

[LB03]   A. Lopes and K. Brodlie. Improving the robustness and accuracy of the marching cubes algorithm for isosurfacing. *IEEE Transactions on Visualization and Computer Graphics*, 9(1):16–29, 2003.

[LC87]   W. E. Lorensen and H. E. Cline. Marching cubes: A high resolution 3d surface construction algorithm. In *SIGGRAPH '87: Proceedings of the 14th annual conference on Computer graphics and interactive techniques*, pages 163–169. ACM Press, 1987.

[LLVT03]   T. Lewiner, H. Lopes, A. W. Vieira, and G. Tavares. Efficient implementation of Marching Cubes cases with topological guarantees. *Journal of Graphics Tools*, 8(2):1–15, 2003.

[MSS94]   C. Montani, R. Scateni, and R. Scopigno. Discretized marching cubes. In *VIS '94: Proceedings of the conference on Visualization '94*, pages 281–287. IEEE Computer Society Press, 1994.

[Nie04]   G. M. Nielson. Dual marching cubes. In *VIS '04: Proceedings of the IEEE Visualization 2004 (VIS'04)*, pages 489–496. IEEE Computer Society, 2004.

[sci04]   science + computing ag. Efficient preprocessing using scFEMod. *http://www.science-computing.de/en/software/scfemod.html*, 2004.

[Sra01]   M. Sramek. High precision non-binary voxelization of geometric objects. In *SCCG '01: Proceedings of the 17th Spring conference on Computer graphics*, page 220. IEEE Computer Society, 2001.

[VKK⁺03]   G. Varadhan, S. Krishnan, Y. J. Kim, S. Diggavi, and D. Manocha. Efficient max-norm distance computation and reliable voxelization. In *SGP '03: Proceedings of the 2003 Eurographics/ACM SIGGRAPH symposium on Geometry processing*, pages 116–126. Eurographics Association, 2003.

[ZZHW91]   J. Z. Zhu, O. C. Zienkiewicz, E. Hinton, and J. Wu. A new approach to the development of automatic quadrilateral mesh generation. *International Journal for Numerical Methods in Engineering*, 32:849–866, 1991.

Structured Grid Generation over NURBS and Facetted Surface Patches by Reparametrization

Sankarappan Gopalsamy, Douglas H. Ross, Yasushi Ito and Alan M. Shih

Enabling Technology Laboratory, Mechanical Engineering Department
University of Alabama at Birmingham, AL, U.S.A.
{sgopals, dhross, yito, ashih}@uab.edu

Abstract

This paper deals with structured grid generation using Floater's parametrization algorithm for surface triangulation. It gives an outline of the algorithm in the context of structured grid generation. Then it explains how the algorithm can be used to generate a structured grid over a singular NURBS surface patch. This is an alternate method to the known carpeting method of reparametrization for structured grid generation over a NURBS surface patch. The paper also explains how to generate a structured grid over a four sided trimmed patch of a facetted surface using the parametrization algorithm. All the procedures are explained using examples.

Keywords: structured grid generation, parametrization, reparametrization, Floater's algorithm, NURBS surface, facetted surface

1 Introduction

Currently common methods of structured grid generation are transfinite interpolation (TFI) and elliptic smoothing [1]. Most of the time, the grid is generated over NURBS or other parametrically defined piecewise smooth surfaces. Direct implementations of TFI and elliptic smoothing work well for simple four-sided surface patches defined over a single surface without singular points — a point on a surface is called a singular point if a directional derivative or normal is zero at that point. To obtain a structured grid over a complex surface patch that is either defined over multiple surfaces or over a surface with singular points, one has to first reparametrize that surface patch by a new surface and then has to generate the grid. The grid points will be finally projected onto the original surfaces if the reparametrization is not exact. One method of reparametrization used by grid generators is by defining a carpet surface [1, 2]. A carpet surface is created by obtaining the four boundary curves, generating a surface from these four curves by TFI, and then projecting this surface onto the underlying original surface patches [1].

In this paper an alternate method of reparametrization is shown for generating structured grids over complex NURBS surface patches. The method also works for generating structured grids over facetted surfaces, which are being used in bio-medical and other applications. The reparametrization is based on Floater's parametrization algorithm [3] for surface triangulations.

Floater's algorithm is widely used in computer graphics applications, but it is not used by many people to generate quality grids for Computational Fluid Dynamics (CFD) and Computational Solid Mechanics (CSM) applications. It has been mentioned in [4] as a method of obtaining valid structured grids in the parametric domain of a surface patch, which will then be optimized to obtain quality grids. However that procedure may not work if the parametric function has singularities. Whereas the method proposed in this paper obtains a structured grid directly over the surface patch.

The reparametrization method is being implemented as part of GGTK – the Geometry and Grid Toolkit [5] developed at the University of Alabama at Birmingham. It is being provided in addition to the existing TFI and elliptic methods. In the following sections, the method is explained and illustrated using examples. Section 2 gives an outline of Floater's parametrization algorithm. Section 3 shows how it can be used for structured grid generation over a trimmed NURBS surface patch. Section 4 shows how it can be used to generate structured grids over facetted surfaces. Finally, Section 5 gives conclusions and mentions further work along this direction.

2 Parametrization Algorithm

This section first explains the notations and terminology used in parametrization and then gives an outline of Floater's parametrization algorithm. It also briefly talks about choice of coordinate values and shape preserving parametrization. Then it mentions how to obtain a structured grid over the surface using the parametrization.

2.1 Notations and Definitions

In [3] a surface triangulation is defined using graph theory notations. But in this paper it will be defined in terms of common notations used in grid generation as follows.

A *surface triangulation* S consists of a set of vertices V, a set of edges E, and a set of triangular faces F satisfying the following properties:

- Each vertex in V is a point in R^3; each edge in E is a line segment joining two vertices; each triangle is formed by three edges.
- Each vertex will be part of at least one edge and each edge will be part of at least one triangle.
- Intersection of any two triangles will be either empty or an edge or a vertex.

An edge of a surface triangulation S is said to be a *boundary edge* if it is shared by only one triangle. A vertex is said to be a *boundary vertex* if it lies on a boundary edge; otherwise it is called an *internal vertex*.

A surface triangulation is said to be *simply connected* if all its boundary edges form a single connected loop. Figure 1 (a) shows a simply connected surface triangulation.

Two surface triangulations are said to be *isomorphic* [3] if there is a one-one correspondence between their vertices, edges and faces in such a way that corresponding edges join corresponding vertices and corresponding triangles are formed by corresponding edges.

A *planar triangulation* is a special case of surface triangulation when all the vertices lie on the *xy* plane R^2.

A surface triangulation S is said to have a *parametrization* if there is a planar triangulation that is isomorphic to S. Figure 1 (b) shows a planar triangulation that is isomorphic to the surface triangulation shown in Figure 1 (a). That is, Figure 1 (b) is a parametrization of Figure 1 (a).

In [3] it is shown that every simply connected surface triangulation has a parametrization.

2.2 Floater's Algorithm

The procedure given in [3] to compute a planar triangulation that is isomorphic to a given simply connected surface triangulation S can be outlined as follows:

Step 1: Let $V = \{P_i\}$ be the set of vertices of S. Label the vertices in such a way that $\{P_1, \dots P_n\}$ are internal vertices and $\{P_{n+1}, \dots P_N\}$ are boundary vertices, where N is the total number of vertices and n is the number of internal vertices.

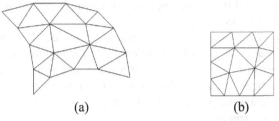

(a)   (b)

Fig. 1. (a) A simply connected surface triangulation; (b) A planar triangulation isomorphic to (a).

Step 2: Select a convex domain D in R^2 and choose points $Q_{n+1}, ...Q_N$ on the boundary of D such that they form an anticlockwise sequence.

Step 3: For each i in $\{1, ..., n\}$ and j in $\{1, ..., N\}$, choose λ_{ij} such that
$\lambda_{ij} = 0$ if P_iP_j is not an edge
$\lambda_{ij} > 0$ if P_iP_j is an edge
For each i, $\sum\limits_{j=1}^{N} \lambda_{ij} = 1$.

Step 4: Form the system of linear equations
$$Q_i = \sum_{j=1}^{N} \lambda_{ij} Q_j, \ i = 1, ..., n,$$

where $Q_1, ...Q_n$ are the unknown interior points to be computed in the domain D and $Q_{n+1}, ...Q_N$ are the known boundary points chosen in Step 2. The above equation can be rewritten as

$$Q_i - \sum_{j=1}^{n} \lambda_{ij} Q_j = \sum_{j=n+1}^{N} \lambda_{ij} Q_j, \ i = 1, ..., n.$$

This can be written in matrix form as $\mathbf{AQ} = \mathbf{B}$ with

$$\mathbf{A} = \begin{bmatrix} a_{11} & .. & a_{1n} \\ : & :: & : \\ a_{n1} & .. & a_{nn} \end{bmatrix}, \ \mathbf{Q} = \begin{bmatrix} Q_1 \\ : \\ Q_n \end{bmatrix}, \ \mathbf{B} = \begin{bmatrix} B_1 \\ : \\ B_n \end{bmatrix},$$

where $a_{ii} = 1$, $a_{ij} = -\lambda_{ij}$ for $i \neq j$, $B_i = \sum\limits_{j=n+1}^{N} \lambda_{ij} Q_j$.

Then the matrix $\mathbf{A}$ is non-singular [3].

Step 5: Solve the above system and obtain the points $Q_1, \ldots Q_n$ in the interior of the domain D.

Step 6: Form a planar triangulation T over D with $\{Q_1, \ldots Q_N\}$ as the set of vertices and having edges and triangles in such a way that Q_iQ_j is an edge in T if and only if P_iP_j is an edge in the given surface triangulation S and $Q_iQ_jQ_k$ is a triangle in T if and only if $P_iP_jP_k$ is a triangle in S.

The planar triangulation T formed as above is a valid triangulation [3] and is isomorphic to the given surface triangulation S. That is, it defines a parametrization of S. For example, the planar triangulation shown in Figure 1 (b) has been obtained by the above procedure as the parametrization of the surface triangulation of Figure 1 (a).

2.3   Choice of λ_{ij}

The values λ_{ij} chosen in Step 3 above are called coordinates of the vertex P_i with respect to its surrounding vertices. The quality of the parametrization will depend on the choice of these coordinates λ_{ij}. In [3] three methods of computing λ_{ij} are discussed resulting in three kinds of parametrization. They are:
1. Uniform parametrization: $\lambda_{ij} = 1/d_i$, where d_i is number of vertices connected to P_i.
2. Weighted least squares of edge lengths: λ_{ij} is proportional to $1/\|P_i - P_j\|$.
3. Shape preserving parametrization: A procedure is given in [3] to compute λ_{ij} that will result in a shape preserving parametrization analogous to the chord length parametrization of a sequence of points.
Another choice for λ_{ij} is the mean value coordinates given in [6].

Our purpose of parametrization is to obtain good quality grids on the surface by mapping good quality grids in the new parametric domain D. Hence we need to choose a parametrization that maps good triangles in the parametric domain to good triangles on the surface. This is the characterizing property of the shape preserving parametrization. In the case of uniform parametrization, whatever the shapes of the triangles in the original surface triangulation, the shapes of the triangles of the parametrization in the domain D are good because they satisfy the Laplacian smoothing equation. That means a good triangle in the parametric domain may be mapped to a bad triangle on the surface if we use uniform parametrization. In [3], it is mentioned that the shape preserving parametrization gives smoother results than the other two. Hence in all results shown in this paper, the shape preserving parametrization has been used.

2.4   Structured Grid Generation

Suppose the surface triangulation S has four corner points. In order to generate a structured grid over S, choose a square or a rectangular domain D in Step 2 and choose boundary points in D in such a way that the four corner points of S are mapped to the four corners of D. Then compute the isometric planar triangulation over D. After that, generate the desired structured grid over D, and map that grid onto S using the parametrization map defined as follows: If Q is a point in D then Q will lie in a triangle $Q_iQ_jQ_k$ of the planar triangulation. Let α, β, γ be the barycentric coordinates of Q with respect to $Q_iQ_jQ_k$. Then Q will be mapped to P on the surface triangulation, where P is the point on the triangle $P_iP_jP_k$ with barycentric coordinates α, β, γ.

For the example parametrization shown in Figure 1, Figure 2 (a) shows a structured grid over the parametric domain D and Figure 2 (b) shows the resulting structured grid over the surface S.

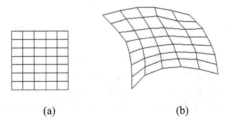

(a)   (b)

Fig. 2. (a) Structured grid over the parametric domain D; (b) Corresponding structured grid over the surface S.

2.5   Computational Issues

The parametrization procedure involves solving a linear system of equations. For small values of n the system can be solved by LU decomposition. Note that the matrix **A** (obtained in Step 4 of the procedure) is diagonally dominant and the matrix elements a_{ij} have the property that $a_{ij} = 0$

if P_iP_j is not an edge. That is, the matrix **A** is of the same kind as a typical finite element matrix. So, for large values of n, one should use other efficient solvers used in finite element methods.

3 Structured Grid over a NURBS Surface Patch

This section shows how to generate a structured grid over a singular NURBS patch using the parametrization described above. The procedure involves two steps: (a) Obtain a surface triangulation over the patch; (b) compute a parametrization using a square domain, and generate the structured grid. These steps are explained below using an example.

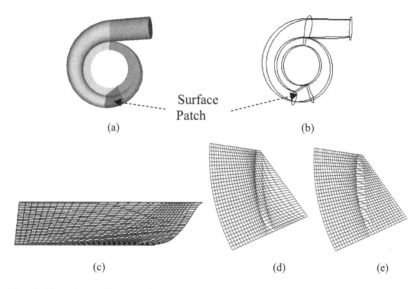

Fig. 3. Singular surface patch of a volute. (a), (b) The patch shown as part of the volute; (c) TFI grid over the trimmed parametric domain of the patch; (d) Corresponding structured grid over the face; (e) Grid obtained by TFI in 3D and projection.

3.1 Example of a singular NURBS patch

An example of a singular NURBS surface patch is shown in Figure 3. The surface patch is one of 26 boundary faces of a *volute model,* shown in Figures 3 (a) and 3 (b). The surface patch is defined by a trimmed NURBS surface [7]. A TFI grid over the trimmed parametric domain of the surface patch is shown in Figure 3 (c). The corresponding TFI grid over the surface patch obtained by the NURBS mapping is shown in Figure 3 (d). Figure 3 (e) shows the grid obtained by TFI in 3D and then projection on to the surface. As can be seen in the Figures 3 (d) and 3 (e), the TFI grids over the surface are not good. This is because the NURBS function has singularities along the line of bending in the interior of the surface. Actually one directional derivative vanishes, and the second order derivative in that direction is discontinuous.

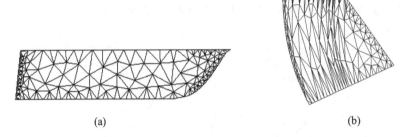

(a)   (b)

Fig. 4. (a) Triangulation of the original parametric domain; (b) Corresponding triangulation of the surface patch.

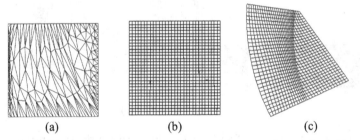

(a)   (b)   (c)

Fig. 5. (a) Triangulation of the new parametric domain obtained by parametrization of the surface triangulation shown in Fig 4 (b); (b) Structured grid on the new parametric domain; (c) Corresponding structured grid on the surface obtained using the new parametrization mapping.

3.2 Surface Triangulation

To obtain a triangulation of the surface patch, the boundary is first discretized and the corresponding grid points are obtained in the original parametric domain of the NURBS surface. Care is taken to ensure that the set of boundary grid points contains the four corner points of the patch. The parametric domain of the patch is triangulated using an automatic 2D triangulation routine such as *Triangle* [8] that preserves the boundary grid points and adds new interior grid points. The parametric domain triangulation is then mapped on to the surface using the NURBS function. This will result in a surface triangulation S of the surface patch.

Figure 4 (a) shows a triangulation of the parametric domain of the example surface patch and Figure 4 (b) shows the corresponding surface triangulation on the patch. Note that the quality of the surface triangulation is not important for the reparametrization.

3.3   Reparametrization and Structured Grid

For the surface triangulation S computed as above, a shape preserving parametrization is computed using the Floater's algorithm explained in Section 2. Since the purpose is to compute a structured grid, a square is chosen as the domain D for the parametrization. This will result in an isometric triangulation over D and will define a map from D on to S. In order to obtain a structured grid on the surface patch, corresponding structured grid is generated on the square parametric domain D and mapped onto the surface triangulation S. The grid points will lie on the triangles of S and may not lie on the original surface. So they are finally projected onto the original surface.

For the example surface patch, Figure 5 (a) shows triangulation of the new parametric domain obtained by the parametrization; Figure 5 (b) shows a structured grid on the new parametric domain; Figure 5 (c) shows the final structured grid on the surface patch obtained by mapping the structured grid on the new parametric domain onto the surface triangulation using the parametrization and then projection.

4   Structured Grid Over a Faceted Surface Patch

This section shows how to generate a structured grid over a portion of a facetted surface. A facetted surface is a surface defined by a collection of polygons. They are used to represent surfaces reconstructed from scanned data of objects such as MRI/CT scan of internal human body parts [9]. In this paper it is assumed that a facetted surface is defined by a surface tri-

angulation, that is, it is collection of triangles. Figure 6 shows a human
lung model defined as facetted surface consisting of triangles.

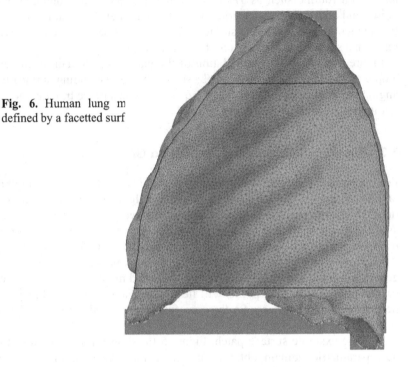

Fig. 6. Human lung m
defined by a facetted surf

4.1 Surface Triangulation

The main issue in generating a structured grid over a facetted surface is defining a four sided surface patch and obtaining a surface triangulation for
the surface patch. In many cases, a facetted surface may not have boundary
curves on the surface defining four sided patches. In such cases, one can
define the required curves by creating curves in 3D space and projecting
them onto the surface or by intersecting the surface with planes. These
curves may be intersecting the triangles. So in order to obtain surface triangulation of the required patch, one has to retriangulate the surface along
these curves and ensure that the curves pass along the edges of the triangulation. Then a surface triangulation of the patch can be obtained by collect-
ing all the triangles inside the patch. Figure 7 shows a surface triangulation
over a four sided patch of the lung model obtained as above.

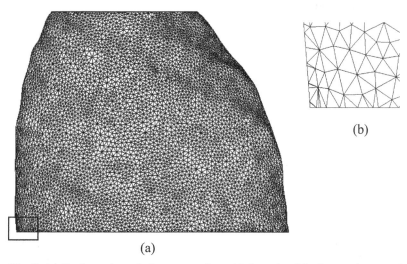

(b)

(a)

Fig. 7. (a) Surface triangulation over a four-sided patch of the human lung model; (b) Zoomed-in view of the lower left corner.

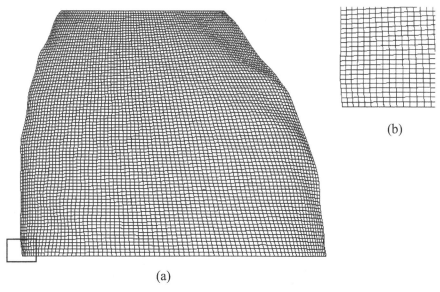

(b)

(a)

Fig. 8. (a) Structured grid over the four-sided patch of the human lung model obtained using the parametrization; (b) zoomed-in view of the lower left corner.

4.2 Structured Grid

Once a surface triangulation is obtained for the surface patch, a structured surface grid can be obtained by parametrization as explained in Section 2. Figure 8 shows a structured grid over the surface patch of the human lung model obtained by this procedure.

5 Conclusions and Further Work

This paper has shown how Floater's parametrization algorithm can be used to generate structured grids over trimmed patches of NURBS as well as facetted surfaces. In the case of NURBS surfaces, this provides an alternate method of reparametrization to the known method of carpeting. As examples, the paper has used a trimmed singular NURBS surface patch of a volute model and a four sided facetted surface patch of a human lung model. The results obtained on these examples show that this is a good alternative method for structured grid generation. In this paper, only one kind of coordinates λ_{ij} producing shape preserving parametrization has been used. There seems to more scope in the choice of the coordinates λ_{ij} to see if they can be computed differently to produce a parametrization that will result in structured orthogonal grids. More work can be done along this direction.

Acknowledgements

This research was supported in part by the following projects: (a) Improvements in Turbomachinery Grid Generation Tools, sponsored by NASA Marshall Space Flight Center; (b) Automated and Parametric Dynamic Geometry-Meshing Tools and Technology for Thrust Chamber Assemblies, sponsored by NASA CUIP Program; and (c) Mesh Improvement Algorithms and Software Technology, sponsored by DOD MSRC PET Project.

References

1. Joe F. Thompson, "A Reflection on Grid Generation in the 90s: Trends, Needs, and Influences," *Proceedings of the 5th International Conference on Numerical Grid Generation in Computational Field Simulations*, Mississippi State University, pp. 1029-1110, April 1996.
2. B. A. Jean and B. Hamann, "Interactive Techniques for Correcting CAD Data," *Proceedings of the 4th International Grid Conference*, Pineidge Press Limited, Swansea Wales (UK), 1994.
3. M. S. Floater, "Parametrization and smooth approximation of surface triangulations," *Computer Aided Geometric Design*, vol. 14, pp. 231-250, 1997.
4. D. Shikhare, S. Gopalsamy and S. P. Mudur, "A two phase technique for optimal tessellation of complex geometric models," *Proceedings of the eighth international conference on engineering computer graphics and descriptive geometry*, Austin, Texas, USA, July 31-August 3, 1998.
5. GGTK Documentation, http://www.eng.uab.edu/me/ETLab/Software/GGTK/manual/index.html
6. M. S. Floater, "Mean value coordinates," *Computer Aided Geometric Design*, vol. 20, pp. 19-27, 2003.
7. L. Piegl and W. Tiller, *The NURBS Book*, 2nd Edition, Springer, 1997.
8. J. R. Shewchuk, "Triangle – A two dimensional quality mesh generation and Delaunay triangulator," http://www-2.cs.cmu.edu/~quake/triangle.html
9. Y. Ito, P. C. Shum, A. M. Shih, B. K. Soni, and K. Nakahashi, "Medical Image-Based Unstructured Mesh Generation," submitted for publication.

A fews snags in mesh adaptation loops

Frederic Hecht[1]

Universit Pierre et Marie, Laboratoire Jacques-Louis Lions, Paris
`frederic.hecht@umpc.fr`

Abstract

The first stage in an adaptive finite element scheme (cf. [CAS95, bor1]) consists in creating an initial mesh of a given domain Ω, which is used to perform an initial computation (for example a flow solver). A size specification field is deduced (e.g. at the vicinity of each mesh vertex, the desired mesh size is specified), based on the numerical results. If the mesh does not satisfy the size specification field, then a new constrained mesh, governed by this field, is constructed. The size specification field is usually obtained via an error estimate [FOR, VER96]. Actually, the estimation gives a discrete size specification field. Using an adequate size interpolation over the mesh elements, a continuous field is then obtained.

Metrics are commonly used to normalize the mesh size specification to one in any direction (cf. [VAL92]), and are defined as a symmetric positive definite matrix associated to any point of the domain.

A classical adaptation loop is:

0 Build a initial mesh $\mathcal{T}_h^0$
1 loop $i = 0, \ldots$
   - Solve your problem on mesh $\mathcal{T}_h^i$
   - Compute an error indicator , and if the error is small enough then stop.
   - Compute a metric $\mathcal{M}^{i+1}$,
   - Bound, regularize the metric $\mathcal{M}^{i+1}$,
   - Compute a new unit mesh $\mathcal{T}_h^{i+1}$ with respect to the new metric.

In this kind of algorithm, there are two problematic cases:

One) if the minimal mesh size is reached then we generally lose the anisotropy of the mesh in this region.

Two) In the adaptation loop, we use a hidden scheme to evaluate the metric, so some-times the mesh size to compute a good approximation of the solution is incompatible with the scheme to get a good approximation of the metric.

First, we do the numerical experiment to show this two snags. All the experiments are done with **FreeFem++** software , see [freefempp, DAN03].

In this article we present the classical mesh adaptation with metric in section 2. And in section 3 we present the first trouble and some way to solve it. In section 4, a second problem is described and we explain when it occurs.

1 Metric and anisotropic mesh adaptation

In this section we recall the notion of mesh *adapted* to a control space or a metric map. Let Ω be a domain of R^d ($d = 2$ or 3) and $\mathcal{M}_d(\Omega)$ be a *continuous* field of metrics associated with the points of Ω. The metric at a point P of Ω characterizes the desired edge- or element-size in the vicinity of P. By normalizing the desired size to *one*, the metric at P can be defined as a symmetric positive definite matrix of order d, denoted as $\mathcal{M}_d(P)$,

Under this metric definition, in the Riemannian space the length $\mathcal{L}$ of a curve Γ of $\mathbb{R}^d$, parametrized by $\gamma(t)_{t=0..1}$, is

$$\mathcal{L} = \int_0^1 \sqrt{{}^t\gamma(t)'\mathcal{M}(\gamma(t))\gamma(t)'}dt \tag{1}$$

The metric given is a tool to govern the local mesh size, and so we build a unit mesh in the metric. A unit mesh is such that the size of all the edges in a metric is close to 1.

A way to build a metric with a method of order 2 in space, is to use the Hessian $\mathcal{H}$ as a good error indicator. Details on the ingredients used in the metric definition for inviscid and viscous laminar and turbulent flows involving shocks and boundary layers can be found in [missi, bor2, CAS00, HEC97, GEO99, HAB].

The Key idea is that: the order of the error is given by of the interpolation error, and for a P^1 Lagrange discretization of a variable u, the interpolation error is bounded by:

$$\mathcal{E} = ||u - \Pi_h u||_\infty \leq \frac{1}{2} \sup_{T \in \mathcal{T}_h} \sup_{x,y,z \in T} |\mathcal{H}(x)|(y - z).(y - z) \tag{2}$$

where $\Pi_h u$ is the P^1 interpolate of u, $|\mathcal{H}(x)|$ is the Hessian of u at point x after being made positive definite, and where . is the dot product.

The formular (2) give a way to evaluted the interpolation $\mathcal{E}_{pq}$ error on an edge (p,q)

$$\mathcal{E}_{pq} = |\mathcal{H}(x)|(q - p).(p - q) \simeq \int_0^1 |\mathcal{H}(tp + (1 - t)q)|(q - p).(p - q)dt$$

Now, we generate, using a Delaunay procedure as a mesh generation method, a mesh with edges close to unit length in the metric

$$\mathcal{M} = \frac{|\mathcal{H}|}{(\mathcal{E})}. \tag{3}$$

The interpolation error $\mathcal{E}_{pq}$ is equi-distributed with $\mathcal{E}$ over the edges pq of the mesh. More precisely, we have

$$(pq)^T \mathcal{M}(pq) = \frac{1}{\mathcal{E}}(pq)^T |\mathcal{H}|(pq) \leq 1. \tag{4}$$

so the interpolation error is less than $\mathcal{E}$. Remark, we forgot the coefficient $\frac{1}{2}$ because the two screw $\mathcal{E}$ and $\frac{\mathcal{E}}{2}$ are equivalent.

For systems, the previous approach leads to a metric for each variable. For two metrics $\mathcal{M}_1$ and $\mathcal{M}_2$, we define a metric intersection $\mathcal{M} = \mathcal{M}_1 \cap \mathcal{M}_2$, such that the unit ball of $\mathcal{M}$ is included in the intersection of the two unit balls of metric $\mathcal{M}_2$ and $\mathcal{M}_1$ with the procedure defined in [GEO99, chap 10.3.3] for example.

To evalute the Hessian of u_h a P_1 finite element function, we can use two formulas like:

$$\mathcal{H}^k_{h\,ij} = -\frac{1}{3|\Omega_k|} \int_{\Omega_k} \partial_i u_h \partial_j w^k \tag{5}$$

where w^k is the finite element basis function associated to the vertex k and $|\Omega_k|$ is the measure of the support of this basis function.
Remark, this approximation generally does not converge when the size of the triangle becomes close to zero.

Or a more stable technique is to make a double projection like:

$$\mathcal{H}_h = I_{L^2}\left(\nabla\big(I_{L^2}(\nabla u)\big)\right) \tag{6}$$

where I_{L^2} is the L^2 projection on the P_1 Lagrange finite element space.
Remark, we have no convergence proof, but result is better with using(6, see [alauzet, remark 1.14] for more detail. For the convergence problem, just take the following example approximation of $f(x,y) = x^2 + y^2$. If we take a mesh pattern around point $x_0 = \mathbf{0}$, and we do a homothety with matrix $\left(\begin{smallmatrix} h & 0 \\ 0 & h \end{smallmatrix}\right)$ on this pattern. The value of approximate Hessian at mesh point x_0 given by the two method on the transform pattern are independent of h, because f is quadratic.

1.1 Mesh generation and adaptation

We use the mesh generation tools developed at INRIA-Gamma project [bamg, GEO99]. One novelty however in this work in the Delaunay mesh generation part is to introduce an extra criteria keeping the new mesh nodes and connectivities unchanged as much as possible compared to the previous mesh where the mesh prescribed by the metric is similar to the previous mesh. This is of course suitable for time dependent simulations and reduces the perturbation introduced by remeshing and solution interpolation from the background over the new mesh.

The algorithm of adapted mesh generation is a Delaunay-like method ([GEO99, Sec. 7.3.1]) where the insert point procedure is fully described in [GEO99, Sec. 7.6.1].

To introduce the idea above, we need to use the points of the background mesh. The algorithm is:

1. Create a bounding box mesh.
2. Discretize the boundary with respect to the metric, Let $\mathcal{P}$ be a set equal $\mathcal{S}_B$ the set of boundary point.
3. Sequentially add the internal points of the background mesh to $\mathcal{P}$, if a point is not too close (distance in the metric less than $\frac{1}{\sqrt{2}}$) to a point of $\mathcal{P}$.

4. Insert the points of $\mathcal{P}$ one by one.
5. Enforce the boundary mesh.
6. Remove outside triangles.
7. DO: far field point creation $\mathcal{S} = \emptyset$.
   - Split all the edges which are too long such that the lengths of sub-edges 'in the metric' are close to one. Add the new splitting points in $\mathcal{S}$.
   - Insert all the points of $\mathcal{S}$,if they are not too close 'in the metric' to an existing point.
8. if $\mathcal{S}$ is not empty, repeat 7.
9. Do some small regularization.

Finally, the adaptation loop scheme is:

0) Build a initial mesh T_h^0
1) loop $i = 0, ...$
    1. Solve your problem on mesh T_h^i
    2. Compute an error indicator , and if the error is small enough then stop.
    3. Compute a metric $\mathcal{M}$,
    4. Bound, regularize the metric $\mathcal{M}$,
    5. Compute a new mesh unit mesh T_h^{i+1} with respect to the new metric.

Remark, all these tools are available in the **FreeFem++** software [freefempp].

2 Lose of Anisotropy

In case of a solution with shock layer, the minimal mesh size goes to zero perpendicularly to the shock during the adaptation loop, so the user wants a method to bound computations in size or in time. To do this, he gives a minimal mesh size cutoff value to limit the number of mesh points.

We see the problem immediately, if the adaptation loop tries to build an anisotropic mesh with smaller a mesh size than the minimal mesh size cutoff value.

First, to focus on this problem, we just try to build an optimal mesh with respect to the function

$$f(x, y) = yx^2 + y^3 + \tanh(10(sin(5y) - 2x)).$$

This function has a smooth shock line. the equation of the line is $sin(5y) = 2x$. Figure 3 shows a three dimensional representation of this function f.

The adaptation algorithm in **FreeFem++** [freefempp] language is:
```
// interpolation error: eps = E, see formula (2)
real eps = 0.002;
the user parameter to set the miminal mesh size hmin
real hmin=0.005;//                                      see figure 1
// or hmin=0.000005 see figure 2
func f = y*x*x+y*y*y+h*tanh(10*(sin(5.0*y)-2.0*x));
```

```
border cercle(t=0,2*pi) x=cos(t);y=sin(t);
mesh Th=bzildmesh(cercle(20));

for (int i=0;i<6;i++)
{
Th=adaptmesh(Th,f,hmin=hmin,err=eps,nbvx=100000);
plot(Th,ps="la-Th-"+hmin+".eps");
}
```

 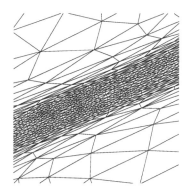

Fig. 1. Adaptated mesh with $hmin = 0.005$, the left picture seems good but on the picture on the right , we see the lose of anisotropy (Number of Triangles = 11456, Number of Vertices 5806)

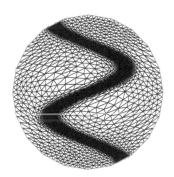 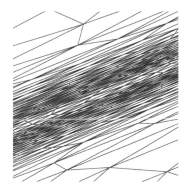

Fig. 2. Adaptated mesh with $hmin = 0.00005$, everything is correct (Nb of Triangles = 14054, Nb of Vertices 7110)

The natural question is: *"Why does this problem of losing anisotropy occur?"*

The answer is given in the three dimensional plot of the f function on a coarse grid with shading. As you can see thanks to the shading in the shock region it is dented, irregular because the surface is badly approximated. In a tangent direction to the shock layer, the second derivative is high and thus the method puts lots of points in this direction, and the mesh become isotropic.

Fig. 3. the 3d representation of function f on a grid which is too coarse, with $hmin = 0.05$

Now, how to solve this problem?

- Change the mesh generator or do some optimization, to force points to be on extremal curvature lines to get super convergence propriety (see paper with J.-F. Lage in preparation).
- Change the error indicator, in a shock region, see the work of F. Alauzet [alauzet].
- Smooth the solution before the computation of the metric with for example a convolution with a regularizing real function Φ_h of $\mathbb{R}^d$ to be sure that the built metric respects the constraint on $hmin$.

  Remember that, the convolution operator $\star$ of two functions is defined by

$$(f \star g)(x) = \int f(x - y)g(y)dy.$$

The regularizing function must clearly satisfy:

$$\int_{\mathbb{R}^d} \Phi_h(x)dx = 1, \quad \text{and } ||(\partial_{ij}(\Phi_h))_{ij \in (1,..,d)^2}|| \leq \frac{1}{h^2} \tag{7}$$

where here h is the minimal mesh size cutoff value $hmin$.
The function can be radial and can be defined with $\Phi_h(x) = \phi_h(||x||)$. The real function ϕ_h has a small support, must verify

$$\int_0^\infty r^{d-1}\phi_h(r)dr = \frac{1}{S_d}, \quad |\phi_h''| \leq 1/h.$$

where S_d is the measure of the sphere in dimension d ($S_2 = 2\pi$, $S_3 = 4\pi$).
Let us introduce the function ρ defined by

$$\rho(x) = \begin{cases} 1 - \frac{1}{2}x^2 & \text{if } x \in [0, 1], \\ \frac{1}{2}(x-2)^2 & \text{if } x \in [1, 2], \\ 0 & \text{otherwise } x \in [2, \infty]. \end{cases}$$

and denote $\alpha_d = \int_0^\infty x^{d-1}\rho(x)dx$, we have $\alpha_2 = 7/12$ and $\alpha_3 = 1/2$ and we can remark that $|\rho''| = 1$ on $]0, 2[$.
We define the function ϕ_h by

$$\phi_h(x) = \frac{1}{\tilde{h}^d \alpha_d S_d}\rho(\frac{x}{\tilde{h}}).$$

On the support $[0, 2\tilde{h}[$ of $\phi_h(x)$ we have

$$|\phi_h''| = \frac{1}{\tilde{h}^{d+2}\alpha_d S_d} = \frac{1}{h^2} \quad \text{thus } \tilde{h} = (\frac{h^2}{\alpha_d S_d})^{\frac{1}{d+2}}$$

So finally the function is

$$\Phi_h(x) = (\alpha_d S_d)^{-\frac{2}{d+2}} h^{-\frac{2d}{d+2}}\rho((\frac{h^2}{\alpha_d S_d})^{-\frac{1}{d+2}}||x||)$$

The same example with this technique solves the problem, as you can see on figure 4.

Remark, we preserve the anisotropy but we lose a little bit in the thickness of the shock layer. This is due to the size of the support of function Φ_h.

3 Problem of convergence in adaptation

As I say in the introduction, there is a hidden scheme to compute the metric. To compute the metric we use formula (3). Here, to evaluate the Hessian of a discrete function u_h, we use (6). This hidden scheme has some criterion of convergence, this criterion is hard to define exactly. But anyway, it is possible to show some convergence problems.

Let take the same adaptation loop with a more harder function f.

$$f = 0.1 * (y * x * x + y * y * y) + h * tanh(200 * (sin(3y) - 2x));$$

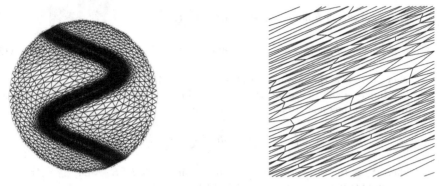

Fig. 4. Adaptated mesh with $hmin = 0.005$, with the convolution technique, Nb of Triangles = 9659, Nb of Vertices 4905

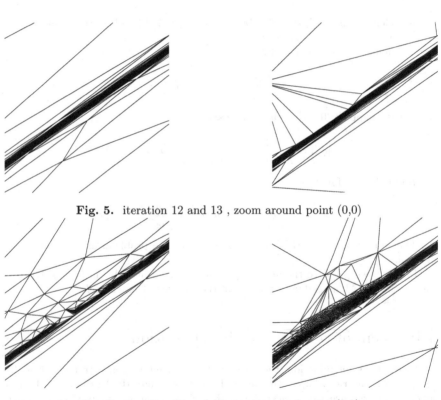

Fig. 5. iteration 12 and 13 , zoom around point (0,0)

Fig. 6. iteration 14 and 16, zoom around point (0,0)

The shock on this function is straighter, due to the coefficient set to 200.

```
real eps = 0.01;
real hmin=0.000005; //
int nbiter=40;
func f = 0.1*(y*x*x+y*y*y)+tanh(200*(sin(3*y)-2*x));
```

// same loop and initial mesh as in the previous // freefem++ example at page 2.

We see that at iteration 12 the mesh is correct, but in the next three steps the meshes are bad because the metric is cleary wrong. The reason is: We do not have enough points to compute the second derivative to evaluate the metric of the function f correctly, from the interpolation of f on the current mesh.

A classical analysis of a scheme to compute a second order derivative, shows that the error depends of the third derivative. In the example the third derivative is huge compare to the second.

First, This problem happens rarely, if we change the norm to evaluate the error , the L^∞ norm is changed to a more geometrical error (the Hausdorff distance). In this case, the metric associed to function u is defined by

$$\mathcal{M} = \frac{|\mathcal{H}|}{(\sqrt{1+||\nabla u||^2 \mathcal{E}})}, \tag{8}$$

where $\mathcal{H}$ is the Hessian of u.

With this metric, computational results are often better (see figure 7 and 8) , but the problem can still occur, and I have no another solution that completely solves this problem.

Conclusion:

This kind of adaptation scheme gives very good results but sometimes, as you have seen they are problems. Generally, we can explain why the problem occurs, by analysing an example. Sometimes it is more difficult to correct, but anyway mesh adaptation can help you to solve a lot of problems.

Acknowledgments: Thank to the helpful discution and comment with Frédéric Alauzet, Pascal Frey, Jean-françois Lagüe and Antoine Le Hyaric,

[alauzet] F. ALAUZET, Adaptation de maillage anisotrope en trois dimensions. Application aux simulations instationnaires en manique des fluides, *The de doctorat de l'UniversitMontpellier II*, 2003.

[missi] H. BOROUCHAKI H., M.J. CASTRO-DIAZ, P.L. GEORGE , F. HECHT AND B. MOHAMMADI, Anisotropic adaptive mesh generation in two dimensions for CFD *5th Inter. Conf. on Numerical Grid Generation in Computational Field Simulations*, Mississipi State Univ., 1996.

[bor1] H. BOROUCHAKI, P.L.GEORGE, F.HECHT, P.LAUG AND E.SALTEL, Delaunay Mesh Generation Governed by Metric Specifications. Part I: Algorithms., *Finite Elements in Analysis and Design*, 25, pp. 61–83, 1997.

[bor2] H.BOROUCHAKI, P.L.GEORGE AND B.MOHAMMADI, Delaunay Mesh Generation Governed by Metric Specifications. Part II: Applications., *Finite Elements in Analysis and Design*, 25, pp. 85–109, 1997.

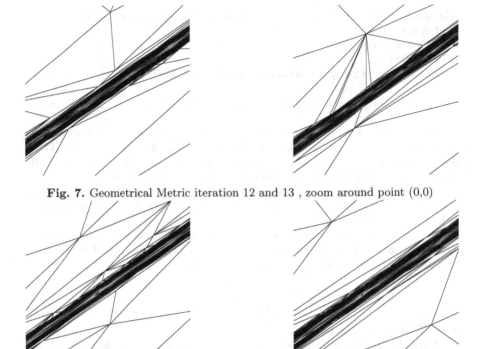

Fig. 7. Geometrical Metric iteration 12 and 13 , zoom around point (0,0)

Fig. 8. iteration 14 and 16, zoom around point (0,0)

[CAS95] M.J.CASTRO-DÍAZ, F.HECHT, AND B.MOHAMMADI, New progress in anisotropic grid adaptation for inviscid and viscid flows simulation, *4th International Mesh Roundtable*, Albuquerque, New-Mexico, october 1995.

[CAS00] M.J. CASTRO-DIAZ , F. HECHT F. AND B. MOHAMMADI Anisotropic Grid Adaptation for Inviscid and Viscous Flows Simulations *IJNMF*, Vol. 25, 475-491, 2000.

[DAN03] I. Danaila, F. Hecht, and O. Pironneau. *Simulation numérique en C++.* Dunod, Paris, 2003.

[GEO99] FREY P.J. ET GEORGE P.L., *Maillages*, Hermès, Paris, 1999.

[FOR] M. FORTIN, M.G. VALLET, J. DOMPIERRE, Y. BOURGAULT AND W.G. HABASHI Anisptropic Mesh Adaption: Theory, Validation and Applications, *Eccomas 96, PARIS*, CFD book, pp 174–199.

[HAB] W. HABASHI ET AL. A Step toward mesh-independent and User-Independent CFD *Barriers and Challenges in CFD*, pp. 99-117, Kluwer Ac. pub.

[HEC97] F. HECHT AND B. MOHAMMADI Mesh Adaptation by Metric Control for Multi-scale Phenomena and Turbulence *AIAA*, paper 97-0859, 1997.

[bamg] F. HECHT The mesh adapting software: bamg. INRIA report 1998. http://www-rocq.inria.fr/gamma/cdrom/www/bamg/eng.htm.

[freefempp] F. Hecht, K. Ohtsuka, and O. Pironneau. *FreeFem++ manual*. Universite Pierre et Marie Curie, 2002–2005. on the web at http://www.freefem.org/ff++/index.htm.

[VAL92] M.G. VALLET, Génération de maillages Éléments Finis anisotropes et adaptatifs, *thèse Université Paris VI*, Paris, 1992.

[VER96] R. VERFÜRTH, A review of a posteriori error estimation and adaptive refinement techniques, *Wiley Teubner*, 1996.

On discrete boundaries and solution accuracy in anisotropic adaptive meshing

Konstantin Lipnikov[1] and Yuri Vassilevski[2]

[1] Los Alamos National Laboratory, `lipnikov@lanl.gov`
[2] Institute of Numerical Mathematics, `vasilevs@dodo.inm.ras.ru`

1 Introduction

Large-scale simulations in engineering applications are most effective when they are combined with adaptive methods. The adaptive methods reduce greatly the demand for larger number of unknowns and improve accuracy of the simulations via grid adaptation near fine-scale features of the solution. In this paper, we consider a tensor metric based adaptive methodology [AG99, HE98, BU97, HA95, LI03, VA03]. The metric is induced by an approximate Hessian (matrix of second derivatives) of a discrete solution. The focus of this paper is on treatment of curved (non-planar) internal and boundary surfaces.

In many applications, exact parameterization of curved surfaces may be unknown. In this case, the surfaces are described by triangular meshes (e.g., meshes coming from CAD systems) which reduce performance of adaptive methods due to the limited surface resolution. To resolve geometrical features of a model, the trace of an optimal adaptive mesh on a curved surface should be close to the given surface triangulation which is not always the case. One of the possible solutions is to fix only nodes of the surface triangulation (see [VA05] for a more adequate solution) which imposes simple constraints for mesh adaptation algorithms. The fixed-node constraints are easily embedded into the metric-based adaptive methodology.

If the underlying surfaces are sufficiently smooth (or piecewise smooth), the original triangular meshes carry additional information about these surfaces. In this paper, we use this fact to design a new surface reconstruction method and analyze it both theoretically and numerically. In principle, the reconstructed surface can be triangulated to use again the metric-based adaptive methodology with the fixed-node constraints.

There are many methods for higher order reconstruction of piecewise linear surfaces (see [GA05, MI97, ME02, MO01] and references therein). In [MI97, MO01] the surface is parameterized and the desired surface characteristics are computed from the derivatives of functions specifying the parameterization. In [GA05, MI97], the discrete surface is approximated by a piecewise quadratic surface using the best fit algorithm. The method proposed in this paper uses technique of the discrete differential geometry to compute an approximate Hessian of a piecewise quadratic function representing the reconstructed surface. The Hessian is computed in a weak sense by analogy with the finite element methods. The developed method is exact for quadratic surfaces.

The reconstruction method is local, and therefore it can be easily parallelized. Its computational cost is proportional to the number of surface triangles.

We demonstrate efficiency of the reconstruction method for the solution of a convection-diffusion problem simulating transport phenomena around a spherical obstacle. The solution has a boundary layer along a part of the obstacle boundary. As the result, accuracy of the discrete solution depends on the accuracy of the boundary representation and is significantly improved on the reconstructed surfaces.

The paper outline is as follows. In Section 2, we describe briefly the Hessian based adaptation methodology. In Section 3, we propose and analyze a new surface reconstruction method. In Section 4, we illustrate our adaptive methodology with numerical tests.

2 Hessian Based Mesh Adaptation

2.1 Quasi-Optimal Meshes

Let Ω_h be a mesh with $N(\Omega_h)$ elements and u_h be a discrete piecewise linear solution computed at mesh nodes with some numerical method which we denote by $\mathcal{P}_{\Omega_h}$. We shall simply write that $u_h = \mathcal{P}_{\Omega_h} u$ where u is an unknown exact solution. The ideal goal would be to find a mesh (probably anisotropic) which minimizes the maximal norm of the discretization error $\|u - \mathcal{P}_{\Omega_h} u\|_\infty$. In many problems, this error can be majorized by the interpolation error, $\|u - \mathcal{I}_{\Omega_h} u\|_\infty$, where $\mathcal{I}_{\Omega_h}$ is the linear interpolation operator on mesh Ω_h. It gives us the following mesh optimization problem:

$$\Omega_h^{opt} = \arg \min_{N(\Omega_h) < N_{\max}} \|u - \mathcal{I}_{\Omega_h} u\|_\infty \qquad (1)$$

where $N_{\max}$ is the maximal number of mesh elements (tetrahedra) defined by the user. This problem was analyzed both theoretically and numerically in [AG99, VA03]. In fact, problem (1) was replaced by a simpler problem which provides a constructive way for finding an approximate solution of (1) which we refer to as a *quasi-optimal* mesh. This mesh is quasi-uniform in the metric $|H^h|$ derived from the discrete Hessian H^h of u_h. The generation of quasi-uniform meshes is based on the notion of a *mesh quality*.

Let G be a metric generated by a symmetric positive definite 3×3 matrix whose entries depend on point $\mathbf{x} \in \Omega$. For an element e in Ω_h, we denote by $|e|_G$ its volume in metric G and by $|\partial\partial e|_G$ the total length of its edges (also in metric G). We define the mesh quality as

$$Q(\Omega_h) = \min_{e \in \Omega_h} Q(e) \qquad (2)$$

where $Q(e)$ is the quality of a single element e,

$$Q(e) = 6\sqrt[4]{2}\frac{|e|_G}{|\partial\partial e|_G^3} F\left(\frac{|\partial\partial e|_G}{6h^*}\right), \quad 0 < Q(e) \leqslant 1. \qquad (3)$$

Here h^* is the mesh size in the G-uniform mesh with N_{max} elements and $F(t)$ is a continuous smooth function, $0 \leqslant F(t) \leqslant 1$, with the only maximum at point 1,

$F(1) = 1$, and such that $F(0) = F(+\infty) = 0$. The last factor in (3) controls the size of the element, whereas the remaining factors control its shape.

The optimization of the mesh Ω_h with respect to the mesh quality (2) results in a G-quasi-uniform grid. Since the mesh quality is as good as the quality of the worst element, the mesh optimization can be achieved with a series of *local* operations applied to this element. The list of such operations includes alternation of topology with node deletion/insertion, edge/face swapping and node movement (see Fig. 1 for 2D analogs of local operations and [AG99] for more details).

The local operations such as node deletion/insertion and edge/face swapping are well described in the literature while implementation of the node movement requires additional comments. It is driven by minimization of the smooth functional $\mathcal{F} : \mathbb{R}^3 \to \mathbb{R}$, of the node position $\mathbf{x}$, defined as a reciprocal of the mesh quality (2), i.e. $1 \leqslant \mathcal{F} < \infty$. Some restrictions have to be imposed on mesh modifications to keep the mesh unfolded and to preserve internal and boundary surfaces.

2.2 Adaptive Iterative Algorithm

We use the following loop to build a quasi-optimal mesh:

- Generate any initial tetrahedrization $\Omega_h^{(1)}$ of the computational domain.
- For $k = 1, 2, \ldots$, repeat
  - Compute the discrete solution u_h and generate the discrete Hessian-based metric $|H^h|$ which is the symmetric positive definite matrix given by

$$H^h = W_h \Lambda_h W_h^T, \quad |H^h| = W_h |\Lambda_h| W_h^T$$

  where W_h is the orthonormal matrix, $\Lambda_h = \text{diag}\{\lambda_1, \lambda_2, \lambda_3\}$ is the diagonal matrix, and

$$|\Lambda_h| = \text{diag}\{\max\{|\lambda_1|; \varepsilon\}, \max\{|\lambda_2|; \varepsilon\}, \max\{|\lambda_3|; \varepsilon\}\}$$

  with $\varepsilon > 0$ being a user defined tolerance.
  - Terminate the adaptive loop if the mesh quality in metric $|H^h|$ is bigger than Q_0 which is the user defined number (e.g., $Q_0 \sim 0.4$).
  - Generate the mesh $\Omega_h^{(k+1)}$ which is quasi-uniform in the metric $|H^h|$ and is such that $Q(\Omega_h^{(k+1)}) > Q_0$. To do this, we use local operations such as node deletion/insertion, edge/face swapping and node movement (see Fig. 1).

It is proved in [AG99] that quasi-optimal meshes in polyhedral domains result in the asymptotically optimal estimate:

$$\|u - \mathcal{I}_{\Omega_h} u\|_\infty \sim N(\Omega_h)^{-2/3}. \tag{4}$$

In Section 3, we demonstrate numerically that (4) holds in a more general case of curved boundaries. We also show that the optimal estimate is violated when these boundaries are represented by triangular meshes.

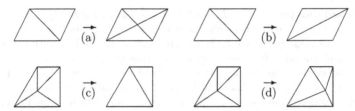

Fig. 1. Local topological operations for 2D triangular meshes: (a) node insertion, (b) edge swapping, (c) node deletion, and (d) node movement

2.3 Treatment of Surface Constraints

The distinctive geometrical features of any model are internal and boundary surfaces (*feature surfaces*) and their intersections (*feature edges*). Let us consider a particular feature surface $\Gamma \subset \mathbb{R}^3$ and a feature edge $\Theta \subset \mathbb{R}^3$. If the analytical formulae for Γ and Θ were available, they could be used in the metric-based adaptive methodology. For example, a node living on the feature surface could be easily projected on the analytic surface.

Often, the analytic representation may not be available for all geometric features constituting the model. In this case, the geometric features are modeled with faces and edges of the original mesh $\Omega_h^{(1)}$. It is highly desirable if the trace of the quasi-optimal mesh on a geometric feature is close to the given feature triangulation. One of the possible solutions is discussed below.

Let the discrete feature surface Γ_h be the triangulated surface of the original mesh $\Omega_h^{(1)}$ approximating Γ with triangular faces Γ_t, $\Gamma_h = \cup_t \Gamma_t$, and the discrete feature edge Θ_h be a polyline formed by the edges of $\Omega_h^{(1)}$ approximating Θ. In this paper, we fix (freeze) nodes living on Γ_h and Θ_h. This imposes simple constraints on the local mesh modifications and leaves enough freedom for realization of mesh modifications with surrounding tetrahedra. Still, the fixed-nodes constraints may result in unnecessary fine mesh in domains where solution u_h is very smooth. The more adequate treatment of discrete boundaries is described in [VA05].

In Section 5, we shall demonstrate that accuracy of boundary representation makes significant impact on accuracy of the discrete solution. The accuracy may be improved if we assume that the underlying surfaces are sufficiently smooth or piecewise smooth. Then the discrete feature surface Γ_h carries additional information about Γ. Our surface reconstruction method is described in the next section.

3 Piecewise Quadratic Extrapolation of Piecewise Linear Surfaces

In this section, we consider again the feature surface Γ. To simplify the presentation, we assume that Θ is its boundary. We assume also that nodes of Γ_h and Θ_h belong to Γ and Θ, respectively, although this assumption is not necessary in practice.

The piecewise quadratic extrapolation $\tilde{\Gamma}_h$ of Γ_h is defined as the continuous surface being the closure of a union of open non-overlapping pieces $\tilde{\Gamma}_t$ of local quadratic extrapolations over faces Γ_t.

The local extrapolation $\tilde{\Gamma}_t$ is described by a quadratic function $\varphi_{2,t}$. Hereafter, we shall omit the superscript t whenever it does not result in confusion. For our purposes, it will be convenient to describe the function φ_2 in a local coordinate system (ξ_1, ξ_2) associated with the plane of Γ_t. In this coordinate system, the 2D multi-point Taylor formula [CI71] for a quadratic function φ_2 with the Hessian $H^{\varphi_2} = \{H_{pq}^{\varphi_2}\}_{p,q=1}^2$ reads

$$\varphi_2(\boldsymbol{\xi}) = -\frac{1}{2} \sum_{i=1}^{3} (H^{\varphi_2}(\boldsymbol{\xi} - \mathbf{a}_i),\ (\boldsymbol{\xi} - \mathbf{a}_i))\, p_i(\boldsymbol{\xi}) \tag{5}$$

where $\mathbf{a}_1, \mathbf{a}_2, \mathbf{a}_3$ are vertices of the triangle Γ_t and $p_i(\boldsymbol{\xi})$ is a piecewise linear function such that $p_i(\mathbf{a}_j) = \delta_{ij}$.

In order to recover the Hessian H^{φ_2}, we *first* assume that numbers $\alpha_i = (H^{\varphi_2}\ell_i, \ell_i)$, $i = 1, 2, 3$, representing the projection of this Hessian on edges ℓ_i of Γ_t, are given. Hereafter, we use ℓ_i for both the mesh edge and the corresponding vector. In the local coordinate system, vectors ℓ_i are described by two coordinates, $\ell_i = (l_1^i, l_2^i)$. We assume that the vector ℓ_i begins at vertex $\mathbf{a}_i$ and ends at vertex $\mathbf{a}_{i+1}$ with $\mathbf{a}_4 = \mathbf{a}_1$. Then, the definition of α_i gives

$$\left(\begin{pmatrix} H_{11}^{\varphi_2} & H_{12}^{\varphi_2} \\ H_{12}^{\varphi_2} & H_{22}^{\varphi_2} \end{pmatrix} \begin{pmatrix} l_1^i \\ l_2^i \end{pmatrix},\ \begin{pmatrix} l_1^i \\ l_2^i \end{pmatrix} \right) = \alpha_i$$

which in turn results in the system of three linear equations for the unknown entries of the matrix H^{φ_2}:

$$l_1^i l_1^i H_{11}^{\varphi_2} + l_2^i l_2^i H_{22}^{\varphi_2} + 2\, l_1^i l_2^i H_{12}^{\varphi_2} = \alpha_i,\ i = 1, 2, 3. \tag{6}$$

Lemma 1. *The matrix of the system (6) is non-singular.*

Proof. Let us denote the coefficient matrix of system (6) by B. Note that $\ell_1 + \ell_2 + \ell_3 = 0$. Using this fact in direct calculations of the determinant of matrix B, we get

$$|\det B| = 2|l_1^1 l_2^2 - l_1^2 l_2^1|^3 = 16|\Gamma_t|^3 > 0 \tag{7}$$

where $|\Gamma_t|$ is the area of the triangle Γ_t. This proves the assertion of the lemma. $\square$

Second, we use results of [AG99] where the algorithm for computing the discrete Hessian $H^h(\mathbf{a}_i)$ of a continuous piecewise linear function is presented and analyzed. We define α_i as the average of two nodal approximations,

$$\alpha_i = ((H^h(\mathbf{a}_i)\ell_i, \ell_i) + (H^h(\mathbf{a}_{i+1})\ell_i, \ell_i))/2, \tag{8}$$

associated with the edge ℓ_i. There are two exceptions from this rule. If $\mathbf{a}_i \in \Theta_h$ and $\mathbf{a}_{i+1} \notin \Theta_h$, then α_i is equal to $(H^h(\mathbf{a}_{i+1})\ell_i, \ell_i)$. If $\mathbf{a}_i \in \Theta_h$ and $\mathbf{a}_{i+1} \in \Theta_h$, then $\alpha_i = 0$. This implies that the nodal approximation of the Hessian is not recovered at feature edges and therefore the traces of Γ_h and $\tilde{\Gamma}_h$ on Θ_h coincide.

It remains to describe how we recover $H^h(\mathbf{a}_i)$ for every interior node $\mathbf{a}_i$ of Γ_h. We begin by introducing a few additional notations. For each $\mathbf{a}_i$, we define the superelement σ_i as a union of all triangles of Γ_h sharing $\mathbf{a}_i$. Then, we define a plane approximating in the least square sense the nodes of this superelement and associate this plane with a local coordinate system (ξ_1, ξ_2). Let $\hat{\sigma}_i$ be the projection of σ_i onto the $\xi_1\xi_2$-plane. Further, let $\varphi(\xi_1, \xi_2)$ be the continuous function representing locally Γ, and $\varphi_h^i(\xi_1, \xi_2)$ be the continuous piecewise linear function representing σ_i. We assume that both functions are single-valued over $\hat{\sigma}_i$. Finally, we denote the Hessian of φ by H^φ.

The components H_{pq}^h, $p, q = 1, 2$, of the discrete Hessian H^h are defined in a weak sense by

$$\int_{\hat{\sigma}_i} H_{pq}^h(\mathbf{a}_i)\psi_h \, dS = -\int_{\hat{\sigma}_i} \frac{\partial \varphi_h^i}{\partial \xi_p} \frac{\partial \psi_h}{\partial \xi_q} \, dS, \qquad (9)$$

which holds for any continuous piecewise linear function ψ_h vanishing on $\partial \hat{\sigma}_i$. Note that the discrete Hessian $H^h(\mathbf{a}_i)$ is a geometric characteristic of the feature surface Γ at point $\mathbf{a}_i$ (related to its curvature) and therefore is invariant of the position of the projection plane associated with the superelement σ_i. In other words, the value $(H^h(\mathbf{a}_i)\ell_i, \ell_i)$ is independent of the local transformation of the coordinate system.

In addition to the above invariance and the obvious uniqueness of H^h, the presented extrapolation is exact for quadratic surfaces as long as the triangle Γ_t has no edges on Θ_h. Indeed, for a quadratic function φ, the recovery method (9) is exact, i.e. $H_{pq}^h(\mathbf{a}_i) = H_{pq}^\varphi(\mathbf{a}_i)$. Therefore, for all $\mathbf{a}_i \notin \Theta_h$,

$$(H^\varphi \ell, \ell) = (H^h(\mathbf{a}_i)\ell, \ell)$$

for every edge $\ell \subset \Gamma_h \setminus \Theta_h$ and $H^{\varphi_2} = H^\varphi$ follows from (8) and Lemma 1.

The proposed reconstruction method is local and therefore it can be easily parallelized. Its computational cost is proportional to the number of surface triangles. It is pertinent to note that this cost is negligent compared to the cost of anisotropic mesh adaptation.

Now we consider the approximation property of our extrapolation method. For every triangle Γ_t, we define a superelement σ^t as union of superelements σ_i corresponding to vertices $\mathbf{a}_i$ of Γ_t. Again, we use the local coordinate system (ξ_1, ξ_2) associated with the triangle Γ_t. Let $\hat{\sigma}^t$ (resp., $\hat{\Gamma}_t$) be the projection of σ^t (resp., Γ_t) onto the $\xi_1\xi_2$-plane. We define the constant tensor $H_{\sigma^t}^\varphi$ for the superelement $\hat{\sigma}^t$ as

$$H_{\sigma^t}^\varphi = H^\varphi(\arg \max_{\boldsymbol{\xi} \in \hat{\sigma}^t} |det H^\varphi(\boldsymbol{\xi})|). \qquad (10)$$

Theorem 1. *Let edges of a triangle Γ_t be interior edges of Γ_h and $\hat{\sigma}_t$ be a quasi-uniform triangulation with size h. Let $\varphi(\xi_1, \xi_2)$ be a $C^2(\hat{\sigma}^t)$ function representing locally Γ and $\varphi_h = \mathcal{I}_{\hat{\sigma}^t}\varphi$ be a continuous piecewise linear function representing σ^t. Moreover, let H^φ and H^h be the differential and discrete Hessians of φ and φ_h, respectively, such that*

$$\|H_{pq}^\varphi - H_{\sigma^t, pq}^\varphi\|_{L_\infty(\hat{\sigma}^t)} < \delta, \qquad (11)$$

$$\|\nabla(\varphi - \mathcal{I}_{\hat{\sigma}^t}\varphi)\|_{L_2(\hat{\sigma}^t)} < \epsilon. \qquad (12)$$

Then, the quadratic function φ_2 describing $\tilde{\Gamma}_t$ and defined by (5), (6), (8) and (9) satisfies

$$\|\varphi - \varphi_2\|_{L_\infty(\hat{\Gamma}_t)} \leqslant C(\epsilon + \delta h^2) \qquad (13)$$

where constant C is independent of δ, ϵ, h and φ.

<u>Proof.</u> Hereinafter, we shall use C and C_i for generic constants having different values in different places. The definition (9) of the discrete Hessian implies that

$$\int_{\hat{\sigma}_i} (H_{pq}^\varphi - H_{pq}^h(\mathbf{a}_i)) \psi_h \, dS = -\int_{\hat{\sigma}_i} \frac{\partial(\varphi - \varphi_h)}{\partial \xi_p} \frac{\partial \psi_h}{\partial \xi_q} \, dS$$

for any $\psi_h \in P_1(\hat{\sigma}_i)$ vanishing on $\partial \hat{\sigma}_i$. Now, using the triangle inequality and then the Cauchy inequality, we get

$$\int_{\hat{\sigma}_i} |H_{\sigma^t, pq}^\varphi - H_{pq}^h(\mathbf{a}_i)||\psi_h| \, dS \leqslant \left\| \frac{\partial(\varphi - \varphi_h)}{\partial \xi_p} \right\|_{L_2(\hat{\sigma}_i)} \left\| \frac{\partial \psi_h}{\partial \xi_q} \right\|_{L_2(\hat{\sigma}_i)}$$
$$+ \int_{\hat{\sigma}_i} |H_{\sigma^t, pq}^\varphi - H_{pq}^\varphi||\psi_h| \, dS.$$

Let us evaluate all terms in the above inequality for a particular choice of ψ_h such that $\psi_h(\mathbf{a}_i) = 1$. The term in the left hand side is estimated from below as follows:

$$\int_{\hat{\sigma}_i} |H_{\sigma^t, pq}^\varphi - H_{pq}^h(\mathbf{a}_i)||\psi_h| \, dS \geqslant C_1 |H_{\sigma^t, pq}^\varphi - H_{pq}^h(\mathbf{a}_i)| \, |\hat{\sigma}_i|.$$

The terms in the right hand side may be easily estimated from above using quasi-uniformity of $\hat{\sigma}^t$ and assumption (11):

$$\left\| \frac{\partial \psi_h}{\partial \xi_q} \right\|_{L_2(\hat{\sigma}_i)} \leqslant C_2, \qquad \int_{\hat{\sigma}_i} |H_{\sigma^t, pq}^\varphi - H_{pq}^\varphi||\psi_h| \, dS \leqslant C_3 \delta |\hat{\sigma}_i|.$$

Combining the above inequalities, we get

$$|H_{\sigma^t, pq}^\varphi - H_{pq}^h(\mathbf{a}_i)| \leqslant \frac{C_2}{C_1 |\hat{\sigma}_i|} \epsilon + \frac{C_3}{C_1} \delta. \tag{14}$$

Let H^{φ_2} be the Hessian of the quadratic function φ_2. The next step in the proof is to estimate the discrepancy between $H_{\sigma^t}^\varphi$ and H^{φ_2}. For this purpose, we use the perturbation analysis and Lemma 1. Since both Hessians $H_{\sigma^t}^\varphi$ and H^{φ_2} are constant, they are uniquely defined by the right hand side of system (6) and edges of triangle Γ_t. Let α_1, α_2 and α_3 be the entries of the right hand side, $H_{pq}^{\varphi_2}$ be the solution of (6), and let $\beta_i = (H_{\sigma^t}^\varphi \ell_i, \ell_i)$, $i = 1, 2, 3$. Using definition (8), inequality (14), a linear algebra estimate for eigenvalues of a 2×2 matrix, and the assumption of quasi-uniformity of $\hat{\sigma}^t$, we get

$$|\alpha_i - \beta_i| = \frac{1}{2} |(H^h(\mathbf{a}_i)\ell_i, \ell_i) + (H^h(\mathbf{a}_{i+1})\ell_i, \ell_i) - 2(H_{\sigma^t}^\varphi \ell_i, \ell_i)|$$
$$\leqslant 2 \left(\frac{C_2 \epsilon}{C_1 \min_{i=1,2,3} |\hat{\sigma}_i|} + \frac{C_3}{C_1} \delta \right) (\ell_i, \ell_i) \leqslant C(\epsilon + \delta h^2).$$

The perturbation analysis states that

$$|H_{pq}^{\varphi_2} - H_{\sigma^t, pq}^\varphi| \leqslant C |\lambda_{\min}^{-1}(B)| \max_{i=1,2,3} |\alpha_i - \beta_i|$$

where the matrix B is defined in Lemma 1 and $\lambda_{\min}(B)$ is its closest to zero eigenvalue. The application of the Gershgorin theorem and the quasi-uniformity assumption give the estimate for the maximal eigenvalue of B:

$$\lambda_{\max}(B) \leqslant 2 \max_{1\leqslant i\leqslant 3} |\ell_i|^2 \leqslant Ch^2.$$

Therefore, due to (7)

$$|\lambda_{\min}(B)| \geqslant \frac{|\det B|}{\lambda_{\max}^2(B)} = \frac{16|\Gamma_t|^3}{\lambda_{\max}^2(B)} \geqslant Ch^2.$$

Using the last estimate, we get easily that

$$|H_{pq}^{\varphi_2} - H_{\sigma^t,pq}^{\varphi}| \leqslant C(\epsilon/h^2 + \delta). \tag{15}$$

Finally, by virtue of the multi-point Taylor formula for a general function φ whose linear interpolant φ_h vanishes on the triangle $\hat{\Gamma}_t = \Gamma_t$, we have:

$$\varphi(\boldsymbol{\xi}) = -\frac{1}{2} \sum_{i=1}^{3} (H^{\varphi}(\boldsymbol{\xi}_i^*)(\boldsymbol{\xi} - \mathbf{a}_i), (\boldsymbol{\xi} - \mathbf{a}_i)) \, p_i(\boldsymbol{\xi})$$

where $\boldsymbol{\xi}_i^*(\boldsymbol{\xi})$ is a point inside $\hat{\Gamma}_t$, $\boldsymbol{\xi} \in \Gamma_t$. Together with formula (5), it gives

$$|\varphi(\boldsymbol{\xi}) - \varphi_2(\boldsymbol{\xi})| = \frac{1}{2} \left| \sum_{i=1}^{3} ([H^{\varphi}(\boldsymbol{\xi}_i^*) - H^{\varphi_2}](\boldsymbol{\xi} - \mathbf{a}_i), (\boldsymbol{\xi} - \mathbf{a}_i)) \, p_i(\boldsymbol{\xi}) \right|$$
$$\leqslant C(\epsilon + \delta h^2).$$

This proves the assertion of the theorem. $\qquad\qquad\square$

Generally speaking, the values of ϵ and δ depend on the derivatives of φ. If φ is sufficiently smooth, for example it is in $C^3(\hat{\sigma}^t)$, then $\epsilon \sim h^3$ [CI78], $\delta \sim h$ and we get the expected result

$$\|\varphi - \varphi_2\|_{L_\infty(\hat{\Gamma}_t)} \leqslant Ch^3.$$

4 Numerical Experiments

As the model problem, we consider the convection-diffusion equation

$$-0.01\Delta u + \mathbf{b} \cdot \nabla u = 0 \quad \text{in} \quad \Omega \tag{16}$$
$$u = g \quad \text{on} \quad \Gamma_{in}$$
$$\frac{\partial u}{\partial n} = 0 \quad \text{on} \quad \Gamma_{out}$$
$$u = 0 \quad \text{on} \quad \partial\Omega \setminus (\Gamma_{in} \cup \Gamma_{out}).$$

Here $\mathbf{b} = (1,0,0)^T$ is the velocity field, $\Omega = (0,1)^3 \setminus B_{0.5}(0.18)$ is the computational domain with $B_{0.5}(r) = \{\mathbf{x}: \sum_{i=1}^{3}(x_i - 0.5)^2 \leqslant r^2\}$, $\Gamma_{in} = \{\mathbf{x} \in \partial\Omega: x_1 = 0\}$, $\Gamma_{out} = \{\mathbf{x} \in \partial\Omega: x_1 = 1\}$, and $g(x_2, x_3) = 16x_2(1 - x_2)x_3(1 - x_3)$ is the standard Poiseille profile of the entering flow.

The solution u to (16) possesses a boundary layer along the upwind side of the spherical obstacle $B_{0.5}(0.18)$ and is very smooth in the shadow region of this obstacle. Since the exact solution is not known, in our experiments, we replace it with the piecewise linear finite element solution u_* computed on a very fine adaptive (quasi-optimal) mesh containing more than 1.28 million tetrahedra (see Fig. 2, left

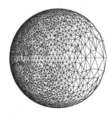

Fig. 2. Isolines of solution u_* in the plane passing through the center of the obstacle and parallel to the x_1x_2-plane (left picture) and the trace of a typical quasi-optimal mesh at the obstacle with analytic boundary (right picture)

picture). To built the adaptive mesh, we used the analytical representation of $\partial\Omega$. The trace of the adapted mesh on the surface of the obstacle shows coarsening in the shadow region and refinement in the upwind region (see Fig. 2, right picture).

In the first set of experiments (the top-left picture in Fig. 3), we demonstrate the asymptotic result (4) with u_* instead of u. The L_∞ error fits the analytic curve $60\,N(\Omega_h)^{-2/3}$.

In the second set of experiments (the top-right picture in Fig. 3), the boundary $\Gamma = \partial B_{0.5}(0.18)$ is approximated with a quasi-uniform mesh Γ_h. We measure the L_∞ error as a function of $N(\Omega_h)$ for three different values of h. Figure 3 presents saturation of this error due to the limited boundary resolution. We observe that the saturated error ε_h is almost reciprocal to h^2: $\varepsilon_{0.05} = 0.20$, $\varepsilon_{0.025} = 0.067$, and $\varepsilon_{0.0125} = 0.021$. This is probably related to the second order approximation of the smooth boundary Γ by the piecewise linear manifold Γ_h.

In these experiments, we fixed the nodes on Γ_h. The fixed-node constraints result in unnecessary fine mesh only in the shadow region of the obstacle (see Fig. 4, right picture). Therefore, the mesh is too stretched there in contract to the case of analytical representation of the obstacle boundary (see Fig. 4, left picture, and Fig. 2 for the mesh trace). This results in mesh elements with a lower quality in the shadow region. However, the excessive refinement and the low quality of these elements do not affect the value of the saturated error. The number of extra elements is small compared to $N(\Omega_h)$ and the solution is very smooth in these elements.

In the third set of experiments, we study the effect of the piecewise quadratic extrapolation $\tilde{\Gamma}_h$ of Γ_h on accuracy of the discrete solution. We compare saturation errors for three surface meshes: $\Gamma_{0.025}$, $\Gamma_{0.0125}$ and $\Gamma^*_{0.0125}$. The third mesh is obtained from $\Gamma_{0.0125}$ by projection its mesh nodes onto $\tilde{\Gamma}_{0.025}$. This mesh must provide the saturation error ε^*_h which is between saturation errors on the other two meshes. This is illustrated in the bottom picture of Fig. 3 where $\varepsilon_{0.0125} = 0.021$, $\varepsilon_{0.025} = 0.067$, and $\varepsilon^*_{0.0125} = 0.043$.

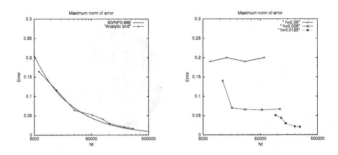

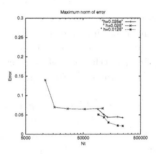

Fig. 3. Convergence analysis: using analytic representation of the obstacle boundary (top-left picture), using three discrete models $\Gamma_{0.05}$, $\Gamma_{0.025}$, and $\Gamma_{0.0125}$ for $\partial B_{0.5}(0.18)$ (top-right picture), using the piecewise quadratic extrapolation $\tilde{\Gamma}_{0.025}$ (bottom picture)

Another approach for building a piecewise linear surface $\Gamma_{0.0125}^*$ is based on the uniform refinement of $\Gamma_{0.025}$ with subsequent projection of new mesh nodes onto $\tilde{\Gamma}_{0.025}$. We use the first approach because it gives the most rigorous comparison of saturation errors on meshes $\Gamma_{0.0125}$ and $\Gamma_{0.0125}^*$.

In practice, the surface reconstruction should be dynamic and driven by the size of mesh elements. For the convection-diffusion problem (16), the surface extrapolation is required only in the upwind part of the obstacle boundary. We shall address this problem in the future.

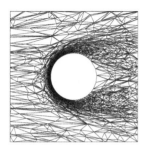 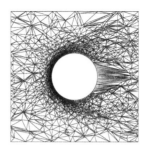

Fig. 4. The mesh cuts in the plane passing through the center of the obstacle and parallel to the $x_1 x_2$-plane: The left picture corresponds to the case of analytic boundary representation. The right picture corresponds to the case of fixed-node constraints. The meshes have approximately the same number of elements ($\sim 200k$)

5 Conclusion

We have shown that representation of curved surfaces with triangular meshes restricts the use of adaptive methods. For a particular convection-diffusion problem, we have shown numerically that the saturated discretization error is proportional to h^2 where h is the size of the quasi-uniform mesh approximating the curved surface. We have proposed and analyzed theoretically and numerically a new surface reconstruction technique which improves performance of adaptive methods.

Acknowledgments. The work of the second author has been supported by the RFBR grant 04-07-90336 and the academic program "Computational and informatic issues of the solution of large problems".

[AG99] Agouzal A, Lipnikov K, Vassilevski Y (1999) East-West J Numer Math 7:223–244
[BO03] Bonnans JF, Gilbert C, Lemarécha C, Sagastizábal CA (2003) Numerical Optimization. Theoretical and Practical Aspects. Springer-Verlag, Berlin
[HE98] Borouchaki H, Hecht F, Frey PJ (1998) Inter J Numer Meth Engrg 43:1143–1165
[BU97] Buscaglia GC, Dari EA (1997) Inter J Numer Meth Engrg 40:4119–4136
[CI71] Ciarlet PG, Wagschal C (1971) Numer Math 17:84–100
[CI78] Ciarlet PG (1978) The finite element method for elliptic problems. North-Holland
[HA95] Dompierre J, Vallet M-G, Fortin M, Habashi WG, Ait-Ali-Yahia D, Tam A (1995) CERCA Report R95-73
[GA05] Garimella RV, Swartz BK (2003) Los Alamos Report LA-UR-03-8240

[GA04] Garimella R, Knupp P, Shashkov M (2004) Comput Meth Appl Mech Engng 193:913–928
[LI03] Lipnikov K, Vassilevski Y (2003) Comput Meth Appl Mech Engrg 192:1495–1513
[MI97] McIvor AM, Valkenburg RJ (1997) Machine Vision and Applications 10:17–26
[ME02] Meyer M, Lee H, Barr AH, Desbrun M (2002) Journal of Graphics Tools 7:13–22
[MO01] Mokhtarian F, Khalili N, Yuen P (2001) Computer Vision and Image Understanding 83:118–139
[NO99] Nocedal J, Wright SJ (1999) Numerical Optimization. Springer-Verlag, New York
[PE02] Petitjean S (2002) ACM Computing Surveys 32:211–262
[VA03] Vassilevski Y, Lipnikov K (2003) Comp Math Math Phys 43:827–835
[VA05] Vassilevski Y, Dyadechko V, Lipnikov K (2005) Russian J Numer Anal Math Modelling (to appear)

Weighted Delaunay Refinement for Polyhedra with Small Angles

S.-W. Cheng[1], T. K. Dey[2], and T. Ray[2]

[1] Hong-Kong U. of Science and Technology, Hong Kong. scheng@cs.ust.hk
[2] The Ohio State U., Columbus, OH, USA. tamaldey,rayt@cse.ohio-state.edu

Summary. Recently, a provable Delaunay meshing algorithm called QMESH has been proposed for polyhedra that may have acute input angles. The algorithm guarantees bounded circumradius to shortest edge length ratio for all tetrahedra except the ones near small input angles. This guarantee eliminates or limits the occurrences of all types of poorly shaped tetrahedra except slivers. A separate technique called weight pumping is known for sliver elimination. But, allowable input for the technique so far have been periodic point sets and piecewise linear complex with non-acute input angles. In this paper, we incorporate the weight pumping method into QMESH thereby ensuring that all tetrahedra except the ones near small input angles have bounded aspect ratio. Theoretically, the algorithm has an abysmally small angle guarantee inherited from the weight pumping method. Nevertheless, our experiments show that it produces better angles in practice.

Key words: mesh generation, computational geometry, Delaunay refinement, sliver, weighted Delaunay triangulation.

1 Introduction

Meshing a polyhedral domain with well shaped tetrahedra occurs as an important problem in finite element methods. The *aspect ratio* of a tetrahedron is the ratio of its circumradius to its inradius. A tetrahedron has bounded aspect ratio if all the face angles and dihedral angles are greater than a constant threshold. A weaker measure is the *circumradius-edge ratio*: ratio of the circumradius to the shortest edge length. If the tetrahedra have bounded circumradius-edge ratio, only one class of poorly shaped tetrahedra may remain and they are known as *slivers* [CDET99]. Eliminating slivers is a challenge in mesh generation.

Three main paradigms are known for polyhedra meshing: *octree based* [MV00, SG91], *advancing front* [Lo91b, L96], and *Delaunay based* [ACYD05, B89, BGL00, WH94] methods. The Delaunay based methods are popular in practice [O98], perhaps due to their directional independence and good quality meshing in general. In this paper, we focus on the Delaunay based methods that can provide theoretical guarantees on the quality of tetrahedra.

Based on Ruppert's 2D Delaunay refinement paradigm [Rup95], Shewchuk presented an algorithm that constructs a Delaunay mesh with bounded circumradius-edge ratio [Shew98]. Unfortunately, the algorithm may not recover the input boundary when some input angle is acute. Although Shewchuk [Shew00], Murphy, Mount and Gable [MMG00], and Cohen-Steiner, de Verdière and Yvinec [SVY02] subsequently proposed algorithms that can handle acute input angles, no guarantee on the tetrahedral shape was provided. A significant theoretical progress was obtained by Cheng and Poon [CP03], who proposed an algorithm that guarantees bounded circumradius-edge ratio everywhere in the mesh. However, given the complexity of the algorithm, its practicality is doubtful. Recently, Cheng, Dey, Ramos, and Ray [CDRR04] presented a simple algorithm QMESH and an implementation for polyhedra. All tetrahedra have bounded circumradius-edge ratio, except those near small input angles. There are very few such tetrahedra as observed in the experiments. There are several innovations in this algorithm: local feature sizes are only needed at vertices with small input angles, explicit protecting regions for the input edges are no longer needed, and the splitting of non-Delaunay triangles in recovering the input boundary (instead of splitting non-Gabriel triangles alone). These new features ease the implementation tremendously and help to keep the mesh size small. Subsequently Pav and Walkington [PW04] proposed an algorithm for handling non-manifold boundaries with very similar guarantees.

All of the algorithms mentioned so far focused on ensuring bounded circumradius-edge ratio, which eliminates all types of poorly shaped tetrahedra but slivers. The work of Chew [Chew97], Cheng et al. [CDET99], and Edelsbrunner et al. [E00] are the first theoretical results on sliver elimination, albeit for point sets only. Edelsbrunner and Guoy [EG02] experimented with the sliver exudation technique in [CDET99]. They demonstrated that the technique is more effective in practice than what the theory predicts. Later, Li and Teng [LT01], and Cheng and Dey [CD03] presented algorithms for meshing piecewise linear complex with well-shaped tetrahedra. But only non-acute input angles were allowed.

In this paper, we incorporate the weight pumping technique [CD03, CDET99] into QMESH [CDRR04] to purge slivers. In essence, we make the following contributions in this paper.

Theoretical: We present a Delaunay meshing algorithm for polyhedra possibly with acute input angles. It guarantees bounded aspect ratio for all tetrahedra except the ones near small input angles. No Delaunay meshing algorithm with such guarantees is known to date. The new algorithm changes the mesh connectivity using the weighted Delaunay triangulation. Since weight pumping may challenge the input boundary, we first extend the algorithm in [CDRR04] to prepare for it. The extended algorithm offers the same guarantee that remaining skinny tetrahedra are near small input angles. Then the algorithm assigns proper weights to sliver vertices to purge them. Most slivers are eliminated except those near small input angles.

Practical: The weight pumping procedure of Cheng and Dey [CD03] needs the local feature sizes at the sliver vertices which are expensive to compute. We improve upon this by showing that it suffices to work with the nearest neighbor distances of the sliver vertices. This is readily available in a Delaunay mesh and eases the weight pumping implementation substantially.

Experimental: We experimented with an implementation of the proposed algorithm. In our experiments, we set the circumradius-edge ratio threshold at 2.2 and dihedral angle bound for slivers at $3°$. This means that the algorithm tries to eliminate any poorly shaped tetrahedron with circumradius-edge ratio more than 2.2 and any sliver with dihedral angle less than $3°$. Our experiments show that there are extremely few tetrahedra in the output that violate these thresholds. The plot of the distribution of the face and dihedral angles shows that over 90% of the angles are more than $10°$ and many of them are between $15°$ and $30°$.

2 Input domain

The input domain is a polyhedron bounded by a 2-manifold. The 2-manifold is the underlying space of a piecewise-linear-complex called PLC defined as follows. A vertex v is a point in $\mathbb{R}^3$; its boundary ∂v is v itself. An edge e is a closed segment between two vertices v_1 and v_2 where its boundary is $\partial e = \{v_1, v_2\}$. A facet is a subset of the plane bounded by a collection simple polygonal cycles made out of vertices and edges. A PLC is a collection of vertices, edges, and facets that intersect properly, i.e., the intersection of any two elements is either empty or a collection of lower-dimensional elements. Although this definition disallows special cases such as isolated vertex or a dangling edge in the middle of a facet, we believe that, after adding extra steps to deal with special cases, our algorithm works for any input PLC as long as its underlying space remains a 2-manifold.

Also, we make a modification to the input domain by encompassing the original input polyhedron with a large enough bounding box. Our algorithm meshes the interior of the box conforming to the input polyhedron and keeps only the tetrahedra covering the interior of the input polyhedron. Let $\mathcal{B}$ denote the bounding box. We use $\mathcal{P}$ to denote the PLC of the input polyhedron together with $\mathcal{B}$. Two elements in $\mathcal{P}$ are *incident* if one is in the boundary of the other. We call two elements of $\mathcal{P}$ *adjacent* if they intersect.

Our algorithm works with two types of input angles. For any two incident edges of a vertex u, we measure the angle between them. We call such angles *edge-edge angles*. For any edge uv and a facet F incident to u such that uv is neither incident on F nor coplanar with F, the angle between uv and F is $\min\{\angle puv : p \in F, p \neq u\}$. We call such angles *edge-facet angles*. At an edge of $\mathcal{P}$, we measure the internal and external dihedral angles at the edge. Throughout this paper, we use ϕ_m to denote the minimum input angle in $\mathcal{P}$. We call an edge *sharp* if the internal or external dihedral angle at the edge is acute. We call a vertex u *sharp* if an edge-edge angle or an edge-facet angle at u is acute, or if u is an endpoint of a sharp edge.

The *local feature size* $f(x)$ for $\mathcal{P}$ at $x \in \mathbb{R}^3$ is the radius of the smallest ball centered at x intersecting two non-adjacent elements of $\mathcal{P}$.

3 Weighted Delaunay and shape measure

We denote a *weighted point* at location x by $\hat{x}$ and we use X^2 to denote its weight. The *weighted distance* between $\hat{x}$ and $\hat{y}$ is $\pi(\hat{x}, \hat{y}) = \|x - y\|^2 - X^2 - Y^2$. We say that $\hat{x}$ and $\hat{y}$ are *orthogonal* if $\pi(\hat{x}, \hat{y}) = 0$. One can view $\hat{x}$ as a sphere centered at x with radius X. Then $\hat{x}$ and $\hat{y}$ are orthogonal if the two spheres intersect at right angle. An *orthosphere* of k, $2 \leq k \leq 4$, weighted points $\hat{x}_1, \cdots, \hat{x}_k$ is a sphere $\hat{y}$

such that $\pi(\widehat{x}_i, \widehat{y}) = 0$ for $1 \leq i \leq k$. The orthosphere is unique for four weighted points in general position. The center and radius of the orthosphere are known as *orthocenter* and *orthoradius*.

Given a weighted point set, a tetrahedron spanning four points is *weighted De-launay* if the weighted distance between its orthosphere and any other weighted point is non-negative. The *weighted Delaunay triangulation* is the collection of all weighted Delaunay tetrahedra along with their triangles, edges and vertices. It can be built using an incremental algorithm [ES96] and CGAL has a library function for its construction [cgal]. We will be interested in the special case where the weights are zero or small. Specifically, the weight of $\widehat{v}$ can take on a value between 0 and $\omega_0 N(v)$, where $\omega_0 \in [0, 1/3)$ is a constant to be defined later and $N(v)$ is the nearest neighbor distance of v. In this case, we say that the point set has *weight property* $[\omega_0]$. If all weights are zero for a vertex set $\mathcal{V}$, we get the standard unweighted Delaunay triangulation $\mathrm{Del}\,\mathcal{V}$.

The volume of a tetrahedron Δt in a normalized sense captures its shape quality. We define the *orthoradius-edge ratio* $\rho(\Delta t) = R/L$ and the *normalized volume* $\sigma(\Delta t) = \mathrm{vol}(\Delta t)/L^3$, where R and L are the orthoradius and the shortest edge length of Δt, respectively. These two values determine the shape of Δt as indicated in the following lemma. In other words, the quality measure in the weighted mesh is sufficient for actual quality, if the weight property $[\omega_0]$ holds for sufficiently small ω_0.

Lemma 1. *Assume that vertices of Δt have weight property $[\omega_0]$. If $\rho(\Delta t) \leq \rho_1$ and $\sigma(\Delta t) > \sigma_0$ for some constants ρ_1 and σ_0, all angles of Δt are greater than some constant threshold.*

Proof. Let pq be the longest edge of Δt. Let $\widehat{z}$ be the orthosphere of Δt. Since both $\widehat{p}$ and $\widehat{q}$ overlap with $\widehat{z}$, $Z + P \geq \|p - z\|$ and $Z + Q \geq \|q - z\|$. The weight property implies that $\max\{P, Q\} \leq \omega_0 \cdot \|p - q\|$. Thus, $2Z \geq \|p - z\| + \|q - z\| - 2\omega_0 \cdot \|p - q\| \geq (1 - 2\omega_0) \cdot \|p - q\|$. Then $\rho(\Delta t) \leq \rho_1$ implies that $L \geq Z/\rho_1 \geq \frac{1 - 2\omega_0}{2\rho_1} \cdot \|p - q\|$. That is, the length of any two edges of Δt are within a constant factor. It follows that the area of any face of Δt is $O(L^2)$. So any height of Δt is at least $\sigma_0 L^3/O(L^2) = \Omega(L)$. The sine of any dihedral angle of Δt is at least the ratio of a height to the longest edge length. This implies that all dihedral angles of Δt are greater than some constant. So are the angles of the triangles of Δt.

4 Algorithm

The algorithm has four distinct phases, INITIALIZE, CONFORM, REFINE, and PUMP. The phases INITIALIZE, CONFORM, and REFINE are from [CDRR04]. Some modifications are incorporated into REFINE to prepare for PUMP. The result of the first three phases is a Delaunay mesh with bounded circumradius-edge ratio, except near the small input angles. In PUMP, we assign weights to sliver vertices to remove the slivers.

The algorithm maintains a vertex set $\mathcal{V}$. The edges of $\mathcal{P}$ are divided into *subsegments* by inserted vertices. The subsegments inside the vertex balls are protected by the vertex balls, so we are not concerned about them. We call a subsegment *sharp* if it lies outside the vertex balls and on a sharp edge. It is *non-sharp* otherwise.

A *circumball* of a subsegment is a finite ball with the subsegment endpoints on its boundary. The *diametric ball* of a subsegment is its smallest circumball. A point p *encroaches* a subsegment e if p is not an endpoint of e and lies on or inside its diametric ball. This encroachment definition is stricter than usual. When p is weighted, we say $\hat{p}$ encroaches e if $\pi(\hat{p}, \hat{z}) \le 0$ where $\hat{z}$ is the diametric ball of e.

When no subsegment is encroached by any vertex in $\mathcal{V}$, the facets are decomposed into *subfacets* defined as follows. For each facet F of $\mathcal{P}$, consider the 2D Delaunay triangulation of the vertices in $\mathcal{V} \cap F$. The Delaunay triangles in F are its subfacets. Thanks to the stricter definition of subsegment encroachment, the circumcenters of subfacets lie strictly in the interior of F[CD03]. A *circumball* of a subfacet h is a finite ball with the vertices of h on its boundary. The *diametric ball* of h is its smallest circumball. A point p *encroaches* h if p lies inside its diametric ball. When p is weighted, we say $\hat{p}$ encroaches h if $\pi(\hat{p}, \hat{z}) < 0$ where $\hat{z}$ is the diametric ball of h.

There are several subroutines used by our algorithm. We describe them in Sections 4.1 and 4.2 and then give the algorithm in Section 4.3. Some of these subroutines are exactly same as those described in [CDRR04]. We include them here for completeness.

4.1 Sharp vertex protection

We protect each sharp vertex u with a *vertex ball* centered at u with radius $f(u)/4$. The distance of u from each non-adjacent input vertex, edge and facet is computed to determine $f(u)$. We denote the protecting ball of u by $\hat{u}$. The points where the boundary of $\hat{u}$ intersects edges of $\mathcal{P}$ are inserted into the vertex set $\mathcal{V}$. We protect a subset of $\hat{u} \cap \mathcal{P}$ using the SOS method of Cohen-Steiner, de Verdière and Yvinec [SVY02] (see [BG02] for a generalized variant of this operation). At any generic step of the algorithm, $\mathcal{V}$ contains vertices on the arc where a facet F incident to u intersects the boundary of $\hat{u}$. The segments connecting consecutive points on such an arc form *shield subsegments*. Let ab be a shield subsegment. If the angle of the sector aub on F is at least π, we insert the midpoint x on the arc between a and b on the boundary of $\hat{u} \cap F$. The subsegment ab is replaced with two shield subsegments ax and bx; see Figure 1(a). If the angle of the sector aub is less than π and ab is encroached, we also insert x to split ab; see Figure 1(b).

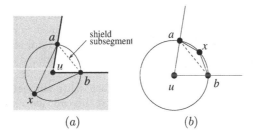

Fig. 1. Shield subsegment and SOS splitting.

When no shield subsegment corresponds to a sector at u with angle π or more, the shield subsegments around u create a set of *shield subfacets* incident to u. It turns out that the diametric ball of a shield subfacet lies in the union of the vertex ball $\widehat{u}$ and the diametric ball of the corresponding shield subsegment. Since $\widehat{u}$ is kept empty throughout the algorithm, it is sufficient to keep the shield subsegments non-encroached to ensure that shield subfacets appear in Del $\mathcal{V}$.

In INITIALIZE, we only insert the points where the incident edges of u intersect the boundary of $\widehat{u}$, and split shield subsegments that correspond to sectors with angles π or more. The encroachment of shield subsegments is handled in the next phase CONFORM.

4.2 Edge, Facet, Ball and Tetra Splitting

Edges are split in both the CONFORM and REFINE phases to recover the edges of $\mathcal{P}$ as union of Delaunay edges. Any subsegment (sharp, non-sharp or shield) that is encroached is split using SPLITE until no such segment exists.

SPLITE(e)
If e is a shield subsegment, split it with SOS else insert the midpoint of e and update Del $\mathcal{V}$.

After we recover all the edges, we start splitting facets so that they appear in Del $\mathcal{V}$ as union of subfacets. Standard Delaunay refinement insists that no encroached subfacet exists. While such a condition can be enforced for subfacets on the boundary of the bounding box $\mathcal{B}$, it may never be satisfied for all subfacets. Instead, we check only that if any subfacet does not appear in Del $\mathcal{V}$. We argue that, for a polyhedron, all subfacets must appear in Del $\mathcal{V}$ after sufficient but finite amount of splitting.

A subfacet h that does not appear in Del $\mathcal{V}$ is split using the procedure SPLITF. Certainly, h cannot be a shield subfacet since there is no encroached shield subsegment when the algorithm reaches the facet splitting step.

SPLITF(h)
(i) Compute the circumcenter c of h;
(ii) Let F be the facet containing h. If c does not encroach any subsegment, insert c and update Del $\mathcal{V}$. Otherwise, reject c and
1. pick a subsegment g encroached by c with preference for those in ∂F or on F (shield subsegment), and
2. call SPLITE(g).

In the CONFORM phase, after all the edges and facets are recovered, further splittings of subsegments and subfacets may be performed to reduce the diametric balls of the sharp subsegments roughly to the order of local feature sizes. In order to avoid the computation of local feature sizes, this is achieved in a roundabout way (see rule 3 in QMESH).

At the end of the CONFORM phase, for each sharp subsegment, we double the radius of its diametric ball with the center fixed and call this a *protecting ball*. These protecting balls and the vertex balls at the sharp vertices constitute the entire set of *protecting balls* for the REFINE phase.

Skinny tetrahedra are split in the REFINE phase by inserting their circumcenters. But we disallow the insertion of the circumcenters of skinny tetrahedra inside any protecting ball. The reason is that once these points are allowed to be inserted, they can cause perpetual splittings of the subsegments or subfacets. This means that some skinny tetrahedra may remain at the end. It has been proved that all such tetrahedra lie close to sharp vertices or edges [CDRR04]. We also prepare for pumping in the REFINE phase. For this we assign a weight to each vertex of a sliver tetrahedron and check if the weighted vertex threatens to destroy the input conformity. If so, we split subsegments and subfacets further. Potentially we can choose to pump all vertices in PUMP and thus can prepare them all in REFINE. But, for efficiency we choose only a subset of tetrahedra that have a small dihedral angle, say $3°$.

4.3 WQMESH

The following algorithm WQMESH triangulates $\mathcal{P}$ with the claimed theoretical guarantees. WQMESH uses the following subroutine ENCROACH to test whether a point c can be inserted and if not, return the appropriate point to be inserted.

> ENCROACH(c)
> 1. If c does not encroach any subsegment or subfacet, return c.
> 2. If c encroaches some subsegment e, reject c and if e is a shield subsegment, return the point given by SOS; else return the midpoint of e.
> 3. If c encroaches some subfacet, one such subfacet h contains the orthogonal projection of c. Reject c. If the circumcenter p of h does not encroach any subsegment, return p; otherwise, reject p and return the point as in case 2.

The following are the descriptions of the four phases of WQMESH($\mathcal{P}$). Recall that $\mathcal{P}$ includes a bounding box $\mathcal{B}$ which encompasses the original polyhedron.

INITIALIZE. Initialize $\mathcal{V}$ to be the set of vertices of $\mathcal{P}$. Compute the vertex balls. Insert the intersections between their boundaries and the edges of $\mathcal{P}$ into $\mathcal{V}$. If any shield subsegment forms a sector with angle π or more, split it with SOS. Compute Del $\mathcal{V}$.

CONFORM. Repeatedly apply a rule from the following list until no rule is applicable. Rule i is applied if it is applicable and no rule j with $j < i$ is applicable.

> Rule 1. If there is an encroached subsegment e, call SPLITE(e).
> Rule 2. If there is a subfacet $h \subset \mathcal{B}$ that is encroached, or if $h \not\subset \mathcal{B}$ and h does not appear in Del $\mathcal{V}$, call SPLITF(h).
> Rule 3. Let s be a sharp subsegment on an edge e. If the midpoint of s encroaches a subsegment or subfacet h, where h and e are contained in disjoint elements of $\mathcal{P}$, split h accordingly using SPLITE(h) or SPLITF(h).

At the end of CONFORM, we double the sizes of the diametric balls of sharp subsegments. These expanded balls and the vertex balls are the *protecting balls*. The sharp subsegments may be split further in the next phase, but the locations and sizes of these protecting balls do not change.

We call a subfacet *guarded* if its diametric ball lies inside some protecting ball.

REFINE. Repeatedly apply a rule from the following list until no rule is applicable. Rule i is applied if it is applicable and no rule j with $j < i$ is applicable. The parameter $\rho_0 > 2/(1 - \tan(\pi/8))$ is a constant chosen a priori.

> Rule 4. If there is an encroached subsegment e, call SPLITE(e).

Rule 5. If there is a non-guarded subfacet h that is encroached or a guarded subfacet h that does not appear in $\text{Del}\,\mathcal{V}$, then call $\text{SPLITF}(h)$.

Rule 6. Assume that there is a tetrahedron with circumradius-edge ratio exceeding ρ_0. Let z be its circumcenter. If z does not lie on or inside any protecting ball, then compute $p := \text{ENCROACH}(z)$ and insert p into $\mathcal{V}$.

Rule 7. Take a vertex v of a tetrahedron Δt with a dihedral angle of $3°$ or less. Let $\hat{v}$ be the weighted vertex v with weight $L_v^2/9$ where L_v is the length of the shortest incident edge of v. We ignore v if $\hat{v}$ encroaches the two times expansion of some protecting ball. Otherwise, we take action only in the following cases: (i) if $\hat{v}$ encroaches some non-guarded subfacet, we split by SPLITF the one that contains the projection of v; (ii) if $\hat{v}$ makes some guarded subfacet disappear from the 3D weighted Delaunay triangulation, we split the subfacet by SPLITF.

We always maintain an unweighted Delaunay triangulation in REFINE and $\hat{v}$ is used only for checking encroachments. Notice that, as claimed in the introduction, the weight pumping depends on nearest neighbor distances as opposed to local feature sizes in [CD03].

$\underline{\text{PUMP}}$. We examine all vertices v incident on a tetrahedron with a dihedral angle of $3°$ or less such that $\hat{v}$ does not encroach the two times expansion of any protecting ball. For each such vertex v, we assign to v the weight in the interval $[0, L_v^2/9]$ that maximizes the minimum dihedral angle of the tetrahedra incident to v. We maintain the weighted Delaunay triangulation during the weight pumping. We claim that no pumped vertex encroaches upon any weighted-subsegment and weighted-subfacet.

Note: We used $3°$ as a threshold on the dihedral angles of tetrahedra to select vertices to pump. The choice of $3°$ is dictated by our experiments which shows that in most cases pumping eliminates tetrahedra with dihedral angles smaller than $3°$. In theory, we show that pumping eliminates all tetrahedra with dihedral angle below a positive threshold. Although this threshold is much lower than $3°$, the algorithm as stated works. Only that, sometimes tetrahedra with dihedral angles between the theoretical threshold and $3°$ may not get eliminated.

Let $\mathcal{M}$ denote the unweighted Delaunay mesh at the end of REFINE. Let U denote the set of vertices of tetrahedra in $\mathcal{M}$ with circumradius-edge ratio greater than ρ_0. It has been proved that all vertices in U are close to a small input angle [CDRR04]. Our main result is that all sliver vertices left after PUMP also satisfies similar property. We say that a vertex v is in the E-neighborhood of U if v is within E edges in $\mathcal{M}$ from some vertex of U. We say that Δt is in the E-neighborhood of U if some vertex of Δt is in the E-neighborhood of U. Recall that $\rho(\Delta t)$ and $\sigma(\Delta t)$ denote the orthoradius-edge ratio and the normalized volume of Δt, respectively. The main theorem we prove is:

Theorem 1. *There are constants $\rho_1, \sigma_0, E > 0$ such that for every tetrahedron Δt in the output mesh of WQMESH, $\rho(\Delta t) \leq \rho_1$ and $\sigma(\Delta t) > \sigma_0$ unless Δt is in the E-neighborhood of U or Δt is within distance $O(f(x))$ from some point x on some sharp edge.*

5 Analysis

In this section, we first prove that the first three phases terminate with a graded Delaunay mesh and all tetrahedra have bounded circumradius-edge ratio, except those near small input angles. Then we show that the phase PUMP eliminates slivers, except those that are near skinny tetrahedra or the protecting balls.

The main purpose of the bounding box $\mathcal{B}$ is to disallow any point to be inserted outside $\mathcal{B}$ so that one can claim termination by applying a packing argument within a bounded domain. The set of points, say P, that are inserted or rejected while conforming to $\mathcal{B}$ maintain a lower bound on their distances to all other existing points by the arguments in [CD03] because all angles of $\mathcal{B}$ are $\pi/2$. In the analysis we skip explicit arguments about P and focus on the set of points, say Q, that are inserted or rejected for conforming to the edges and facets of the original polyhedron. Of course, we do not lose any generality by doing so as the lower bounds on distances for P are dominated by those for Q.

5.1 Termination and conformity

In this section, we inductively prove that the inter-vertex distances remain above certain thresholds in the first three phases. The inductive argument makes use of a predecessor relation defined as follows. Let x be a vertex inserted or rejected by WQMESH. The *predecessor* of x is an input vertex or a vertex inserted or rejected by WQMESH. If x is a vertex of $\mathcal{P}$ or a vertex inserted during INITIALIZE, its predecessor is undefined. Otherwise, the predecessor p is defined as follows.

- Suppose that x splits a subsegment (shield or non-shield) or a subfacet h. Let B be the diametric ball of h.
  If B contains some vertex in $\mathcal{V}$, p is the encroaching vertex nearest to x. If B is empty and WQMESH is going to reject a vertex inside B for encroaching h, p is that vertex. Otherwise, B is empty and the encroachment must be due to a weighted vertex outside the ball in rule 7. Then p is that weighted vertex.
- If x is the circumcenter of a skinny tetrahedron Δt, then p is one of the endpoints of the shortest edge of Δt. Between the two endpoints of the shortest edge, p is the one that appears in $\mathcal{V}$ later.

It is still possible that x has no predecessor. This happens when x is inserted to split a subsegment in rule 3. When a subfacet is split by x in rule 3, x may or may not have a predecessor depending on whether the diametric ball is empty or not.

The *neighbor radius* r_x of x is the distance from x to its nearest neighbor in the current $\mathcal{V}$ when x is inserted or rejected. So if x is an input vertex, then $r_x \geq f(x)$. The following result shows that $r_x = \Omega(f(x))$ and that the protecting balls are not too small at the end of CONFORM.

Lemma 2. [CDRR04] *At the end of* CONFORM, *there are constants* $\mu_1, \mu_2 > 0$ *such that*

(i) *For each vertex* $x \in \mathcal{V}$, $r_x \geq \mu_1 f(x)$.
(ii) *For any point* z *on a sharp subsegment, a ball centered at* z *with radius* $\mu_2 f(z)$ *lies inside some protecting ball.*

So we focus on analyzing REFINE. We will show by induction that the rules 4–7 in REFINE preserve the following property. Let $\omega_0 \in [0, 1/3)$ be a constant to be defined later.

VERTEX GAP PROPERTY: For each vertex $p \in \mathcal{V}$, the nearest neighbor distance of p is at least $\omega_0 f(p)$ throughout REFINE.

We need the following result, Lemma 3, about REFINE. It is the same as Lemma 4.16 in [CDRR04](extended version). The original proof almost works here but some change is needed due to our extension of REFINE to prepare for PUMP. The only case that is not handled in the original proof is that the weighted encroachment in rule 7 may split a guarded subfacet h, although h belongs to the 3D unweighted Delaunay triangulation. If we can argue that the circumradius of h is $\Omega(f(x))$ where x is its circumcenter, the original argument goes through. We can invoke Lemma 4.10 in [CDRR04] to show that the circumradius of h is $\Omega(f(x))$, provided that we can find a circumball B of h such that B contains a vertex u on some element of $\mathcal{P}$, u and the center z of B lie on different sides of the plane of h, and radius$(B) \geq cf(x) - \|x - z\|$ for some constant $c > 0$. Since some weighted vertex $\hat{v}$ makes h disappear from the 3D weighted Delaunay triangulation in rule 7 and $\hat{v}$ is relatively far from any protecting ball, it can be checked that such a ball B exists. Thus Lemma 3 holds.

Lemma 3. *Assume that no subsegment is encroached by a vertex. Let x be the circumcenter of a subfacet inserted or rejected during REFINE and p be its predecessor. If p is a vertex in $\mathcal{V}$ lying on some element of $\mathcal{P}$, then $r_x = \Omega(f(x))$.*

Lemma 4. *Let x be a vertex inserted or rejected during REFINE. Assume that x splits a subfacet h and its predecessor p is defined. If p does not encroach h, then $\|p - x\| \geq \min\{\mu_3 f(x), \mu_4 r_p\}$ for some constants $\mu_3, \mu_4 > 0$.*

Proof. Since p does not encroach h, the diametric ball of h centered at x is empty. Moreover, the encroachment happens in rule 7 when p is pumped. So $p \in \mathcal{V}$. By rule 7, p lies outside all protecting balls and h is a non-guarded subfacet. Go back to the time t when p was first inserted into $\mathcal{V}$. Let F be the input facet containing h.

Let abc be the subfacet on F that contains the projection of p at that time. If p lies outside the diametric ball of abc, it can be shown that the distance from p to F is at least $r_p/\sqrt{2}$ [CD03]. So is $\|p-x\|$. If p lies inside the diametric ball of abc, then p lies on some input element F'. Otherwise, p would be the circumcenter of a tetrahedron and be rejected by the algorithm. If F' is disjoint from F, then $\|p - x\| \geq f(x)$. Otherwise, F and F' meet at a sharp vertex or a sharp edge. Let z be the point in $F \cap F'$ closest to p. Since p lies outside all protecting balls, Lemma 2(ii) implies that $\|p - z\| \geq \mu_2 f(z)$. The Lipschitz condition implies that $\|p - z\| \geq \mu_2 f(p)/(1 + \mu_2)$. Thus, $\|p - x\| \geq \|p - z\| \sin \phi_m \geq \mu_2 \sin \phi_m f(p)/(1 + \mu_2)$. Invoking the Lipschitz condition again yields $\|p - x\| \geq \mu_2 \sin \phi_m f(x)/(1 + \mu_2 + \mu_2 \sin \phi_m)$.

Now we can show that the splitting vertex x triggered by the weighted encroachment in rule 7 has neighbor radius $\Omega(f(x))$, if the predecessor of x is outside the diametric ball centered at x.

Lemma 5. *Suppose that the vertex gap property is satisfied. Let x be a vertex that splits a subfacet h. Assume that its predecessor p is defined. If p lies outside the diametric ball of h, then $r_x \geq \mu_5 f(x)$ for some constant $\mu_5 > 0$.*

Proof. Since p is the predecessor of x but p does not encroach h, the encroachment occurs in rule 7 when p is pumped. Let L_p be the nearest neighbor distance of p at that time.

We claim that $L_p \leq 2\|p - x\|$. If the claim is false, the ball B centered at p with radius L_p contains x. Moreover, the distance from x to the boundary of B is greater than $L_p - \|p - x\| \geq L_p/2$. By the definition of L_p, B is empty. So the vertices of h lies outside B, which implies that the diametric ball of h (centered at x) has radius at least $L_p/2 > \|p - x\|$. But then p must encroach h, contradicting our assumption. This proves the claim.

By the weighted encroachment, $r_x^2 + L_p^2/9 \geq \|p - x\|^2$. So our claim implies that $r_x \geq \frac{\sqrt{5}}{3}\|p - x\|$. By Lemma 4, $\|p - x\| \geq \min\{\mu_3 f(x), \mu_4 r_p\}$. If $\|p - x\| \geq \mu_3 f(x)$, then $r_x \geq \frac{\sqrt{5}\mu_3}{3} f(x)$. If $\|p - x\| \geq \mu_4 r_p$, then $\|p - x\| \geq \mu_4 L_p \geq \mu_4 \omega_0 f(p)$ by the vertex gap property. Thus, $f(x) \leq f(p) + \|p - x\| = O(\|p - x\|) = O(r_x)$.

We are ready to apply induction to prove the vertex gap property holds throughout REFINE and to bound the inter-vertex distances from below by the local feature sizes.

Lemma 6. *Let x be a vertex such that x exists in $\mathcal{V}$ before the invocation of* REFINE *or x is inserted or rejected by* WQMESH *during* REFINE. *The following invariants hold.*

(i) *$r_x \geq f(x)/C$ for some constant $C > 0$.*
(ii) *For any other vertex y currently in $\mathcal{V}$, $\|x - y\| \geq \max\{f(x)/C, f(y)/(1+C)\}$.*
(iii) *Assume that $\omega_0 = 1/(1+C)$. If x is inserted, the vertex gap property holds afterwards.*

Proof. By Lemma 2(i), if x exists in $\mathcal{V}$ before the invocation of REFINE, then $r_x \geq \mu_1 f(x)$. Assume that x is inserted or rejected during REFINE. If x splits a subsegment, we say x has type 4; if x splits a subfacet, x has type 5; and if x splits a tetrahedron, x has type 6. We prove a stronger statement by induction: if x has type i, then $r_x \geq f(x)/C_i$ for some constants $C_4 > C_5 > C_6 > 1$. Let p be the predecessor of x.

Case 1: x has type 4. Let ab be the subsegment split by x. Since only subfacets are split in rule 7, p must be unweighted and p encroaches ab. If p has type 4, then p must be inserted and it is a subsegment endpoint. In this case, it has been proved that $r_x \geq \mu_6 f(x)$ for some constant $\mu_6 > 0$ [CDRR04]. Suppose that p has type 5 or 6. It has been shown [CDRR04] that $r_x \geq r_p/\beta_1$ and $r_x \geq \|p - x\|/\beta_2$, where $\beta_1 = \frac{\sqrt{2}}{1 - \tan(\pi/8)}$ and $\beta_2 = \frac{1 + \tan(\pi/8)}{1 - \tan(\pi/8)}$. Since $C_5 > C_6$, the induction assumption implies that $r_p \geq f(p)/C_5$ regardless of the type of p. Thus $\frac{f(x)}{r_x} \leq \beta_1 C_5 + \beta_2$.
So the inequalities $\beta_1 C_5 + \beta_2 \leq C_4$ and $C_4 \geq 1/\mu_6$ should hold.

Case 2: x has type 5. If p is a vertex lying on some element of $\mathcal{P}$, then by Lemma 3, $r_x \geq \mu_7 f(x)$ for some constant $\mu_7 > 0$. Otherwise, p must have type 6. Let h be the subfacet split by x. There are two cases.

Case 2.1: p encroaches h. In this case, it has been shown p is not a vertex in $\mathcal{V}$ [CDRR04]. Therefore, we can infer that WQMESH tries to insert p to split a tetrahedron but then it rejects p for encroaching h. And the diametric ball of h is empty. Therefore, $r_x \geq r_p/\sqrt{2}$ and $r_x \geq \|p - x\|$. Thus, $\frac{f(x)}{r_x} \leq \sqrt{2}C_6 + 1$.

Case 2.2: If p does not encroach h, $\widehat{p}$ encroaches h in rule 7. And the diametric ball of h is empty. Lemma 5 implies that $r_x \geq \mu_5 f(x)$.

So the inequalities $\sqrt{2}C_6 + 1 \leq C_5$ and $C_5 \geq \max\{1/\mu_5, 1/\mu_7\}$ need to hold.

Case 3: x has type 6. In this case, it has been proved that either $r_x \geq \frac{\rho_0\mu_1}{\rho_0\mu_1+1} f(x)$ or $\frac{f(x)}{r_x} \leq \frac{f(p)}{\rho_0 \cdot r_p} + 1 \leq \frac{C_4}{\rho_0} + 1$ [CDRR04]. So the inequalities $C_4/\rho_0 + 1 \leq C_6$ and $C_6 \geq \frac{\rho_0\mu_1+1}{\rho_0\mu_1}$ need to hold.

This finishes the case analysis. To satisfy all the above inequalities, we set C_6 to be the maximum of $\frac{\rho_0+\beta_1+\beta_2}{\rho_0-\sqrt{2}\beta_1}$, $\frac{1}{\mu_5}$, $\frac{1}{\mu_6}$, $\frac{1}{\mu_7}$, and $\frac{\rho_0\mu_1+1}{\rho_0\mu_1}$. Then we set $C_5 = \sqrt{2}C_6 + 1$ and $C_4 = \sqrt{2}C_5 + 1$. Notice that $C_i > 0$ for $i = 4, 5, 6$ since $\rho_0 > \sqrt{2}\beta_1 = 2/(1 - \tan(\pi/8))$ as chosen by WQMESH. Finally, we set $C = \max\{C_4, 1/\mu_1\}$. This proves invariant (i).

Consider invariant (ii). For any vertex y that appears in $\mathcal{V}$ currently, $\|x - y\| \geq f(x)/C$. Since $f(x) \geq f(y) - \|x - y\|$, we have $\|x - y\| \geq f(x)/C \geq f(y)/C - \|x - y\|/C$. This implies that $\|x - y\| \geq f(y)/(1 + C)$.

Consider invariant (iii). Invariant (ii) implies that for any vertices $a, b \in \mathcal{V}$, $\|a - b\| \geq f(a)/(1+C)$. The choice of the values of C and ω_0 enforce that $f(a)/(1+C) = \omega_0 f(a)$. This proves that the vertex gap property holds.

By Lemma 6 and a standard packing argument, WQMESH must terminate. Since no protecting ball is encroached by any weighted vertex at the end, the non-shield subsegments appear as weighted Delaunay edges in the end, i.e., the input edges are recovered. The guarded subfacets and subfacets incident to sharp vertices are explicitly kept as weighted Delaunay by WQMESH. The proof of Theorem 7.2 in [CD03] can be used to show that the other subfacets are recovered as the union of weighted Delaunay triangles. Hence, WQMESH terminates with a conforming weighted Delaunay mesh. It has been proved [CDRR04] that at the end of REFINE, the vertices of any tetrahedra with circumradius-edge ratio exceeding ρ_0 are within distance $O(f(x))$ from some point x on some sharp edge.

5.2 Sliver exudation in PUMP

Let $\mathcal{M}$ denote the unweighted Delaunay mesh at the end of REFINE. Let $\mathcal{M}[\omega_0]$ denote a weighted Delaunay mesh obtained after assigning arbitrary weights to the vertices of $\mathcal{M}$ such that the weight property $[\omega_0]$ holds. So $\mathcal{M}[\omega_0]$ may be the mesh obtained at the end of PUMP, but there are many other possibilities for $\mathcal{M}[\omega_0]$ as well. Note that $\mathcal{M}$ and $\mathcal{M}[\omega_0]$ share the same vertex set. Let U denote the set of vertices of tetrahedra in $\mathcal{M}$ with circumradius-edge ratio greater than ρ_0. In this section we complete the proof of Theorem 1.

We need the following result from Talmor's thesis [T97]. We have rephrased it to fit our presentation. Given any vertex p in $\mathcal{M}$, we use V_p to denote its Voronoi cell in the unweighted Voronoi diagram. We say that V_p is ρ_0-round if the circumradius-edge ratio of all tetrahedra in $\mathcal{M}$ incident to p are bounded by ρ_0. In fact, if V_p is ρ_0-round, it has a bounded aspect ratio. It is known that the lengths of adjacent edges in $\mathcal{M}$ differ by a constant factor $L = 2^{2m-1}\rho_0^{m-1}$, where $m = 2/(1 - \cos\frac{\eta}{4})$ and $\eta = \frac{1}{2}\arctan(2\rho_0 - \sqrt{4\rho_0^2 - 1})$ [CDET99]. Let B_p be a ball centered at p with radius $\rho_0 L \cdot N(p)$.

Lemma 7. *Let* xz *be a line segment lying inside* $\bigcup_{p \in \mathcal{V}} V_p \cap B_p$. *Let* $\widehat{z}$ *be the sphere centered at* z *with radius* $\|x - z\|$. *There exists a constant* $K > 0$ *such that if* $\widehat{z}$ *is empty and* xz *intersects* ρ_0*-round Voronoi cells only, then* xz *intersects at most* K *Voronoi cells and* $N(z) \le C \cdot N(x)$ *for some constant* $C > 0$.

Recall the definition of E-neighborhood from section 4.3.

Lemma 8. *There exists a constant* $\rho_1 > 0$ *such that any tetrahedron* Δt *in* $\mathcal{M}[\omega_0]$ *with orthoradius-edge ratio exceeding* ρ_1 *is in the* K*-neighborhood of* U *where* K *is given by Lemma 7.*

Proof. Let $\widehat{z}$ be the orthosphere of some tetrahedron Δt in $\mathcal{M}[\omega_0]$. Let qr be the shortest edge of Δt. Let x be the intersection point $qz \cap \widehat{z}$. Note that x lies inside $\widehat{q}$ and $\widehat{q}$ lies inside V_q. By assumption, z lies inside Conv $\mathcal{V}$, the convex hull of $\mathcal{V}$. So xz lies inside Conv $\mathcal{V}$ by convexity. It has been proved that $V_p \cap$ Conv $\mathcal{V} \subseteq V_p \cap B_p$ [CD03]. So $xz \subseteq \bigcup_{p \in \mathcal{V}} V_p \cap B_p$. Note that $\widehat{z}$ is empty. Walk from x towards z. Stop at z or when we have encountered K Voronoi cells (including V_q). If we encounter a Voronoi cell that is not ρ_0-round, the site v owning the cell is incident to some tetrahedron in $\mathcal{M}$ with circumradius-edge ratio exceeding ρ_0. That is, $v \in U$. So q is in the K-neighborhood of U. Otherwise, Lemma 7 says that we must reach z and $Z \le N(z) \le C \cdot N(x)$, where Z is the radius of $\widehat{z}$. The Lipschitz condition implies that $N(x) \le N(q) + \|q - x\| \le N(q) + \omega_0 N(q)$. Since $N(q) \le \|q - r\|$, we have $N(x) \le (1 + \omega_0) \cdot \|q - r\|$. Thus $Z \le C(1 + \omega_0) \cdot \|q - r\|$ and the lemma is true for $\rho_1 = C(1 + \omega_0)$. $\square$

Let $\mathcal{G}$ be the graph consisting of the edges in all possible meshes $\mathcal{M}[\omega_0]$. Let p be a vertex in $\mathcal{G}$. It has been proved that [CD03] if both p and its neighbors are not incident to any tetrahedron Δt with orthoradius-edge ratio $\rho(\Delta t) > \rho_1$ in any $\mathcal{M}[\omega_0]$, then $\deg(p) \le \delta_0$ for some constant δ_0. Thus, by Lemma 8, the following result holds.

Lemma 9. *For any vertex* p *of* $\mathcal{G}$, *if* p *is not in the* $(K+1)$*-neighborhood of* U, *then* $\deg(p) \le \delta_0$ *for some constant* δ_0.

We are now ready to analyze the effects of weight assignment in PUMP. Let σ_0 be a constant to be specified later. Let $pqrs$ be a tetrahedron with $\sigma(pqrs) \le \sigma_0$ that is not in the $(K+1)$-neighborhood of U. Some vertices may have been assigned some weights already. Suppose that p is pumped with weight P^2 from the interval $[0, L_p^2/9]$. It has been proved [CD03] that for $pqrs$ to remain weighted Delaunay, P^2 must be at most $k\sigma_0 L_p^2$ for some constant k. That is, $pqrs$ define a forbidden weight interval for p and the length of this weight interval is at most $k\sigma_0 L_p^2$. By Lemma 9, there are no more than δ_0^3 tetrahedra incident to p throughout the pumping. Thus the total length of forbidden weight intervals for p is at most $k\delta_0^3\sigma_0 L_p^2$. Since the weight of p is chosen from $[0, L_p^2/9]$, if $\sigma_0 < 1/(9k\delta_0^3)$, p can be assigned a weight such that p is not incident to any tetrahedron Δt with $\sigma(\Delta t) \le \sigma_0$.

It may be the case that we abort the pumping of p because it encroaches the two times expansion of some protecting ball. A vertex ball $\widehat{x}$ has radius $f(x)/4$. By Lemma 2(ii), a protecting ball with center x has radius $\Omega(f(x))$. Thus there is some constant $\mu > 0$ such that $\mu f(x) + L_p/3 \ge \|p - x\|$. It has been shown that $L_p \le 2\sqrt{2}f(p)$ (Lemma 7.5 [CD03]). Thus $\mu f(x) + 2\sqrt{2}f(p)/3 \ge \|p - x\|$. By the

Lipschitz condition, we have $(\mu + 2\sqrt{2}f(x)/3) + (2\sqrt{2}/3) \cdot \|p - x\| \geq \|p - x\|$, which yields $\|p - x\| \leq \frac{3\mu+2\sqrt{2}}{3-2\sqrt{2}} f(x)$.

This completes the proof of all claims in Theorem 1.

6 Experimental results and conclusions

We have seen that WQMESH is a modification of QMESH [CDRR04] (with the step REFINE modified) and an added step for sliver removals. Therefore, we took an implementation of QMESH [soft] and modified it to WQMESH. This required adapting QMESH to weighted Delaunay triangulations. We used CGAL [cgal] for this weighted Delaunay triangulation.

Figure 4 shows the results. Although we proved a bound of $2/(1 - \tan(\pi/8))$ for the circumradius-edge ratio, we do not know the optimal bound on ρ_0 which the algorithm can tolerate. We took $\rho_0 = 2.2$ for the experiments because earlier work proved the guarantees for $\rho_0 = 2$ [PW04, Shew98].

We experimented with the performance of WQMESH on a number of data sets. We made several observations summarized in Table 2 and in the bar graph of Figure 2. Table 2 shows the relevant data about input and output. Observe that most of the tetrahedra have good circumradius-edge ratio. For measuring slivers we put a threshold of $3°$ on dihedral angles, i.e., a tetrahedron is declared sliver if its circumradius-edge ratio is no more than 2.2, but has a dihedral angle less than $3°$. One should note that, in theory, the sliver exudation step can only eliminate slivers with abysmally small dihedral angles. However, our experiments show that it is more effective than the theory confirms. The eighth column of Table 2 shows the number of slivers in different dihedral angle ranges remaining in the output. The distribution of minimum angles (both dihedral and face) for all tetrahedra are shown in Figure 2. It is clear that over 95% of the tetrahedra have minimum angle more than $3°$ while 90% of them have minimum angles more than $10°$.

Slivers with a dihedral angle less than $3°$ and tetrahedra with circumradius-edge ratio greater than 2.2 are shown in the second column of Figure 4. All of them lie near some small input angle. Third column of the figure shows the triangulation of the input polyhedron.

Table 1. Effect of sliver exudation.

model	# slivers before exudation	# slivers after exudation
Simplebox	6	0
Anchor	9	0
Tower	9	2
Hole	14	0
Hammer	30	6
Pawn	183	16

In spite of the abysmally small angle guaranteed by the sliver exudation step in theory, perhaps its most convincing advantage is its ability to remove slivers with

Table 2. Relevant input and output data.

model	# input points	# sharp elements	#points inserted	# tetrahedra with R/l			# tets(min dihed. ang.)		
				0.6-1.4	1.4-2.2	> 2.2	0-3	3-5	5-10
Anchor	28	27	981	2324	1045	35	0	15	48
Box	32	30	1044	2468	832	41	0	15	42
Wiper	72	58	979	1895	573	19	0	4	43
Blade	36	32	367	777	212	3	0	4	16
Tower	33	59	512	969	399	15	2	1	41

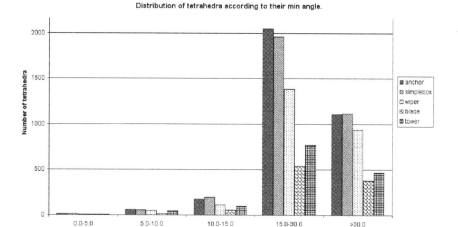

Fig. 2.

dihedral angles much larger than the ones predicted by theory. Table 1 shows the number of slivers remaining at the end of QMESH which does not have any sliver removal step, and the number of slivers after WQMESH. The data clearly shows that slivers are drastically reduced by the sliver exudation step incorporated in WQMESH. Figure 3 also confirms this conclusion.

Of course, from practical viewpoint it remains open if there is any provable method that can produce guaranteed quality meshing of polyhedra with a sufficiently large lower bound on angles. Also, the case for non-polyhedral inputs remains open.

Fig. 3. PAWN: input (left), slivers before PUMP (middle), few slivers survive PUMP (right). Most of the input face angles close to $\frac{\pi}{2}$ on the boundary are evaluated as acute for numerical tolerances. All surviving slivers are near some such acute angles.

Acknowledgments

Research is supported by RGC, Hong Kong, China, under grants HKUST6190/02E, Army Research Office, USA under grant DAAD19-02-1-0347 and NSF, USA under grants DMS-0310642 and CCR-0430735.

[ACYD05] P. Alliez, D. Cohen-Steiner, M. Yvinec and M. Desbrun. Variational tetrahedral meshing. *Proc. SIGGRAPH 2005*, to appear.

[B89] T.J. Baker. Automatic mesh generation for complex three-dimensional regions using a constrained Delaunay triangulation. *Engineering with Computers*, 5 (1989), 161–175.

[BG02] C. Boivin and C. Ollivier-Gooch. Guaranteed-quality triangular mesh generation for domains with curved boundaries. *Intl. J. Numer. Methods in Engineer.*, 55 (2002), 1185–1213.

[BGL00] H. Borouchaki, P.L. George, and S.H. Lo. Boundary enforcement by facet splits in Delaunay based mesh generation. *Numerical Grid Generation in Computational Field Simulations*, 2000, 203–221.

[cgal] Computational Geometry Algorithms Library (CGAL). URL: www.cgal.org

[CD03] S.-W. Cheng and T. K. Dey. Quality meshing with weighted Delaunay refinement. *SIAM J. Comput.*, 33 (2003), 69–93.

[CDET99] S.-W. Cheng, T. K. Dey, H. Edelsbrunner, M. A. Facello and S.-H. Teng. Sliver exudation. *J. ACM*, 47 (2000), 883–904.

[CDRR04] S.-W. Cheng, T. K. Dey, E. A. Ramos, and T. Ray. Quality Meshing for Polyhedra with Small Angles. *Proc. 20th Annu. ACM Sympos. Comput. Geom.*, 2004, 290–299. Extended versions available from authors' web-pages.

[CP03] S.-W. Cheng and S.-H. Poon. Graded conforming tetrahedralization with bounded radius-edge ratio. *Proc. 14th Annu. ACM-SIAM Sympos. Discrete Algorithms* (2003), 295–304.

[Chew97] L. P. Chew. Guaranteed-quality Delaunay meshing in 3D. *Proc. 13th Annu. ACM Sympos. Comput. Geom.* (1997), 391–393.

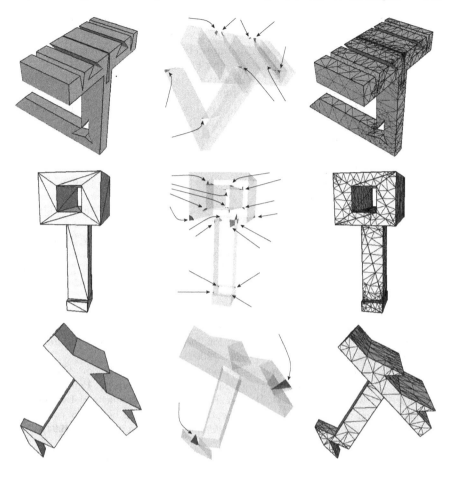

Fig. 4. First column shows the input, second column shows the skinny tetrahedra left (none of the slivers are left after exudation) and third column shows the surface triangulation.

[SVY02] D. Cohen-Steiner, E. C. de Verdière and M. Yvinec. Conforming Delaunay triangulations in 3D. *Proc. Annu. Sympos. Comput. Geom.* (2002), 199–208.

[EG02] H. Edelsbrunner and D. Guoy. An Experimental Study of Sliver Exudation. *Engineering With Computers*, 18 (2002), 229–240.

[E00] H. Edelsbrunner, X.-Y. Li, G. Miller, A. Stathopoulos, D. Talmor, S.-H. Teng, A. Üngör, and N. Walkington. Smoothing cleans up slivers. *Proc. 32th Annu. ACM Sympos. Theory of Computing*, 2000, 273–277.

[ES96] H. Edelsbrunner and N. R. Shah. Incremental topological flipping works for regular triangulations. *Algorithmica*, 15 (1996), 223–241.

342   S.-W. Cheng, T. K. Dey, and T. Ray

[LT01] X.-Y. Li and S.-H. Teng. Generating well-shaped Delaunay meshes in 3D. *Proc. 12th. Annu. ACM-SIAM Sympos. Discrete Algorithm* (2001), 28–37.

[Lo91b] S.-H. Lo. Volume discretization into tetrahedra-II. 3D triangulation by advancing front approach. *Computers and Structures*, 39 (1991), 501–511.

[L96] R. Lohner. Progress in grid generation via the advancing front technique. *Engineering with Computers*, 12 (1996), 186–210.

[MV00] S. A. Mitchell and S. A. Vavasis. Quality mesh generation in higher dimensions. *SIAM J. Comput.*, 29 (2000), 1334–1370.

[MMG00] M. Murphy, D. M. Mount and C. W. Gable. A point placement strategy for conforming Delaunay tetrahedralization. *Intl. J. Comput. Geom. Appl.*, 11 (2001), 669–682.

[O98] S.J. Owen. A survey of unstructured mesh generation technology. *Proc. 7th Intl. Meshing Roundtable* 1998.

[PW04] S. Pav and N. Walkington. A robust 3D Delaunay refinement algorithm. *Proc. Intl. Meshing Roundtable* 2004.

[Rup95] J. Ruppert. A Delaunay refinement algorithm for quality 2-dimensional mesh generation. *J. Algorithms*, 18 (1995), 548–585.

[SG91] M.S. Shephard and M.K. Georges. Three-dimensional mesh generation by finite octree technique. *Intl. J. Numer. Methods in Engineer.*, 32 (1991), 709–749.

[Shew98] J. R. Shewchuk. Tetrahedral mesh generation by Delaunay refinement. *Proc. 14th Annu. ACM Sympos. Comput. Geom.* (1998), 86–95.

[Shew00] J. R. Shewchuk. Mesh generation for domains with small angles. *Proc. 16th Annu. Sympos. Comput. Geom.* (2000), 1–10.

[T97] D. Talmor. Well-spaced points for numerical methods. Report CMU-CS-97-164, Dept. Comput. Sci., Carnegie-Mellon Univ., Pittsburgh, Penn., 1997.

[WH94] N.P. Weatherill and O. Hassan. Efficient three-dimensional Delaunay triangulation with automatic point creation and imposed boundary constraints. *International Journal for Numerical Methods in Engineering*, 37 (1994), 2005–2039.

[soft] http://www.cse.ohio-state.edu/~tamaldey/qualmesh.html

Polygonal Surface Remeshing with Delaunay Refinement

T. K. Dey[1], G. Li[1], and T. Ray[1]

The Ohio State University, Columbus, Ohio, USA
{tamaldey,ligan,rayt}@cse.ohio-state.edu

Summary. Polygonal meshes are used to model smooth surfaces in many applications. Often these meshes need to be remeshed for improving the quality, density or gradedness. We apply the Delaunay refinement paradigm to design a provable algorithm for isotropic remeshing of a polygonal mesh that approximates a smooth surface. The proofs provide new insights and our experimental results corroborate the theory.

Key words: surface meshing, computational geometry, computational topology, Delaunay refinement.

1 Introduction

Polygonal meshes including the triangular ones are often used in many applications of science and engineering to model a smooth surface. Such a mesh is usually designed by some modeling software (CAD software), or is generated by some reconstruction algorithm from a set of sample points provided by a scanning device. These meshes often lack properties that are useful for subsequent processing. For example, the triangle shapes may be poor for subsequent numerical methods [PB01], or the application may need a graded mesh with different levels of density without sacrificing the shape quality. To address these requirements, the input mesh needs to be remeshed, that is, they need to be sampled and triangulated appropriately.

Because of its application needs, the problem of remeshing has been a topic of research in many areas. We refer the readers to [Alliez03, Sifri] and the references therein. In this work, we apply the Delaunay refinement technique to design a provable algorithm for remeshing polygonal surfaces. The Delaunay refinement, originally pioneered by Chew [Chew89] is a powerful paradigm for meshing. First, it produces a Delaunay mesh as output which is often favored over other meshes due to its isotropic nature. Second, the paradigm offers a very simple mechanism to guarantee the quality of the mesh. It works on the following "furthest-point" principle. Whenever some desired criterion of the mesh is not satisfied, the algorithm inserts a point within the domain which is locally furthest from all other existing points. Then, the desired condition is automatically satisfied when the algorithm terminates. The main challenge entails to guarantee the termination. The Delaunay refinement paradigm has been successfully used for meshing two and three dimensional domains [CDRR204, PW04, Rup95, Shew98].

Researchers have also explored the Delaunay refinement technique for surface meshing. Chew [Chew93] proposed the first surface meshing algorithm with this technique though without any guarantee. Cheng et al. [CDES01] combined the sampling theory of Amenta and Bern [AB98] for surface reconstruction with the Delaunay refinement for producing a mesh for skin surfaces. Boissonnat and Oudat [BO03] gave an elegant algorithm for general surfaces assuming that the feature sizes of the surface can be computed. Cheng, Dey, Ramos and Ray [CDRR04] gave a different algorithm for surface meshing which replaced the feature size computations with critical point computations. All these algorithms are meant for smooth surfaces and not for polygonal surfaces. Although the authors of [BO03] and [CDRR04] indicate that their algorithms work well for polygonal surfaces in practice, no guarantee is proved.

Of course, treating a polygonal mesh as an input polyhedron one can use any of the polyhedra meshing algorithms [CDRR204, PW04] to obtain a meshing of the polygonal surfaces. Unfortunately, the output mesh produced with this approach may have some undesirable properties. These algorithms respect the edges and vertices of the input polygonal mesh so that the underlying space of the output is exactly the same as that of the input. As a result small input angles are not eliminated. In particular, if applied to a triangular mesh, the algorithm tries to compensate for acute angles invariably present in each triangle and produces too many sample points. What we are looking for is a remeshing of the input polygonal mesh where the new points are constrained to be on the input mesh though the underlying space of the output is allowed to differ from the input.

Results. Given an input polygonal mesh G, our algorithm samples G with a Delaunay refinement approach and produces an output mesh that has the same topology and approximate geometry of G. Moreover, the output triangles have bounded aspect ratios. These guarantees are proved assuming that G satisfies certain conditions. Specifically, we show that if G approximates a smooth surface both point-wise and normal-wise closely, then the Delaunay refinement running with the desired conditions terminates. It is only the proofs that use a hypothetical smooth surface Σ approximated by G, but Σ plays no role in the algorithm. In practice, there are many situations where such assumption is valid. For example, G might be a polygonal surface reconstructed from a dense sample of a smooth surface [Dey03], or a designed surface approximating a smooth surface closely.

Overview. Our algorithm has two distinct phases. In the first phase it recovers the topology of G. In the second phase, it refines further to recover geometry and ensures quality of the triangles. For topology recovery we follow the approach of Cheng et al. [CDRR04] to build the mesh as the restricted Delaunay triangulation of a set of points sampled from the input polygonal surface. This restricted Delaunay triangulation has the same topology of the input surface if a property called topological ball holds [ES97]. We prove that one can find a point on the input surface that is far-away from all existing sampled points if this property does not hold. Such a point is sampled to drive the Delaunay refinement. For geometry recovery we present similar results that lead to a new refinement algorithm. Section 2 and 3 describe the algorithms for topology and geometry recovery respectively. Section 4 and 5 build all necessary lemmas for the termination proof in section 6. Finally, in section 7 we show that meshes computed from dense point cloud data satisfy the conditions required for our proofs.

2 Delaunay refinement for topology

The Delaunay refinement algorithm first concentrates on getting the topology right. Before we describe the algorithm we briefly set up our notations for Delaunay and Voronoi diagrams. The Delaunay triangulation of a point set $P \subset \mathbb{R}^3$ is denoted as $\text{Del}\,P$ and its dual Voronoi diagram as $\text{Vor}\,P$. A Voronoi cell for a point $p \in P$ is denoted as V_p. Skipping the details we just mention that $\text{Del}\,P$ is a simplicial complex where a k-simplex is dual to a Voronoi face of dimension $3 - k$ which is the intersection of $3 - k$ Voronoi cells. Zero-,one- and two- dimensional Voronoi faces are called Voronoi vertices, Voronoi edges and Voronoi facets respectively.

The topology is recovered in two phases. First, in the manifold recovery phase, a Delaunay mesh is computed which is guaranteed to be a 2-manifold. Then, further refinement is carried out so that the 2-manifold has the same topology as that of the input. In both phases we grow a sample Q, generated from G iteratively. The notion of restricted Delaunay triangulation plays an important role in the algorithms and the proofs.

Definition 1. *Let $v(\sigma)$ denote the set of vertices of a simplex σ and $V_q|_G = V_q \cap G$. The Delaunay complex*

$$\text{Del}\,Q|_G = \{\sigma \in \text{Del}\,Q \mid \bigcap_{q \in v(\sigma)} V_q|_G \neq \emptyset\}$$

is called the restricted Delaunay triangulation of Q with respect to G.

Basically, $\text{Del}\,Q|_G$ contains a dual Delaunay simplex for every Voronoi face intersected by G. The triangulation $\text{Del}\,Q|_G$ plays a key role in the topology recovery phase. However, we need only a subcomplex of $\text{Del}\,Q|_G$ to guarantee the manifold property. Let T be the set of triangles in $\text{Del}\,Q$ dual to the Voronoi edges that intersect G. The *edge restricted Delaunay triangulation* $\text{EDel}\,Q|_G$ is the simplicial complex made by T and its edges and vertices. The manifold recovery phase MFLRE-COV computes $\text{EDel}\,Q|_G$ while the topology recovery phase TOPORECOV computes $\text{Del}\,Q|_G$.

The refinement routines check if the Voronoi diagram $\text{Vor}\,Q$ satisfies certain conditions. If not, they insert a point from G into Q which is far away from its nearest neighbor and hence from all points in Q. Specifically, MFLDRECOV enforces a manifold property explicitly and TOPORECOV enforces a topological ball property. The routines are similar in spirit with those used for smooth surface meshing by Cheng et al. [CDRR04]. However, they differ in crucial details. We take the liberty of computing the furthest point in $G \cap V_q$ from q since G is polygonal and these computations do not require any optimization steps as opposed to the smooth surface meshing case. Also, the critical point computations in various phases of the algorithm of Cheng et al. are completely eliminated which results into new subroutines like VCELL with its new justification.

2.1 Manifold Recovery

The manifold recovery phase explicitly enforces the manifold condition. Two subroutines, VEDGE and DISK are called by the manifold recovery algorithm. VEDGE checks if any Voronoi edge, say in a Voronoi cell V_q, intersects G in more than one point. If so, it outputs the intersection point, say p, furthest from q. Clearly, p cannot be closer to any point in Q than q. Let T be the set of triangles in EDel $Q|_G$, i.e., the triangles dual to the Voronoi edges in Vor Q intersected by G. Denote the set of triangles incident to q in T as Δt_q. DISK checks if Δt_q is a topological disk. If not, it returns the point in $G \cap V_q$ furthest from q. MFLDRECOV inserts the points returned by VEDGE and DISK into Q and updates Vor Q. It starts with a vertex from each component in G and finally returns T when no more points need to be inserted.

Observation 2.1 *When* MFLDRECOV *terminates, T is a 2-manifold as Δt_q for each vertex q is a 2-disk.*

VEDGE($e \in V_q$)
    If e intersects G tangentially or at least in two points, return the point furthest from q among them, otherwise return nzll.

DISK(q)
    If Δt_q is not a topological disk, return the furthest point in $G \cap V_q$.

MFLDRECOV(G)
    1. Initialize Q with a vertex from each component of G.
    2. Compute Vor Q.
    3. If there is a Voronoi edge $e \in$ Vor Q so that VEDGE(e) returns a point p, insert p into Q and go back to step 2.
    4. If there is a point $q \in Q$ so that DISK(q) returns a point p, insert p into Q and go back to step 2.
    5. Output the set T of triangles dual to the Voronoi edges intersecting G.

2.2 Topology recovery

In the topology recovery phase we insert points into Q till Vor Q satisfies the following property. We say Vor Q satisfies the *topological ball property* if each Voronoi face of dimension k, $0 < k \le 3$, intersects Σ in a topological $(k-1)$-ball if it intersects G at all. For $k = 0$, this intersection should be empty. The motivation for ensuring the topological ball property comes from the following result of Edelsbrunner and Shah [ES97].

Theorem 1. *The underlying space of the complex* $\mathrm{Del}\,Q|_G$ *is homeomorphic to* G *if* $\mathrm{Vor}\,Q$ *has the topological ball property.*

We use this theorem to guarantee the topology of the output. The VEDGE subroutine can be used to enforce the topological ball property for the Voronoi edges. A Voronoi facet $F \in V_q$ can violate this property by intersecting G in

(i) more than one topological interval, and/or
(ii) in one or more cycle.

If (i) happens, either a Voronoi edge of F intersects G more than once, or the dual Delaunay edge $\mathrm{dual}(F)$ has more than two triangles incident to it in $\mathrm{EDel}_{Q|G}$. The topological disk condition for Δt_q will be violated where q is any of the end vertices of $\mathrm{dual}(F)$. So, this violation can be handled by the subroutine DISK. For (ii) we introduce a subroutine FCYCLE to check the condition and to identify the point in $F \cap G$ furthest from q. This furthest point will be a point of intersection between F and an edge of G.

FCYCLE($F \in V_q$)

　　If $F \cap G$ has a cycle, return the point in $F \cap G$ furthest from q.

So far we have subroutines for checking the topological ball property for Voronoi edges and facets. The Voronoi cell also needs a separate check. For a Voronoi cell V_q, the subroutine VCELL checks if $W = V_q \cap G$ is a topological disk. By the time VCELL is called in the algorithm, W is ensured to be a 2-manifold with a single boundary. Also, W has a single component as Q is initialized with a vertex from each component of G. It is easy to check if such a surface is a disk by computing the Euler characteristic that is given by the alternating sums of the number of vertices, edges and polygons in W. Of course, this requires one to compute the boundary edges and vertices of W. However, we can simply ignore those vertices and edges as they cancel out in the alternating sum.

VCELL(q)

　　Determine the number $\#v$ of vertices, $\#e$ of edges and $\#g$ of polygons in G intersecting V_q. If $\#v - \#e + \#g \neq 1$, return the point in $G \cap V_q$ furthest from q.

Now we have all ingredients to recover the topology of G into the new mesh T.

TOPORECOV(G, Q)

　　1. If $Q = \emptyset$ initialize Q with a vertex from each component of G.
　　2. Compute $\mathrm{Vor}\,Q$.
　　3. If any of VEDGE, DISK, FCYCLE and VCELL, necessarily in this order, succeeds and returns a point p, insert p into Q and go back to step 2.
　　5. Output the set T of triangles dual to the Voronoi edges intersecting G.

3 Delaunay refinement for geometry

It is often not enough to recover only the topology of G. The remeshed surface T should also follow the geometry of G. For this, we need to sample G more densely

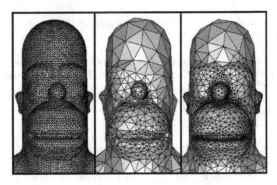

Fig. 1. The input mesh is uniformly dense everywhere. The output is a graded mesh at two different levels of density.

than the topology recovery requires. The level of density is controlled by an user parameter λ. Figure 1 shows an example of remeshing at two different levels of density.

The algorithm for geometry recovery uses the structure of the Voronoi cells to determine if the mesh should be refined locally. For a point $q \in Q$, the set $W = V_q \cap G$ is a 2-disk after the topology recovery phase. This disk separates V_q into two subsets one on each side of W. Let V_q^+ and V_q^- denote these subsets and let q^+ and q^- be the Voronoi vertices in V_q^+ and V_q^- respectively furthest from q. If any of V_q^+ and V_q^- is unbounded, the corresponding furthest vertex is taken at infinity. The points q^+ and q^- are like poles of the Voronoi cell V_q as defined by Amenta, Bern [AB98] for smooth surfaces. When points are sampled from a smooth surface densely, it is known that the poles of a point remain far away from it. Although we deal with a non-smooth surface G, we can claim (Lemma 15) a similar property if G approximates a smooth surface closely enough.

Geometry recovery checks if all triangles Δt_q incident to a vertex q are small enough compared to the pole distance $h_q = \min\{\|q-q^+\|, \|q-q^-\|\}$. Let $r(t)$ denote the circumradius of a triangle in Δt_q. For an user parameter λ, if $r(t)/h_q$ is larger than 12λ, we insert the point c where the dual Voronoi edge of t, $\mathrm{dual}(t)$, intersects G. Clearly, the value of λ denotes the level of refinement.

A similar procedure can be used to guarantee the quality of the triangles. Let $\rho(t)$ denote the ratio $r(t)/\ell(t)$ where $\ell(t)$ is the length of the shortest edge of t. It is known that triangles with bounded ρ value have bounded aspect ratios. In the algorithm, we check if a triangle t has $\rho(t)$ more than $(1 + 8\lambda)$, and, if so, we insert the point $\mathrm{dual}(t) \cap G$ into Q.

The particular choice of 12λ and $(1 + 8\lambda)$ comes from our proofs.

GEOMRECOV(G, λ)

1. $Q := \emptyset$
2. $T := \text{TOPORECOV}(G, Q)$, $Q :=$ vertices of T
3. For each $q \in Q$, if there is a $t \in \Delta t_q$ with $c = \text{dual}(t) \cap G$ so that either
   (i) $\rho(t) > (1 + 8\lambda)$, or
   (ii) $r(t)/h_q > 12\lambda$
   then insert c into Q and go back to step 2.
4. Output T.

Notice that, although GEOMRECOV requires an user supplied λ, MFLDRECOV and TOPORECOV require no such λ. We choose $\lambda = 0.02$ in GEOMRECOV for our experiments. Figure 5 shows some results of this experiment.

4 Prelude to proofs

The refinement routines may not terminate for arbitrary polygonal meshes but we prove that when the input mesh G approximates a smooth surface closely enough, they necessarily terminate. We need several definitions to state this approximation precisely.

Let Σ be a smooth, compact surface without boundary which is approximated by G. In general, G and hence Σ may have more than one connected component. The output is a triangulation T whose vertex set lies in G. Abusing the notations slightly we will use G and T to mean their underlying complexes and spaces as well.

4.1 Definitions

Distances: For a point $x \in \mathbb{R}^3$ and a set $X \subseteq \mathbb{R}^3$, let $d(x, X)$ denote the Euclidean distance of x from X, i.e.,

$$d(x, X) = \inf_{y \in X} \|x - y\|.$$

A ball $B_{c,r}$ is the set of points whose distance to c is no more than r.

Medial axis and feature size: The *medial axis* M of Σ is the closure of the set $X \subset \mathbb{R}^3$ so that, for each point $x \in X$, $d(x, \Sigma)$ is realized by two or more points. Alternatively, M is the loci of the centers of the maximal balls whose interiors are empty of points from Σ. These balls, called *medial balls*, are tangent to Σ at one or more points. At each point $x \in \Sigma$, there are two such tangent medial balls. Define a function $f : \Sigma \to \mathbb{R}$ where $f(x) = d(x, M)$. The value $f(x)$ is called the *local feature size* of Σ at x [ABE98].

Projection map: Often we will use a projection map $\nu : \mathbb{R}^3 \setminus M \to \Sigma$ where $\tilde{x} = \nu(x)$ is the closest point in Σ, i.e., $d(x, \Sigma) = \|x - \tilde{x}\|$.

Oriented normals: The normals of the three spaces, Σ, G and T play an important role in the proofs. These normals need to be oriented. We denote the normal of Σ at a point x as $\tilde{\mathbf{n}}_x$. These normals are oriented, i.e., the normal $\tilde{\mathbf{n}}_x$ points to the bounded component of $\mathbb{R}^3 \setminus \Sigma'$ where x is in the connected component $\Sigma' \subseteq \Sigma$. Similarly, we define an oriented normal $\mathbf{n}_g$ for each polygon g in G. Let g be in the connected component $G' \subseteq G$. The normal $\mathbf{n}_g$ points to the bounded component of $\mathbb{R}^3 \setminus G'$. We will also orient the normals of the Delaunay triangles computed by our algorithms. For a Delaunay triangle t, we orient its normal $\mathbf{n}_t$ so that $\angle \mathbf{n}_t, \tilde{\mathbf{n}}_{\tilde{p}} < \pi/2$ where p is a vertex subtending the largest angle in t.

Angle notation: The angle between two oriented normals $\mathbf{a}$ and $\mathbf{b}$ is denoted as $\angle \mathbf{a}, \mathbf{b}$.

Thickening and approximation: We require that G approximate Σ pointwise, i.e., it resides within a small thickening of Σ. In particular, for $0 \le \delta < 1$, we introduce a thickened space

$$\delta\Sigma = \{x \in \mathbb{R}^3 | d(x, \Sigma) \le \delta f(\tilde{x})\}.$$

We will also require that G approximates Σ normal-wise. To specify this point-wise and normal-wise approximation we define:

Definition 2. *G is (δ, μ)-flat with respect to Σ if the two conditions hold:*

(i) For $\delta < 1$, the closest point $\tilde{x} \in \Sigma$ of each point $x \in G$ is within $\delta f(\tilde{x})$ distance, and conversely, each point $x \in \Sigma$ has a point of G within $\delta f(x)$ distance.

(ii) For any point x in a polygon $g \in G$, the angle $\angle \mathbf{n}_g, \mathbf{n}_{\tilde{x}}$ is at most μ for some $\mu < 1$.

4.2 Consequences

We assume G to be (δ, μ)-flat with respect to Σ. The following lemmas are direct consequences of this assumption.

Lemma 1. *(i) $G \subset \delta\Sigma$. (ii) Let $\mathbf{n}_g$ and $\mathbf{n}'_g$ be the normals of two adjacent polygons $g, g' \in G$. We have $\angle \mathbf{n}_g, \mathbf{n}'_g \le 2\mu$.*

Lemma 2. *Let p and q be two points in $\delta\Sigma$ with $\|p - q\| \le \lambda f(\tilde{p})$. Then, $\|\tilde{p} - \tilde{q}\| \le 2(\lambda + \delta)f(\tilde{p})$.*

Proof. We have

$$\|\tilde{p} - \tilde{q}\| \le \|\tilde{p} - p\| + \|p - q\| + \|q - \tilde{q}\|$$
$$\le 2(\|\tilde{p} - p\| + \|p - q\|) \le 2(\lambda + \delta)f(\tilde{p}).$$

The medial balls of Σ are empty of points from Σ but not from G. This is a reason why the proofs for Σ as detailed in [CDRR04] do not extend to G. First, we clear this obstacle by showing that the medial balls after appropriate shrinking can be made empty of points from $\delta\Sigma$ and hence G. This claim is formalized in the following lemma. Let L_x^+ and L_x^- denote the rays originating at $x \in \Sigma$ in the directions of $\tilde{\mathbf{n}}_x$ and $-\tilde{\mathbf{n}}_x$ respectively, see Figure 2 (proof in the extended version [Dey05]).

Lemma 3. *Let $0 < \delta < 1/4$. For each point $x \in \Sigma$ there are two balls $B = B_{c,r}$ and $B' = B_{c',r'}$ with*

(i) $c \in L_x^+$ and $c' \in L_x^-$,
(ii) $r = r' = (1 - 4\delta)f(x)$,
(iii) $d(x, B) = d(x, B') = 4\delta f(x)$,
(iv) $\delta\Sigma \cap B = \emptyset$ and $\delta\Sigma \cap B' = \emptyset$.

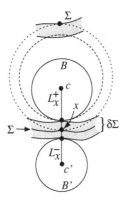

Fig. 2. A medial ball tangent to Σ at x (shown with dotted boundary) is shrunk radially first to the ball with dashed boundary. Then it is shrunk further to be empty of $\delta\Sigma$ to the ball with solid boundary.

5 Normals and conditions

The normals to the triangles and edges that the Delaunay refinement produces play an important role in the analysis. For convenience we define the following two functions for any $\lambda > 0$.

$$\alpha(\lambda) = \frac{\lambda}{1 - 4\lambda} \text{ and}$$

$$\beta(\lambda) = \arcsin \lambda + \arcsin \left(\frac{2}{\sqrt{3}} \sin(2 \arcsin \lambda) \right).$$

For smooth surfaces it is known that the triangles with small circumradius lie almost parallel to the surface. This follows from the following lemma proved by Amenta, Choi, Dey and Leekha [ACDL00] and the fact that medial balls incident to a point on the surface are relatively large.

Lemma 4. *Let $B = B_{c,r}$ and $B' = B_{c',r'}$ be two balls meeting at a single point p of a smooth surface Σ. Let $t = pqr$ be a triangle where*

(i) p subtends the largest angle of t,
(ii) the vertices of t lie outside of $B \cup B'$, and
(iii) the circumradius of t is no more than $\lambda \min\{r, r'\}$ where $\lambda < \frac{1}{\sqrt{2}}$.[1]

Then the acute angle between the line of the normal $\mathbf{n}_t$ of t and the line joining c, c' is no more than $\beta(\lambda)$.

---

[1] In [ACDL00], the constant is stated smaller, but $\frac{1}{\sqrt{2}}$ is also a valid choice.

This lemma implies that the small Delaunay triangles lie almost parallel to the surface, a key fact used in the Delaunay refinement algorithm of Cheng et al. [CDRR04]. Lemma 7 is a version of this fact for the non-smooth surface G which we prove with our assumption that G follows Σ point-wise.

Another key ingredient used in smooth case is that the normals of a smooth surface cannot vary too abruptly. Precisely, Amenta and Bern [AB98] proved the following lemma.

Lemma 5. *Let x and y be any two points in Σ with $\|x - y\| \leq \lambda f(x)$ and $\lambda < 1/4$. Then, $\angle \tilde{\mathbf{n}}_x, \tilde{\mathbf{n}}_y \leq \alpha(\lambda)$.*

Lemma 1 is the non-smooth version of the above lemma which holds because G follows Σ normal-wise.

5.1 Triangle and edge normals

Now we focus on deriving the flatness property of the triangles and edges that are produced by the Delaunay refinement of G. The Delaunay refinement of G generates a point set Q on G. Thus, necessarily $Q \subset \delta\Sigma$ as $G \subset \delta\Sigma$. It is easy to see that the triangles and edges with such points as vertices may have normals in any direction no matter how small they are. A key observation we make and formalize is that when vertices have a suitable sparsity condition, i.e., a lower bound on their mutual distances, the triangles lie almost parallel to the surface Σ. This is proved in Lemma 7 with the help of large empty balls guaranteed by Lemma 3 and the technical Lemma 6 (proof in [Dey05]).

Lemma 6. *Let $\ell > \sqrt{\delta} > 0$ and $\delta \leq 1/4$. Let $B = B_{a,R}$ and $B' = B_{b,r}$ be two balls whose boundaries ∂B and $\partial B'$ intersect in a circle C with the following conditions.*

(i) Let p be the point where the line joining a and b intersects $\partial B'$ outside B. The distance of p from any point on C is at least ℓR.
(ii) The distance of p from ∂B is no more than δR.

Then,

$$r \geq \frac{R}{9}.$$

Lemma 7. *For $0 \leq \delta < \lambda < 1/48$ and $\ell > \sqrt{6\delta}$, let t be a triangle and q be any of its vertices where*

(i) vertices of t lie in $\delta\Sigma$,
(ii) q is at least $\ell f(\tilde{q})$ distance away from all other vertices of t,
(iii) the circumradius of t is at most $\lambda f(\tilde{q})$.

Then, $\angle \mathbf{n}_t, \tilde{\mathbf{n}}_{\tilde{q}} \leq \beta(12\lambda) + \alpha(6\lambda)$.

Proof. Let p be the vertex of t subtending the largest angle. First, we prove that there are two balls of radius at least $\frac{f(\tilde{p})}{12}$ being tangent at p and with centers on $L_{\tilde{p}}^+$ and $L_{\tilde{p}}^-$ respectively.

If $\delta = 0$, the vertices of t lie on Σ and the two medial balls at p satisfy the condition. So, assume $\delta \neq 0$. Consider a ball $B = B_{c,(1-4\delta)f(\tilde{p})}$ as stated in Lemma 3 for the point $\tilde{p}$. This ball is empty of any point from $\delta\Sigma$ and therefore does not contain any vertex of t. Consider a ball D with the center p and radius $\ell f(\tilde{p})$ where $\ell > \sqrt{6\delta}$. This ball also does not contain any vertex of t by the condition (ii). Let C be the circle of intersection of the boundaries of B and D. Let $B' = B_{w,r}$ be the ball whose boundary passes through C and p (Figure 3). No vertex of t lies inside B' as $B' \subset B \cup D$ and both B and D are empty of the vertices of t. We claim that $r \geq f(\tilde{p})/12$.

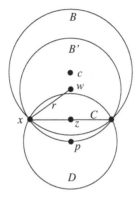

Fig. 3. Illustration for Lemma 7.

Let x be any point on the circle C whose center is z. The radius R of B is equal to $(1 - 4\delta)f(\tilde{p})$. So, $\|x - p\| = \frac{\ell}{1-4\delta}R$. The distance $d(p, B) \leq \|p - \tilde{p}\| + d(\tilde{p}, B)$. Since p lies in $\delta\Sigma$, $\|p - \tilde{p}\| \leq \delta f(\tilde{p})$, and $d(\tilde{p}, B) \leq 4\delta f(\tilde{p})$ by Lemma 3(iii). So,

$$d(p, B) \leq 5\delta f(\tilde{p}) \leq \frac{5\delta}{1 - 4\delta}R \leq 6\delta R \text{ for } \delta < \tfrac{1}{24}.$$

Since $\ell > \sqrt{6\delta}$ we can apply Lemma 6 to B and B' to get

$$r \geq \frac{R}{9} = \frac{f(\tilde{p})(1 - 4\delta)}{9} > \frac{f(\tilde{p})}{12}$$

using $(1 - 4\delta) > \frac{3}{4}$ for $\delta \leq \frac{1}{16}$.

Applying the above argument to the other ball $B_{c',(1-4\delta)f(\tilde{p})}$ as guaranteed by Lemma 3, we get another empty ball B'' with radius at least $\frac{f(\tilde{p})}{12}$ touching p. The centers of both B' and B'' lie on the line of the normal $\tilde{\mathbf{n}}_{\tilde{p}}$. Notice that both B' and B'' meet at a single point p. Shrink both B' and B'' keeping them tangent at p till their radius is equal to $f(\tilde{p})/12$. Now the balls B', B'' and the triangle t satisfy the conditions of Lemma 4. First, the vertices of t lie outside B' and B''. Secondly, the circumradius of t is at most $\lambda f(\tilde{p})$, i.e., 12λ times the radii of B' and B''. Therefore, the acute angle between the lines of $\mathbf{n}_t$ and $\tilde{\mathbf{n}}_{\tilde{p}}$ is at most $\beta(12\lambda)$. For $\lambda < 1/48$, $\beta(12\lambda) < \pi/2$. This implies that the upper bound of $\beta(12\lambda)$ also holds for the oriented normals, i.e.,

$$\angle \mathbf{n}_t, \tilde{\mathbf{n}}_{\tilde{p}} \leq \beta(12\lambda).$$

Now, consider any vertex q of t. Since the circumradius of t is no more than $\lambda f(\tilde{q})$, $\|p - q\| \leq 2\lambda f(\tilde{q})$. By Lemma 2, $\|\tilde{p} - \tilde{q}\| \leq 2(2\lambda + \delta)f(\tilde{q}) \leq 6\lambda f(\tilde{q})$ for $\delta < \lambda$. Then, by Lemma 5 $\angle \tilde{\mathbf{n}}_{\tilde{p}}, \tilde{\mathbf{n}}_{\tilde{q}} \leq \alpha(6\lambda)$ provided $6\lambda < 1/4$, or $\lambda < 1/24$. Therefore,

$$\angle \mathbf{n}_t, \tilde{\mathbf{n}}_{\tilde{p}} \leq \beta(12\lambda) + \alpha(6\lambda).$$

Similar to triangles, small edges with vertices on G also lie almost parallel to Σ (proof in [Dey05]).

Lemma 8. *For $0 \leq \delta < \lambda < 1/48$ and $\ell > \sqrt{6\delta}$, let pq be an edge where*

(i) p and q lie in $\delta\Sigma$,
(ii) $\ell f(\tilde{q}) < \|p - q\| < \lambda f(\tilde{q})$.

Then, $\angle \mathbf{q}p, \tilde{\mathbf{n}}_{\tilde{q}} \geq \frac{\pi}{2} - \arcsin 6\lambda$.

5.2 Conditions

We use Lemma 7 and Lemma 8 to prove the correctness of the remeshing algorithms. These results depend on certain conditions, i.e., the values of λ and δ have to satisfy some constraints. They would in turn suggest some condition on the sparsity of the sampling by the Delaunay refinement.

Condition on sparsity

Lemma 7 and Lemma 8 will be applied to the Delaunay triangles and edges for a sample Q that the Delaunay refinement generates. It will be required that Q maintain a lower bound on the distances between its points.

Definition 3. *A point set Q is λ-sparse if each point $q \in Q$ is at least $\frac{\lambda}{(1+8\lambda)}f(\tilde{q})$ distance away from every other point in Q.*

The particular choice of the factor $\frac{\lambda}{1+8\lambda}$ will be clear when we argue about termination. One condition of lemmas 7 and 8 says that the length of an edge pq has to be more than $\sqrt{6\delta}f(\tilde{q})$. When Q is λ-sparse, this condition is satisfied if

$$\sqrt{6\delta} < \frac{\lambda}{1 + 8\lambda}. \tag{1}$$

This means that the Delaunay refinement has to maintain a λ-sparse sample Q where λ satisfies the inequality (1).

Bounding conditions

Inequality (1) says that λ needs to satisfy a lower bound in terms of δ. We will see later that it also needs to satisfy some upper bounds for guaranteeing termination of the algorithms. For reference to these conditions on λ, we state them with Condition 1 and 2 and refer them together as Bounding condition on λ.

$$\text{Condition } 1 : \sqrt{6\delta} < \frac{\lambda}{1 + 8\lambda} \text{ and } \lambda < \frac{1}{48}.$$
$$\text{Condition } 2 : \beta(12\lambda) + \alpha(6\lambda) + \alpha(4\lambda) + 3\mu < \frac{\pi}{2}.$$

Observation 5.1 *If Condition 1 holds, $\delta < \lambda$.*

Recall that the map ν takes a point to its closest point on Σ. It turns out that the map ν restricted to G induces a homeomorphism between G and Σ if δ and μ are sufficiently small(proof in [Dey05]).

Observation 5.2 *If G is (δ, μ)-flat with respect to Σ where δ and μ satisfy the Bounding conditions for some $\lambda > 0$, then G is homeomorphic to Σ and the map ν restricted to G is a homeomorphism.*

6 Termination proofs

The main theorems we prove are:

Theorem 2. *Let G be a polygonal mesh that is (δ, μ)-flat with respect a smooth surface. If there exists a $\lambda > 0$ so that the Bounding conditions hold, then*

(i) MFLDRECOV *terminates and outputs a manifold Delaunay mesh whose vertex set is λ-sparse,*

(ii) TOPORECOV *terminates and outputs a Delaunay mesh homeomorphic to G whose vertex set is λ-sparse.*

Theorem 3. *Let G be a polygonal mesh that is (δ, μ)-flat with respect a smooth surface. If the chosen λ satisfies the Bounding conditions, then* GEOMRECOV *terminates and outputs a Delaunay mesh homeomorphic to G whose vertex set is λ-sparse.*

The key to the success of the topology recovery phase is that, for sufficiently small δ and μ, there exists a λ satisfying the Bounding conditions. For example, if $\delta = 4 \times 10^{-5}$ and $\mu = 0.1$, one can choose $\lambda = 0.02$. Notice that the requirement on δ is rather too small. First of all, this is an artifact of our proofs. Secondly, when G is a reconstructed mesh from a dense point sample, we will see later that δ will be $O(\varepsilon^2)$ where $\varepsilon < 1$ measures the sampling density and thus the requirement on ε will be less stringent. For geometry recovery phase we explicitly need that the user supplied λ satisfy the Bounding conditions.

We use several lemmas about the Voronoi diagram of Q to prove the above two theorems. A common theme in these lemmas is that if a Voronoi face does not intersect G appropriately, there is a point in G far away from all existing points in Q. Recall that algorithmically we used this result by inserting such a far-away point to drive the Delaunay refinement. A similar line of arguments was used by Cheng, Dey, Ramos and Ray [CDRR04] for meshing smooth surfaces. However, as we indicated before, some of the proofs need different reasoning since G is not smooth. We skip those proofs that can be adapted from Cheng et al. [CDRR04] with only minor changes and include the ones that need fresh arguments. In particular, Lemma 9, which needs new arguments, is an essential ingredient for other lemmas. In what follows we assume G to be (δ, μ)-flat with respect to a smooth surface Σ for some appropriate $\delta < 1$ and $\mu < 1$. All lemmas in this section involve faces of Vor Q where Q is λ-sparse for a λ satisfying the Bounding conditions.

Lemma 9. *Let $e \in V_q$ be a Voronoi edge that intersects G either (i) tangentially at a point, or (ii) transversally at two or more points. Let x be the point among these intersection points which is furthest from q. Then, x is at least $\lambda f(\tilde{q})$ away from q.*

Proof. Suppose that contrary to the lemma $\|q - x\| < \lambda f(\tilde{q})$. Observe that

$$\|\tilde{x} - \tilde{q}\| \leq 2(\lambda + \delta)f(\tilde{q}) \text{ (Lemma 2)}$$
$$\leq 4\lambda f(\tilde{q}) \text{ by Observation 5.1.}$$

By Lemma 5, $\angle \tilde{\mathbf{n}}_{\tilde{x}}, \tilde{\mathbf{n}}_{\tilde{q}} \leq \alpha(4\lambda)$.

Orient e along $\mathbf{n}_{pqr}$ where pqr is the dual Delaunay triangle of e. The conditions (i) and (ii) of Lemma 7 hold for pqr since $G \subset \delta\Sigma$, Q is λ-sparse and the Bounding condition 1 holds. The circumradius of pqr is no more than $\|q - x\| \leq \lambda f(\tilde{q})$ satisfying the condition (iii) of Lemma 7. So, we have

$$\angle e, \tilde{\mathbf{n}}_{\tilde{x}} \leq \angle \mathbf{n}_{pqr}, \tilde{\mathbf{n}}_{\tilde{q}} + \angle \tilde{\mathbf{n}}_{\tilde{q}}, \tilde{\mathbf{n}}_{\tilde{x}}$$
$$\leq \beta(12\lambda) + \alpha(6\lambda) + \alpha(4\lambda) \text{ (Lemma 7, 5).}$$

Let g be a polygon in G containing x. Since $\angle \mathbf{n}_g, \tilde{\mathbf{n}}_{\tilde{x}} \leq \mu$, oriented e makes an angle of at most $\beta(12\lambda) + \alpha(6\lambda) + \alpha(4\lambda) + \mu$ with $\mathbf{n}_g$.

Suppose, e intersects G tangentially at x. Then, e makes at least $(\pi/2) - 2\mu$ angle with the normal of one of the polygons containing x since the normals of adjacent polygons in G make at most 2μ angle (Lemma 1). We reach a contradiction if

$$\beta(12\lambda) + \alpha(6\lambda) + \alpha(4\lambda) + \mu < \frac{\pi}{2} - 2\mu,$$

which is satisfied by the Bounding condition 2 proving (i).

Because of the previous argument we can assume that e intersects G only transversally. Let y be an intersection point next to x on e and $g' \in G$ be a polygon containing y. The distance $\|q - y\|$ is at most $\lambda f(\tilde{q})$ as the furthest intersection point x from q is within $\lambda f(\tilde{q})$ distance from it. Then, applying the same argument as for x, we get that e makes an angle of at most $\beta(12\lambda) + \alpha(6\lambda) + \alpha(4\lambda) + \mu$ with $\mathbf{n}_{g'}$. The oriented e leaves the bounded component of $\mathbb{R}^3 \setminus G$ at one of x and y. At this exit point, e makes more than $\frac{\pi}{2}$ angle with the oriented normal of the corresponding polygon. This means we reach a contradiction if

$$\beta(12\lambda) + \alpha(6\lambda) + \alpha(4\lambda) + \mu < \frac{\pi}{2}.$$

This inequality is satisfied if the Bounding condition 2 holds.

Next lemma says that if a Voronoi facet does not intersect G properly, one can find a far away point to insert. Its proof depends on Lemma 9 and is very similar to Lemma 8 of [CDRR04].

Lemma 10. *Let F be a Voronoi facet in V_q where $F \cap G$ contains at least two closed topological intervals. Furthermore, assume that each Voronoi edge intersects G in at most one point. The furthest point in $F \cap G$ from q which lies on a Voronoi edge of V_q is at least $\lambda f(\tilde{q})$ away from q.*

Next three lemmas deal with different cases of the boundaries of the manifold in which a Voronoi cell intersects G. We skip the proof of Lemma 11 since it is same as that of Lemma 11 of [CDRR04] and refer to [Dey05] for the proofs of others.

Lemma 11. *For a vertex q in $\mathrm{Vor}\, Q$ let $W = V_q \cap G$ is a manifold with at least two boundaries both of which intersect Voronoi edges of V_q. Then the point $x \in W$ furthest from q is within $\lambda f(\tilde{q})$ distance.*

Lemma 12. *Let $F \subset V_q$ intersect G in a cycle. The point in $F \cap G$ furthest from q is at least $\lambda f(\tilde{q})$ distance away from q.*

Lemma 13. *For a point $q \in Q$ let $W = V_q \cap G$ intersect no Voronoi edge. Then, the point $x \in W$ furthest from q is at least $\lambda f(\tilde{q})$ away from q.*

The next lemma will ensure that the point inserted by VCELL cannot be very close to all other points in Q (proof in [Dey05]).

Lemma 14. *Let x be a point in G and $W \subset G$ be a subset so that $x \in W$ and $\|x - y\| \le \lambda f(\tilde{x})$ for each point $y \in W$. Furthermore, W has a single boundary. Then, W is a 2-disk when $\delta < 1/5$ and $\lambda < 1/4$.*

Observation 6.1 *Let p and q be any two points in $\delta\Sigma$ with $\|p-q\| \ge \lambda f(\tilde{q})$. Then, for $\delta < \lambda$, $\|p - q\| \ge \frac{\lambda}{(1+4\lambda)} f(\tilde{p})$.*

Proof. If $\|p-q\| \ge \lambda f(\tilde{p})$, there is nothing to prove. So, we assume $\|p-q\| < \lambda f(\tilde{p})$. By Observation 2, $\|\tilde{p} - \tilde{q}\| \le 2(\lambda + \delta)f(\tilde{p}) \le 4\lambda f(\tilde{p})$. By Lipschitz property of $f()$, $f(\tilde{q}) \ge \frac{1}{1+4\lambda}f(\tilde{p})$ which applied to the given inequality $\|p-q\| \ge \lambda f(\tilde{q})$ yields $\|p - q\| \ge \frac{\lambda}{1+4\lambda}f(\tilde{p})$.

Now we have all ingredients to prove Theorem 2.

Proof. (Theorem 2) We show that the vertex set Q remains λ-sparse for a $\lambda > 0$ throughout MFLDRECOV and TOPORECOV. Termination of these algorithms is immediate since only finitely many points can be accommodated in the bounded domain $\delta\Sigma$ with non-zero nearest neighbor distances.

Initially Q is λ-sparse trivially since it contains a single point from each component of G. Let p be any point inserted by any of the subroutines called by MFLDRECOV and TOPORECOV.

We claim that p is at least $\lambda f(\tilde{q})$ distance away from all other points in Q where q is a nearest point to p.

If VEDGE inserts p, the claim is true by Lemma 9. If DISK inserts p, then there was an existing point $q \in Q$ so that $p \in V_q$ and Δt_q was not a disk. If Δt_q were empty, $G \cap V_q$ did not intersect any Voronoi edge. Then, by Lemma 13 the claim is true. If Δt_q were not empty, either (i) there was an edge e of Δt_q not having two triangles incident to it, or (ii) there were two topological disks pinched at q. For (i) let F be the dual Voronoi facet of e. If e had a single triangle incident to it, G intersected a Voronoi edge in F either tangentially or at least twice. Both of these cases would have been caught by VEDGE test. So, e had three or more incident triangles. Hence F intersected G in more than one topological interval and the claim follows from Lemma 10. For (ii) observe that $G \cap V_q$ had two or more boundaries. The claim follows from Lemma 11. If p is inserted by FCYCLE, apply Lemma 12 for the claim. For VCELL observe that if it inserts a point, the subset $W = V_q \cap G$ is not a 2-disk. Also, since it is called after all other tests, W has a single boundary. Then, Lemma 14 is violated which implies the claim.

Applying Observation 6.1, we get that p is at least $\frac{\lambda}{1+4\lambda} f(\tilde{p})$ away from all other points of Q. Applying Observation 6.1 once more to any other point $s \in Q$, we get that

$$\|s - p\| \geq \frac{\lambda}{1 + 8\lambda} f(\tilde{s})$$

proving Q remains λ-sparse after insertion of p.

Next, we prepare to prove Theorem 3. Recall that q^+ and q^- are the two poles defined for a vertex q. First, we show that these poles are far away from q.

Lemma 15. *If Q is λ-sparse and the Bounding conditions hold, then for each vertex $q \in Q$, $\min\{\|q - q^+\|, \|q - q^-\|\} \geq \frac{f(\tilde{q})}{12}$.*

Proof. Following the proof of Lemma 7, we get two empty balls that are tangent to each other at q whose radii are at least $\frac{f(\tilde{q})}{12}$. The centers of these empty balls reside inside V_q. Also, they are separated locally within V_q by G. The points q^+ and q^- are even further from q than these centers. The lemma follows.

Proof. (Theorem 3) Since GEOMRECOV calls TOPORECOV it is sufficient to argue that if Q is λ-sparse, then it remains so after inserting a point c in the steps 3(i) and 3(ii) of GEOMRECOV. First, consider step (i). Since Q is λ-sparse, $\ell(t) > \frac{\lambda}{1+8\lambda} f(\tilde{q})$ where q is a vertex of the shortest edge in t. Then c is at least $\lambda f(\tilde{q})$ distance away from q. Next, consider setp (ii). The radius $r(t)$ is more than $12\lambda h_q$ which by Lemma 15 is at least $\lambda f(\tilde{q})$. Therefore, the point c is at least $\lambda f(\tilde{q})$ distance away from q.

Observe that in both cases q is also a nearest point of c in Q. Therefore, following the proof of Observation 6.1, we get that Q remains λ-sparse after the insertion of c.

7 Input Meshes

We have already seen that when δ and μ are sufficiently small, there exists a $\lambda > 0$ satisfying the Bounding conditions. This means that, for a mesh G that is (δ, μ)-flat with respect to a smooth surface for sufficiently small values of δ and μ, the manifold recovery and topology recovery terminate. Figure 4 shows two examples of polygonalized surfaces on which our remeshing algorithm is applied.

An interesting and perhaps the most important input for our algorithms would be the polygonal meshes created from point cloud data. When G is such a mesh we show that it is necessarily (δ, μ)-flat with respect to the surface Σ from which the point cloud is drawn. Of course, the point cloud should be sufficiently dense. A point set $P \subset \Sigma$ is called an ε-sample if $d(x, P) \leq \varepsilon f(x)$ for each point x of Σ [ABE98]. When G is reconstructed from an ε-sample P of Σ, it becomes (δ, μ)-flat where δ and μ depend on the sampling density ε. In general we can assume that any of the provable reconstruction algorithms [Dey03] is applied to create G from P. What is important is that all these algorithms produce triangles in G with small circumradius. For precision we assume that G is created from P using the COCONE algorithm of Amenta, Choi, Dey and Leekha [ACDL00]. Then, the following fact holds.

Fact 7.1 *Each triangle $t \in G$ has a circumradius of $\frac{1.15\varepsilon}{1-\varepsilon} f(p)$ where p is any vertex of t.*

We can derive bounds on δ and μ from the above fact. It turns out that $\mu = O(\varepsilon)$ while $\delta = O(\varepsilon^2)$. The proofs appear in [Dey05].

Lemma 16. *Let x be any point in a triangle $t \in G$. We have $\angle \mathbf{n}_t, \tilde{\mathbf{n}}_{\tilde{x}} \leq \beta(\varepsilon) + \alpha(4.6\varepsilon')$ where $\varepsilon' = \frac{\varepsilon}{1-\varepsilon}$.*

Lemma 17. *Let x be any point in a triangle $t \in G$. Then $\|x - \tilde{x}\| \leq \left(\frac{4.6\varepsilon'}{1-4.6\varepsilon'}\right)^2 f(\tilde{x})$.*

From the above two lemmas, we find that $\delta \leq \left(\frac{4.6\varepsilon'}{1-4.6\varepsilon'}\right)^2$ and $\mu \leq \beta(\varepsilon) + \alpha(4.6\varepsilon')$ for G. If $\varepsilon \leq 0.001$, we get $\delta = 2 \times 10^{-5}$ and $\mu = 0.009$. This allows to choose $\lambda = 0.02$ to satisfy the Bounding conditions. Figure 4 shows examples of reconstructed meshes that are remeshed.

Acknowledgements. We acknowledge the support of Army Research Office, USA under grant DAAD19-02-1-0347 and NSF, USA under grants DMS-0310642 and CCR-0430735.

[Alliez03] P. Alliez, E. C. de Verdière, O. Devillers and M. Isenburg. Isotropic surface remeshing. *Proc. Shape Modeling Internat.* (2003).

[AB98] N. Amenta and M. Bern. Surface reconstruction by Voronoi filtering. *Discr. Comput. Geom.* **22** (1999), 481–504.

[ABE98] N. Amenta, M. Bern and D. Eppstein. The crust and the β-skeleton: combinatorial curve reconstruction. *Graphical Models and Image Processing*, **60** (1998), 125-135.

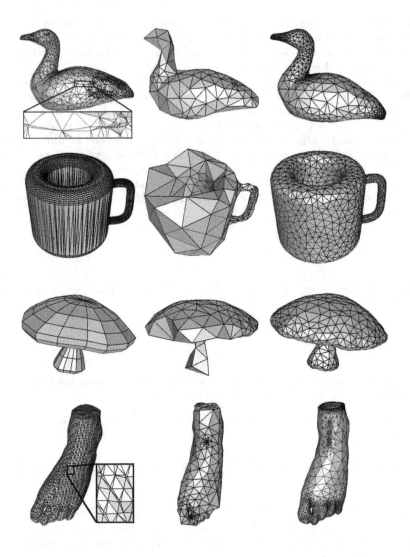

Fig. 4. The input meshes are shown in the first column. Second and third columns show the output meshes for two different levels of refinements. (a) First row shows remeshing of a reconstructed surface from a point cloud. Notice the skinny triangles present in the input. (b) Second and third rows show remeshing of a designed triangular and non-triangular surface respectively. (c) Fourth row shows remeshing of an iso-surface extracted by a marching cube algorithm from volume data. Notice that the artificial small feature created in the iso-surface is meshed densely.

[ACDL00] N. Amenta, S. Choi, T. K. Dey and N. Leekha. A simple algorithm for homeomorphic surface reconstruction. *Internat. J. Comput. Geom. Applications* **12** (2002), 125–141.

[BO03] J.-D. Boissonnat and S. Oudot. Provably good surface sampling and approximation. *Eurographics Sympos. Geom. Process.* (2003), 9–18.

[CDES01] H.-L. Cheng, T. K. Dey, H. Edelsbrunner and J. Sullivan. Dynamic skin triangulation. *Discrete Comput. Geom.* **25** (2001), 525–568.

[CDRR04] S.-W. Cheng, T. K. Dey, E. A. Ramos and T. Ray. Sampling and meshing a surface with guaranteed topology and geometry. *Proc. 20th Annu. Sympos. Comput. Geom.* (2004), 280–289.

[CDRR204] S.-W. Cheng, T. K. Dey, E. A. Ramos and T. Ray. Quality meshing for polyhedra with small angles. *Proc. 20th Annu. Sympos. Comput. Geom.* (2004), 290–299.

[Chew89] L. P. Chew. Guaranteed-quality triangular meshes. Report TR-98-983, Comput. Sci. Dept., Cornell Univ., Ithaca, New York, (1989).

[Chew93] L. P. Chew. Guaranteed-quality mesh generation for curved surfaces. *Proc. 9th Annu. ACM Sympos. Comput. Geom.*, (1993), 274–280.

[Dey03] T. K. Dey. Curve and surface reconstruction. Chapter in *Handbook on Discrete and Computational Geometry*, 2nd Edition (2004), eds. J. Goodman and J. O'Rourke, CRC press, Boca Raton, Florida.

[Dey05] T. K. Dey, G. Li and T. Ray. Polygonal surface remeshing with Delaunay refinement. Extended version, 2005. http://www.cse.ohio-state.edu/∼tamaldey/papers.html.

[ES97] H. Edelsbrunner and N. Shah. Triangulating topological spaces. *Internat. J. Comput. Geom. Appl.* **7** (1997), 365–378.

[PW04] S. Pav and N. Walkington. A robust 3D Delaunay refinement algorithm. *Proc. Intl. Meshing Roundtable* (2004).

[PB01] P. P. Pébay and T. J. Baker. Comparison of triangle quality measures. *Proc. 10th Internat. Meshing Roundtable*, Sandia National Laboratories, (2001), 327–340.

[Rup95] J. Ruppert. A Delaunay refinement algorithm for quality 2-dimensional mesh generation. *J. Algorithms*, **18**, (1995), 548–585.

[Shew98] J. R. Shewchuk. Tetrahedral mesh generation by Delaunay refinement. *Proc. 14th Annu. ACM Sympos. Comput. Geom.*, (1998), 86–95.

[Sifri] O. Sifri, A. Sheffer and C. Gotsman. Geodesic-based surface remeshing. *Proc. Internat. Meshing Roundtable* (2003).

An All-Hex Meshing Strategy for Bifurcation Geometries in Vascular Flow Simulation

Chaman Singh Verma[1], Paul F. Fischer[1], Seung E. Lee[2], and F. Loth[3]

[1] Mathematics and Computer Science Division, Argonne National Laboratory, Argonne, IL 60439
[2] Dept. of Mechanical Engineering, Massachusetts Institute of Technology, Cambridge, MA 02319
[3] Dept. of Mechanical Engineering, University of Illinois, Chicago, IL 60607

Summary. We develop an automated all-hex meshing strategy for bifurcation geometries arising in subject-specific computational hemodynamics modeling. The key components of our approach are the use of a natural coordinate system, derived from solutions to Laplace's equation, that follows the tubular vessels (arteries, veins, or grafts) and the use of a tripartitioned-based mesh topology that leads to balanced high-quality meshes in each of the branches. The method is designed for situations where the required number of hexahedral elements is relatively small ($\sim$ 1000–4000), as is the case when spectral elements are employed in simulations at transitional Reynolds numbers or when finite elements are employed in viscous dominated regimes.

1 Introduction

We develop an automated all-hex meshing strategy for bifurcation geometries arising in subject-specific computational hemodynamics modeling. Our strategy is designed for the spectral element method (SEM) but works equally well for hex-based finite volume or finite element methods. Similar to the finite element method, the SEM is a high-order weighted residual technique featuring isoparametric hexahedral (hex) elements that are globally assembled into an unstructured mesh. A distinguishing feature of spectral element meshes is that they generally require orders of magnitude fewer elements than their finite-element counterparts because each spectral element typically contains 100s to 1000s of gridpoints. This reduction often poses significant challenges for conventional approaches to automated hex-mesh generation because there are relatively few interior elements over which to apply smoothing in order to absorb topological corrections that arise from, say, merging advancing fronts. It is of interest, therefore, to construct meshes having intrinsically compatible geometries and topologies. For most hemodynamic flow domains, which primarily comprise tubes and bifurcations, there are decompositions that are readily tessellated by hexahedra using sweeping methods [LPL$^+$00, Lee02]. The challenge, however, is to develop robust, fast, and fully automated schemes for patient-specific geometries, which often feature tortuous passages with sharp curvature and (in the presence of stenoses) rapid variation in diameter. An additional complication, separate from the question of meshing is that the geometry is not well defined but must generally be inferred from slice-based medical images that are often highly pixelated with respect to the vessel diameter. Certainly, the geometry definition, which involves image registration and segmentation, is an integral part of the automated procedure, but we do not discuss it further here. We also note that, while the mesh generation techniques proposed here are designed specifically for vascular flow geometries, they could be adapted to work equally well in other internal flow configurations involving bifurcating channels.

2 Background and Flow Modeling

During the past two decades, the role of hemodynamics, or fluid mechanics of blood flow, has been implicated in the development of arterial disease and in the regulation of cellular biology in both normal and diseased arteries [FKS$^+$95, PCG93]. Vascular disease, including atherosclerosis, aneurysms, and plaque disruption is one of the leading causes of death in the United States. A number of methods are being used to investigate the hemodynamic forces in the vascular system, with computational fluid dynamics (CFD) becoming the most prevalent because of its ability to provide more detailed flow information than either in vivo or in vitro experiments. Although significant insight has been gained from CFD simulations in idealized vascular geometries, geometry clearly has a dominant influence on the local hemodynamics and there is a consequent need for subject-specific vascular flow modeling [Ku97].

While the natural flow state in the vasculature is laminar, it is possible to have a transition to a weakly turbulent state in the presence of stenoses (blockages) or subsequent to surgical procedures such as arteriovenous graft implantation [LAF$^+$03]. The transition to turbulence induces a sudden change in the range of spatial and temporal scales in the solution, resulting in a need for two to three orders of magnitude increase in computational resources for the same physical-time simulation. The flow physics in the turbulent case is dominated by convection of momentum with relatively little diffusion. The non-dimensional ratio of these two processes is denoted as the Reynolds number, which is typically $Re \sim 350$ in healthy vessels but can reach as high as $Re=1000$–3000 in the transitional cases. For simulations in the high Reynolds number regime where physical dissipation is small, it is beneficial to use high-order numerical discretizations that have minimal numerical dispersion and dissipation per grid point [DFM02].

Our numerical approach is based on the SEM, which is a high-order weighted residual technique that combines the geometric flexibility of the finite element method (FEM) with the rapid convergence and tensor-product efficiencies of global spectral methods. In the SEM, the domain is decomposed into E curvilinear hexahedral elements, and the solution within each element is represented as an Nth-order tensor-product nodal-based (Lagrange) polynomial. For three-dimensional problems, there are approximately EN^3 gridpoints in the entire domain.

We remark that high-order methods do not circumvent the need to resolve flow structures such as boundary layers and vortices. Both low- and high-order methods must capture the predominant structures. In the FEM the requisite resolution is attained by varying E, whereas in the SEM the resolution is attained through a combined variation of E and N, with $N=8$–12 being typical. For the same resolution (i.e., number of gridpoints) the number of elements for an SEM discretization would be two to three orders of magnitude smaller than its FEM counterpart. For example, typical SEM discretizations of a bifurcation geometry involve 1000–4000 elements, whereas typical FEM discretizations involve $> 10^5$ elements. Because there are so few elements, it is critical that the SEM mesh topology and geometry be relatively free of the point and line dislocations that can result when two or more mesh fronts converge when using advancing front or similar generalized meshing methods. For the FEM, achieving this topology is less of a problem because mesh smoothing can be used to effectively spread the geometric penalty arising from such topological defects over a large number of elements. For the SEM, mesh smoothing is also important but is more constrained by the (topological) proximity of the boundaries. Fortunately, for the class of geometries that we are considering here, which comprises blood vessels and bifurcations, high-quality all-hex decompositions exist, and we are able to exploit this domain-specific information to develop a robust and fully automated meshing procedure that works equally well for the SEM and, through mesh refinement, for the FEM.

3 Vessel Surface Definition

Currently, we are using noncompact radial basis functions (RBFs) to represent the vessel surface [CFB97]. We opted for noncompact RBF for the following reasons: (i) the initial point cloud was noise-free, (ii) the number of points was reasonable to be handled by ordinary Linux machines, and (iii) noncompact thin-plate RBFs provide smooth surfaces. The RBF-based implicit surface construction requires a set of known interpolating points that we construct as follows. Given surface points $\mathbf{x}_j$, $j = 1, \ldots, n$, we identify a set of external points, y_k, $k = 1, \ldots, m$, near the surface using an oct-tree algorithm that marches inward with successively smaller boxes from the edges of a bounding box. With the known oct-tree topology, we identify as external points the centers of small empty boxes adjacent to those containing surface points. From these, we define a signed distance and construct an interpolating RBF, $\phi(\mathbf{x})$, that vanishes at all $\mathbf{x}_j$ and matches the signed distance at all y_k. The surface where $\phi(\mathbf{x}) \equiv 0$ defines the vessel wall.

The radial basis function is evaluated on a set of structured points, and the $\phi = 0$ surface is then triangulated by using the marching cubes algorithm [LC87]. This surface triangulation is then smoothed by using Taubin's nonshrinking smoothing algorithm [Tau95, TZG96]. It was observed that RBF-based surface is not accurate near the end-caps. These sites are locations where artificial boundary conditions are applied in the blood-flow simulations and are thus not required to be physically correct. We therefore cut the end-caps and triangulated the surface cuts using Shewchuck's *Triangle* software [She97].

4 Conduction-Based Sweeping

To introduce the basic elements of our bifurcation meshing procedure, we consider a simpler model problem, namely, the meshing of a curvilinear tube representative of an artery or vein. The strategy is to employ a sweeping algorithm, in which one projects a templated quadrilateral O-grid onto specified cross-sections of the tube and connects corresponding vertices in adjacent cross-sections to form a hex mesh. The central element to our meshing strategy is to develop a natural coordinate system that follows the undulating vessel and satisfies two criteria:

Criterion 1: no two cross-sections intersect, and
Criterion 2: each cross-section is orthogonal to the vessel wall.

A robust approach to producing such a coordinate system is to solve Laplace's equation in the vessel with Dirichlet boundary conditions at the tube ends and homogeneous Neumann conditions along the tube wall. This corresponds to a steady thermal conduction problem with, say, temperature $\alpha = 0$ at one end, $\alpha = 1$ at the other, and $\nabla\alpha\cdot\mathbf{n} =$ along the sidewalls, where $\mathbf{n}$ is the outward pointing normal. Once α is known, selecting a set of isosurfaces for a monotonically increasing sequence of temperatures $0 = \alpha_1 < \alpha_2 < \cdots < \alpha_m = 1$ will produce a set of cross-sections that satisfy the desired non-intersecting and orthogonality conditions. These surfaces often have double curvature like a potato chip and are therefore termed *chips*. The α_is can be chosen such that chips have the desired separation in Euclidean space. A typical set of chips and a 12-element O-grid template are shown in Fig. 1.

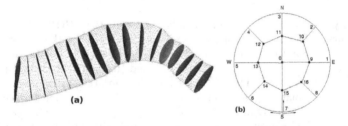

Fig. 1. (a) Isosurfaces of α in an isolated vessel. (b) Level-0 O-grid template.

We note that potential-based grid generation is a well-established technique [TWM85], particularly for external flow configurations, where analytical solutions to Laplace's equation can readily be computed. In the present case, one needs a base mesh in order to solve the conduction (i.e., potential) problem. Fortunately, Laplace's equation is very well conditioned and virtually any reasonable mesh/discretization pairing is up to the task, particularly because we are not concerned with accuracy but only with satisfying the non-intersection and orthogonality criteria. We therefore employ linear finite elements for the conduction problem. Starting with a surface triangulation, we build a tetrahedral mesh using the advancing front algorithm [Mar00]. We then construct the finite element Laplacian, apply the requisite boundary conditions, and solve the sparse linear system by using conjugate gradient iteration. The surface triangulation is smoothed by using Taubin's non-shrinking algorithm [Tau95]. With α known, the O-grid is swept through the

Table 1. Boundary Conditions for Bifurcation Conduction Problems

Problem	Γ_A	Γ_B	Γ_C
A: $\nabla^2\alpha = 0$	$0 = \nabla\alpha \cdot \mathbf{n}$	$\alpha = -\alpha_P$	$\alpha = 1 - \alpha_P$
B: $\nabla^2\beta = 0$	$\beta = 1 - \beta_P$	$0 = \nabla\beta \cdot \mathbf{n}$	$\beta = -\beta_P$
C: $\nabla^2\gamma = 0$	$\gamma = -\gamma_P$	$\gamma = 1 - \gamma_P$	$0 = \nabla\gamma \cdot \mathbf{n}$

domain and projected onto chips where $\alpha(\mathbf{x}) = \alpha_i$. The O-grid template is aligned by minimizing the Euclidian distance between the "S" points (Fig. 1b) when the template is mapped onto adjacent chips. The remaining edge points are distributed according to arclength. The interior point distribution is parameterized by using local polar coordinates, with arclength acting as the azimuthal variable, and the radial coordinate normalized by the distance from the chip centroid to the perimeter.

5 Bifurcation Geometries

Our bifurcation meshing strategy is based on a tripartition of the domain into three tubes, each of which is meshed using a variant of the sweeping algorithm described in the preceding section. This approach yields minimal elemental deformation and balanced resolution in each branch. (The initial mesh can be further improved through mesh smoothing techniques [FP00, Mun04] and the local resolution enhanced through p- or h-type refinement, e.g., [FKL02].) Our original implementation of the tripartition algorithm was based on a series of cross-sections that were orthogonal to the plane containing the bifurcation [LPL+00, Lee02]. Such an approach, however, is not robust because it does not guarantee Criteria 1 and 2. We therefore extend this tripartition approach to incorporate the natural coordinate system introduced in Section 4.

We demonstrate that, in addition to satisfying Criteria 1 and 2, the Laplacian isosurfaces provide a robust and natural tripartition of the bifurcation geometry. We begin by noting that any two of the three branches may be swept by solving Laplace's equation with $\alpha = 0$ on one end-cap, $\alpha = 1$ on another, and $\nabla\alpha \cdot \mathbf{n} = 0$ elsewhere. The third branch can be swept by cyclically permuting these boundary conditions and solving a second conduction problem. For symmetry and for completeness of the algorithm, we solve three conduction problems in total, each having, in turn, one branch insulated while the other two have end-caps at different fixed values. Denoting the branches as A, B, and C, with respective end-caps Γ_A, Γ_B, and Γ_C, we solve conduction problems for α, β, and γ satisfying the boundary conditions listed in Table 1. For each case, the Dirichlet values have been shifted by a constant (denoted with subscript P) so that the temperature at the center of the insulated end-cap is zero. The isosurfaces that (nominally) emanate from these centerpoints are referred to as the *principal isosurfaces* and, with the shift, correspond to $\alpha = 0$, $\beta = 0$, and $\gamma = 0$.

The solutions for problems A–C are shown in Fig. 2 for a stenosed carotid artery bifurcation. A critical observation is that, to the level of truncation error, the principal isosurfaces intersect in a unique curve, as illustrated in Fig. 2d. (Uniqueness is further discussed in Appendices A and B.) These isosurfaces consequently serve to tripartition the bifurcation, as required for our sweeping strategy. In addition, they provide the bisecting "plane" that is required to orient the O-grid so that element boundaries are aligned with the cusp of each branch where it connects to the bifurcation. Thus six subdomains are defined by the principal isosurfaces: three branches, each having two halves. Denoting the half of branch A that connects to B by A_B, and so on for each half, we formally identify "half-chips" in subdomains A_B and B_A with isosurfaces of γ, those in B_C and C_B with isosurfaces of α, and those in C_A and A_C with isosurfaces of β. As in Section 4, the chips defining adjacent slabs of hex elements are found by choosing α-isosurfaces (or β- or γ-) that yield a desired Euclidean separation. (Hereon, unless otherwise indicated, we will restrict our discussion to the problem of meshing the A-branch using isosurfaces of β and γ, with the understanding that the procedure for the other branches follows from symmetry.)

Refinements to the Approach

In infinite-precision arithmetic, the approach outlined above would suffice to mesh each branch in its entirety. A modest complication arises because the insulated branch assumes a nearly uniform temperature as one moves away from the bifurcation. Choosing α_P so that the temperature in the insulated branch is close to zero and then *recomputing* α provides several significant digits that would otherwise be lost and allows one to more precisely identify the principal isosurface. Nevertheless, it is often difficult to identify the principal isosurface more than two diameters away from the bifurcation, as is clear from Fig. 2. For two reasons, this situation is not a particular problem. First, we note that identifying the principal isosurface *near* the bifurcation is a very robust procedure. In exact arithmetic, all isosurfaces in the insulated branch collapse *exponentially fast* to the principal isosurface as one approaches the bifurcation. Thus, starting from any point in the insulated branch, one can use a marching tetrahedra algorithm and arrive at the principal isosurface. Second, away from the bifurcation, the principal isosurface merely serves as a guide to orient the O-grid and is not really needed. In the extremal regions, the O-grid can be mapped to either constant-β or constant-γ surfaces (for the A-branch), as the two families are indistinguishable. Thus, away from the bifurcation, we simply pick one of two solutions (e.g., β or γ) and sweep toward the extremity (A) following the procedure of Section 4.

Summary of Approach

With the basic concepts in hand, we summarize the steps of the procedure as follows:

1. Starting from a noncompact RBF representation, construct a triangulation of the vessel surface.
2. Apply Taubin smoothing [Tau95] to the surface triangulation.
3. Use the advancing front algorithm to construct a preliminary tet mesh of the domain volume.

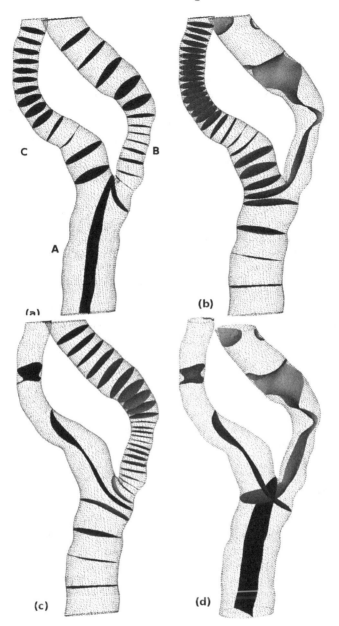

Fig. 2. Isosurfaces for conduction problems A (a), B (b), and C (c) of Table 1; (d) intersection of principal isosurfaces.

4. Using linear finite elements and conjugate gradient iteration, solve conduction problems for α, β, and γ (Table 1) with $\alpha_P = \beta_P = \gamma_P = 0$.

5. Set α_P, β_P, and γ_P to the negative of their respective insulated end-cap centerpoint values and recompute α, β, and γ. (Iterate on Step 5, if needed.)

6. Identify the $\alpha = 0$ principal isosurface with marching tets, and compute β and γ on this surface using the corresponding marching-tet interpolants.

7. Use marching triangles to identify the $\beta = 0$ and $\gamma = 0$ curves on the $\alpha = 0$ principal isosurface. Average these curves to determine a common representation of the trisection curve. Identify the centerpoint (measured in arc-length) of the trisection curve as the origin, O.

8. On the A-branch, march outward from O on the $\alpha = 0$ surface a user-specified distance δ (typ., $\delta \approx 1/4D$, where D is the local vessel diameter). Denote the corresponding point as $\mathbf{x}_1$.

9. Find the values $\beta_1 := \beta(\mathbf{x}_1)$ and $\gamma_1 := \gamma(\mathbf{x}_1)$, and compute the corresponding isosurfaces $\beta(\mathbf{x}) = \beta_1$ and $\gamma(\mathbf{x}) = \gamma_1$ with marching tets. Identify the first pair of half-chips with these isosurfaces.

10. Repeat Steps 8 and 9, generating points $\mathbf{x}_i$ and corresponding isosurfaces until the dihedral angle of the $\beta(\mathbf{x}) = \beta_i$ and $\gamma(\mathbf{x}) = \gamma_i$ chips at $\mathbf{x}_i$ is $< 10°$; then switch to the single-chip algorithm of Section 4 using either β or γ as the coordinate.

11. Starting at O, project templated O-grids onto each chip or half-chip pair.

12. Smooth the quadrilateral surface mesh using Taubin smoothing, and project the smoothed points onto the triangulated chip surface.

13. Repeat Steps 8–12 for branches B and C.

14. Consistently order each contour on each chip with respect to the principal isosurface half-chips.

15. Construct the hexahedral volume mesh taking two adjacent chips at a time.

Figure 3 shows the resulting templated chip set for the stenosed carotid bifurcation. A close up of the half-chips on the principal isosurfaces that trisect the bifurcation are shown in Fig. 3b.

6 Results

We have successfully used an earlier variant of this approach, developed in [LPL$^+$00, Lee02], to build several meshes for the study of transition in vascular flows [LFL$^+$05, LAF$^+$03], This earlier mesh construction approach involved several disjoint pieces of software and required significant user expertise to produce an acceptable mesh. Typical turnaround time was on the order of five to ten days. Our current focus is on streamlining the entire procedure so that it is fully automated and requires minimal user input.

The current development code is in C++ and uses the Matrix Template Library [SL] to solve the linear equations for the RBF and heat conduction problems. The RBF coefficients were computed by using direct solves, whereas the conduction problems were solved with Jacobi-preconditioned conjugate gradient iteration. Table 2 shows preliminary performance results on a Linux 2.4 GHz Pentium IV machine with a breakdown of the CPU time for each phase. We stress that this initial development code has not been optimized. The turnaround for the entire procedure has nonetheless been reduced to a matter of minutes, rather than days.

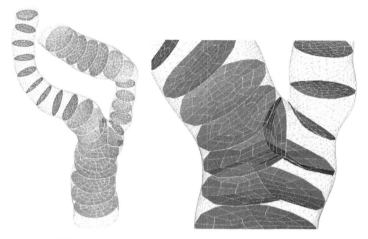

Fig. 3. Templated chips in a bifurcation geometry.

Table 2. Timing for Automated Generation of Carotid Bifurcation Mesh

Function	Time (in Seconds)
RBF coefficients evaluation	450
RBF function evaluation on box	90.0
Marching cube for $\phi = 0$ evaluation	2
Taubin surface fairing	1
Trimming end sections	0.5
Solution of three heat equations	400
Principal chips construction	1
Split chips construction	2
Simple chips construction	4
Mesh template	0.5
Surface smoothing	0.001
Volume mesh	0.0001

7 Future Directions

Our goal is to provide rapid generation of a quality hex mesh for vascular flow geometries, starting with a sliced-based stack of medical images (i.e., CT-scan or MRI). presently, we are using commercial software to segment the images. This will be replaced by a snake algorithm that is incorporated as a module in our mesh generation software. With all the modules in place, we will tackle optimization of the algorithms described in the preceding sections. In particular, we expect significant performance gains by using multigrid to precondition the thermal conduction problems. Similar performance gains are expected by improving the RBF approach. For instance, we have been investigating the use of multilevel RBFs that yield sparse coefficient matrices.

Acknowledgments

This work was supported by the National Institutes for Health, RO1 Research Project Grant (2RO1HL55296-04A2) and by the Mathematical, Information, and Computational Sciences Division subprogram of the Office of Advanced Scientific Computing Research, U.S. Department of Energy, under Contract W-31-109-Eng-38.

Appendix A

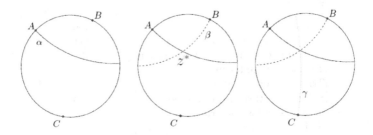

Fig. 4. Two-dimensional model problem showing common intersection point z^* for isopotentials α, β, and γ.

Our assertion that the three principal isosurfaces intersect in a unique one-dimensional curve is based primarily on observation of tens of two- and three-dimensional cases. Here, we demonstrate that uniqueness holds rigorously for a special class of two-dimensional geometries, namely, for an arbitrary distribution of source, sink, and isocontour points on a circle.

Consider the two-dimensional potential problem in the unit-disk having a point source at B and a point sink of equal strength at C, as illustrated in Fig. 4. The general solution in the complex z-plane is

$$u + iv = \ln \frac{z - B}{z - C} + c,$$

with u and v real-valued, $i := \sqrt{-1}$, and c arbitrary. The isopotentials (corresponding to isothermal surfaces in Fig. 2) are given by isocontours of the real part of the solution, u, and in this case are circular arcs. The isosurface passing through A is given by

$$\alpha := \left\{ z : \left| \frac{z - B}{z - C} \right| = \left| \frac{A - B}{A - C} \right| \right\}. \tag{1}$$

Cyclically permuting the role of A, B, C yields a new problem with isocontour β passing through B given by

$$\beta := \left\{ z : \left| \frac{z - C}{z - A} \right| = \left| \frac{B - C}{B - A} \right| \right\}. \tag{2}$$

The intersection of α and β is given by the point z^* that simultaneously satisfies (1) and (2). Inserting z^*, multiplying (1) and (2), and inverting the result shows that z^* also satisfies

$$\left| \frac{z^* - A}{z^* - B} \right| = \left| \frac{C - A}{C - B} \right|, \tag{3}$$

which implies that z^* is also an element of γ, the third principal isocontour obtained by again permuting the roles of A, B, and C. Thus, the intersection of the principal isocontours α, β, and γ is unique.

Note that this class of solutions can be immediately extended to any domain Ω that whose boundaries are given by isocontours of v. Such boundaries are orthogonal to the isopotential lines and an arbitrary choice of these boundaries does not affect the solution u. Thus, the result is immediately applicable to football and crescent shaped domains. Numerical evidence suggests that the result holds for any simply connected domain in $\mathbb{R}^2$, but we've yet to rigorously establish this generalization.

Appendix B

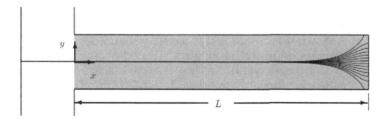

Fig. 5. Two-dimensional model problem illustrating exponential convergence of end-cap isosurfaces to the principal isosurface.

The model problem of Appendix A employed idealized point sources and sinks to establish the unique intersection point of the three principal isosurfaces. In our meshing application, we have set the entire end-caps to be either a fixed temperature or insulated. Strictly speaking, we can no longer expect uniqueness of the intersection point because there is an infinity of isosurfaces emanating from the insulated end-cap. The following example, however, illustrates that these isosurfaces collapse exponentially fast to the principal isosurface. Consider the computational domain Ω with $(x, y) \in [0, L] \times [-\frac{1}{2}, \frac{1}{2}]$ indicated by the gray region in Fig. 5, which is a two-dimensional model of an insulated branch having unit diameter D. To first order, we can approximate the potential at the left edge of the domain as $T(0, y) = \sin \pi y$, for which the solution is $T = \sin \pi y \cosh \pi (x - L) / \cosh \pi L$. An isosurface emanating from the right at (L, y_0) is given by

$$ y = \sin^{-1} \left[\frac{\sin \pi y_0}{\cosh \pi (L - x)} \right], $$

which establishes the exponential convergence to the principal isosurface. Several such isosurfaces are shown in the figure. As an example, if $L = 5D$, then the maximum separation of any isosurface pair emanating from the end-cap is $< 5.e - 7$ at the bifurcation point.

[CFB97]   J.C. Carr, W.R. Fright, and R.K. Beatson. Surface interpolation with radial basis functions for medical imaging. *IEEE Trans Med. Imaging*, 16:96–107, 1997.

[DFM02]   M.O. Deville, P.F. Fischer, and E.H. Mund. *High-order methods for incompressible fluid flow*. Cambridge University Press, Cambridge, 2002.

[FKL02]   P.F. Fischer, G.W. Kruse, and F. Loth. Spectral element methods for transitional flows in complex geometries. *J. Sci. Comput.*, 17:81–98, 2002.

[FKS⁺95]   M. H. Friedman, B. D. Kuban, P. Schmalbrock, K. Smith, and T. Altan. Fabrication of vascular replicas from magnetic resonance images. *J. Biomech. Eng.*, 117:364–366, 1995.

[FP00]   L. Freitag and P. Plassmann. Local optimization-based simplical mesh untangling and improvement. *Int. J. Numer. Methods Eng.*, 48:109–125, 2000.

[Ku97]   D. N. Ku. Blood flow in arteries. *Annu. Rev. Fluid Mech.*, 29:399–434, 1997.

[LAF⁺03]   F. Loth, N. Arslan, P. F. Fischer, C. D. Bertram, S. E. Lee, T. J. Royston, R. H. Song, W. E. Shaalan, and H. S. Bassiouny. Transitional flow at the venous anastomosis of an arteriovenous graft: Potential relationship with activation of the erk1/2 mechanotransduction pathway. *ASME J. Biomech. Engr.*, 125:49–61, 2003.

[LC87]   W.E. Lorensen and H. E. Cline. Marching cubes: A high resolution 3d surface construction algorithm. In *SIGGRAPH '87: Proceedings of the 14th annual conference on Computer graphics and interactive techniques*, pages 163–169, New York, NY, USA, 1987. ACM Press.

[Lee02]   S.E. Lee. Solution method for transitional flow in a vascular bifurcation based on in vivo medical images. Master's thesis, Univ. of Illinois, Chicago, 2002. Dept. of Mechanical Engineering.

[LFL⁺05]   S.W. Lee, P.F. Fischer, F. Loth, T.J. Royston, J.K. Grogan, and H.S. Bassiouny. Flow-induced vein-wall vibration in an arteriovenous graft. *ASME J. Fluid-Structures (to appear)*, 2005.

[LPL⁺00]   S.E. Lee, N. Piersol, F. Loth, P. Fischer, G. Leaf, Smith B., Yedevalli R., A. Yardimci, N. Alperin, and L. Schwartz. Automated mesh generation of an arterial bifurcation based upon in vivo mr images. *Proceedings of the 2000 World Congress on Medical Physics and Bioengineering*, CD ROM, 2000.

[Mar00]   D.L. Marcum. Efficient generation of high quality unstructured surface and volume grids. October 2-5,2000.

[Mun04]   T. S. Munson. Mesh shape-quality optimization using the inverse mean-ratio metric. Preprint ANL/MCS-P1136-0304, Argonne National Laboratory, Argonne, Illinois, 2004.

[MW95]   D. L. Marcum and N. P. Weatherill. Unstructured grid generation using iterative point insertion and local reconnction. *AIAA Journal, AIAA,,* 33, Num 9:1619–1625, September 1995.

[PCG93]   Giddens D. P., C.K.Zarins, and S. Glagov. The role of fluid mechanics in the localization and detection of atherosclerosis. *J. Biomech. Eng.,* 115:588–594, 1993.

[She97]   J. Shewchuk. *Delaunay Refinement Mesh Generation.* PhD thesis, School of Computer Science, Carnegie Mellon University, Pittsburgh, Pennsylvania, May 1997. Available as Technical Report CMU-CS-97-137.

[SL]   J. Siek and A. Lumsdaine. Generic programming for high performance numerical linear algebra. http://www.osl.iu.edu/research/mtl.

[Tau95]   G. Taubin. A signal processing approach to fair surface design. In *SIGGRAPH '95: Proceedings of the 22nd annual conference on Computer graphics and interactive techniques*, pages 351–358, New York, NY, USA, 1995. ACM Press.

[TWM85]   J.E. Thompson, Z.U.A. Warsi, and C. W. Mastin. *Numerical Grid Generation.* Elsevier Science Publisging Co., 1985.

[TZG96]   G. Taubin, T. Zhang, and G. H. Golub. Optimal surface smoothing as filter design. In *ECCV '96: Proceedings of the 4th European Conference on Computer Vision-Volume I*, pages 283–292, London, UK, 1996. Springer-Verlag.

AN INTERIOR SURFACE GENERATION METHOD FOR ALL-HEXAHEDRAL MESHING

Tatsuhiko Suzuki[1], Shigeo Takahashi[2], Jason Shepherd[3, 4]

[1] Digital Process Ltd., 2-9-6, Nakacho, Atsugi City, Kanagawa, Japan.
tsuzuki@dipro.co.jp
[2] The University of Tokyo, 5-1-5 Kashiwanoha, Kashiwa-city, Chiba, Japan.
takahashis@acm.org
[3] University of Utah, Scientific Computing and Imaging Institute, SLC, Utah,
USA. jfsheph@sci.utah.edu   [4]Sandia National Lab., Computational Modeling
Sciences, Albuquerque, NM, USA. jfsheph@sandia.gov

ABSTRACT

This paper describes an interior surface generation method and a strategy for all-hexahedral mesh generation. It is well known that a solid homeomorphic to a ball with even number of quadrilaterals bounding the surface should be able to be partitioned into a compatible hex mesh, where each associated hex element corresponds to the intersection point of three interior surfaces. However, no practical interior surface generation method has been revealed yet for generating hexahedral meshes of quadrilateral-bounded volumes. We have deduced that a simple interior surface with at most one pair of self-intersecting points can be generated as an orientable regular homotopy, or more definitively a sweep, if the self-intersecting point types are identical, while the surface can be generated as a non-orientable one (i.e. a Möbius band) if the self-intersecting point types are distinct. A complex interior surface can be composed of simple interior surfaces generated sequentially from adjacent circuits, i.e. non-self-intersecting partial dual cycles partitioned at a self-intersecting point. We demonstrate an arrangement of interior surfaces for Schneiders' open problem, and show that for our interior surface arrangement Schneiders' pyramid can be filled with 146 hexahedral elements. We also discuss a possible strategy for practical hexahedral mesh generation.
Keywords: all-hexahedral mesh generation, interior surface arrangement, Schneiders' pyramid

1. INTRODUCTION

Product development in every industrial field is in a severe competitive phase, and FEM analysis plays an important role in product development. It is known that, in many FEM applications, hexahedral meshes give better and more effective results than tetrahedral ones. However, due to the difficulty in generating hexahedral meshes, industrial interests are shifting to hex-dominant meshes [Ow-99][Ow-01], even though all-hex meshes are still demanded because they give better results than hex-dominant ones in FEM analysis.

It is desirable to generate an all-hexahedral mesh for a model of arbitrary topological type, rather than generating a sweep-type hexahedral mesh, as is more commonly done. However, a generalized algorithm for creating such meshes does not currently exist. Techniques such as plastering [BM-93], whisker weaving [TM-95][TMB-96][FM-98], and dual cycle elimination [Mh-98] have been developed, in attempt to realize such an algorithm. Unfortunately, none of them are considered to be reliable in heavy practical use, because they can handle only a limited class of solids, and have no guarantees on the quality of the resulting meshes.

In 1995, Mitchell proposed a *hexahedral mesh existence theorem* based on an arrangement of dual-cycle-extending surfaces in the interior of solids [Th-93][Mi-95][Ep-96]. This theorem states that any simply-connected three-dimensional domain, with an even number of quadrilateral boundary faces, can be partitioned into a hexahedral mesh respecting the boundary.

However, no practical interior surface generation method has been revealed yet, and despite the proof we (and many others) have encountered several severe deadlocks in creating the topology of hexahedral mesh that conforms to a given quadrilateral boundary mesh. In this paper we present an interior surface classification theory, along with a method, that could be utilized to create actual interior surfaces, with some discussion as to how these surfaces represent a hexahedral mesh. As our later examples show, even if a topological solution exists, it may not always be an acceptable solution. Therefore, we will also discuss strategies to avoid inappropriate quadrilateral meshes when generating a hex mesh for practical analysis.

In this paper we focus on interior surfaces' properties, and leave the representation of interior surface arrangement as a future problem. The remainder of this paper is organized as follows. In Section 2, we review interior surfaces and their arrangements, and deduce the requirements for these interior surfaces for sound hexahedral mesh generation. In section 3, the classification of self-intersecting point types is discussed. In section 4 we describe a method for generating simple interior surfaces that have at most one pair of self-intersecting points. Section 5 contains a discussion of a method for composing more general interior surfaces using the simple ones. In section 6, we demonstrate a dual space solution of Schneiders' open problem based on the methods described in the previous sections. In Section 7, we conclude with a brief discussion of some future developments needed in order to generate acceptable hex meshes.

2. INTERIOR SURFACE ARRANGEMENT

2.1 Dual representation

In this paper we propose a technique using the notion of an *interior surface arrangement* [Mi-95] following the developers of the whisker weaving technique. For simplicity, we discuss hex meshing of solids homeomorphic to a ball, hereafter.

We denote a set of vertices, edges, quads and hexes as $V=\{v\}$, $E=\{e\}$, $Q=\{q\}$ and $H=\{h\}$, respectively.

For a planar graph G of a quad mesh M_Q, a *dual graph* $G^*=G^*(V^*,E^*)$ is composed as follows. A *dual vertex* v^* is placed in each quad q, and a *dual edge* e^* is placed in each edge e incident to the adjacent two dual vertices. A vertex v of a *primal graph* G is represented as a *dual face* q^* surrounded by dual edges in dual graph G^*. An edge e of a primal graph corresponds to a dual edge e^* (Fig. 1a). A sequence of dual edges connecting opposite edges in a quad (Fig. 1 b) is always closed, since the opposite edge of a quad is uniquely determined and the number of edges is finite. Thus the sequence of dual edge is called a *dual cycle* [Mh-98] (Fig. 1b). A dual cycle may self-intersect. A dual vertex is the intersection of two (local) dual cycles (Fig. 1a).

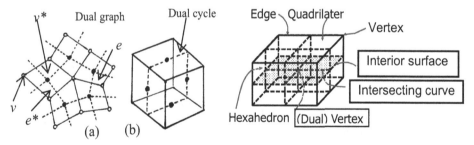

Fig. 1. (a) A Dual graph and (b) a dual cycle   **Fig. 2.** Dual representation of a hex mesh

Similarly for a graph $G=G(V,E)$ of a hex mesh M_H, a dual graph $G^*=G^*(V^*,E^*)$ is constructed as follows. A dual vertex v^* is placed in each hex h, and a dual edge e^* is placed through each quad q incident to the adjacent two hexes. A hex h, quad q, edge e and vertex v of a primal graph G corresponds respectively to a dual vertex v^*, dual edge e^*, a dual face q^*, and dual polygon h^* enclosed by dual faces in dual graph G^*. A topological representation by a dual vertex $v*$, dual edge $e*$, *dual face $q*$*, and dual polygon $h*$ is called a *dual representation M_H** of a hex mesh M_H.

A layer of hexahedra corresponds to an *interior surface* [Ep-96]. A line, or column, of hexahedra corresponds to the intersection of two interior surfaces. A hexahedron is the dual to a vertex at the intersection of three interior surfaces in the dual representation of the hex mesh (Fig. 2).

2.2 Hex mesh existence theorem

The hexahedral mesh existence theorem is described as follows [Ep-96]:

Any simply connected three-dimensional domain with an even number of quadrilateral bounding faces can be partitioned into a hexahedral mesh respecting the boundary.

In an all-hex mesh, quads on the surface are always coupled with another quad on the surface. Therefore, a necessary condition for an all-hex mesh is that the surface be covered with even number of quads. Thus it can be shown that the number of self-intersections of dual cycles must also be even, since the intersection of distinct two dual cycles makes a pair of quads.

The proof steps of the hex mesh existent theorem are as follows [Mi-95]:
1. The surface mesh of the object is mapped onto a sphere preserving the quadrilateral mesh connectivity,
2. The quadrilateral mesh on the spherical surface forms an arrangement of dual cycles,
3. The arrangement of dual cycles is extended to an arrangement of 2D manifolds through the interior of the ball, using the theorem on "regular curve on Riemannian manifold" based on the homotopy theory [Sm-58].
4. Additional manifolds that are closed, and completely within the solid are inserted, if necessary,
5. The arrangement of 2D manifolds is dualized back to induce a hexahedral mesh.

The hex mesh existence theorem, however, addresses only the existence and does not describe how to extend dual cycles to the interior surfaces in the above-mentioned step 3. This paper attempts to propose a method to create an arrangement of interior surfaces from dual cycles, and provide a novel guideline to generate all hex meshes that attempts to use not only topological information but also geometric information. We give a solution for Schneiders' open problem as an example of utilizing an arrangement of interior surfaces.

2.3 Requirements for interior surfaces

2.3.1 Boundary

An interior surface can be bounded by one or more dual cycles, or the interior surface can also be entirely closed with no dual cycle boundary.

2.3.2 Regularity

For a collection of continuous mappings, $F(u,v)=f_v(u)$, $u \in I=[0,1]$ is called homotopy if $F(u,v)=f_v(u)$ is continuous. A parametric *regular homotopy, F,* is a homotopy where at every parameter location F is a regular curve ($F(u,v)=f_v(u)$, $u \in I$), and keeps end points and direction fixed and such that tangent vector moves continuously with the homotopy [Sm-58]. Regularity is a necessary condition to create a hex element at any point on the interior surface, and we will endeavor to create interior surfaces based on regular homotopies. If there is a *singular point* in a 3D space, a hex element layer cannot be formed, and the resulting topological structure of the hex mesh is not valid.

This can be illustrated by a simple example. If two circles C_1 and C_2, which are oriented to different directions: one is clockwise and another is counter-clockwise, on two parallel planes π_1 and π_2 are given, then a non-orientable homotopy H formed between the planes π_1 and π_2 cannot be regular (Fig. 3a), because there exists a self-intersecting curve L with the end points P_1 and P_2 on the homotopy H, and the end points P_1 and P_2 are singular. The self-intersecting curve L and an arbitrary surface ω will form a dual vertex v^*, at the intersection point. The singular points P_1 and P_2 cannot be dual vertices. Thus, there exists at least one dual vertex connected to the singular points P_1 and P_2 with a self-intersecting curve. Therefore, one intersecting curve from a dual vertex is not connected with any other vertex (Fig. 3a), resulting in an invalid topology of a hex mesh.

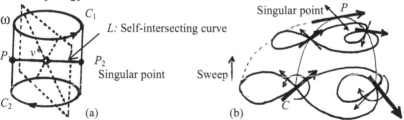

Fig. 3. A singular point makes the mesh topologically invalid.

Figure 3b depicts an orientable surface created by sweeping the curve C. The top point P is a singular point, where the tangent becomes discontinuous.

Assume that the point P is a triple point and, thus, is a dual vertex. The number of the dual edges emanating from the dual vertex P is 2, which is invalid.

2.3.3 Self-intersecting point pair connectivity

Two self-intersecting points on the boundary of an interior surface should always be paired, and connected by a self-intersecting curve. This feature is called *self-intersecting point-pair connectivity*.

Regular homotopies are compatible with self-intersecting point-pair connectivity. Some self-intersecting curves of regular homotopies, however, have complicated structure. For example, the self-intersecting curve of Fig. 4 disappears from the scope at the arrow, continuing to the 'back'. The sectional transition of the regular homotopy on the left of Fig. 4 is illustrated on the right in Fig. 4, where the self-intersecting curve is represented by a dotted line. The advancing direction changes several times. Such an interior surface will undoubtedly reduce the quality of the hex mesh, especially when the radius of curvature is small. These kinds of homotopies should be avoided.

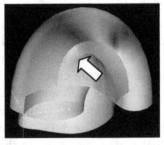 

Fig. 3. A self-intersecting curve of a regular homotopy goes out of the view.

2.4 Requirement for interior surface arrangement

2.4.1 Convexity

Hex elements used in FEM must be convex, and the hexes shown in Fig. 5 are not allowed. A hex mesh is *topologically convex*, if any two quads are not incident to common edges (Fig. 5). The condition that a hex mesh M_H is convex can be described by showing the dual graph $G_F^*(F^*, E^*)$ between the dual face set F^* and the dual edge set E^* of M_H is simple. In Fig. 6 examples of convexity-lost elements in a dual space, where two dual faces f_1^*, f_2^* incident to two common dual edges e_1^*, e_2^* are illustrated.

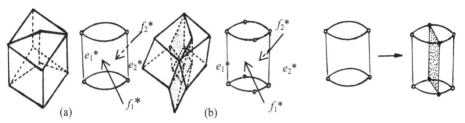

Fig. 4. Non-convex hexes   **Fig. 5.** Convexity recovery

Note that the simplicity of a dual graph $G_V^*(V^*,E^*)$ between the dual vertex set V^* and the dual edge set E^* of M_H does not always ensure the simplicity of the dual graph $G_F^*(F^*,E^*)$, where simple means that a graph has no self-loops or multi-edges. A convexity recovery operation is depicted in Fig. 6. It should also be noted that an all-hex mesh bounded by a convex quad mesh is not always topologically convex.

2.4.2 Hex element existence and connectivity

There must be at least one dual vertex on an intersecting curve between interior surfaces. Any two hex elements in a hex mesh must be connected through quads. In other words a dual graph of a hex mesh must be connected.

3. SELF-INTERSECTING POINT TYPE

Two curves with even number of self-intersecting points on a sphere are topologically deformable (Fig. 7). However, the two surfaces bounded by a curve with even number of self-intersecting points (for example in Fig. 7 the surface bounded by the first is orientable and one by the last curve is non-orientable) are not always deformable into each other. In this section the identity of self-intersecting point types are discussed in order to classify interior surface generation methods. Not only orientable surfaces but also non-orientable[1] surfaces such as a Möbius band can be interior surfaces [SZ-03], and in actuality, the identity of self-intersecting point types is closely related to orientability of interior surfaces.

---

[1] While orientability is basically defined for only a closed surface, we can inherit and consequently define the 'orientability' of an open surface by gluing topological disks along the circles that bound the open surface to refer to the orientability of its corresponding closed surface.

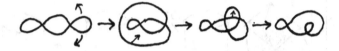

Fig. 6. Two curves with even number of self-intersecting points are deformable.

The vertices and edges of a convex polyhedron can be mapped onto a plane through a face, which is called a *window* (Fig.8), and can be transformed into a plane graph G and its dual graph G^*.

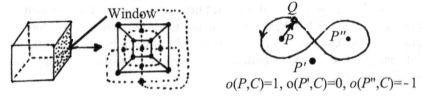

$o(P,C)=1, o(P',C)=0, o(P'',C)=-1$

Fig. 7. Mapping to a plane graph   **Fig. 8.** Order of a point for a curve

Suppose that a plane π contains a closed curve C and a point P as shown in Fig. 9. The *order o(P,C)* of the point P for the curve C is defined as the (signed) number of the vector PQ's rotation around the point P along the curve C in the predefined direction by the parameter (Fig. 9). A dual cycle mapped on a plane divides the plane into several regions. There are 4 regions around a self-intersecting point partitioned by the curve, and the order of any point in a region is identical.

Fig. 10 and Fig. 11 show dual cycles with a pair of self-intersecting points and the orders of the regions around the self-intersecting points. The orders of the points in the regions around a self-intersecting point can be represented as (i, $i+1$, $i+2$, $i+1$), where i is called the *minimum order*. If the minimum orders of the paired self-intersecting points are equal, then the *self-intersecting point-type of the (simple) dual cycle* is said to be identical (Fig. 10), otherwise distinct (Fig. 11).

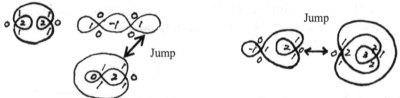

Fig. 9. Identical self-intersecting point-type   **Fig. 10.** Distinct self-intersecting point-type

There are three important points to note. A portion of a dual cycle can "jump" over the two self-intersecting points without crossing over it on the surface (see

Fig. 10,11). This does not change the self-intersecting point-type. Secondly, a dual edge "jumps" if the window crosses over the edge composing the dual cycle. This also does not change the self-intersecting point-type. Thirdly, the self-intersecting point-type is changed if an edge crosses over one of the self-intersecting points (see Fig. 12).

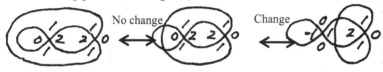

Fig. 11. Edge crossing causes a point-type change.

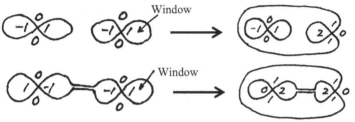

Fig. 12. Two dual cycles bounding an identical interior surface should be connected.

Two dual cycles, having a self-intersecting point respectively bounding an identical interior surface, should be connected in order to determine the self-intersecting point-type. This is due to the fact that the point-type is determined for the dual cycles respectively, and thus if the selection of the window is not appropriate, potentially erroneous results may be obtained, because an inappropriate window may "jump" only one dual cycle. A correct result will be obtained if the two dual cycles are connected into one (Fig. 13).

4. SIMPLE INTERIOR SURFACES

In this section we will describe a technique to create "simple" interior surfaces from the dual cycles containing, at most, a single pair of self-intersecting points. In the next section we will describe a technique to generate interior surfaces of two or more pairs of self-intersecting points by sequentially connecting the "simple" interior surfaces described in this section.

4.1 Simple interior surfaces bounded by a single dual cycle

In this subsection, we describe a technique to create a *simple interior surface* bounded by a single dual cycle. We define an interior surface as 'simple' if its bounding dual cycle(s) has (have) at most a single pair of self-intersecting points.

If the boundary of a simple interior surface is comprised of a single dual cycle, then the number of the self-intersecting points of the dual cycle has to be 0 or 2. The self-intersecting point-type is classified according to the following criteria into the following (a), (b), or (c).

(a) No self-intersection,
(b) Identical self-intersecting point type, or
(c) Different self-intersecting point type.

Based on Smale's 1958 theorem [Sm-58], Mitchell pointed out [Mi-95] that there exists a regular homotopy between a closed curve with even number of self-intersecting points and a closed curve without a self-intersection (Fig. 21).

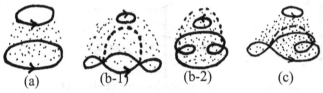

(a)   (b-1)   (b-2)   (c)

Fig. 13. Interior surface by Mitchell's proof

From an implementation standpoint, however, even if the existence of an interior surface might be proven, it is not always appropriate to create a homotopy using the method described in the proof. For example our ability to control the shape of the homotopy shown in Fig. 17 is poorer than sweep-based method. Therefore, we will consider other implementation techniques to create simple interior surfaces bounded by a single dual cycle.

However, before we consider methods for generating the interior surfaces, let us first consider some helpful constraints that must, or should, be satisfied in the creation of interior surfaces with respect to hexahedral mesh generation:

1. The interior surface bounded by a dual cycle should be nearly orthogonal to the solid surface at the boundary,
2. The interior surface should be smooth,
3. The interior surface must be completely enclosed within the solid,
4. The self-intersecting curve connecting a pair of self-intersecting points should be smooth. It is also desirable that this curve will not alternate direction multiple times (e.g. Fig. 4 is not desirable).
5. If the interior surface self-intersects, the angle of self-intersection should be nearly a right angle along the self-intersecting curve.
6. No self-intersections within the interior surface are allowed (with the exception of the possible self-intersecting curve connecting a pair of self-intersecting points).

7. The absolute value of the radius of surface curvature should be as large as possible.

Following the classification of the self-intersecting point type identity for the simple interior surface (i.e. (a), (b), or (c) from Fig. 14), we will study the implementation of generating interior surfaces to satisfy the above-mentioned requirements.

(a) Case of no self-intersecting points

First, we will define some helpful terminology. If the ruled surface created by connecting the barycenter of a dual cycle and every point of the dual cycle forms a nearly-planar surface, then the dual cycle is defined to be *planar*. Otherwise, it is defined to be *hemispherical* (Fig. 15).

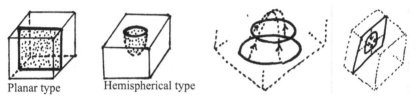

Planar type   Hemispherical type

Fig. 14. Planar type and hemispherical one   **Fig. 15.** Advancing closed curve

If the dual cycle is located along a closed sequence of sharp edges, a surface that is parallel to the surface of the solid is selected as the interior surface [Mh-98].

Otherwise, we can use an "*advancing closed curve method*" that repeats advancing the closed curve to the nearly normal direction of the surface, adjusting it, and finally filling the disk when the closed curve becomes small enough (Fig. 16). This method can be applied to both the planar and the hemispherical types.

(b) Case of dual cycle with identical self-intersecting point-type

For dual cycles with identical types of self-intersecting points, the homotopies used in Mitchell's proof of hex mesh existence are illustrated in Fig. 17. It is difficult to control the shapes of these homotopies, so, here we propose another form of homotopy, for which it is easier to control the shape.

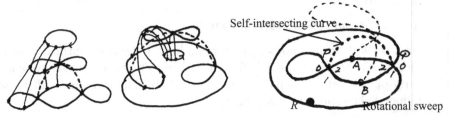

Fig. 16. Homotopy to a self-intersection free loop   **Fig. 17.** Sweep type homotopy

If the points P and Q, have identical self-intersecting point types as shown in Fig. 18, then their orders are symmetric with respect to a curve passing through the points A and B on the two *paths* connecting the self-intersecting points (as shown in Fig. 18). Since the interior surface can be represented as a *sweep* of a sectional curve whose end points lie on the two paths, the corresponding swept surface is orientable. A special type of surface is a *rotational sweep* whose rotational axis passes through the points A and B (*pivots*) on the two paths.

Note that if a dual cycle is simple and the self-intersecting point-type is identical, the two points on a dual cycle cannot always be connected in an arc-wise manner without intersecting the self-intersecting curve. For example, in Fig. 18 the points A and B; R and A; R and B cannot be connected as above-mentioned manner respectively.

(c) Case of distinct self-intersecting point type

Finally, we consider the difficult case of an interior surface bounded by a curve with distinct types of self-intersecting points (Fig. 14c). Using a Möbius band instead of an orientable surface, as depicted in Fig. 19, which is homeomorphic to Fig. 14c, we can obtain an interior surface that is a regular homotopy when it is embedded in a 3D space [Fr-87]. Fig. 20 demonstrates a method to create a Möbius band, which is expressed as a regular homotopy such that the points P_0, P_1 are $f_0(0)=f_1(1)$; $f_1(0)=f_0(1)$ respectively.

For this case, it is not easy to satisfy the interior surface generation requirements by a simple method where only the end points and tangential vectors are supplied. Our experience has shown that it is quite difficult to obtain a Möbius band with such a neat transition diagram shown in Fig. 21 by a simple method automatically, as can be seen in Fig. 4, where the self-intersecting curve changes the vertical direction 3 times. Utilizing the sectional transition from the left shown in Fig. 21, however, we can reduce the number of vertical direction changes to 1.

Even though there are, indeed, topological solutions, non-orientable interior surfaces, which contains Möbius bands, should be avoided as much as possible.

Note that if a dual cycle is simple and the intersecting point types are not identical, the two points on a dual cycle cannot always be connected in an arc-wise manner without intersecting the self-intersecting curve.

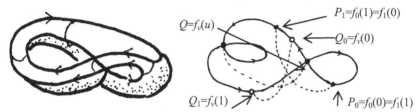

Fig. 18. Möbius band   **Fig. 19.** Möbius band generation

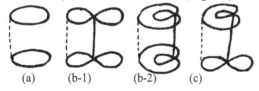

Fig. 20. Sectional transition diagram of a Möbius band to suppress the directional alternation

4.2 Simple interior surfaces bounded by two dual cycles

We will now discuss a method of generating simple interior surfaces bounded by two dual cycles, where the number of self-intersections is 0 or 2.

If an interior surface is bounded by two dual cycles, then they must also be connected by the interior surface (Fig. 22).

(a)   (b-1)   (b-2)   (c)

Fig. 21. Connected Dual Cycles

Consequently, the two dual cycles can be connected into one using a smooth transition. (This corresponds to the statement in Section 3 that two dual cycles with a self-intersecting point respectively bounding an interior surface should be connected prior to determining the type of the self-intersecting point.) The resulting types of self-intersecting points for generating a surface bounded by two dual cycles are the same as those in Fig. 10 and 11 of the previous section. Once the two dual cycles are converted to a closed curve, a regular homotopy can be obtained by the method described in the subsection 4.1.

5. GENERAL INTERIOR SURFACE GENERATION

We can now propose a technique to "constructively" compose a general interior surface whose dual-cycle is bounded by two or more pairs of self-intersecting points. This will be accomplished by connecting simple surfaces with at most one pair of self-intersecting points successively to form the more complex interior surfaces.

5.1 Simple interior surface decomposition

In order to create an interior surface by connecting simple interior surfaces, it is necessary to identify the pairs of self-intersecting points. In this subsection we discuss the problem of determining self-intersecting point-type combinations.

5.1.1 Circuit, triple-circuit, and basic interior surface

We define a *circuit* to be a partial dual cycle split at a self-intersecting point until it does not contain any other self-intersections. We call the splitting point a *base point* (Fig. 23). (Note that intersections between different circuits are allowed, and that the selections of circuits are not unique.) Removing specified circuits, new circuits can be specified recursively (Fig. 24).

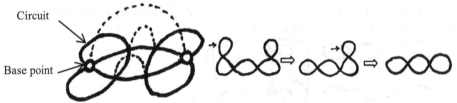

Circuit

Base point

Fig. 22. Circuit and base points   **Fig. 23.** Recursive specification of circuits

The simple interior surfaces described in the previous section have, at most, two self-intersecting points. Therefore, the interior surface generation method in the previous section will work for dual cycles with at most two base points. Let's call the tuple of three connected circuits a *triple-circuit* (Fig. 25). A tuple may have more then two self-intersecting points, while a simple interior surface has at most two.

We can extend a simple interior surface to a *basic interior surface* (Fig. 25) that is created by a triple-circuit, where the self-intersecting curve connecting the two base points is called the *basic self-intersecting curve* (Fig. 25). A basic interior surface may be either an "orientable" or a "non-orientable" type.

Because the triple-circuit is not always determined uniquely (see Fig. 26), the interior surface of the dual cycle shown in Fig. 26 might be orientable (left) or non-orientable (right) depending on the self-intersecting point type and the base-point selected. If the selected base points types are identical it will be orientable, otherwise it will be non-orientable.

Two circuits are considered co*incident* if they share a base point. If the circuit C_1 and C_2 are coincident, and C_2 and C_3 are coincident, then we call the two circuits C_1 and C_3 *adjacent*. For example in Fig. 25 the circuits L and R are called adjacent.

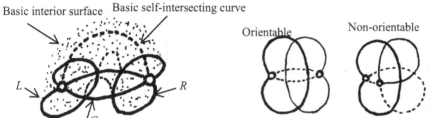

Fig. 24. Triple-circuit   Fig. 25. Selection of a triple-circuit is not unique.

5.1.2 Interior Surface creation

We will call the operation to map a triple-circuit to a basic interior surface a *"basic interior surface creation"*. Any interior surface can be created by iterating through successive basic interior surface creation operations. In other words, a basic interior surface can be created for the triple-circuit formed by cutting off existing basic interior surfaces, where we let the two circuits be coincident at the base points P and Q, by the on-surface curves connecting the two base points P and Q (Fig. 27). The on-surface curves connecting the two base points P and Q cannot intersect with existing basic self-intersecting curves (Fig. 28), because at the intersecting point four local interior surfaces meet together, which will not result in a valid hexahedral mesh.

Fig. 26. Basic interior surface creation   **Fig. 27.** Two basic self-intersecting curves cannot intersect each other.

5.2 Other self-intersecting curves

When simple interior-surface connections have been completed and the self-intersecting curves between a pair of base points have been created, it may be necessary to create other self-intersecting curves. The shape of interior surface is fixed by the creation of the other self-intersecting curves and then by dividing the faces as shown in Fig. 29.

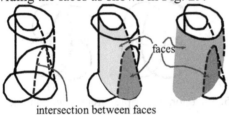

intersection between faces

Fig. 28. Intersection between basic interior surfaces

6. SOLUTION OF SCHNEIDERS' OPEN PROBLEM

Schneiders presents [Sch-www] a problem regarding whether, or not, there exists a hexahedral mesh whose boundary exactly matches a pyramid with a prescribed surface mesh as shown in Fig. 30, and hereafter called *"Schneiders' pyramid"*. (Schneiders also presents the problem of whether, or not, there exists a mesh of hexahedral elements whose boundary matches a pre-specified mesh of quadrilateral faces. Hereafter, we'll refer to this second problem as *'Schneiders' general open problem'*.) Though, several solutions have been published for Schneiders' pyramid (see [Ca-www] [YS-01]), we will attempt to solve the first problem by utilizing our "interior surface direct arrangement technique" by creating the interior surfaces directly in a dual space.

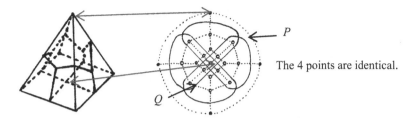

The 4 points are identical.

Fig. 29. Schneiders' pyramid   **Fig. 30.** Triple-circuit for Schneiders' pyramid

The dual cycles of Schneiders' pyramid are shown in Fig. 31, and we let the points P and Q be selected as the base points of the tri-circuit for the dual cycle depicted with solid line. (Note that this is the same as the dual cycle depicted in Fig. 26.) The base points are selected such that the self-intersecting point-types are identical so that we can obtain an orientable surface. Then, the interior surface can be created as a rotational sweep.

Let the sweep-section rotate from the base point P to the base point Q, and let the points A and B be the pivots in Fig. 32. To obtain a regular homotopy, the sweep-vector is oriented upward at the starting point (left half), horizontally at the intermediate point (center; red), and downward at the end point (right half). Therefore, the trajectory of the point X on the sweep-section becomes a curve with a loop in blue. Fig. 33 shows the interior surface created with this curve (left whole; right section).

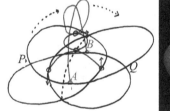

Fig. 31. Rotational sweep   **Fig. 32.** Interior surface for Schneiders' pyramid

This interior surface has four features. The first is that it is regular (if the trajectory has no loop, it will make a singular point). The second is that that it is orientable. The third is that it has two triple points. (However, because this surface is orientable, it is not a Boy's surface [BE-02][Fr-87].) The fourth feature is "*through hole*" whose section is shown on the right of Fig. 33, which makes self-loops and increases the number of hexes needed to fill the volume considerably. Note that the through hole appears due to the intersection of rotational sweep section.

Fig. 33 shows a section of the interior surfaces for Schneiders' problem. Each of the triple-intersection points represents a dual vertex, and the intersecting curves are the dual edges in the dual graph of the resulting mesh. Because of the four self-loops, several of the resulting hexes require their convexity to be recovered.

In order to recover the hex convexity, we insert a cylindrical (closed) surface (depicted with dotted line as a closed interior surface in Fig. 34). The resulting intersecting curves are added to the dual graph, along with the other symmetric half of the original surface, which produces 20 dual vertices (hexes). The dual graph was represented with a prototype 3D cell-complex model, and converted to a primal graph, whose NASTRAN format data is shown in Tab.1 in the appendix together with the node numbers on the surface.

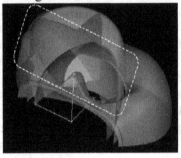 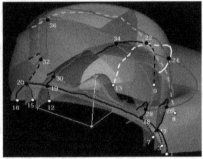

Fig. 33. Section of the interior surfaces   **Fig. 34.** Intersection between the interior surfaces for Schneiders' pyramid

Topologically, there are four dual vertices having double edges, namely 26, 28, 30, 32 (see Fig. 35). We can recover the convexity for these dual vertices by adding another cylindrical surface, resulting in an additional 18 hexes. Furthermore, still there exist 9 convexity-taking edges, and this recovery requires 9 additional cylinders, which consequently create additional 90 hexes. For our interior surface arrangement it is confirmed that 146 dual vertices (hexes) complete the hexahedral mesh topology.

As shown above, a hex mesh can be generated from a closed quad mesh with complicated dual cycles using an "interior surface direct arrangement technique".

Schneiders' pyramid demonstrates to us that there exist quad meshes where there is an obvious interplay between interior surfaces that result in a poorer quality mesh. The process of recovering the convexity of the elements may intrinsically require the addition of many hex elements, which inevitably reduces the industrial value of the solution.

7. CONCLUSION

7.1 Our contribution

Indeed, the hex mesh existence theory, that "Any simply connected three-dimensional domain with an even number of quadrilateral bounding faces can be partitioned into a hexahedral mesh respecting the boundary (using interior surfaces extended from dual cycles into solids)" is well known. However, a practical algorithm for generating the meshes indicated by the proof has not yet materialized. Our contribution to all-hex meshing is as follows:

We have introduced the notion of self-intersecting point-type identity for a dual cycle, and deduced that if the types of self-intersecting point pair are identical, then the interior surface will be an orientable homotopy (and can be generated via a sweep), otherwise it will be a non-orientable homotopy (i.e. a Möbius band will be contained in the interior surface).

Let a circuit be specified as partial dual cycle split by a self-intersecting point, such that it does not self-intersect. A "general" interior surface can be comprised of "simple" interior surfaces generated from adjacent circuits sequentially.

We propose to create interior surfaces by the above-mentioned method and to apply the interior surface direct arrangement technique. We claim that this is one of the solutions of Schneiders' general open problem. We have created an interior surface for Schneiders' pyramid, and showed that it can be filled by 20 hexes with convexity lost derived from the topological arrangement, or with 146 hex elements when the convexity is recovered from the topological arrangement (this does not guarantee geometric convexity).

7.2 Future Problems

7.2.1 Unstructured all-hex meshing

The following three questions with respect to unstructured all-hex mesh generation are important, where an algorithm that positively answers all three will have industrially viable solution.

1. Is the topological solution realized?
2. Can the convexity of all hexes be recovered from the topological solution in a reasonable number of hexes?
3. Are the interior surfaces of sufficient geometric quality that the hexahedral mesh also has sufficient quality?

The solution presented in this paper demonstrates an answer to the first question for the example of Schneiders' pyramid. Further research and development will be needed for a general solution.

The second question of convexity recovery for the hexes can lead to large increases in the number of hex elements, and is especially apparent with multiply self-intersecting interior surfaces. These problems have been avoided in the past by enforcing *restricted unstructured hex mesh generation* techniques that generate hex meshes in much smaller geometric and topologic domains.

The third question of how the geometric quality of the interior surface results in hex mesh quality deterioration has had little study, especially on the relationship between surface quad meshes and the interior surface generation requirements in subsection 4.1. This is also an area of future research.

7.2.2 Implementation

The surfaces generated as an example in this paper were generated without referring to the given volumetric space, and the induced topology was then mapped back into the volume. Because of the difficulty of generating interior surfaces in pre-defined volumetric boundaries (which would be required for sampling quality metrics of the resulting mesh that such interior surfaces might generate), a technique for directly arranging interior surfaces will have some difficult implementation issues. For example, surface creation with boundary constraints, geometrical surface evaluation, topological representations of the induced interior surface arrangement, etc. should be investigated. These problems are also areas for future research.

Acknowledgements

The authors would like to thank Scott Mitchell (Sandia National Laboratories), Soji Yamakawa (Carnegie Mellon University), and Hiroshi Sakurai (Colorado State University) for their advice and insights on hexahedral meshing.

REFERENCES

[BE-02] Marshall Bern, David Eppstein; Flipping Cubical Meshes, ACM Computer Science Archive June 29, 2002
[BM-93] Ted D. Blacker, Ray J. Meyer; Seams and Wedges in Plastering: A 3-D Hexahedral Mesh Generation Algorithm, Engineering with Computers (1993) 9, 83-93
[Ca-www] Carlos D. Carbonera;
http://www-users.informatik.rwth-aachen.de/~roberts/SchPyr/index.html

[Ep-96] David Eppstein; Linear Complexity Hexahedral Mesh Generation, Computational Geometry '96, ACM (1996)

[FM-98] Nathan T. Fowell, and Scott A. Mitchell; Reliable Whisker Weaving via Curve Contraction, Proceedings, 7th International Meshing Roundtable, (1998)

[Fr-87] G. K. Francis, *A Topological Picturebook*, Springer-Verlag, New York, 1987

[Mh-98] Matthias Müller-Hannemann; Hexahedral Mesh Generation with Successive Dual Cycle Elimination: Proc. of 7th International Meshing Roundtable, pp.365-378 (1998)

[Mi-95] Scott A. Mitchell; A Characterization of the Quadrilateral Meshes of a Surface Which Admits a Compatible Hexahedral Mesh of Enclosed Volume, 5th MSI WS. Computational Geometry, 1995
Online available from ftp://ams.sunysb.edu/pub/geometry/msi-workshop/95/samitch.ps.gz

[Mu-95] Murdoch, P. J. ,"The Spatial Twist Continuum: A Dual Representation of the All-Hexahedral Finite Element Mesh", Doctoral Dissertation, Brigham Young University, December, 1995.

[Ow-99] Steve J. Owen; Constrained Triangulation: Application to Hex-Dominant Mesh Generation, Proceedings, 8th International Meshing Roundtable, pp.31-41 (1999)

[Ow-01] Steve J. Owen; Hex-dominant mesh generation using 3D constrained triangulation, CAD, Vol. 33, Num 3, pp.211-220, March 2001

[Sch-www] Robert Schneiders; http://www-users.informatik.rwth-aachen.de/~roberts/open.html

[Sm-58] S. Smale, Regular Curves on Riemannian Manifolds, Tr. of American Math. Soc. vol. 87, pp. 492-510, (1958)

[SZ-03] Alexander Schwartz, Günter M. Ziegler, Construction techniques for cubical complexes, add cubical 4-polytopes, and prescribed dual manifolds

[TBM-96] Timothy J. Tautges, Ted Blacker, and Scott A. Mitchell; The Whisker Weaving Algorithm: A Connectivity-Based Method for Constructing All-Hexahedral Finite Element Method, Draft submitted to the International Journal of Numerical Methods in Engineering, (Mar. 12, 1996)

[Th-93] W. Thurston; Hexahedral Decomposition of Polyhedra, Posting to Sci. Math, 25 Oct 1993

[TM-95] Timothy L. Tautges, Scott A. Mitchell; Whisker Weaving: Invalid Connectivity Resolution and Primal construction Algorithm, Proceedings, 4th International Meshing Roundtable, pp.115-127 (1995)

[YS-01] Soji Yamakawa, Kenji Shimada; Hexpoop: Modular Templates for Converting a Hex-dominant Mesh to An All-hex Mesh, 10th International Meshing Round Table (2001)

APPENDIX

Hex	N 1	N 2	N 3	N 4	N 5	N 6	N 7	N 8
17	1	2	3	4	5	6	7	8
18	9	10	11	12	1	2	3	4
19	13	14	15	16	9	10	11	12
20	17	18	19	8	13	14	15	16
21	20	3	11	15	8	7	21	19
22	22	17	8	5	9	13	23	1
23	7	8	20	3	24	25	26	27
24	5	1	23	8	24	27	26	25
25	28	4	3	27	25	8	7	24
26	27	1	4	28	24	5	8	25
27	29	12	11	30	28	4	3	27
28	30	9	12	29	27	1	4	28
29	31	16	15	32	29	12	11	30
30	32	13	16	31	30	9	12	29
31	25	8	19	33	31	16	15	32
32	33	17	8	25	32	13	16	31
33	3	20	15	11	27	26	32	30
34	1	9	13	23	27	30	32	26
35	19	15	20	8	33	32	26	25
36	17	8	23	13	33	25	26	32

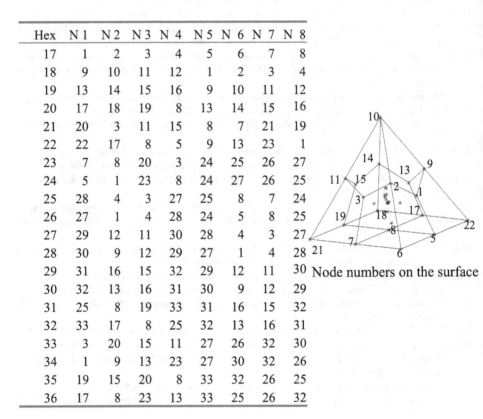

Node numbers on the surface

Table 1. The 20-Hex mesh (convexity lost) by NASTRAN CHEXA Format

Unconstrained Paving & Plastering: A New Idea for All Hexahedral Mesh Generation

Matthew L. Staten, Steven J. Owen, Ted D. Blacker

Sandia National Laboratories[1], Albuquerque, NM, U.S.A.,
mlstate@sandia.gov, sjowen@sandia.gov, tdblack@sandia.gov

Summary: Unconstrained Plastering is a new algorithm with the goal of generating a conformal all-hexahedral mesh on any solid geometry assembly. Paving[1] has proven reliable for quadrilateral meshing on arbitrary surfaces. However, the 3D corollary, Plastering [2][3][4][5], is unable to resolve the unmeshed center voids due to being over-constrained by a pre-existing boundary mesh. Unconstrained Plastering attempts to leverage the benefits of Paving and Plastering, without the over-constrained nature of Plastering. Unconstrained Plastering uses advancing fronts to inwardly project unconstrained hexahedral layers from an *unmeshed* boundary. Only when three layers cross, is a hex element formed. Resolving the final voids is easier since closely spaced, randomly oriented quadrilaterals do not over-constrain the problem. Implementation has begun on Unconstrained Plastering, however, proof of its reliability is still forthcoming.

Keywords: mesh generation, hexahedra, plastering, sweeping, paving

1 Introduction

The search for a reliable all-hexahedral meshing algorithm continues. Many researchers have abandoned the search, relying upon the widely available and highly robust tetrahedral meshing algorithms [6]. However, hex meshes are still preferable for many applications, and depending on the solver, still required. This paper introduces a new method for hexahedral mesh generation called Unconstrained Plastering.

1.1 Previous Research

For all-quadrilateral meshing, Paving [1] and its many permutations [7][8] have proven reliable. Paving starts with pre-meshed boundary edges which are classified into fronts and

---

1 Sandia is a multiprogram laboratory operated by Sandia Corporation, a Lockheed Martin Company, for the United States Department of Energy under Contract DE-AC04-94AL85000

The submitted manuscript has been authored by a contractor of the United States Government under contract. Accordingly the United States Government retains a non-exclusive, royalty-free license to publish or reproduce the published form of this contribution, or allow others to do so, for United States Government purposes

advanced inward. As fronts collide, they are seamed, smoothed, and transitioned until only a small unmeshed void remains (usually 6-sided or smaller). Then a template is inserted into this void resulting in quadrilaterals covering the entire surface.

Paving's characteristic of maintaining high quality, boundary-aligned rows of elements is what has made it a successful approach to quad meshing. In addition, because of its ability to transition in element size, Paving is able to match nearly any boundary edge mesh.

There have been many attempts to extend Paving to arbitrary 3D solid geometry. While valuable contributions to the literature, these attempts have not resulted in reliable general algorithms for hexahedral meshing. Plastering [2][3][4][5] was one of the first attempts. In Plastering, the bounding surfaces of the solid are quad meshed, fronts are determined and then advanced inward. However, once opposing fronts collide, the algorithm frequently has deficiencies. Unless the number, size, and orientation of the quadrilateral faces on opposing fronts match, Plastering is rarely able to resolve the unmeshed voids.

Many creative attempts have been made to resolve this unmeshed void left behind by plastering. Since arbitrary 3D voids can be robustly filled with tets, the idea of plastering in a few layers, followed by tet-meshing the remaining void was attempted [9][10]. Transitions between the tets and hexes were done with Pyramids [11] and multi-point constraints. The Geode-Template [12] provided a method of generating an all-hex mesh by refining both the hexes and tets. However, this required an additional refinement of the entire mesh resulting in meshes much larger than required. In addition, the Geode-Template was unable to provide reasonable element quality.

A draw-back of paving is the need for expensive intersection calculations. An alternative to Paving called Q-Morph [7] eliminated the need for intersection calculations by first triangle meshing the surface. This triangle mesh is then "transformed" into a quad mesh. Using a similar advancing front technique to paving, triangles are locally reconnected, repositioned, and combined to form quads. Q-Morph is able to form high-quality quadrilateral elements with similar characteristics to paving. Q-Morph has proven to be a robust and reliable quad meshing algorithm in common use in several commercial meshing packages.

An attempt at extending Q-Morph to a hex-dominant meshing algorithm was done with H-Morph [13]. This algorithm takes an existing tetrahedral mesh and applies local connectivity transformations to the elements. Groups of tetrahedral are then combined to form high-quality hexahedra. The advancing front approach was also used for ordering and prioritizing tetrahedral transformations. Although H-Morph had the desirable characteristics of regular layers near the boundaries, it was unable to reliably resolve the interior regions to form a completely all-hex mesh since it also attempted to honor a pre-meshed quad boundary.

Recognizing the difficulty of defining the full connectivity of a hex mesh using traditional geometry-based advancing front approaches, the Whisker-Weaving algorithm [14][15][16] attempted to address the problem from a purely topological approach. It attempts to first generate the complete dual of the mesh, from which the primal, or hex elements, are readily obtainable. Although wisker-weaving can in most cases generate a successful dual topology, resulting hex elements are often poorly shaped or inverted.

Plastering, H-Morph, Whisker Weaving and all of their permutations are classified as Outside-In-Methods. They start with a pre-defined boundary quad mesh and then attempt to use that to define the hex connectivity on the inside. Another class of Hex meshing algorithms can be classified as Inside-Out methods [17][18][19]. These algorithms fill the inside of the solid with elements first, often using an octree-based grid. This grid is then adapted to fit the boundary. These methods place high quality elements on the interior of the volume, however, they typically generate extremely poor quality elements on the boundary. In addition, traditional Inside-Out methods are unable to mesh assemblies with conformal meshes. These inside-out methods seem particularly popular with the metal forming industry, but of less appeal in structural mechanics applications.

Sweeping based methods [20][21][22] are among the most widely used hexahedral based meshing algorithms in industry today. Sweeping, however, applies only to solids which are 2.5D, or solids which can be decomposed into 2.5D sub-regions. There has been a considerable amount of research in sweeping and many successful implementations have been published. It is typically quite simple to decompose and sweep simple to medium complexity solids. However, as more complexity is added to the solid model, the task of decomposing the solids into 2.5D sub-regions can be daunting, and in some regards, an art-form requiring significant creativity and experience.

Advancing front methods have proven ideal for triangle, quadrilateral and even tetrahedral meshes. They have been successful in these arenas because of the smaller number of constraints imposed by the connectivity of these simple element shapes. Hexahedral meshes, on the other hand, must maintain a connectivity of eight nodes, 12 edges, and six faces per element, with strict constraints on warping and skewness. As a result, unlike tetrahedral meshes, minor local changes to the connectivity of a hex mesh can have severe consequences to the global mesh structure. For this reason, current hexahedral advancing front methods where the boundary is prescribed apriori have rarely succeeded for general geometric configurations.

Current advancing front methods, while having the high ideal of maintaining the integrity of a prescribed boundary mesh, frequently fail because the very boundary mesh they are attempting to maintain over-constrains the problem, creating a predicament which can be intractable.

To resolve this issue, we introduce a new concept, known as Unconstrained Plastering. With this approach, we relax the constraint of prescribing a boundary apriori quad mesh. While still maintaining the desirable characteristics of advancing front meshes, Unconstrained Plastering is free to define the topology of its boundary mesh as a consequence of the interior meshing process. It is understood that not prescribing an apriori boundary quad mesh can have implications on the traditional bottom-up approach to mesh generation. These implications, however, are significantly outweighed by the prospect of automating the all-hex mesh generation process through a more top-down approach to the problem that Unconstrained Plastering offers.

1.2   Unconstrained Plastering, an Unproven Concept

Unconstrained Plastering is a new approach that is unique from all the others. Although it contains similarities to other existing algorithms, primarily Plastering, it should not be considered an extension of Plastering.

Finally, Unconstrained Plastering is not a finished work. Rather, Unconstrained Plastering is an idea that holds promise for hex meshing researchers and should be studied further. Implementation has begun on a prototype and there is reason to be optimistic that it has a greater potential for success than others. However, evidence of it being robust enough to handle general purpose mesh generation for industrial applications is forthcoming.

2   Unconstrained Paving

To best understand the general concept behind Unconstrained Plastering, we first examine the 2D corollary, Unconstrained Paving.

2.1   Advancing Unconstrained Rows

Fig. 1 shows a geometric surface ready for quad meshing. If we were to pave this surface, we would first mesh each of the surfaces boundary curves, after which we would advance a row of quads along one of the boundary curves as can be seen in Fig. 2. In this case four quads were added because the curve along which the row was paved was pre-meshed with four mesh edges.

If, instead, the surface was being meshed with Unconstrained Paving, the boundary curves would not be pre-meshed with edges. Advancing, or paving, an unconstrained row would result in Fig. 3. In this case a row of quads have been inserted, however, we do not know how many quads will be in that row. The number of quads in this row is determined as adjacent rows cross it. Fig. 4 shows what the mesh looks like after a second row is advanced. Since the second row advanced crossed the first row advanced, a single quad is formed (shaded) in the corner where the two rows cross. However, both of the rows still have an undetermined number of quads in them.

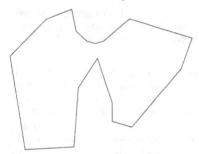

Fig. 1 Example Surface

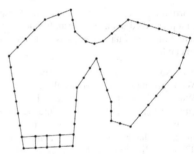

Fig. 2 Example Surface with meshed boundary and one row paved

Fig. 3 One row advanced with Unconstrained Paving

Fig. 4 Two rows advanced with Unconstrained Paving

In Fig. 5, several additional rows have been advanced and 12 quadrilateral elements have been formed where the various unconstrained rows have crossed. At this point, the un-meshed portion of the surface has been subdivided into two sub-regions (sub-region A & B). It is important to note that both of these unmeshed regions are completely uncon-strained. For example, sub-region A is bound by five edges, however, none of these edges has been meshed. Sub-region A is free to be meshed with as many divisions as needed along all of these edges.

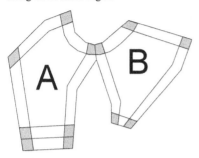 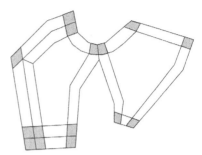

Fig. 5 Additional rows advanced with Unconstrained Paving

Fig. 6 Transition row inserted based on large angle between two adjacent rows

2.2   Transitioning Unconstrained Rows

Like traditional Paving, Unconstrained Paving has the ability to insert irregular nodes (nodes with more or less than four adjacent quads) in order to transition and fit the shape of the surface. In traditional Paving, this is done by assigning states to the fronts based on an-gles with adjacent fronts. Unconstrained Plastering is no different. The start and end of an advanced unconstrained row likewise depends upon states and angles. Fig. 6 shows the ad-vancement of an additional row, which, because of angles is the advancement of two previ-ously advanced rows.

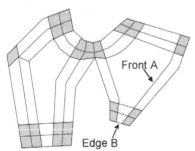 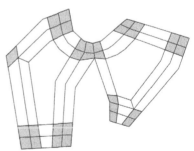

Fig. 7   Front A cannot be advanced normally because Edge B is too short

Fig. 8   Transition row inserted based on front sizes

Fig. 7 shows an additional case where rows must be advanced with care. *Front A* is the next front to advance, however, *Edge B* is too short even though angles indicate that the advancement of *Front A* should extend to *Edge B*. In this case, *Front A* can be advanced as shows in Fig. 8.

Unconstrained rows continue to advance as previously described. Rows bend through the mesh as required to maintain proper quadrilateral connectivity ensuring that all quadrilateral elements created are of proper size. In addition, Paver-like row smoothing and seaming, along with the insertion of tucks and wedges [1] are additional operations that can be performed on the unconstrained rows. Fig. 9 illustrates the example surface and how it may look after several more rows are advanced. All edges on the unmeshed sub-regions A and B are less than two times the desired element size, and so we stop advancing fronts. At this point, it is time to resolve the unmeshed voids.

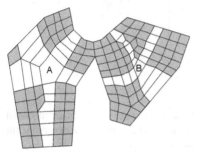

Fig. 9 Unconstrained rows advanced leaving only small unmeshed voids and connecting tubes; Quad are shaded, connecting tubes are white.

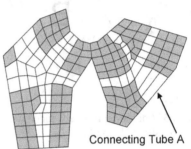

Connecting Tube A

Fig. 10 Unmeshed voids have been meshed

2.3 Resolving Unmeshed Voids

In general, the unmeshed sub-regions will be any general polygon, with any number of sides. It is assumed that each polygon will be convex. If it is not convex, that would suggest that an additional row needs to be advanced before resolving the unmeshed void. It is also assumed that the size of the polygon is roughly one-to-two times the desired element size. If it is larger than this, then additional unconstrained rows should be advanced until the remaining polygon is one-to-two times the desired element size.

Also, note that the unmeshed region is completely unconstrained. Each of the edges on the unmeshed polygons are connected to the boundary of the mesh through "connecting tubes". Connecting tubes are the white regions in Fig. 9 that have been crossed by only a single row. The edges of each polygon can be meshed with any number of edges, which will be propagated back to the boundary through the connecting tubes.

At this point, the polygon is meshed with a template quad mesh similar to the templates used to fill the voids during Paving [1]. The template inserted is based on the relative lengths of edges, and angles between edges. In the general case, any convex polygon can be meshed with midpoint subdivision [23]. Midpoint subdivision meshes convex polygons by adding a node at the centroid of the polygon and connecting it to nodes added at the center of each polygon boundary edge. The number of new quads formed is equal to the number of points defining the polygon. Although midpoint subdivision can always be used to mesh the void, simpler templates are often possible.

In Fig. 9, since sub-region B is already four-sided and is of proper size and shape, it can be converted into a single quadrilateral element. However, sub-region A is meshed with midpoint subdivision since it has five sides. The resulting mesh is illustrated in Fig. 10.

Before Unconstrained Paving is finished, the connecting tubes must be examined for size. In Fig. 10 *Connecting Tube A* is much too wide. This can be fixed by advancing a few more rows until the proper size is obtained as shown in Fig. 11. Traditional quadrilateral cleanup operations and smoothing can then be performed to finalize the mesh connectivity and quality as shown in Fig. 12.

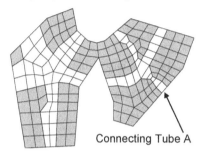

Connecting Tube A

Fig. 11   Sizes in connecting tubes have been resolved

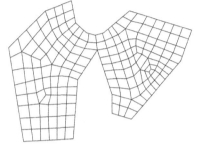

Fig. 12   Final quad mesh after cleaning and smoothing

2.4   Unconstrained Paving with Multiple Surfaces

In real world models, rarely is the geometry confined to a single surface. For example, sheet metal parts in the auto industry representing automobile hoods often contain thousands of surfaces. Each of these surfaces must share nodes and element edges with its neighboring surfaces across its boundary edges in order to ensure a conformal mesh.

Typically, algorithms that do not pre-mesh the curves of surface before meshing have difficulty ensuring a conformal mesh [17][18][19]. However, Unconstrained Paving can be extended to ensure conformal meshes between any number of surfaces. The penalty, however, is that all of the surfaces must be meshed at the same time. For example, Fig. 13 illustrates four adjacent surfaces which require a conformal mesh. Fig. 14 shows the same model with one unconstrained row advanced. The row was advanced in three of the surfaces. Fig. 15 shows several additional rows inserted and the formation of a tuck in surface 2. Notice that curves which are shared by more than one surface are double-sided fronts advancing into both adjacent surfaces.

After additional rows are advanced, Fig. 16 shows the small unmeshed voids and the connecting tubes. It is important to note that when meshing multiple surfaces at once, the connecting tubes impose additional constraints on how the unmeshed voids can be meshed. However, these constraints can always be satisfied with midpoint subdivision since this would split each edge in every connecting tube exactly once. Fig. 17 shows what the mesh may look like before a final pass through cleanup and smoothing.

In the current research, Unconstrained Paving is used only as a thought experiment to help illustrate the concepts of Unconstrained Plastering. While implementation of Unconstrained Paving may be beneficial current unstructured quadrilateral meshing techniques satisfy FEA needs. For this reason the current research focuses implementation and prototyping efforts on the 3D Unconstrained Plastering problem.

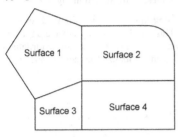

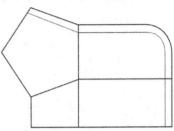

Fig. 13 Multiple adjacent surfaces requiring a conformal mesh

Fig. 14 An unconstrained row has been advanced extending through multiple surfaces

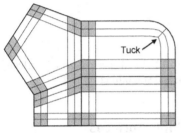

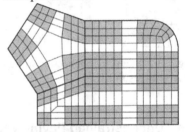

Fig. 15 Additional rows are advanced including a tuck

Fig. 16 Only small voids and connecting tubes remain

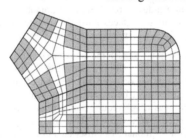

Fig. 17 Unmeshed voids and connecting tubes are meshed

3   Unconstrained Plastering

The basic principles of Unconstrained Paving extend to 3D as Unconstrained Plastering. The basic algorithm is as follows and is described in the following sections.

1. Start with a solid assembly with unmeshed volume boundaries.

2. Define fronts, which initially are the surfaces of the volumes.

3. While the unmeshed voids of the solids are larger than twice the desired element size:

   a. Select a front to advance,

   b. Based on sizes of fronts, and angles with adjacent fronts, determine which adjacent fronts should be advanced with the current front.

   c. Advance the fronts

   d. Form unconstrained columns of hexahedra where 2 layers cross.

   e. Form actual hexahedral elements where 3 layers cross.

   f. Perform layer smoothing and seaming.

   g. Insert tucks and wedges as needed.

4. Identify unmeshed voids, connecting tubes, and connecting webs.

5. Define constraints between unmeshed voids through connecting tubes.

6. Mesh the interior voids with either midpoint subdivision or T-Hex.

7. Sweep the connecting tubes between voids and out to the boundary.

8. Split connecting webs as needed.

9. Smooth all nodes to improve element quality.

3.1   Advancing Unconstrained Layers

The model in Fig. 18 will be used an example of Unconstrained Plastering. Fig. 19 shows a single unconstrained hexahedral layer advanced. The new surface displayed in Fig. 19 represents the top of the layer of hexes which will be adjacent to the advanced surface. The region between the boundary and the new surface represents an unconstrained layer of hexahedra. It is still unknown how many hexes will be in this layer, however, we do know that it will contain a single layer of hex elements.

In Fig. 20, a second layer is advanced, which crosses the first layer advanced previously. When two layers cross, a column of hexahedra is formed, however, the size and number of hexahedra in this column will not be determined until additional hex layers cross this column.

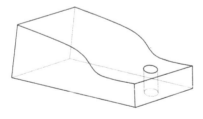

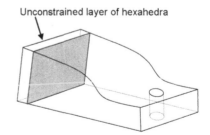

Unconstrained layer of hexahedra

Fig. 18 Unconstrained Plastering example model

Fig. 19 One unconstrained hex layer has been advanced

Unconstrained column of hexahedra

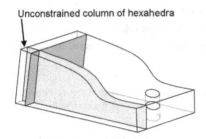

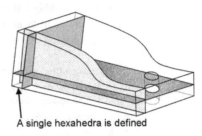

A single hexahedra is defined

Fig. 20   When two layers cross, an unconstrained column of hexes is defined

Fig. 21   When three layers cross a hexahedral element is defined

In Fig. 21, a third layer is advanced, which crosses both of the previously defined layers. Where ever three layers cross, a hexahedral element is formed. In Fig. 21, a single hex element is defined in the lower left corner. Note that until now, no final decisions have been imposed on placement of hexahedra. It is only when three orthogonal layers intersect that hex placement becomes finalized.

The process continues in Fig. 22, Fig. 23, and Fig. 24. Each time two layers cross a column of hexahedra is defined. Each time a third layer cross a column, a single hexahedral element is defined. During this process, there will be transition layers inserted with logic similar to that described in section 2.2. Layers are advanced until the unmeshed void is approximately twice the desired element size.

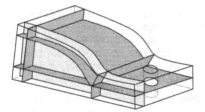

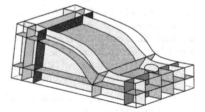

Fig. 22   Additional layers are advanced

Fig. 23   Additional layers are advanced

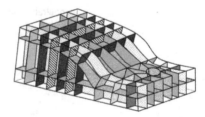

Fig. 24 Additional layers are advanced until the unmeshed void is small

3.2   Front Processing Order

A front to advance is a group of one or more adjacent surfaces which are advanced together. The order that fronts are processed in Unconstrained Plastering is very important. This is an area where additional research is required. However, factors to consider when choosing the next front to advance include:

1.   Number of layers away from the boundary the front is. Fronts closer to the boundary should be processed first.

2.   If the front is "complete" or not. A complete front is a group of surfaces which are completely surrounded by what are referred to as "ends" in Paving and Submapping[24][25]. For example, in Fig. 25, Surface 1 is complete since its boundary is adjacent to a cylindrical face which is perpendicular to Surface 1. In contrast, Surface 2 is incomplete since it is bounded on one of its loops by an "end", but is bound on its other loop with a "corner" [24]. The best way to proceed would be advance Surface 1 several times until it becomes even with Surface 2, at which the front from Surface 1 and Surface 2 would be combined and advanced as a single front.

3.   The size of the front. Smaller fronts should probably be processed first.

4.   How much distance there is ahead of the front before a collision will occur. Fronts with a lot of room to advance should probably be processed first.

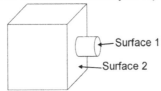

Fig. 25 Surface 1 is a "complete" front, while Surface 2 is "incomplete"

3.2   Resolving Unmeshed Voids

Like Unconstrained Paving, there will be unmeshed voids at the center of each volume being meshed. The unmeshed voids can be easily identified because they are the regions in space that have not been crossed by any hex layers. Fig. 26 illustrates the unmeshed void at the center of the example model. In general, these voids define general polyhedra. It is assumed that these polyhedra are convex. If they are not convex, that suggests that an additional layer should be advanced before resolving the voids. Although we have no theoretical basis to prove that advancing additional rows will always ensure a convex polyhedra, experience has shown that it does. Likewise, it is assumed that all edges and faces of the polyhedra are approximately twice the desired element size, or less. If they are larger than this, it suggests that an additional layer should be advanced.

In addition to identifying the unmeshed voids, we must also identify the connecting tubes and connecting webs. Connecting tubes are those regions in space which have been crossed by only a single hex layer as illustrated in Fig. 27. In order to define a hex, three layers must cross, which gives the connecting tubes two degrees of freedom, allowing them to be swept as a one-to-one sweep[22]. The direction of the sweep is perpendicular to the single layers which already cross the connecting tube. The number of layers in the sweep is the same as the number of hex layers the connecting tube crosses between the unmeshed polyhedra and the boundary surface.

Connecting webs are those regions in space which have been crossed by only two hex layers as illustrated in Fig. 28. Connecting webs only have a single degree of freedom. Essentially, connecting webs represents a layer of hexahedra which will be split the same number of times that the adjacent connecting tubes are split. In this example model, there is one small connecting web section which is not attached to any connecting tubes or unmeshed void. This will happen in locations where seaming has been performed, since seaming will often eliminate or split the unmeshed void. In the example model, the front that was advanced from the front-right surface was seamed with the fronts extending from the hole.

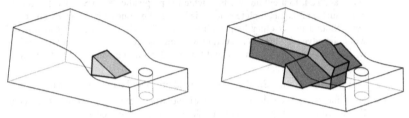

Fig. 26 The unmeshed void   Fig. 27 The connecting tubes

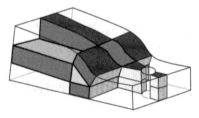

Fig. 28 The connecting webs

After the unmeshed voids, connecting tubes, and connecting webs have been identified, the unmeshed void is meshed using either midpoint subdivision[23], or T-Hex. Midpoint subdivision is the preferable method since it generates higher quality elements. To determine if midpoint subdivision is possible, a simple count of the number of curves connected to each vertex on the unmeshed polyhedra is done. If there are any vertices which have four or more connected curves, then midpoint subdivision is not possible. The unmeshed void in Fig. 26 can be meshed with midpoint subdivision as illustrated in Fig. 29.

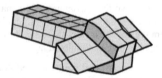

Fig. 29 Midpoint subdivision of the   Fig. 30   The connecting tubes are
unmeshed void   swept

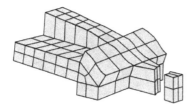

Fig. 31 The connecting webs are split

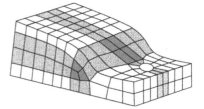

Fig. 32 Final mesh using midpoint subdivision of voids; connecting tubes are shaded; connecting webs are cross-hatched

After the unmeshed void is meshed, the connecting tubes are swept as shown in Fig. 30 using the mesh from the unmeshed void as the source. Finally, the connecting webs can be split as illustrated in Fig. 31. The final mesh on the example model, after some global smoothing, is shown in Fig. 32. The connecting tubes exposed to the boundary of the mesh are shaded dark and the exposed connecting webs are cross-hatched.

If the polyhedra cannot be meshed with midpoint subdivision, it is meshed with the T-Hex template instead. To do this, we first take each non-triangular polygon on each unmeshed polyhedra and split it into triangles. If the polygon being split is connected to other unmeshed polyhedra through connecting tubes, we must be careful that the face is split the same on both polyhedra so the sweeper can match them up through the connecting tubes. To ensure that they are split the same, a node can be added at the center of the face and triangles are formed using each edge on the polygon and the newly created center node. After each face is split into triangles, the polyhedra are meshed with tets. Since we are assuming that the unmeshed void is 1-2 times the desired element size, we would like to mesh these polyhedra without introducing any nodes interior to the polyhedra. Not putting any new nodes in the polyhedra will also help with element quality since T-Hex meshes are worst when T-Hexing around a node surrounded completely by tet elements. After the polyhedra are tet meshed, each tet is split into four hexahedral elements using the T-Hex template shown in Fig. 33. The T-Hex mesh for the polyhedra in the example problem is shown in Fig. 34 and the mesh on the connecting tubes in Fig. 35. Since, in this example, we were meshing only a single solid, there were no constraints between multiple unmeshed voids. Therefore, the quadrilateral face on the top of the polyhedra was split into only two triangles before tetrahedralization. Fig. 36 shows the final mesh after some global smoothing. The connecting tubes exposed to the boundary of the mesh are shaded dark and the exposed connecting webs are cross-hatched.

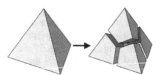

Fig. 33 T-Hex template

Fig. 34 T-Hex on unmeshed void

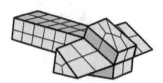

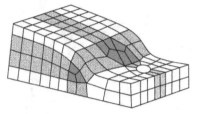

Fig. 35 The connecting tubes are swept with T-Hex mesh

Fig. 36 Final mesh using T-Hex on voids; connecting tubes are shaded; connecting webs are cross-hatched

T-Hex has long been known as a guaranteed way to get an all hexahedral mesh on nearly any solid geometry. However, the quality of the elements that result is rarely sufficient for most solver codes. Critics of Unconstrained Plastering will point to the use of T-Hex on interior voids as a major downfall of Unconstrained Plastering. However, before that judgment can be made, the following should be considered:

1.   T-Hex is only used when interior voids have a vertex with a valence of four or more. In most cases, the interior voids can be meshed with midpoint subdivision.

2.   The worst quality hexahedra in T-Hex meshes are found adjacent to nodes which were completely surrounded by tets in the initial tet mesh. This is because a tet mesh can have nodes with a valence of 15 or more, which results in the same number of hexahedra when the T-Hex template is applied. This case should not appear during Unconstrained Plastering, since we assume that enough unconstrained layers have been advanced to make the interior voids small enough to be tet meshed with no interior nodes.

3.   The T-Hex looking elements that are swept to the boundary through connecting tubes are not poor in quality since a swept T-Quad mesh is much higher quality than a traditional T-Hex mesh.

4.   Any poor quality hexahedra that are formed by Unconstrained Plastering will be in the interior voids which should be on the deep interior of the volumes, with the exception of thin parts which require only one or two layers of hexahedra through the thickness.

3.4   Unconstrained Assembly Meshing

Even though Unconstrained Plastering does not seem capable of honoring existing boundary quad meshes, it can be used to mesh assemblies of solids and still get a conformal mesh. Like Unconstrained Paving, however, all of the volumes in the assembly must be meshed at once. Fig. 37 illustrates a simple assembly model to be meshed. Fig. 38 and Fig. 39 show the same model after 3 unconstrained layers have been advanced. Fig. 39 shows that the unconstrained layers have been extended and advanced through both of the solids in the assembly. Like Unconstrained Paving, surfaces which are shared by both volumes will behave as double-sided fronts advancing into both volumes.

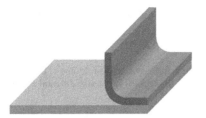

Fig. 37 Assembly model

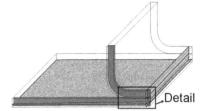

Fig. 38 Assembly model with three advanced unconstrained plastering layers

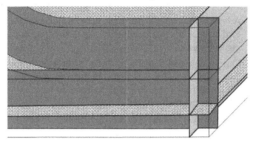

Fig. 39 Detail of Fig. 38

3.5 Element Quality with Unconstrained Plastering

Implementation of Unconstrained Plastering has not progressed far enough to make any claims on element quality. However, like other advancing front algorithms, Unconstrained Plastering will have the tendency to put the highest quality element near the boundary. Subsequent publications on Unconstrained Plastering will report element quality findings as the research matures.

One limitation that Unconstrained Plastering will have compared to Unconstrained Paving is the lack of hexahedral cleanup operations. Unlike quadrilateral cleanup, hexahedral cleanup operations are limited due to the highly constrained nature of hexahedra [26][27]. As a result, Unconstrained Plastering will be required to create hexahedral topology that will permit good element quality rather than relying on a post-processing cleanup step to fix poor elements.

4   Implementation Details

As stated earlier, implementation has begun on a full 3D implementation of Unconstrained Plastering. The authors have chosen to use a faceted surface based approach. The basic algorithm that is being followed is:

1. Triangle mesh all of the boundary surfaces using an element size approximately equal to the desired hexahedral element size.

2. Traverse through this triangle mesh to eliminate any unnecessary CAD artifacts (i.e. small angles, slivers, etc.

3. Divide the triangles up into groups which form Surfaces or Fronts. They are grouped together considering the original CAD topology, but also dihedral angles between the original CAD surfaces.

4. For each volume in the assembly being meshed, form a "cell". Each cell has three associated layer ids. These initial cells get (UNDEFINED, UNDEFINED, UNDEFINED) as their initial layer ids.

5. While any cell is larger than twice the desired element size:

    a. Choose a set of Surfaces to advance to form a new layer.

    b. Advance the triangle mesh on these surfaces into the volume. A faceted surface is created offset by the desired element size to the advancing surfaces.

    c. Form a new cell between each newly created surface its corresponding front surface. This new Cell inherits the layer ids from the cell being advanced into. It is also assigned a new layer id which represents the layer just created.

    d. Smooth and seam the newly advanced faceted surface with its neighboring surfaces.

6. Create constraints between the unmeshed voids through the connecting tubes.

7. Mesh each unmeshed void with either midpoint subdivision or T-Hex.

8. Sweep the connecting tubes.

9. Split the connecting webs as needed.

10. Send the entire mesh to a smoother for global smoothing. Smoothing is needed on curves and surfaces, in addition to the nodes on the interior of the volumes.

5   Conclusions

The concept of advancing unconstrained rows of quads and layers of hexahedra has been introduced through the algorithms of Unconstrained Paving and Unconstrained Plastering. The concept is most relevant with Unconstrained Plastering since it eliminates the problems of resolving highly constrained unmeshed voids which is common with most other advancing front hexahedral meshing algorithms.

The algorithms presented are able to mesh assembly models with conformal meshes with the penalty that all of the volumes/surfaces in the model must be meshed at the same time. Meshing all of the volumes in an assembly at once increases memory requirements since the mesh on the entire assembly will need to be in the mesher's internal datastructures at once, which are typically larger than mesh storage datastructures.

Implementation of Unconstrained Plastering has begun and the authors are optimistic that it will be more successful than other free hex meshing algorithms. However, additional research is required before any claims will be made.

Unconstrained Paving is also presented which is a potential improvement upon traditional advancing front quadrilateral meshing algorithms. However, since the quadrilateral meshing problem already has several solutions, the priority of researching and implementing Unconstrained Paving is lower than that of Unconstrained Plastering.

References

1   T. D. Blacker, M. B. Stephenson. "Paving: A New Approach to Automated Quadrilateral Mesh Generation", *International Journal for Numerical Methods in Engineering*, 32, 811-847 (1991).

2   S. A. Canann, "Plastering: A New Approach to Automated 3-D Hexahedral Mesh Generation", *American Institute of Aeronautics and Astronics* (1992).

3   J. Hipp, R. Lober, "Plastering: All-Hexahedral Mesh Generation Through Connectivity Resolution", *Proc. 3rd International meshing Roundtable* (1994).

4   S. A. Canann, "Plastering and Optismoothing: New Approaches to Automated 3D Hexahedral Mesh Generation and Mesh Smoothing", Ph.D. Dissertation, Brigham Young University, Provo, Utah, USA (1991).

5   T. D. Blacker, R. J. Meyers, "Seams and Wedges in Plastering: A 3D Hexahedral Mesh Generation Algorithm", *Engineering With Computers*, 2, 83-93 (1993).

6   P.-L. George, H. Borouchaki, "Delaunay Triangulation and Meshing: Application to Finite Elements", © Editions HERMES, Paris, 1998.

7   S. J. Owen, M. L. Staten, S. A. Canann, S. Siagal, "Q-Morph: An Indirect Approach to Advancing Fron Quad Meshing", *International Journal for Numerical Methods in Engineering*, 44, 1317-1340 (1999).

8   D. R. White, P. Kinney, "Redesign of the Paving Algorithm: Robustness Enhancements through Element by Element Meshing", *Proc. 6th Int. Meshing Roundtable*, 323-335 (1997).

9   D. Dewhirst, S. Vangavolu, H. Wattrick, "The Combination of Hexahedral and Tetrahedral Meshing Algorithms", *Proc. 4th International Meshing Roundtable*, 291-304 (1995).

10   R. Meyers, T. Tautges, P. Tuchinsky, "The 'Hex-Tet' Hex-Dominant Meshing Algorithm as Implemented in CUBIT", *Proc. 7th International Meshing Roundtable*, 151-158 (1998).

11   S. J. Owen, S. Canann, S. Siagal, "Pyramid Elements for Maintaining Tetrahedra to Hexahedra Conformability", *Trends in Unstructured Mesh Generation*, AMD Vol 220, 123-129, ASME (1997).

12   R. W. Leland, D. Melander, R. Meyers, S. Mitchell, T. Tautges, "The Geode Algorithm: Combining Hex/Tet Plastering, Dicing and Transition Elements for Automatic, All-Hex Mesh Generation", *Proc 7th International Meshing Roundtable*, 515-521 (1998).

13   S. J. Owen, "Non-Simplical Unstructured Mesh Generation", Ph.D. Dissertation, Carnegie Mellon University, Pittsburgh, Pennsylvania, USA (1999).

14   T. J. Tautges, T. Blacker, S. Mitchell, "The Whisker-Weaving Algorithm: A Connectivity Based Method for Constructing All-Hexahedral Finite Element Meshes", *International Journal for Numerical Methods in Engineering*, 39, 3327-3349 (1996).

15   P. Murdoch, S. Benzley, "The Spatial Twist Continuum", *Proc. 4th International Meshing Roundtable*, 243-251 (1995).

16   N. T. Folwell, S. A. Mitchell, "Reliable Whisker Weaving via Curve Contraction," *Proc. 7th International Meshing Roundtable*, 365-378 (1998).

17   R. Schneiders, R. Schindler, R. Weiler, "Octree-Based Generation of Hexahedral Element Meshes", *Proc. 5th International Meshing Roundtable*, 205-217 (1996).

18   P. Kraft, "Automatic Remeshing with Hexahedral Elements: Problems, Solutions and Applications", *Proc. 8th International Meshing Roundtable*, 357-368 (1999).

19   G. D. Dhondt, "Unstructured 20-Node Brick Element Meshing", *Proc. 8th International Meshing Roundtable*, 369-376 (1999).

20   T. D. Blacker, "The Cooper Tool", *Proc. 5th International Meshing Roundtable*, 13-29 (1996).

21   Mingwu Lai, "Automatic Hexahedral Mesh Generation by Generalized Multiple Source to Multiple Target Sweeping", Ph.D. Dissertation, Brigham Young University, Provo, Utah, USA (1998).

22   M. L. Staten, S. Canann, S. Owen, "BMSweep: Locating Interior Nodes During Sweeping", *Proc. 7th International Meshing Roundtable*, 7-18 (1998).

23   T. S. Li, R. M. McKeag, C. G. Armstrong, "Hexahedral Meshing Using Midpoint Subdivision and Integer Programming", *Computer Methods in Applied Mechanics and Engineering*, Vol 124, Issue 1-2, 171-193 (1995).

24   D. R. White, L. Mingwu, S. Benzley, "Automated Hexahedral Mesh Generation by Virtual Decomposition", *Proc. 4th International Meshing Roundtable*, 165-176 (1995).

25   D. R. White, "Automatic, Quadrilateral and Hexahedral Meshing of Pseudo-Cartesian Geometries using Virtual Subdivision", Master's Thesis, Brigham Young University, Provo, Utah, USA (1996).

26   M Bern, D. Eppstein, "Flipping Cubical Meshes," *Proc. 10th International Meshing Roundtable,* 19-29 (2001).

27   P. Knupp and S. A. Mitchell, "Integration of mesh optimization with 3D all-hex mesh generation," *Tech. Rep. SAND99-2852*, Sandia National Laboratories, 1999, http://citeseer.ist.psu.edu/ knupp99integration.html.

Adaptive Sweeping Techniques

Michael A. Scott[1], Matthew N. Earp[2], Steven E. Benzley[3] and
Michael B. Stephenson[4]

[1] Brigham Young University Provo, UT U.S.A. mas88@et.byu.edu
[2] Brigham Young University Provo, UT U.S.A. mne2@et.byu.edu
[3] Brigham Young University Provo, UT U.S.A. seb@byu.edu
[4] M. B. Stephenson & Associates, Provo, UT U.S.A. mbsteph@sandia.gov

Summary. This paper presents an adaptive approach to sweeping one-to-one and many-to-one geometry. The automatic decomposition of many-to-one geometry into one-to-one "blocks" and the selection of an appropriate node projection scheme are vital steps in the efficient generation of high-quality swept meshes. This paper identifies two node projection schemes which are used in tandem to robustly sweep each block of a one-to-one geometry. Methods are also presented for the characterization of one-to-one geometry and the automatic assignment of the most appropriate node projection scheme. These capabilities allow the sweeper to adapt to the requirements of the sweep block being processed. The identification of the two node projection schemes was made after an extensive analysis of existing schemes was completed. One of the node projection schemes implemented in this work, BoundaryError, was selected from traditional node placement algorithms. The second node projection scheme, SmartAffine, is an extension of simple affine transformations and is capable of efficiently sweeping geometry with source and/or target curvature while approximating the speed of a simple transform. These two schemes, when used in this adaptive setting, optimize mesh quality and the speed that swept meshes can be generated while minimizing required user interaction.

Key words: decomposition, hexahedra, mesh generation, node projection, smoothing, sweeping, automatic

1 Introduction

With the use of Finite Element Analysis (FEA) growing across many disciplines, the need for accurate underlying discretizations or meshes has become increasingly important. Because of the inherent difficulty in generating an all-hexahedral mesh on general three-dimensional geometry, methods of producing meshes on special classes of geometry have become popular. One of the most common methods is "sweeping" or "projecting." Sweeping requires that the geometry in question be two-and-one-half dimensional or decomposable into two-and-one-half dimensional sub-geometries (e.g. generalized cylinders). This requirement ensures that the geometry can be meshed with all-hexahedral finite elements [DTJ00].

1.1 One-to-One Sweeping

Sweeping of "one-to-one" geometry begins by identifying "source", "target", and connected "linking" surfaces. The source surface is then usually meshed with quadrilaterals using an unstructured scheme such as paving [T.91]. Each linking surface must also be meshed with a mapped [WW82] or submapped [DLSG95] mesh. The surface mesh on the source is then swept or extruded one layer at a time along the mapped mesh on the linking surfaces toward the target mesh, which may or may not be meshed. This type of sweep is termed "one-to-one" because of the one-to-one correspondence between the source and target surface.

1.2 Many-to-One or Many-to-Many Sweeping

Because of sweeping's strict requirements, few geometries satisfy the topological constraints required to generate a swept mesh. This problem has resulted in the exploration of various methods that decompose more complex geometry into two-and-one-half dimensional sweep "blocks" or "barrels" which can then be swept [T.96, LSGT96, JSPD00, P.98a, MSS, DSS04].

Two such methods, many-to-many and many-to-one, use an internal decomposition approach to generate a series of individual one-to-one sweepable topologies which are then processed by a one-to-one sweep engine. A many-to-many geometry is characterized by multiple source and target surfaces. A many-to-one geometry is characterized by multiple source surfaces and a single target surface. It is important to note that actual decomposition of the geometry does not occur, only an internal characterization of sweep blocks. Sweepable geometries or geometries that may be decomposed into sweepable parts can be detected automatically with a fair amount of success [DTJ00].

1.3 Interior Node Placement

A vital component of the sweep process is the accurate placement of interior points in the volume. Although a number of projection schemes have been proposed for sweeping [LSGT96, P.98a, MSS98, P.99], the problem of efficiently and accurately placing nodes in all types of two-and-one-half dimensional volumes remains elusive. Each scheme has strengths and weaknesses that need to be understood by the user before a sweeping operation can be performed. Requiring the user to understand the details associated with each scheme is a major weakness in current sweeping algorithms. The problems associated with interior point placement schemes can be grouped into three general categories.

Ineffective Capture of Source and Target Curvature If the curvature associated with the source and target surfaces is different, it is important that each interior point reflect this difference in curvature throughout the sweep. If a sweeping algorithm begins at the source and blindly moves toward the target, unconscious of target curvature, a mesh that poorly represents the curved geometry may be produced. Figure 1 shows an example where a node projection scheme did not account for the differences in source and target curvature. Resultant hexahedral elements on the last layer of the sweep (Figure 1(b)) are shown in this example. Notice that the elements reflect the curvature of the source but not the target.

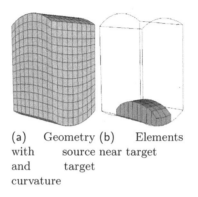

(a)   Geometry (b)   Elements
with   source near target
and   target
curvature

Fig. 1. Example of ineffective capture of source and target curvature

Ineffective Capture of Changes in Linking Curvature Perhaps an even more serious shortcoming in a node projection scheme is the inability to effectively capture changes in linking curvature. Figure 2 shows a geometry that has a drastic change in linking curvature along the sweep axis. The sweep axis originates at the right of the figure and moves toward the left. A node projection scheme that is unable to resolve these changes in linking curvature as the sweep progresses may project nodes outside the volume and produce degenerate elements.

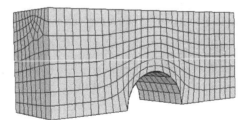

Fig. 2. Geometry with changes in linking curvature

Efficiency Problems The more advanced node projection schemes may suffer from performance problems due to the large number of calculations required to accurately place interior nodes. For example, a basic node projection scheme may be used in connection with a layer smoothing routine [P.98b, P.99]. As each layer of nodes is placed, smoothing is used to reorient the points into a more optimal configuration. These smoothing routines are characteristically slow and on large difficult meshes may render the sweeper useless due to the tremendous slow-down imposed by the smoother. Efficiency also becomes a concern when a geometry is composed of many one-to-one sweep blocks. As the geometry is decomposed and each block is swept, one block may be successfully swept with an efficient basic node projection algorithm while another block may require a computationally intense algorithm because of different, more complex geometric constraints. If a node projection scheme must be selected by the user up front, then the worst case scenario will control and a less efficient, more advanced scheme will be selected. This scheme will mesh both the complicated block as well as the blocks that could be meshed with a faster projection algorithm, resulting in an overall time loss.

1.4 Adaptive Sweeping

The adaptive sweeping approach described in this paper addresses the three general shortcomings described above through the development and implementation of two node projection schemes, SmartAffine and BoundaryError. Both these schemes are used in a many-to-one context, or, in other words, a many-to-one sweep tool was built around them that decomposes a given geometry into one-to-one sweep blocks which are then passed to the most appropriate projection scheme. Each scheme is only used on one-to-one sweep blocks that have been carefully characterized and found well-suited to the relative strengths of that algorithm. In this way, the sweeping algorithm is able to adapt and apply the most appropriate projection scheme on a block-by-block basis. This process is formalized in this paper and called the AdaptiveSweeper.

2 Analysis of Existing Methods

A careful analysis of existing node projection schemes was conducted to understand the relative strengths and weaknesses of each and to develop appropriate schemes for the AdaptiveSweeper. The analysis centered on two main aspects of each algorithm. First, the speed of each algorithm was compared. Second, the quality of the resulting hexahedral elements was compared. Five projection schemes were implemented and analyzed. They are:

- LinearAffine
- Faceted
- BoundaryError
- Smoothing
- Auto

2.1 LinearAffine

The LinearAffine method, developed by Knupp [P.98a], uses an affine transformation matrix and the centroidal locations of the current and next boundary loops to place interior nodes. This type of transformation robustly handles the translation, rotation, and scaling of interior nodes in each sweep layer until the target is reached. A brief explanation of how the transformation is developed and used to place interior nodes follows. For a complete treatment of the subject see [P.98a].

A 3×3 non-singular linear transformation T is computed between current and next boundary loop nodes, x_k and $\tilde{x}_k$ where $k = 1, 2, \cdots, K$ with $K \geq 3$. This transformation T is used with the loop center points, c and $\tilde{c}$, to project interior nodes from the current layer to the next. Equation (1) shows how the transform is used with the loop center points c and $\tilde{c}$ computed in (2) and (3). Notice that x_k and $\tilde{x}_k$ may be replaced with current and next layer interior points.

$$\tilde{x}_k - \tilde{c} = T(x_k - c) \tag{1}$$

$$c = \frac{1}{K} \sum_{k=1}^{K} x_k \tag{2}$$

$$\tilde{c} = \frac{1}{K} \sum_{k=1}^{K} \tilde{x}_k \tag{3}$$

Because, in general, a single T may not exist between arbitrary loops, a least-squares fit to the bounding loop data is performed by minimizing the non-negative function (4) where $u_k = x_k - c$ and $\tilde{u}_k = \tilde{x}_k - \tilde{c}$

$$F(T) = \frac{1}{2} \sum_{k=1}^{K} |\tilde{u}_k - T u_k|^2 \tag{4}$$

Notice that if T is the identity matrix, then loop translation is achieved. Also, if F is not zero at the minimum, then T does not necessarily send all x_k to $\tilde{x}_k$.

2.2 Faceted

The Faceted method is based on the one-to-one BMSweep method introduced by Staten et. al [MSS98], but extended to be usable in a many-to-one setting [MSS]. The Faceted method assumes that a topologically similar mesh is on both the source and target surfaces. It then calculates a faceted mesh using the source surface boundary loops. That faceted representation is used in two ways. First, it is used to calculate the barycentric coordinates of each source interior node on its closest facet. Second, the distance or offset between the newly calculated barycentric coordinate and the source interior node's actual location is determined. These two calculations are also performed for the topologically similar target mesh. The sweep begins at the source surface and the facets are transferred to the next layer of boundary nodes on the linking surfaces. The sweep direction is calculated using the vertices of the facet on the current layer and the vertices of the facet on the next layer. Using the sweep direction, the interpolated barycentric coordinate of the point being projected on the next facet and the interpolated offset information for the next layer, the node can be projected to the next layer. This process is repeated for each node on each layer until the sweep reaches the target surface, which is already meshed.

2.3 BoundaryError

The BoundaryError method, introduced in [T.96] and described in [DSS04], places nodes using the LinearAffine algorithm described above and a subsequent least-squares residual error correction. The BoundaryError method, to successfully capture source and target curvature, calculates the residual error twice, once sweeping from the source surface and terminating at the target surface, and then sweeeping from the target and terminating at the source. These two error distances are then interpolated for final interior node placement. The critical step in this method is the calculation of the residual error, E, defined by (5) which is applied to the current interior node being projected. Once the affine transformation is computed between the current and next layer it is used to project each of the nodes in the current boundary layer to where it believes it should be placed on the next layer. Because the actual location of the boundary node on the next layer is known, the difference, e, between the actual location and the computed location can be computed. This process is repeated for each node in the current boundary list. The next step is to compute the corrected location of each interior node on the next layer. This is done by first projecting the interior node using the affine transformation calculated above. The error, e, associated with each boundary node, and the distance, d, from the current interior node to each boundary node is then used to calculate a least-squares weighted error, E, which is added to the location of the current interior node. Note that n is the number of nodes in the boundary loop.

$$\mathbf{E} = \sum_{i=1}^{n} \frac{\mathbf{e}_i}{\sum_{j=1}^{n} \frac{d_i}{d_j}} \tag{5}$$

The final error, $\mathbf{E}_{final}$, is the linear interpolation of the error, $\mathbf{E}_s$, from the source and the error, $\mathbf{E}_t$, from the target and is defined by (6). The value n_{layers} is the total number of layers in the sweep and i is the layer we are currently creating.

$$\mathbf{E}_{final} = \mathbf{E}_s\left(1 - \frac{i}{n_{layers}}\right) + \mathbf{E}_t\left(\frac{i}{n_{layers}}\right) \tag{6}$$

2.4 Smoothing

The Smoothing method introduced by Knupp [P.99], also uses the LinearAffine method for initial interior node placement. Once the interior points are placed using a simple transformation a traditional structured node smoothing scheme known as weighted Winslow smoothing is used to calculate final point placement. The weight functions associated with this smoother are calculated from the initial source mesh and strive to ensure a faithful copy of the source mesh on all subsequent layers until the target is reached. Specifically, without an attempt at weighting the initial source mesh, any initial biasing of the source mesh would be destroyed by the smoother. In this method the smoothing algorithm is run on every layer of the sweep.

2.5 Auto

The Auto sweeping method is a natural enhancement of the Smoothing method [P.99]. The Smoothing method performs a layer smoothing operation on each layer and is therefore inefficient in terms of algorithmic speed. Because the quality of the affine transformation is determined by how successfully (4) is minimized, a scaled F factor can be computed on a layer by layer basis. This factor will specify the quality of the impending transformation, with zero being a perfect translation that will not require any layer smoothing. This factor may be used to, in effect, turn layer smoothing on and off, depending on the anticipated quality of the affine transformation.

2.6 Timing Results

Because the LinearAffine method is, by far, the least computationally complex of the five methods it was always the fastest algorithm on all the tests conducted. For this reason, it will not be included in the results that follow. It may be assumed that it always outperformed its counterparts in terms of speed. The other four methods were run on thirty test models three different times. Each time the average number of elements required to mesh the model was increased. The timing results were then recorded for all the tests and trends were identified. The results of the testing are show in Table 1.

Table 1. Overall average timing results (seconds/1000 elements created)

Run	Auto	Smooth	Boundary	Faceted
Run 1	0.210	0.515	0.254	0.194
Run 2	0.259	0.568	0.229	0.123
Run 3	0.368	0.767	0.279	0.104

These results can be attributed to the difference in computational intensity of the four algorithms. For example, the Faceted scheme only needs to be aware of the current interior node's location with respect to its matching node on the source and target surface and the locations of the vertices of the current and next facet. The BoundaryError scheme, while less computationally expensive than Smoothing still needs to be aware of the location of all the boundary nodes on the current and next layer and use this information in interior node placement. It is interesting to note that the Auto scheme, while fast when only linear transformations are needed, slowed down considerably because of the need, at times, to perform smoothing on the current layer of interior nodes on complex or highly curved geometry.

2.7 Element Quality Results

Perfect element quality in a hexahedral mesh can only be obtained if all the elements are perfect cubes [P.03]. Because this is impossible to obtain in all cases, the element quality is instead maximized as much as possible to produce a mesh still suitable for finite element analysis. It should be noted that for all types of shape quality metrics, a perfect element will return a quality metric value of one and a degenerate element will return zero. As was done in the timing tests, the results from LinearAffine are not included because it will never outperform the other methods due to the fact it is used as a basic building block in the other four schemes. The minimum element quality was recorded for all the tests during three different runs. The number of elements was increased in each run.

Minimum Element Quality Results

The minimum element quality results help identify where the "weak link" element is in the mesh. If the result is zero, then a negative Jacobian element was produced. The results for all the tests were analyzed and the results are presented in this section.

Table 2 shows the averaged minimum results for the tests. The BoundaryError method generated the highest minimum element quality. The Smoothing and Faceted schemes were within a small percentage of one another.

Table 2. Minimum element quality results

Run	Auto	Smooth	Boundary	Faceted
Run 1	0.331	0.423	0.436	0.394
Run 2	0.299	0.397	0.425	0.387
Run 3	0.287	0.385	0.412	0.377

The BoundaryError method outperformed the other three methods in general. The results for Smoothing and Faceted were within ten percent of the BoundaryError method in every case.

3 Development of Adaptive Node Projection Schemes

The results obtained in Section 2 were used to select and develop the node projection schemes to be implemented in the AdaptiveSweeper. Two schemes were implemented in the sweeper. BoundaryError was selected from the list of existing methods. The second scheme, SmartAffine, is a modification of the LinearAffine method and is presented in Section 3.2. The ideas governing the selection and development process are presented in this section.

3.1 Selection Criteria for BoundaryError

A surprising result of the analysis performed on the node projection methods was the ability of the BoundaryError method to equal or outperform the Smoothing method in terms of the minimum quality of elements generated. This is an important result because it allows for the use of BoundaryError on difficult sweeps instead of much slower smoothing routines. Table 1 shows that, as far as speed is concerned, BoundaryError is, on average, twice as fast as the Smoothing method describe in 2.4. BoundaryError is also able to capture changes in source and target curvature as opposed to Smoothing which is not. Figure 3 shows two cross-sections of a mesh swept with Smoothing and BoundaryError. Notice the difference in the shape of the hexahedral elements on the layers closest to the source and target. Smoothing produces long rectangular elements of low quality while BoundaryError produces high quality elements throughout the mesh. The Faceted method was not chosen over BoundaryError because it did not produce as high quality elements as BoundaryError even though it does have an advantage in terms of speed. Resulting minimum element quality was the controlling factor in the selection of BoundaryError.

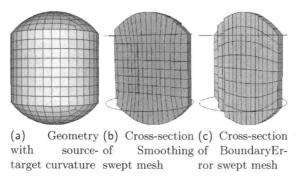

(a)   Geometry (b) Cross-section (c) Cross-section
with   source- of   Smoothing of BoundaryEr-
target curvature swept mesh   ror swept mesh

Fig. 3. Differences in quality of source-target curvature capture between Smoothing and BoundaryError

3.2 The SmartAffine Method

In this section, the SmartAffine method is introduced. The SmartAffine method closely approximates the speed of LinearAffine and is able to robustly sweep much more general problems than the simple translations, rotations and scalings of the source surface that the LinearAffine method is designed to handle. SmartAffine uses simple affine transformations and their inverses to capture source and target curvature. It is also able to, in a limited sense, handle problems that have linking curvature through the interpolation and effective "spreading out" of sweeping errors. Because both SmartAffine and BoundaryError robustly handle changes in source and target curvature it eliminates the need to explicitly make this check during the adaptive sweeping process.

Implementation Details for SmartAffine

The SmartAffine method assumes that the source and target are both meshed with topologically similar meshes. It then performs LinearAffine transformations beginning at the source and target simultaneously. These sweeps will each produce a layer of nodes at the central layer of the volume. The positions of matching nodes on the two layers can then be analyzed to determine the relative error between the source and target sweeps. Let, $\mathbf{p}_s$ and $\mathbf{p}_t$, be the positions of matching nodes at the central layer of the sweep. The point $\mathbf{p}_s$ is the point from the sweep that originated at the source and $\mathbf{p}_t$ is the point from the sweep that originated at the target. We can now calculate two interpolating factors, $n_s = \frac{n_{total}}{2}$ and $n_t = n_{total} - n_s$, where n_{total} is the total number of projections needed to sweep the geometry. It will always be one less than the number of layers in the sweep. This step needs to be performed carefully because n_{total} may be odd.

The error terms to be applied to each half of the total sweep, $\mathbf{e}_s$ and $\mathbf{e}_t$, are calculated in (7) and (8). There will be one such error term associated with each node in both the central source and target layers.

$$\mathbf{e}_s = (\mathbf{p}_t - \mathbf{p}_s)\frac{n_s}{n_{total}} \tag{7}$$

$$\mathbf{e}_t = (\mathbf{p}_s - \mathbf{p}_t)\frac{n_t}{n_{total}} \tag{8}$$

These errors are now applied to each layer of their respective sweeps starting at the mid-layer and working backward to either the source or target surface. Because the error terms were calculated at the central layer they need to be scaled appropriately on each layer of the sweep. This can be accomplished by inverting the affine transformation matrix used to initially place the nodes and interpolating the error term as each layer is processed. The adjusted error terms are presented in (9) and (10), where i represents the current layer being adjusted.

$$\mathbf{e}_{s,adj} = T^{-1}(\mathbf{e}_s \frac{i}{n_s}) \tag{9}$$

$$\mathbf{e}_{t,adj} = T^{-1}(\mathbf{e}_t \frac{i}{n_t}) \tag{10}$$

SmartAffine Results

The SmartAffine method allows for the sweeping of one-to-one geometry that does not meet the simple translation, rotation or scaling requirements of LinearAffine. It also creates meshes nearly as fast as LinearAffine due to the fact that it only proceeds half-way through the geometry before reversing and heading back to where it started. Notice that SmartAffine and LinearAffine are of the same order in terms of algorithmic complexity. Most of the additional computational complexity in SmartAffine can be attributed to inverting the 3×3 affine transformation matrices. Figure 4 shows an example of a geometry with changes in linking curvature. In Figure 4(a) LinearAffine has projected nodes outside of the geometry and produced a degenerate mesh. In Figure 4(b) SmartAffine has produced a mesh with a minimum shape metric of 0.3813, which is very good considering the highly variable curvature in the linking surfaces.

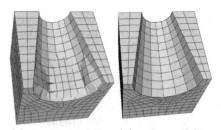

(a)   LinearAffine (b)   SmartAffine projects   points produces a high outside of geometry quality mesh

Fig. 4. Differences in mesh quality between LinearAffine and SmartAffine

Figure 5 shows a mesh with changes in curvature in the linking, source and target surfaces. SmartAffine produced a good quality mesh with a minimum shape metric of 0.4144 and LinearAffine produced a degenerate mesh.

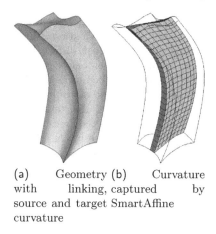

(a)   Geometry (b)   Curvature
with   linking, captured   by
source and target SmartAffine
curvature

Fig. 5. SmartAffine captures linking, source and target curvature

SmartAffine and LinearAffine are comparable in terms of speed, with SmartAffine being, on average, within fifteen to twenty five percent of LinearAffine. For example, a test was run on a model with one million elements and the time spent projecting nodes in LinearAffine was 3.21 seconds while it took SmartAffine 5.51 seconds to perform projections on the same model. This demonstrates that, even on very large models, there is not a sharp divergence in speed between the two.

4 Development of the AdaptiveSweeper

Now that both projection methods are developed and implemented in a many-to-one setting, the next step is the development of routines that automate the task of selecting them so that the strengths of each are captured. To accomplish this, three steps must be followed. First, a close copy of the source mesh must be placed on the target surface. Second, the linking surfaces must be characterized to determine how drastic the changes in linking curvature are from one layer of the sweep to the next. Third, using the information generated in steps one and two, the projection scheme best suited to the geometry is selected. This automated selection of projection schemes allows the sweeper to "adapt" to each sweep block as a many-to-one or one-to-one geometry is processed.

4.1 Creation of the Target Mesh

A high quality copy of the source mesh to the target surface is vital to the overall effectiveness of the projection schemes. If large discrepancies exist between the source and target surface meshes then these errors will be propagated throughout the mesh as the sweeper places interior points. The method used to create the target mesh is based on the mesh morph/copy work of Knupp [P.99]. An affine transformation is computed between the source and target boundary loops. The source interior nodes are then projected as near to the target surface as possible. Because the nodes most likely will not be on the target surface or even near it if the target surface is highly curved, a simple vector calculation is used to iteratively move the nodes in the sweep direction until the target surface is reached. This will ensure that the nodes are in a near optimal position before smoothing is invoked. With the nodes on the surface, they are then smoothed using a weighted Winslow smoothing routine (see 2.4) which preserves biasing and attempts to produce a near copy of the source surface. It is important to note that the iterative vector moving procedure will greatly reduce the number of iterations required to smooth the target mesh and ensure a more faithful copy of the source mesh. On a highly curved target surface, without the iterative vector moving procedure, the smoother would begin with a highly distorted representation of the source mesh and may be unable to recover and produce a close match to the source surface mesh.

4.2 Characterization of Linking Curvature

A more difficult operation is the characterization of changes in linking curvature. To do this (4) is used to measure the quality of the transformation from the current layer to the next layer [P.98a]. As was stated, if F is zero simple loop translation exists, and the larger the value of F the less accurate the affine transformation will be. A layer-by-layer approach is not used in the AdaptiveSweeper because of the efficiency of all the projection schemes. Instead, all the layers are evaluated according to the criteria just described and the worst case transformation is used in projection scheme selection. An additional check that is made is how large the F_{st} (source-to-target) value is when calculated from the source boundary layer to the target boundary layer. It is possible, especially on a very fine mesh, that the layer-by-layer F values remains small while the overall change in shape from the source to the target surface is large. Both checks need to be made in order to fully capture what is occurring on the linking surfaces throughout the sweep.

4.3 Selection of Projection Scheme

Using the results from 4.2, an appropriate projection scheme can be selected for the current sweep block. Conservative threshold values of F were determined experimentally and the results are presented in Table 3.

Table 3. Threshold values of F

F Value	Projection Scheme
$0 \leqslant F < 0.03$ and $F_{st} \leqslant 10$	SmartAffine
$0.03 \leqslant F < \infty$ or $F_{st} \geqslant 10$	BoundaryError

The values in Table 3, coupled with the source and target curvature capture capabilities of each projection method fully define a way to select both node projection schemes in the AdaptiveSweeper on a block-by-block basis. BoundaryError will be used if the F or F_{st} threshold values of the sweep block are exceeded. SmartAffine will be used if neither the F or F_{st} values are exceeded. Notice that, through this method, a complicated many-to-one geometry with multiple blocks can be meshed efficiently without user intervention on a block-by-block basis.

5 Results

The following three examples demonstrate the capabilities of the AdaptiveSweeper. One-to-one and many-to-one geometries were selected that demonstrate each algorithms capability to robustly handle the three types of general problems most common in sweeping (see Section 1.3). Two one-to-one and one many-to-one geometries were selected. Notice that the Faceted method was not included in the tests that follow. It has been demonstrated that, in almost all cases, Faceted produces lower quality elements than BoundaryError (see Table 2). Smoothing was included because traditionally, smoothing schemes coupled with linear projections have been the default method for sweeping difficult geometry. The comparisons that follow between Smoothing and BoundaryError demonstrate why BoundaryError may safely replace Smoothing as a projection method. LinearAffine is included in the tests to demonstrate the similarities in speed between LinearAffine and SmartAffine and SmartAffines superior ability to mesh general one-to-one geometry.

5.1 Test Case 1

Test Case 1 (Figure 6) exhibits changing curvature on a linking surface (Note: The sweep axis is from the top to the bottom of the page). There are no changes in curvature on the source and target, but the target surface is not a scaled version of the source due to the linking curvature change.

Fig. 6. Test Case 1: Variable curvature on a linking surface

Table 4 shows the results for the sweep of Test Case 1. The results demonstrate the ability of SmartAffine to mesh more general geometry than LinearAffine and to approximate the results of the more computationally intense BoundaryError and Smoothing routines. It should also be noted that SmartAffine approximates the results of LinearAffine in terms of speed, and produces a valid mesh while LinearAffine does not. Smoothing produced the highest quality mesh but required 22.36 seconds. The methods in the AdaptiveSweeper for automatic scheme selection chose BoundaryError for Test Case 1 because of a worst-case F value of 0.121 and a F_{st} value of 17.72. 17,525 hexahedral elements were produced for this geometry.

5.2 Test Case 2

Test Case 2 (Figure 7) has variable curvature on the source and target surfaces. Notice that the curvature is slight and the linking surfaces do not have variable curvature.

Table 4. Results of projection methods on Test Case 1

Scheme	Min	Ave	Speed(sec)
Linear	0.0000	0.9071	0.28
Smart	0.3262	0.9089	0.31
Boundary	0.3261	0.9087	1.09
Smooth	0.3814	0.9168	22.36

Fig. 7. Test Case 2: Variable curvature on source and target surfaces

The weaknesses in LinearAffine and Smoothing are highlighted in this example. Even with only slight curvature on the source and target, the resultant quality of the mesh in both cases is very poor. Also, Smoothing required 54.41 seconds to sweep the volume. This is much more than the other methods required. SmartAffine and BoundaryError produced the same mesh with SmartAffine completing in 0.68 seconds as opposed to 3.52 seconds for BoundaryError. As expected, the AdaptiveSweeper selected SmartAffine as the projection scheme of choice to take advantage of the time-savings it provides in this example. The worst-case F value is 0.0 and the F_{st} value is 0.0. 61, 422 hexahedral elements were produced in this example.

Table 5. Results of projection methods on Test Case 2

Scheme	Min	Ave	Speed(sec)
Linear	0.05178	0.9129	0.55
Smart	0.6404	0.9368	0.68
Boundary	0.6404	0.9368	3.52
Smooth	0.1017	0.9456	54.41

5.3 Test Case 3

Test Case 3 shows the results on a many-to-one geometry. As can be seen in Figure 8, Test Case 3 is a many-to-one geometry with six sweep blocks. Notice there are five blocks protruding from the top of the geometry. In the results that follow, the larger of the five protruding blocks will be called block one, the four smaller identical protrusions will be called blocks two through five and the large base will be block six.

Table 6 shows the results for the tests run on Test Case 3. The time required to complete a sweep on each block as well as the overall time were recorded. The overall

Fig. 8. Test Case 3: Many-to-one sweep

minimum and average quality are also reported. The automation schemes in the AdaptiveSweeper assigned BoundaryError to the first five blocks and SmartAffine to block six. In this way the more complicated sweeps one through five took advantage of BoundaryError's residual error correction and block six, which is a simple sweep, benefited from the speed of SmartAffine. As can be seen in Table 6, if BoundaryError had been used on block six, it would have required 34.37 seconds to complete just that block. The AdaptiveSweeper successfully balanced speed and quality in this example. Test Case 3 was meshed with $202,688$ elements.

Table 6. Results for Test Case 3 (Time in seconds)

Scheme	Block 1 Time	Block 2–5 Time	Block 6 Time	Total Time	Min Qual	Ave Qual
Linear	0.11	0.01	1.22	1.40	0.0000	0.9178
Smart	0.14	0.01	1.67	1.85	0.4613	0.9529
Boundary	0.63	0.015	33.68	34.37	0.4618	0.9529
Smooth	12.99	0.22	40.3	54.17	0.01381	0.9348
Adaptive	0.63	0.01	1.61	2.28	0.4618	0.9529

6 Conclusion

The main contribution of this work is the effective automation of node projection schemes on a block-by-block basis. This allows for the optimization of both speed and quality in the generation of finite element meshes on many-to-one geometry. Through the careful characterization of existing projection techniques, two node projection schemes, each having unique strengths were implemented in the AdaptiveSweeper. BoundaryError effectively replaces more computationally intense smoothing routines, while maintaining the quality of the mesh and providing a speed advantage. It is also capable of capturing source and target curvature. SmartAffine, introduced in this work, approximates the speed of LinearAffine while robustly capturing source, target and, to a limited extent, linking curvature. Because both schemes capture source and target curvature, only linking curvature needs to be characterized. This is done through the use of empirically determined F threshold values for the two node projection schemes. All these techniques enhance the automation, speed and quality of swept meshes.

[DLSG95] White D., Mingwu L., Benzley S., and Sjaardema G. Automated Hex-ahedral Mesh Generation by Virtual Decomposition. In *Proceedings 4th International Meshing Roundtable*, pages 165–176. Sandia National Laboratories, 1995.

[DSS04] White D., Saigal S., and Owen S. CCSweep: Automatic Decomposition of Multi–Sweep Volumes. *Engineering with Computers*, 20:222–236, 2004.

[DTJ00] White D., Tautges T., and Timothy J. Automatic Scheme Selection for Toolkit Hex Meshing. *International Journal for Numerical Methods in Engineering*, 49:127–144, 2000.

[JSPD00] Shepherd J., Mitchell S., Knupp P., and White D. Methods for MultiSweep Automation. In *Proceedings 9th International Meshing Roundtable*, pages 77–87. Sandia National Laboratories, 2000.

[LSGT96] Mingwu L., Benzley S., Sjaardema G., and Tautges T. A Multiple Source and Target Sweeping Method for Generating All Hexahedral Finite Element Meshes. In *Proceedings 5th International Meshing Roundtable*, pages 217–225. Sandia National Laboratories, 1996.

[MSS]   Scott M., Benzley S., and Owen S. Improved Many-to-One Sweeping. Submitted for publication.

[MSS98] Staten M., Canaan S., and Owen S. BMSweep: Locating Interior Nodes During Sweeping. In *Proceedings 7th International Meshing Roundtable*, pages 7–18. Sandia National Laboratories, 1998.

[P.98a] Knupp P. Next–Generation Sweep Tool: A Method for Generating All–Hex Meshes On Two–And–One–Half Dimensional Geometries. In *Proceedings 7th International Meshing Roundtable*, pages 505–513. Sandia National Laboratories, 1998.

[P.98b] Knupp P. Winslow Smoothing on Two–Dimensional Unstructured Meshes. In *Proceedings 7th International Meshing Roundtable*, pages 449–457. Sandia National Laboratories, 1998.

[P.99]   Knupp P. Applications of Mesh Smoothing: Copy, Morph, and Sweep on Unstructured Quadrilateral Meshes. *International Journal for Numerical Methods in Engineering*, 45:37–45, 1999.

[P.03]   Knupp P. Algebraic Mesh Quality Metrics for Unstructured Initial Meshes. *Finite Elements in Analysis and Design*, 39:217–241, 2003.

[T.91]   Blacker T. A New Approach to Automated Quadrilateral Mesh Generation. *International Journal for Numerical Methods in Engineering*, 32:811–847, 1991.

[T.96]   Blacker T. The Cooper Tool. In *Proceedings 5th International Meshing Roundtable*, pages 13–29. Sandia National Laboratories, 1996.

[WW82]   Cook W. and Oakes W. Mapped Methods for Generating Three Dimensional Meshes. *Computers in Mechanical Engineering*, pages 67–72, 1982. CIME Research Supplement.

A new least-squares approximation of affine mappings for sweep algorithms

Xevi Roca, Josep Sarrate and Antonio Huerta

Laboratori de Càlcul Numèric, ETSE de Camins Canals i Ports de Barcelona, Universitat Politècnica de Catalunya, Edifici C2, Jordi Girona 1-3, E-08034 Barcelona, Spain
xevi.roca@upc.edu jose.sarrate@upc.edu antonio.huerta@upc.edu **

1 Introduction

The Finite Element Method is currently used to simulate and analyze a wide range of problems in applied sciences and engineering. There are several 3D applications where hexahedral elements are preferred. Hence, the general interest in unstructured hexahedral discretizations has increased. Since an all-hexahedral mesh generation algorithm for any arbitrary geometry is still an unreachable goal, research efforts have been focused on algorithms that decompose the entire geometry into several simpler volumes. In particular, during the last decade significant progress has been made in developing fast and robust sweeping algorithms [Knu98, Knu99, Sta99, Roc04]. Nowadays, the original sweep methods have been modified in order to mesh more complicated geometries allowing multiple source and target geometries [Bla96, Miy00], and multiple axis geometries [Min96].

Given an extrusion volume, the common task of all sweeping algorithms is to identify the *source surfaces*, the corresponding *target surfaces*, and the set of surfaces that join them, called *linking sides*. The source surfaces can be meshed using any structured or unstructured quadrilateral surface mesh generator [Bla91, Cas96, Sar00a, Sar00b]. On the contrary, the linking sides are meshed using any standard structured quadrilateral meshing algorithm, for instance, transfinite interpolation [Tho99]. Then, the source surface meshes are extruded along the sweep direction until they reach the target surfaces. Note that the target surfaces may or may not be previously meshed.

In general, the inner nodes are placed, layer by layer, along the sweep direction. Each layer is delimited by loops of nodes that belong to the structured meshes of the linking-sides. Several algorithms have been developed in order to generate the inner layer of nodes. Most of them generate the new nodes by means of a projection of the source surface mesh onto the inner layers. That is, the inner nodes are located using a least-squares approximation of a linear transformation (the homogeneous

** This work was partially sponsored by the *Ministerio de Educación y Ciencia* under grants DPI2004-03000 and CGL2004-06171-C03-01/CLI, and *E.T.S. d'Enginyers de Camins, Canals i Ports de Barcelona*

part of an affine mapping) between the boundary nodes of the source surfaces and the boundary nodes of the inner layer [Knu98, Bla96].

From the construction point of view, two main strategies are used to compute the position of the inner nodes. The first one computes, starting from the source surface, the position of the new layer from the previous one in an *advanced front manner* [Knu98, Min96]. The second strategy first projects the source surface mesh onto the target surface. Then, the position of the inner layers is computed using a weighted interpolation of the projection of the cap surface meshes [Sta99, Roc04, Bla96].

Most of the proposed algorithms to map meshes between surfaces involve an orthogonal projection of nodes onto the target surface [Goo97]. These projections are expensive from a computational point of view since it is necessary to solve as many root finding problems as internal points are on the mesh of the source surface. In order to address this shortcoming, Roca and co-workers [Roc04] presented a method such that the projection of the source mesh onto the target surface is determined by means of a least-squares approximation of an affine mapping. This affine mapping is defined between the parametric representations of the loops of boundary nodes of the cap surfaces. Once the new mesh is obtained on the parametric space of the target surface, it is mapped up according to the target surface parameterization. It is important to point out that projection algorithms can be used to project meshes both in 2D (parametric space) and 3D geometries (physical space).

The projection algorithms cited above are based on a least-squares approximation of a linear transformation. To this end, several functionals are defined, see sections 2 and 3. However, the minimization of these functionals does not always generate an acceptable projected mesh. For several geometric configurations (for instance, source surfaces with a planar boundary, which are extremely usual in real applications) the minimization of one of the proposed functionals leads to a set of normal equations with a singular system matrix. Nevertheless, these normal equations may be solved using a singular value decomposition algorithm. In this case, if a mesh over a non-planar surface with a planar boundary is projected on an inner layer defined by a planar set of nodes, then a planar mesh is obtained. It is possible to overcome this shortcoming minimizing an alternative functional. However, its minimization may induce a skewness effect on the cross-section of the projected surface mesh. Moreover, its minimization may also lead to a set of normal equations with a singular system matrix, see section 3.

In this paper we propose a new functional that overcomes the drawbacks of the previous formulations. Moreover, we prove that its minimization has a unique solution. Finally, we present several examples in order to assess the robustness of the new formulation and compare it with the previously proposed functionals.

2 Problem Statement

Let $X = \{\mathbf{x}^i\}_{i=1,\dots,m} \subset \mathbb{R}^n$ be a set of source points, and $Y = \{\mathbf{y}^i\}_{i=1,\dots,m} \subset \mathbb{R}^n$ be a set of target points with $m \geq n$. Our goal is to find a mapping $\phi : \mathbb{R}^n \rightarrow \mathbb{R}^n$ such that

$$\mathbf{y}^i = \phi(\mathbf{x}^i), \quad i = 1, \dots, m. \tag{1}$$

We approximate ϕ by an affine mapping φ from $\mathbb{R}^n$ to $\mathbb{R}^n$. This affine mapping is determined by a least-squares fitting of the given data. Thus, we want to find φ

such that minimizes the functional

$$E(\varphi) := \sum_{i=1}^{m} \|\mathbf{y}^i - \varphi(\mathbf{x}^i)\|^2. \tag{2}$$

The above minimization has a clear geometrical meaning: the optimal affine mapping is such that the sum of the square of distances between the target points and the image of the source points is minimized. We also define

$$\mathbf{c}^X := \frac{1}{m} \sum_{i=1}^{m} \mathbf{x}^i \quad \text{and} \quad \mathbf{c}^Y := \frac{1}{m} \sum_{i=1}^{m} \mathbf{y}^i \tag{3}$$

as the geometric centers of the sets X and Y, respectively.

Remark 1. One of the most important practical applications of the least-squares fitting of affine mappings is the projection of a given mesh in a sweeping tool. In these applications X is the set of nodes of the source surface boundary, and Y consists of the loop of nodes that define a given inner layer or the boundary nodes of the target surface discretization.

It is well known that any affine mapping φ from $\mathbb{R}^n$ to $\mathbb{R}^n$ can be written as

$$\varphi(\mathbf{x}) = \mathbf{A}\mathbf{x} + \mathbf{b}',$$

where $\mathbf{A} \in \mathcal{L}(\mathbb{R}^n)$ is a linear transformation (the homogeneous part), and $\mathbf{b}' \in \mathbb{R}^n$ is the affine part. If we consider $\mathbf{b} := \mathbf{b}' + \mathbf{A}\mathbf{c}^X$, then we can write φ as

$$\varphi(\mathbf{x}) = \mathbf{A}(\mathbf{x} - \mathbf{c}^X) + \mathbf{b}. \tag{4}$$

Therefore, without loss of generality, we can write the initial least-squares problem (2) as the minimization of the functional

$$E(\mathbf{A}, \mathbf{b}) := \sum_{i=1}^{m} \|\mathbf{y}^i - \mathbf{A}(\mathbf{x}^i - \mathbf{c}^X) - \mathbf{b}\|^2, \tag{5}$$

where $\mathbf{A} \in \mathcal{L}(\mathbb{R}^n)$ and $\mathbf{b} \in \mathbb{R}^n$.

Remark 2. It is straightforward to prove that if $(\mathbf{A}^E, \mathbf{b}^E) \in \mathcal{L}(\mathbb{R}^n) \times \mathbb{R}^n$ is the optimal solution of (5), then the affine part of (5) is the geometric center of Y, i.e. $\mathbf{b}^E = \mathbf{c}^Y$. Therefore, using (4), the optimal solution maps the center $\mathbf{c}^X$ to the center $\mathbf{c}^Y$.

3 Alternative Formulations

According to Remark 2, the solution of the minimization of E maps the center $\mathbf{c}^X$ to the center $\mathbf{c}^Y$. This property induces, as Knupp does in [Knu98], the definition of the new coordinates $\overline{\mathbf{x}} = \mathbf{x} - \mathbf{c}^X$ and $\overline{\mathbf{y}} = \mathbf{y} - \mathbf{c}^Y$. These new coordinates can be interpreted as translating the sets of points X and Y to the origin, see figure 1(a). Using these new coordinates we have

$$F(\mathbf{A}) := \sum_{i=1}^{m} \|\mathbf{y}^i - \mathbf{c}^Y - \mathbf{A}(\mathbf{x}^i - \mathbf{c}^X)\|^2 = \sum_{i=1}^{m} \|\overline{\mathbf{y}}^i - \mathbf{A}\overline{\mathbf{x}}^i\|^2. \tag{6}$$

Therefore, we are looking for a linear mapping $\mathbf{A}$ such that approximately transforms, in the least-squares sense, $\overline{X} = \{\overline{\mathbf{x}}^i\}_{i=1,\dots,m}$ to $\overline{Y} = \{\overline{\mathbf{y}}^i\}_{i=1,\dots,m}$. Functional (6) is used in [Knu98] in order to reduce the number of degrees of freedom involved in the minimization of functional (5).

However, functional (6) has an important drawback: if the set of source points determines a plane in 3D geometries, or a straight line in 2D applications, then the matrix of the normal equations corresponding to the minimization of functional (6) is singular. Note that this is a usual situation in practical CAD models, see figure 1(b). In order to formalize the analysis of functional F we introduce the following definitions, lemmas and propositions. All of them apply to the general case of $\mathbb{R}^n$. Therefore, we will use the term *hyperplanar* to denote a linear variety of dimension $n - 1$ (a plane for $n = 3$ and a straight line for $n = 2$). From the practical point of view, two cases are important: $n = 2$ used to project sets of points between parametric spaces [Roc04], and $n = 3$ used to project sets of points in the physical space [Knu98, Roc04].

Definition 1 (Hyperplanar set). *A set of points $X = \{\mathbf{x}^i\}_{i=1,\dots,m}$ is hyperplanar if there exists only one hyperplane through all the points in X.*

Remark 3. The definition of hyperplanar set has an interesting geometrical interpretation. It states that there exist n points in X that are linearly independent as affine points. In other words, if we take any point of X, the differences between the rest of points of X and the selected point determine a vectorial subspace of dimension $n - 1$.

Definition 2 (Unitary normal vector). *Let X be a set of points. A unitary normal vector to X is a vector $\mathbf{n}^X \in \mathbb{R}^n$ with $\|\mathbf{n}^X\| = 1$ such that*

$$< \mathbf{n}^X, \mathbf{x}^i > = c, \qquad i = 1, \dots, m, \tag{7}$$

for some $c \in \mathbb{R}$.

Definition 3 (Homogeneous hyperplane). *Let X be a hyperplanar set of points. The homogeneous hyperplane of X is the subspace of vectors*

$$\mathbb{H} = \{\mathbf{v} \in \mathbb{R}^n | < \mathbf{n}^X, \mathbf{v} > = 0\},$$

where $\mathbf{n}^X \in \mathbb{R}^n$ is a normal vector to X.

Lemma 1. *If X is a hyperplanar set, then $\mathbf{c}^X$ is such that*

$$< \mathbf{n}^X, \mathbf{c}^X > = c,$$

where $\mathbf{n}^X$ and c are introduced in Definition 2.

Proof. Since X is a hyperplanar set, equations (7) hold. Adding these m equations, and taking into account that $< \cdot, \cdot >$ is bilinear, then

$$< \mathbf{n}^X, \sum_{i=1}^{m} \mathbf{x}^i > = mc.$$

Dividing both terms of the last equation by m, and using the definition of $\mathbf{c}^X$, see (3), we obtain $< \mathbf{n}^X, \mathbf{c}^X > = c$. □

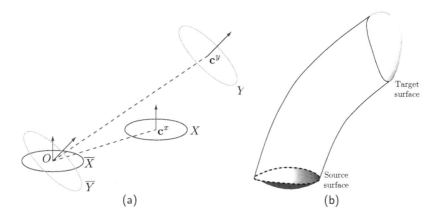

Fig. 1. (a) Geometric representation of the translation of sets X and Y to the origin; (b) Example of a geometry with a source surface defined by a planar boundary loop

Lemma 2. *If X is a hyperplanar set, then*

$$\mathbb{H} = \mathrm{span}(\mathbf{x}^1 - \mathbf{c}^X, \cdots, \mathbf{x}^m - \mathbf{c}^X).$$

Proof. The proof of this Lemma is straightforward from Definition 1, Remark 3, and Lemma 1.

Proposition 1. *If X is hyperplanar, then the minimization of functional F is equivalent to solving n uncoupled overdetermined linear systems of rank $n - 1$.*

Proof. The minimization of functional F is equivalent to imposing the following m constraints

$$\mathbf{A}(\mathbf{x}^i - \mathbf{c}^X) = \mathbf{y}^i, \quad i = 1, \cdots, m. \tag{8}$$

Our unknowns are the coefficients of the $n \times n$ matrix $\mathbf{A}$ which we denote as

$$\mathbf{A} = \begin{pmatrix} a_{1,1} & \cdots & a_{1,n} \\ \vdots & & \vdots \\ a_{n,1} & \cdots & a_{n,n} \end{pmatrix}.$$

Defining

$$\mathbf{X} := \begin{pmatrix} x_1^1 - c_1^X & \cdots & x_1^m - c_1^X \\ \vdots & & \vdots \\ x_n^1 - c_n^X & \cdots & x_n^m - c_n^X \end{pmatrix} \quad \text{and} \quad \mathbf{Y} := \begin{pmatrix} y_1^1 - c_1^Y & \cdots & x_1^m - c_1^Y \\ \vdots & & \vdots \\ y_n^1 - c_n^Y & \cdots & x_n^m - c_n^Y \end{pmatrix},$$

we can write the m constraints (8) as

$$\mathbf{AX} = \mathbf{Y}.$$

Hence, the minimization of F is equivalent to solving

$$\mathbf{X}^T\mathbf{A}^T = \mathbf{Y}^T.$$

This equation is equivalent to solving the following n $(m \times n)$-overdetermined linear systems

$$\mathbf{X}^T\mathbf{a}_k = \mathbf{y}_k, \quad k = 1, \cdots, n,$$

where $\mathbf{a}_k := (a_{k,j})$ for $j = 1, \cdots, n$ and $\mathbf{y}_k = (y_k^l - c_k^Y)$, for $l = 1, \cdots, m$. To conclude, we have to prove that $\mathbf{X}^T$ has rank $n-1$. By Lemma 2, and taking into account that $\dim \mathbb{H} = n - 1$

$$\text{rank}\,\mathbf{X}^T = \dim\,\text{span}(\mathbf{x}^1 - \mathbf{c}^X, \ldots, \mathbf{x}^m - \mathbf{c}^X) = \dim \mathbb{H} = n - 1.$$

$\square$

Remark 4. It is well known that solving a non full-rank overdetermined linear system is equivalent to solving a set of normal equations with singular system matrix [Gil91, Law74].

When X is hyperplanar, we have seen that the minimization of F amounts to solving n uncoupled overdetermined linear systems of rank $n - 1$. Thus, we have n extra degrees of freedom which allow us to find a solution of the minimization of F such that it has $\mathbf{c}^Y - \mathbf{c}^X$ as a fixed vector. This idea, leads to the change of coordinates $\overline{\overline{\mathbf{x}}} = \mathbf{x} - \mathbf{c}^X + \mathbf{c}^Y - \mathbf{c}^X$ and $\overline{\overline{\mathbf{y}}} = \mathbf{y} - \mathbf{c}^X$, see [Knu98] for details. These new coordinates have a clear geometric interpretation: the sets of points X and Y are translated to $\mathbf{c}^Y - \mathbf{c}^X$, see figure 2(a). According to [Knu98], these new coordinates suggest the definition of the following functional

$$G(\mathbf{A}) := \sum_{i=1}^{m} \|\mathbf{y}^i - \mathbf{c}^X - \mathbf{A}(\mathbf{x}^i - \mathbf{c}^X + \mathbf{c}^Y - \mathbf{c}^X)\|^2 = \sum_{i=1}^{m} \|\overline{\overline{\mathbf{y}}}^i - \mathbf{A}\overline{\overline{\mathbf{x}}}^i\|^2. \quad (9)$$

Therefore, we are looking for a linear mapping $\mathbf{A}$ such that it approximately transforms, in the least-squares sense, $\overline{\overline{X}} = \{\overline{\overline{\mathbf{x}}}^i\}_{i=1,\ldots,m}$ to $\overline{\overline{Y}} = \{\overline{\overline{\mathbf{y}}}^i\}_{i=1,\ldots,m}$.

However, functional (9) also leads to normal equations with singular matrix if the vector $\mathbf{c}^Y - \mathbf{c}^X$ lies in the hyperplane determined by the source points, see figure 2(b). Note that this situation is usual in several practical 3D applications if the inner layers are obtained by means of a direct projection from the source surface mesh [Roc04]. On the other hand, this is typically not the case if the position of the new layer is computed from the previous one in an advanced front manner [Knu98].

Proposition 2. *If X is hyperplanar and $\mathbf{c}^Y - \mathbf{c}^X \in \mathbb{H}$, then the minimization of functional G is equivalent to solving n uncoupled overdetermined linear systems of rank $n-1$.*

Proof. This proof only differs from the proof of Proposition 1 on the definitions of matrices $\mathbf{X}$ and $\mathbf{Y}$. In this case, the correspondent matrices are

$$\mathbf{X} := \begin{pmatrix} x_1^1 - c_1^X + c_1^Y - c_1^X & \cdots & x_1^m - c_1^X + c_1^Y - c_1^X \\ \vdots & & \vdots \\ x_n^1 - c_n^X + c_n^Y - c_n^X & \cdots & x_n^m - c_n^X + c_n^Y - c_n^X \end{pmatrix}$$

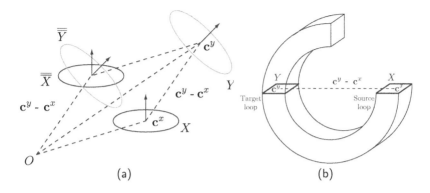

Fig. 2. (a) Geometric representation of the translation of sets X and Y to $\mathbf{c}^Y - \mathbf{c}^X$; (b) Example of a geometry where $\mathbf{c}^Y - \mathbf{c}^X$ lies in the same plane as the source surface and the boundary of the inner layer

and

$$\mathbf{Y} := \begin{pmatrix} y_1^1 - c_1^X & \cdots & x_1^m - c_1^X \\ \vdots & & \vdots \\ y_n^1 - c_n^X & \cdots & x_n^m - c_n^X \end{pmatrix}.$$

To conclude, we have to show that $\mathbf{X}^T$ has rank $n-1$. By assumption $\mathbf{c}^Y - \mathbf{c}^X \in \mathbb{H}$, hence

$$\mathrm{span}(\mathbf{x}^1 - \mathbf{c}^X + \mathbf{c}^Y - \mathbf{c}^X, \ldots, \mathbf{x}^m - \mathbf{c}^X + \mathbf{c}^Y - \mathbf{c}^X) = \mathrm{span}(\mathbf{x}^1 - \mathbf{c}^X, \ldots, \mathbf{x}^m - \mathbf{c}^X).$$

Finally, using this equation, Lemma 2, and taking into account that $\dim \mathbb{H} = n - 1$, we obtain

$$\mathrm{rank}\, \mathbf{X}^T = \dim \mathrm{span}(\mathbf{x}^1 - \mathbf{c}^X + \mathbf{c}^Y - \mathbf{c}^X, \ldots, \mathbf{x}^m - \mathbf{c}^X + \mathbf{c}^Y - \mathbf{c}^X) = \dim \mathbb{H} = n - 1.$$

$\square$

Remark 5. It is also possible to prove that if X is hyperplanar and $\mathbf{c}^Y - \mathbf{c}^X \notin \mathbb{H}$, then the minimization of G is equivalent to solving n uncoupled overdetermined linear systems of rank n.

Remark 6. The minimization of functional G has an additional shortcoming when it is applied to planar sets of points, even in the case of $\mathbf{c}^Y - \mathbf{c}^X \notin \mathbb{H}$. Consider the source surface with a planar boundary and non-planar interior shown in figure 3(a). Assume that we want to project a source surface mesh onto an inner layer (of a sweep volume) defined by a planar boundary, but non-parallel to the source surface. Figure 3(b) shows a cross-section of the source surface, the correspondent cross-section of the computed projection minimizing functional G, and the desired

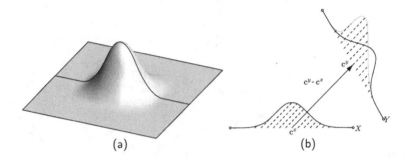

Fig. 3. (a) A source surface with planar boundary and non-planar interior; (b) cross-section view of the source surface and its computed image using the minimization of G (grey line) and the desired solution (black line)

solution. We know that the optimum affine transformation, $\mathbf{A}^G$, has $\mathbf{c}^Y - \mathbf{c}^X$ as fixed vector. Thus, we can observe that the cross-section obtained with $\mathbf{A}^G$ (grey line in figure 3(b)) does not preserve the shape of the original surface.

4 The New Formulation

In order to overcome the drawbacks arising from the minimization of functionals F and G, in this work we propose the following new functional

$$H(\mathbf{A}; \mathbf{u}^X, \mathbf{u}^Y) := \sum_{i=1}^{m} \|\mathbf{y}^i - \mathbf{c}^Y - \mathbf{A}(\mathbf{x}^i - \mathbf{c}^X)\|^2 + \|\mathbf{u}^Y - \mathbf{A}\mathbf{u}^X\|^2, \quad (10)$$

where $\mathbf{u}^X \in \mathbb{R}^n$, and $\mathbf{u}^Y \in \mathbb{R}^n$. It is important to point out that vectors $\mathbf{u}^X$ and $\mathbf{u}^Y$ in (10) can be properly selected in order to obtain several desired properties of functional H.

Proposition 1 states that if X is hyperplanar, the minimization of F is equivalent to solving n uncoupled overdetermined linear systems of rank $n - 1$. Thus, we have n extra degrees of freedom which allow us to find a solution of the minimization of F such that it maps $\mathbf{u}^X$ to $\mathbf{u}^Y$.

Using again the new coordinates introduced by Knupp in [Knu98]: $\overline{\mathbf{x}} = \mathbf{x} - \mathbf{c}^X$ and $\overline{\mathbf{y}} = \mathbf{y} - \mathbf{c}^Y$, we can write

$$H(\mathbf{A}) = \sum_{i=1}^{m} \|\overline{\mathbf{y}}^i - \mathbf{A}\overline{\mathbf{x}}^i\|^2 + \|\mathbf{u}^Y - \mathbf{A}\mathbf{u}^X\|^2.$$

Therefore, we are looking for a linear mapping $\mathbf{A}$ such that it approximately transforms, in the least-squares sense, $\overline{X} = \{\overline{\mathbf{x}}^i\}_{i=1,\ldots,m}$ to $\overline{Y} = \{\overline{\mathbf{y}}^i\}_{i=1,\ldots,m}$, and $\mathbf{u}^X$ to $\mathbf{u}^Y$, see figure 4.

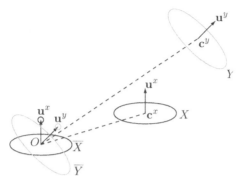

Fig. 4. Geometric representation of the translation of sets X and Y to the origin and the vectors $\mathbf{u}^X$ and $\mathbf{u}^Y$

We will see that the minimization of H, opposite to the minimization of functionals F and G, always leads to a set of normal equations with full rank when X is hyperplanar.

Proposition 3. *If X is a hyperplanar set and $\mathbf{u}^X \notin \mathbb{H}$, then the minimization of functional H is equivalent to solving n uncoupled overdetermined linear systems of rank n.*

Proof. Similar to the proofs of the previous propositions, the minimization of H leads to n uncoupled overdetermined linear systems. In this case we define

$$
\mathbf{X} := \begin{pmatrix} x_1^1 - c_1^X & \dots & x_1^m - c_1^X & u_1^X \\ \vdots & & \vdots & \vdots \\ x_n^1 - c_n^X & \dots & x_n^m - c_n^X & u_n^X \end{pmatrix} \text{ and } \mathbf{Y} := \begin{pmatrix} x_1^1 - c_1^Y & \dots & x_1^m - c_1^Y & u_1^Y \\ \vdots & & \vdots & \vdots \\ x_n^1 - c_n^Y & \dots & x_n^m - c_n^Y & u_n^Y \end{pmatrix}.
$$

By Lemma 2 $\dim \mathbb{H} = \dim \operatorname{span}(\mathbf{x}^1 - \mathbf{c}^X, \dots, \mathbf{x}^m - \mathbf{c}^X) = n - 1$. Since $\mathbf{u}^X \notin \mathbb{H}$, we conclude that $\operatorname{rank} \mathbf{X}^T = \dim \operatorname{span}(\mathbf{x}^1 - \mathbf{c}^X, \dots, \mathbf{x}^m - \mathbf{c}^X, \mathbf{u}^X) = n$.   □

Remark 7. It is also possible to prove that if X generates a linear variety of dimension n (whole $\mathbb{R}^n$) and $\mathbf{u}^X = \mathbf{0}$, then the minimization of H is equivalent to solving n uncoupled overdetermined systems of rank n. From this result, and taking into account Proposition 3, we can conclude that the minimization of H has one and only one solution. Note that we do not consider sets of points X that generate linear varieties of dimension less than $n - 1$. For instance, in $\mathbb{R}^3$ we do not consider source surfaces which degenerate to lines or points, because it does not make sense to sweep them in practical applications.

Remark 8. Vectors $\mathbf{u}^X$ and $\mathbf{u}^Y$ are parameters of functional H. In our implementation we have selected them as:

- X hyperplanar and Y hyperplanar: $\mathbf{u}^X = \mathbf{n}^X$ and $\mathbf{u}^Y = \mathbf{n}^Y$.
- X hyperplanar and Y non-hyperplanar: $\mathbf{u}^X = \mathbf{n}^X$ and $\mathbf{u}^Y = \mathbf{0}$.
- X non-hyperplanar and Y hyperplanar: $\mathbf{u}^X = \mathbf{0}$ and $\mathbf{u}^Y = \mathbf{0}$.
- X non-hyperplanar and Y non-hyperplanar: $\mathbf{u}^X = \mathbf{0}$ and $\mathbf{u}^Y = \mathbf{0}$.

Remark 9. In several applications the source and the target boundaries are not affine. Therefore, it is not possible to obtain an affine transformation that exactly maps X onto Y. In these situations an additional smoothing step is required in order to improve the quality of the final mesh. Hence, our goal is also to obtain a good initial inner node location in order to decrease the number of iterations in the smoothing step. We claim that the minimization of functional H provides better node location than the minimization of functionals F and G. Moreover, this projection algorithm may provide an excellent initial guess for morphing procedures [Knu99].

5 Numerical Examples

In order to assess the advantages and drawbacks of the analyzed functionals used to obtain affine transformations, four examples are presented. These examples are obtained with a sweeping tool that implements the minimizations of functionals F, G, and H. To highlight the analyzed issues, in these examples the inner meshes are obtained projecting directly from the source surface to the inner layers. That is, we have neither used a weighted projection algorithm from both cap surfaces (which we use in practical applications [Roc04]) nor an additional smoothing step to improve the quality of the final mesh. To solve the overdetermined linear systems that do not have full rank, we use a singular value decomposition which supplies the solution with the smallest norm. The set of points X corresponds to the boundary nodes of the source mesh, and the set of points Y corresponds to the boundary nodes of the current inner layer. In all the examples, the source surface has a planar boundary, with non-planar interior. Observe that we have selected source surfaces with planar boundaries in order to force that the minimization of functional F leads to a set of normal equations with singular system matrix. Moreover, the minimization of functional G is only used for source surfaces with planar boundaries. On the contrary, the target surface may be planar or not. Also, in all the examples, the boundary of the source surface is not parallel to the loops of the inner layers. Note that if they were parallel, the minimization of functional G will not produce the skewness effect presented in Remark 6.

In the first example, see figure 5, a C-shaped geometry with circular cross sections is presented. The boundary nodes of the source surface, X, and the boundary nodes of the inner layers, Y, are planar. However, the inner part of the source mesh has curvature. For the minimization of each functional, two views of nine hexahedra layers are provided. The left column is a general view, and the right column is a detail of the fourth, fifth and sixth layers of hexahedra. When we minimize functional F, by Proposition 1, we know that the overdetermined linear system matrix does not have full rank. This implies that the obtained inner layers become flat, despite the source surface has curvature, see figure 5(a). The minimization of G generates inner layers that present curvature on the inner part. However, due to the shape of the geometry the skewness effect appears, see Remark 6. Note that as $\mathbf{c}^Y - \mathbf{c}^X$ tends to the plane defined by X, the skewness effect is more pronounced, see figure 5(b).

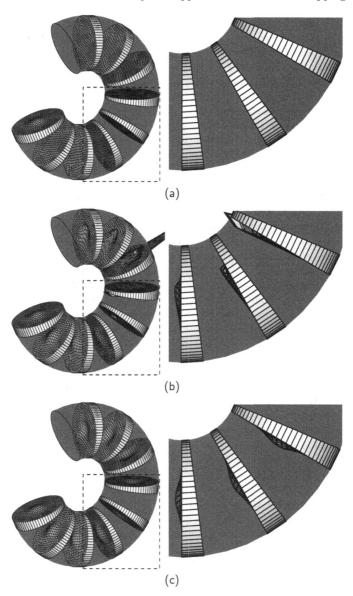

(a)

(b)

(c)

Fig. 5. Projection of a non-planar source surface mesh with planar boundary onto planar inner layers. (**a**) minimizing F; (**b**) minimizing G; and (**c**) minimizing H

In the limit, when the inner layer is on the same plane that the boundary of the source surface, a degenerated projection is obtained (the minimization of G leads to a overdetermined linear system with not full-rank matrix, see Proposition 2). Finally, if we minimize functional H, then the nodes of the inner layers of hexahedra have the desired location and curvature, see figure 5(c). In this example, we see that the minimization of H provides the best location for inner nodes.

The goal of the second example, see figure 6, is to show that the minimization of H gives the best initial inner node location when an additional smoothing step is required. To this end, a square is swept along a semi-circle. The source surface is a planar square with curvature in the inner part, whereas the target surface has a curved boundary. Thus, the inner layers are defined by non-planar loops of nodes that become more curved near the target surface. Four layers of hexahedra are shown from two different points of view. The left column is a general view, and the right column is a detail of the third and the fourth layers of hexahedra. Observe that the source and the target boundaries are not affine. For this reason, the hexahedral elements with nodes on the linking sides present an undesired slope, see figure 6. Therefore, an additional smoothing step will be required, see Remark 9. Like in the first example, minimizing F we obtain layers of hexahedra with flat inner part, see figure 6(a). The minimization of G produces non-flat inner layers of hexahedra, but the skewness effect also appears, see figure 6(b). Finally, minimizing H we obtain an affine transformation that preserves the curvature of the source surface. Since the source and the target surfaces are not affine, the computed solution also presents a slope near the boundary, see figure 6(c). Note that the best initial inner node location is the one provided by the minimization of functional H.

The goal of the third example, see figure 7(a), is to show that the skewness effect introduced by the minimization of functional G may also appear for very simple sweep paths. In particular, this example presents the discretization of an extrusion volume defined by varying cross-sections along a straight and skewed sweep path. These cross-sections are elliptical-shaped with different size, and only the middle of the extrusion path become circular. The source surface is not flat and has a planar boundary, whereas the target surface is planar. For the minimization of each functional, a view of four inner layers of hexahedra elements are presented. Figure 7(b) shows that the minimization of F leads to flat inner layer of elements since the boundary of the source surface is planar. As it is expected, the minimization of G generates non-planar inner layers of hexahedra. However, the shape of the source surface is not well preserved, see the skewness effect in the inner layers presented in figure 7(c). Note that this effect is more pronounced close to the target surface. As in the previous examples, the minimization of H leads to the desired solution, see figure 7(d).

The goal of the last example, see figure 8(a), is to show that if the source and the target surfaces are not affine, the minimization of H provides a better node location than the one obtained with the minimization of F and G (even in the case of geometries simpler than the volume presented in the second example). In this example we discretize an extrusion volume defined by varying cross-sections along a straight and skewed sweep path. The source surface is a planar square with curvature in the inner part. The target surface is planar and its boundary is defined by four arcs. Hence, both surfaces have planar boundaries but not mutually affine. Moreover, the inner layers are defined by planar loops of nodes that become more curved close to the target surface. Note that source surface boundary is not affine to the inner loops of

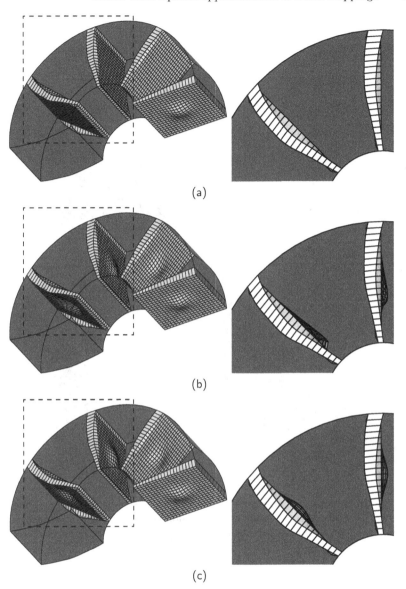

(a)

(b)

(c)

Fig. 6. Projection of a non-planar source surface mesh with planar boundary onto non-planar inner layers. (**a**) minimizing F; (**b**) minimizing G; and (**c**) minimizing H

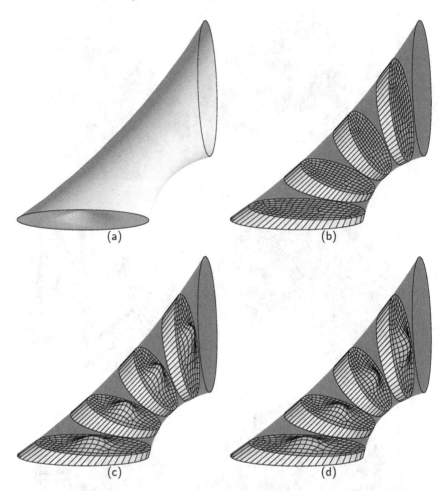

Fig. 7. Projection of a non-planar source surface mesh with planar boundary onto varying shape planar inner layers. (**a**) Extrusion volume to mesh; (**b**) minimizing F; (**c**) minimizing G; and (**d**) minimizing H

nodes. Since the source surface is planar, and similar to the previous examples, the minimization of F generates planar inner layers of hexahedral elements, see figure 8(b). The minimization of G produces skewed layers of elements, see figure 8(c). Finally, the minimization of H preserves the original shape of the source surface, and provides the best initial configuration for the smoothing algorithm.

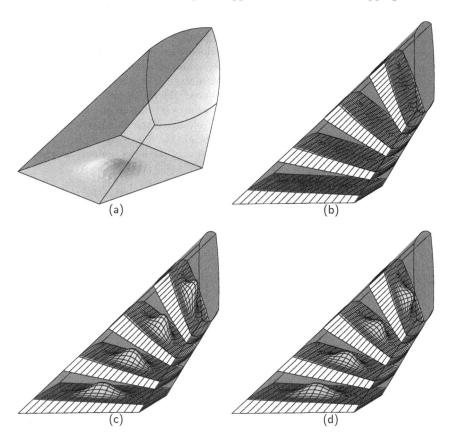

Fig. 8. Projection of a non-planar source surface mesh with planar boundary onto varying shape planar inner layers. The inner loops are not affine to the source surface boundary. (**a**) Extrusion volume to mesh; (**b**) minimizing F; (**c**) minimizing G; and (**d**) minimizing H

6 Concluding Remarks

In this paper we have presented a comparative analysis of several functionals that have been extensively used to project meshes in sweeping procedures. We first stated that the minimization of functional F leads to a set of normal equations with singular system matrix if the source set of points are hyperplanar. We have also proved that the minimization of G leads to normal equations with singular system matrix if, in addition, $\mathbf{c}^Y - \mathbf{c}^X$ lies in the same hyperplane that X. Moreover, we have seen that

448   Xevi Roca, Josep Sarrate and Antonio Huerta

the minimization of G also introduces a skewness effect when the set of points X is hyperplanar, and the sets X and Y are not parallel.

Finally, in order to overcome the previous drawbacks we have proposed the functional H. We have also proved that the minimization of H has one and only one solution. Thus, sets of normal equations with singular system matrix are avoided. Furthermore, if $\mathbf{u}^X$ and $\mathbf{u}^Y$ are properly selected, the minimization of functional H is preferable since it is not affected by the skewness introduced by the minimization of functional G, and tends to preserve the shape of the original source surface. Therefore, it provides suitable node location for the inner layers. In addition, it supplies an excellent initial guess for the position of the inner nodes if an additional smoothing step is required.

[Knu98] Knupp PM (1998) Next-generation sweep tool: a method for generating all-hex meshes on two-and-one-half dimensional geometries. In: 7th Int Meshing Roundtable 505–513

[Knu99] Knupp PM (1999) Applications of mesh smoothing: Copy, morph, and sweep on unstructured quadrilateral meshes. Int J Numer Meth Eng 45:37–45

[Sta99] Staten ML, Canann SA, Owen SJ (1999) BMSweep: Locating interior nodes during sweeping. Eng Comput 15:212–218

[Roc04] Roca X, Sarrate J, Huerta A (2004) Surface mesh projection for hexahedral mesh generation by sweeping. In: 13th Int Meshing Roundtable 169–179

[Bla96] Blacker T (1996) The Cooper Tool. In: 5th Int Meshing Roundtable 13–30

[Miy00] Miyoshi K, Blacker T (2000) Hexahedral mesh generation using multi-axis cooper algorithm. In: 9th Int Meshing Roundtable 89–97

[Min96] Mingwu L, Benzley SE (1996) A multiple source and target sweeping method for generating all-hexahedral finite element meshes. In: 5th Int Meshing Roundtable 217–225

[Bla91] Blacker TD, Stephenson MB (1991) Paving - a new approach to automated quadrilateral mesh generation. Int J Numer Meth Eng 32:811–847

[Cas96] Cass RJ, Benzley SE, Meyers RJ, Blacker TD (1996) Generalized 3-D paving: an automated quadrilateral surface mesh generation algorithm. Int J Numer Meth Eng 39:1475–1489

[Sar00a] Sarrate J, Huerta A (2000) Efficient unstructured quadrilateral mesh generation. Int J Numer Meth Eng 49:1327–1350

[Sar00b] Sarrate J, Huerta A (2000) Automatic mesh generation of nonstructured quadrilateral meshes over curved surfaces in $\mathbb{R}^3$. In: 3rd ECCOMAS

[Tho99] Thompson JF, Soni B, Weatherill N (1999) Handbook of Grid Generation. CRC Press, Boca Raton, London, New York, Washington D.C.

[Goo97] Goodrich D (1997) Generation of All-Quadrilateral Surface Meshers by Mesh Morphing. MA Thesis, Brigham Young University, Utah

[Gil91] Gill PE, Murray W, Wright MH (1991) Numerical Linear Algebra and Optimization. Addison-Wesley, Edwood City

[Law74] Lawson C, Hanson R (1974) Solving Least Squares Problems. Prentice-Hall, Englewood Cliffs

Surface Smoothing and Quality Improvement of Quadrilateral/Hexahedral Meshes with Geometric Flow*

Yongjie Zhang[1], Chandrajit Bajaj[2], and Guoliang Xu[3]

[1] Computational Visualization Center, Institute for Computational Engineering and Sciences, The University of Texas at Austin, USA. jessica@ices.utexas.edu

[2] Computational Visualization Center, Department of Computer Sciences and Institute for Computational Engineering and Sciences, The University of Texas at Austin, USA. bajaj@cs.utexas.edu

[3] LSEC, Institute of Computational Mathematics, Academy of Mathematics and System Sciences, Chinese Academy of Sciences, China. xuguo@lsec.cc.ac.cn

Summary. This paper describes an approach to smooth the surface and improve the quality of quadrilateral/hexahedral meshes with feature preserved using geometric flow. For quadrilateral surface meshes, the surface diffusion flow is selected to remove noise by relocating vertices in the normal direction, and the aspect ratio is improved with feature preserved by adjusting vertex positions in the tangent direction. For hexahedral meshes, besides the surface vertex movement in the normal and tangent directions, interior vertices are relocated to improve the aspect ratio. Our method has the properties of noise removal, feature preservation and quality improvement of quadrilateral/hexahedral meshes, and it is especially suitable for biomolecular meshes because the surface diffusion flow preserves sphere accurately if the initial surface is close to a sphere. Several demonstration examples are provided from a wide variety of application domains. Some extracted meshes have been extensively used in finite element simulations.

Key words: quadrilateral/hexahedral mesh, surface smoothing, feature preservation, quality improvement, geometric flow.

1 Introduction

The quality of unstructured quadrilateral/hexahedral meshes plays an important role in finite element simulations. Although a lot of efforts have been made, it still remains a challenging problem to generate quality quad/hex meshes for complicated structures such as the biomolecule Ribosome 30S shown in Figure 1. We have described an isosurface extraction method to generate quad/hex meshes for

---

*http://www.ices.utexas.edu/~jessica/paper/quadhexgf

arbitrary complicated structures from volumetric data and utilized an optimization-based method to improve the mesh quality [ZBS04] [ZB04] [ZB05], but the surface needs to be smoothed and the mesh quality needs to be further improved.

Geometric partial differential equations (GPDEs) such as Laplacian smoothing have been extensively used in surface smoothing and mesh quality improvement. There are two main methods in solving GPDEs, the finite element method (FEM) and the finite difference method (FDM). Although FDM is not robust sometimes, people still prefer to choosing FDM instead of FEM because FDM is simpler and easier to implement. Recently, a discretized format of the Laplacian-Beltrami (LB) operator over triangular meshes was derived and used in solving GPDEs [MDSB02] [XPB05] [Xu04]. In this paper, we will discretize the LB operator over quadrilateral meshes, and discuss an approach to apply the discretizated format on surface smoothing and quality improvement for quadrilateral or hexahedral meshes.

The main steps to smooth the surface and improve the quality of quadrilateral and hexahedral meshes are as follows:

1. Discretizing the LB operator and denoising the surface mesh - vertex adjustment in the normal direction with volume preservation.
2. Improving the aspect ratio of the surface mesh - vertex adjustment in the tangent direction with feature preservation.
3. Improving the aspect ratio of the volumetric mesh - vertex adjustment inside the volume.

For quadrilateral meshes, generally only Step 1 and Step 2 are required, but all the three steps are necessary for surface smoothing and quality improvement of hexahedral meshes.

Unavoidly the quadrilateral or hexahedral meshes may have some noise over the surface, therefore the surface mesh needs to be smoothed. In this paper, we derive a discretized format of the LB operator, and choose the surface diffusion flow (Equation (1)) to smooth the surface mesh by relocating vertices along their normal directions. The surface diffusion flow is volume preserving and also preserves a sphere accurately if the initial surface mesh is embedded and close to a sphere, therefore it is especially suitable for surface smoothing of biomolecular meshes since biomolecules are usually modelled as a union of hard spheres.

The aspect ratio of the surface mesh can be improved by adjusting vertices in the tangent plane, and surface features are preserved since the movement in the tangent plane doesn't change the surface shape ([Sap01], page 72). For each vertex, the mass center is calculated to find its new position on the tangent plane. Since the vertex tangent movement is an area-weighted relaxation method, it is also suitable for adaptive quadrilateral meshes.

Besides the movement of surface vertices, interior vertices also need to be relocated in order to improve the aspect ratio of hexahedral meshes. The mass center is calculated as the new position for each interior vertex.

Although our relaxation-based method can not guarantee that no inverted element is introduced for arbitrary input meshes, it works well in most cases with the properties of noise removal, feature preservation, mesh quality improvement. Furthermore, it is especially suitable for surface smoothing and quality improvement of biomolecular meshes. As the 'smart' Laplacian smoothing [CTS98] [Fre97], this method is applied only when the mesh quality is improved in order to avoid inverted

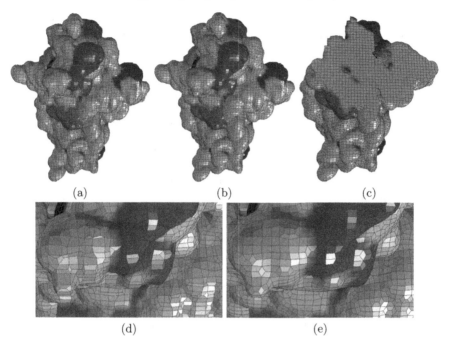

Fig. 1. The comparison of mesh quality of Thermus Thermophilus small Ribosome 30S (1J5E) crystal subunit. The pink color shows 16S rRNA and the remaining colors are proteins. (a) the original quadrilateral mesh (13705 vertices, 13762 quads); (b) the improved quadrilateral mesh; (c) the improved hexahedral mesh (40294 vertices, 33313 hexes); (d) the zoom-in picture of the red box in (a); (e) the zoom-in picture of the red box in (b). The mesh quality is measured by three quality metrics as shown in Figure 2.

elements. This method can also be combined with the optimization-based method to obtain a high quality mesh with relatively less computational cost.

The remainder of this paper is organized as follows: Section 2 reviews the previous related work; Sections 3 discusses the detailed algorithm of the LB operator discretization, surface smoothing and quality improvement of quadrilateral meshes; Sections 4 explains the quality improvement of hexahedral meshes; Section 5 shows some results and applications; The final section presents our conclusion.

2 Previous Work

It is well-known that poor quality meshes result in poorly conditioned stiffness matrices in finite element analysis, and affect the stability, convergence, and accuracy of finite element solvers. Therefore, quality improvement is an important step in mesh generation.

Some quality improvement techniques of triangular and tetrahedral meshes, such as the edge-contraction method, can not be used for quadrilateral and hexahedral meshes because we do not want to introduce any degenerated elements. Therefore, the mesh smoothing methods are selected to improve the quality of quad/hex meshes by adjusting the vertex positions in the mesh while preserving its connectivity. As reviewed in [Owe98] [TW00], Laplacian smoothing and optimization are the two main quality improvement techniques.

As the simplest and most straight forward method for node-based mesh smoothing, Laplacian smoothing relocates the vertex position at the average of the nodes connecting to it [Fie88]. There are a variety of smoothing techniques based on a weighted average of the surrounding nodes and elements [GB98] [ZS00] [SV03]. The averaging method may invert or degrade the local quality, but it is computationally inexpensive and very easy to implement, so it is in wide use. Winslow smoothing is more resistant to mesh folding because it requires the logical variables are harmonic functions [Knu99].

Instead of relocating vertices based on a heuristic algorithm, people utilized an optimization technique to improve mesh quality. The optimization algorithm measures the quality of the surrounding elements to a node and attempts to optimize it [FP00]. The algorithm is similar to a minimax technique used to solve circuit design problems [CC78]. Optimization-based smoothing yields better results but it is more expensive than Laplacian smoothing, and it is difficult to decide the optimized iteration step length. Therefore, a combined Laplacian/optimization-based approach [CTS98] [Fre97] [FOG97] was recommended. Physically-based simulations are used to reposition nodes [LMZ86]. Anisotropic meshes are obtained from bubble equilibrium [SYI97] [BH96].

When we use the smoothing method to improve the mesh quality, it is also important to preserve surface features. Baker [Bak04] presented a feature extraction scheme which is based on estimates of the local normals and principal curvatures at each mesh node. Local parametrization was utilized to improve the surface mesh quality while preserving surface characteristics [GSK02], and two techniques called trapezium drawing and curvature-based mesh improvement were discussed in [SSH04].

Staten et al. [SC97] [Kin97] proposed algorithms to improve node valence for quadrilateral meshes. One special case of cleanup in hexahedral meshes for the whisker weaving algorithm is presented in [MT95]. Schneiders [Sch96] proposed algorithms and a series of templates for quad/hex element decomposition. A recursive subdivision algorithm was proposed for the refinement of hex meshes [BWX02].

3 Quadrilateral Mesh

Noise may exist in quadrilateral meshes, therefore we need to smooth the surface mesh. The quality of some quadrilateral meshes may not be good enough for finite element calculations, and the aspect ratio also needs to be improved.

There are two steps for the surface smoothing and the quality improvement of quadrilateral meshes: (1) the discretization of Laplace-Beltrami opertor and the vertex movement along its normal direction to remove noise, (2) the vertex movement on its tangent plane to improve the aspect ratio while preserving surface features.

3.1 Geometric Flow

Various geometric partial differential equations (GPDEs), such as the mean curvature flow, the surface diffusion flow and Willmore flow, have been extensively used in surface and imaging processing [XPB05]. Here we choose the surface diffusion flow to smooth the surface mesh,

$$\frac{\partial x}{\partial t} = \Delta H(x)\mathbf{n}(x). \tag{1}$$

where Δ is the Laplace-Beltrami (LB) operator, H is the mean curvature and $\mathbf{n}(x)$ is the unit normal vector at the node x. In [EMS98], the existence and uniqueness of solutions for this flow was discussed, and the solution converges exponentially fast to a sphere if the initial surface is embedded and close to a sphere. It was also proved that this flow is area shrinking and volume preserving [XPB05].

In applying geometric flows on surface smoothing and quality improvement over quadrilateral meshes, it is important to derive a discretized format of the LB operator. Discretized schemes of the LB operator over triangular meshes have been derived and utilized in solving GPDEs [MDSB02] [XPB05] [Xu04].

A quad can be subdivided into triangles, hence the discretization schemes of the LB operator over triangular meshes could be easily used for quadrilateral meshes. However, since the subdivision of each quad into triangles is not unique (there are two ways), the resulting discretization scheme is therefore not unique. Additionally in the discretization scheme, the element area needs to be calculated. If we choose to split each quad into two triangles and calculate the area of a quad as the summation of the area of two triangles, then the area calculated from the two different subdivisions could be very different because four vertices of a quad may not be coplanar. Therefore, a unique discretized format of the LB operator directly over quad meshes is required.

3.2 Discretized Laplace-Beltrami Operator

Here we will derive a discretized format for the LB operator over quadrilateral meshes. The basic idea of our scheme is to use the bilinear interpolation to derive the discretized format and to calculate the area of a quad. The discretization scheme is thus uniquely defined.

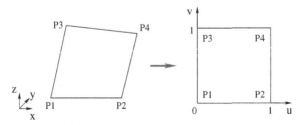

Fig. 2. A quad $[p_1p_2p_4p_3]$ is mapped into a bilinear parametric surface.

Area Calculation: Let $[p_1p_2p_4p_3]$ be a quad in $\mathbb{R}^3$, then we can define a bilinear parametric surface S that interpolates four vertices of the quad as shown in Figure

2:

$$S(u, v) = (1 - u)(1 - v)p_1 + u(1 - v)p_2$$
$$+ (1 - u)vp_3 + uvp_4. \tag{2}$$

The tangents of the surface are

$$S_u(u, v) = (1 - v)(p_2 - p_1) + v(p_4 - p_3), \tag{3}$$
$$S_v(u, v) = (1 - u)(p_3 - p_1) + u(p_4 - p_2). \tag{4}$$

Let ∇ denote the gradient operator about the (x, y, z) coordinates of the vertex P_1, then we have

$$\nabla S_u(u, v) = -(1 - v), \tag{5}$$
$$\nabla S_v(u, v) = -(1 - u). \tag{6}$$

Let A denote the area of the surface $S(u, v)$ for $(u, v) \in [0, 1]^2$, then we have

$$A = \int_0^1 \int_0^1 \sqrt{\| S_u \times S_v \|^2} du dv$$
$$= \int_0^1 \int_0^1 \sqrt{\| S_u \|^2 \| S_v \|^2 - (S_u, S_v)^2} du dv. \tag{7}$$

It may not be easy to obtain the explicit form for integrals in calculating the area, numerical integration quadrature could be used. Here we use the following four-point Gaussian quadrature rule to compute the integral

$$\int_0^1 \int_0^1 f(u, v) du dv \approx \frac{f(q_1) + f(q_2) + f(q_3) + f(q_4)}{4}, \tag{8}$$

where

$$q^- = \frac{1}{2} - \frac{\sqrt{3}}{6}, \quad q^+ = \frac{1}{2} + \frac{\sqrt{3}}{6},$$

$$q_1 = (q^-, \ q^-), \quad q_2 = (q^+, \ q^-),$$
$$q_3 = (q^-, \ q^+), \quad q_4 = (q^+, \ q^+).$$

The integration rule in Equation (8) is of $O(h^4)$, where h is the radius of the circumscribing circle.

Discretized LB Operator: The derivation of the discretized format of the LB operator is based on a formula in differential geometry [MDSB02]:

$$\lim_{diam(R) \to 0} \frac{\nabla A}{A} = \mathbf{H}(p), \tag{9}$$

where A is the area of a region R over the surface around the surface point p, $diam(R)$ denotes the diameter of the region R, and $\mathbf{H}(p)$ is the mean curvature normal.

From Equation (7), we have

$$\nabla A = \int_0^1 \int_0^1 \nabla \sqrt{\| S_u \|^2 \| S_v \|^2 - (S_u, S_v)^2} \, du \, dv$$

$$= \int_0^1 \int_0^1 \frac{S_u(S_v, (v-1)S_v - (u-1)S_u))}{\sqrt{\| S_u \|^2 \| S_v \|^2 - (S_u, S_v)^2}} \, du \, dv$$

$$+ \int_0^1 \int_0^1 \frac{S_v(S_u, (u-1)S_u - (v-1)S_v)}{\sqrt{\| S_u \|^2 \| S_v \|^2 - (S_u, S_v)^2}} \, du \, dv$$

$$= \alpha_{21}(p_2 - p_1) + \alpha_{43}(p_4 - p_3)$$

$$+ \alpha_{31}(p_3 - p_1) + \alpha_{42}(p_4 - p_2), \tag{10}$$

where

$$\alpha_{21} = \int_0^1 \int_0^1 \frac{(1-v)(S_v, (v-1)S_v - (u-1)S_u))}{\sqrt{\| S_u \|^2 \| S_v \|^2 - (S_u, S_v)^2}} \, du \, dv$$

$$\alpha_{43} = \int_0^1 \int_0^1 \frac{v(S_v, (v-1)S_v - (u-1)S_u))}{\sqrt{\| S_u \|^2 \| S_v \|^2 - (S_u, S_v)^2}} \, du \, dv$$

$$\alpha_{31} = \int_0^1 \int_0^1 \frac{(1-u)(S_u, (u-1)S_u - (v-1)S_v)}{\sqrt{\| S_u \|^2 \| S_v \|^2 - (S_u, S_v)^2}} \, du \, dv$$

$$\alpha_{42} = \int_0^1 \int_0^1 \frac{u(S_u, (u-1)S_u - (v-1)S_v)}{\sqrt{\| S_u \|^2 \| S_v \|^2 - (S_u, S_v)^2}} \, du \, dv$$

∇A could be written as

$$\nabla A = \alpha_1 p_1 + \alpha_2 p_2 + \alpha_3 p_3 + \alpha_4 p_4 \tag{11}$$

with

$$\alpha_1 = -\alpha_{21} - \alpha_{31}, \qquad \alpha_2 = -\alpha_{21} - \alpha_{42},$$

$$\alpha_3 = -\alpha_{31} - \alpha_{43}, \qquad \alpha_4 = -\alpha_{43} - \alpha_{42}. \tag{12}$$

Here we still use the four-point Gaussian quadrature rule in Equation (8) to compute the integrals in the α_{ij}. It follows from Equation (12) that $\sum_{i=1}^4 \alpha_i = 0$, we have

$$\nabla A = \alpha_2(p_2 - p_1) + \alpha_3(p_3 - p_1) + \alpha_4(p_4 - p_1). \tag{13}$$

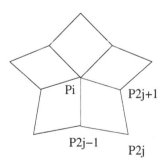

Fig. 3. A neighboring quad $[p_i p_{2j-1} p_{2j} p_{2j+1}]$ around the vertex p_i.

Now let p_i be a vertex with valence n, and p_{2j} $(1 \leqslant j \leqslant n)$ be one of its neighbors on the quadrilateral mesh, then we can define three coefficients $\alpha_2, \alpha_3, \alpha_4$ as in (13). Now we denote these coefficients as α_j^i, β_j^i and γ_j^i for the quad $[p_i p_{2j-1} p_{2j} p_{2j+1}]$ as shown in Figure 3. By using Equation (13), the discrete mean curvature normal can be defined as

$$
\begin{aligned}
\mathbf{H}(p_i) \approx \frac{1}{A(p_i)} \sum_{j=1}^{n} [\alpha_j^i (p_{2j-1} - p_i) \\
+ \beta_j^i (p_{2j+1} - p_i) + \gamma_{j+1}^i (p_{2j} - p_i)] \\
= \sum_{k=1}^{2n} w_k^i (p_k - p_i)
\end{aligned}
\tag{14}
$$

where $\mathbf{H}(p_i)$ denotes the mean curvature normal, $A(p_i)$ is the total area of the quads around p_i, and

$$
w_{2j}^i = \frac{\gamma_j^i}{A(p_i)}, \quad w_{2j-1}^i = \frac{\alpha_j^i + \beta_{j-1}^i}{A(p_i)}, \quad w_{2j+1}^i = \frac{\alpha_{j+1}^i + \beta_j^i}{A(p_i)}.
$$

Using the relation $\Delta x = 2H(p_i)$ ([Wil93], page 151), we obtain

$$
\Delta f(p_i) \approx 2 \sum_{k=1}^{2n} w_k^i (f(p_k) - f(p_i)).
\tag{15}
$$

Therefore,

$$
\begin{aligned}
\Delta H(p_i)\mathbf{n}(p_i) \approx 2 \sum_{k=1}^{2n} w_k^i (H(p_k) - H(p_i))\mathbf{n}(p_i) \\
= 2 \sum_{k=1}^{2n} w_k^i \left[\mathbf{n}(p_i)\mathbf{n}(p_k)^T \mathbf{H}(p_k) - \mathbf{H}(p_i) \right],
\end{aligned}
\tag{16}
$$

where $\mathbf{H}(p_k)$ and $\mathbf{H}(p_i)$ are further discretized by (14). Note that $\mathbf{n}(p_i)\mathbf{n}(p_k)^T$ is a 3×3 matrix.

Figure 4 shows one example of the molecule consisting of three amino acids (ASN, THR and TYR) with 49 atoms. The molecular surface was bumpy as shown in Figure 4(a) since there are some noise existing in the input volumetric data, the surface becomes smooth after the vertex normal movement as shown in Figure 4(b).

3.3 Tangent Movement

In order to improve the aspect ratio of the surface mesh, we need to add a tangent movement in Equation (1), hence the flow becomes

$$
\frac{\partial x}{\partial t} = \Delta H(x)\mathbf{n}(x) + v(x)\mathbf{T}(x),
\tag{17}
$$

where $v(x)$ is the velocity in the tangent direction $\mathbf{T}(x)$. First we calculate the mass center $m(x)$ for each vertex on the surface, then project the vector $m(x) - x$ onto the

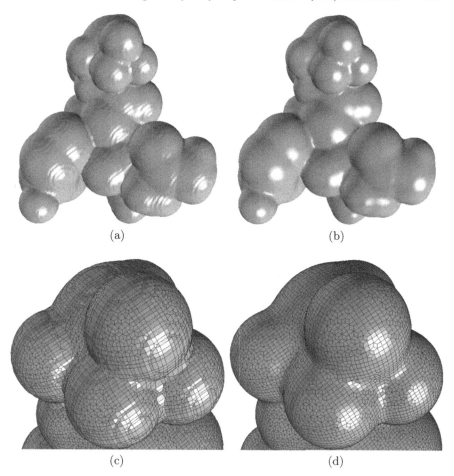

(a)   (b)

(c)   (d)

Fig. 4. Surface smoothing and quality improvement of the molecule consisting of three amino acids (ASN, THR and TYR) with 49 atoms (45534 vertices, 45538 quads). (a) and (c) - the original mesh; (b) and (d) - after surface smoothing and quality improvement.

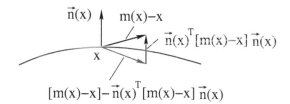

Fig. 5. The tangent movement at the vertex x over a surface. The blue curve represents a surface, and the red arrow is the resulting tangent movement vector.

tangent plane. $v(x)\mathbf{T}(x)$ can be approximated by $[m(x) - x] - \mathbf{n}(x)^T[m(x) - x]\mathbf{n}(x)$ as shown in Figure 5.

Mass Center: A mass center p of a region S is defined by finding $p \in S$, such that

$$\int_S \| y - p \|^2 \, d\sigma = min. \tag{18}$$

S is a piece of surface in $\mathbb{R}^3$, and S consists of quads around vertex x. Then we have

$$\sum(\frac{p_i + p_{2j-1} + p_{2j} + p_{2j+1}}{4} - p_i)A_j = 0, \tag{19}$$

A_j is the area of the quad $[p_i p_{2j-1} p_{2j} p_{2j+1}]$ calculated from Equation (7) using the integration rule in Equation (8). Then we can obtain

$$m(p_i) = \sum_{j=1}^{n}(\frac{p_i + p_{2j-1} + p_{2j} + p_{2j+1}}{4} A_j)/A_{total}^i, \tag{20}$$

where A_{total}^i is the total of quad areas around p_i. The area of a quad can be calculated using Equation (7).

In Figure 4, the vertex tangent movement is used to improve the aspect ratio of the quadrilateral mesh of the molecule consisting of three amino acids. Compared with Figure 4(c), it is obvious that the quadrilateral mesh becomes more regular and the aspect ratio is better as shown in Figure 4(d).

3.4 Temporal Discretization

In the temporal space, $\frac{\partial x}{\partial t}$ is approximated by a semi-implicit Euler scheme $\frac{x_i^{n+1} - x_i^n}{\Delta t}$, where Δt is the time step length. x_i^n is the approximating solution at $t = n\Delta t$, x_i^{n+1} is the approximating solution at $t = (n+1)\Delta t$, and x_i^0 serves as the initial value at x_i.

The spatial and temporal discretization leads to a linear system, and an approximating solution is obtained by solving it using a conjugate gradient iterative method with diagonal preconditioning.

3.5 Discussion

Vertex Normal Movement: The surface diffusion flow can preserve volume. Furthermore, it also preserves a sphere accurately if the initial mesh in embedded and close to a sphere. Suppose a molecular surface could be modelled by a union of hard spheres, so it is desirable to use the surface diffusion flow to evolve the molecular surface. Figure 4 shows one example, the molecular surface becomes more smooth and features are preserved after surface denoising.

Vertex Tangent Movement: If the surface mesh has no noise, we can only apply the tangent movement $\frac{\partial x}{\partial t} = v(x)\mathbf{T}(x)$ to improve the aspect ratio of the mesh while ignoring the vertex normal movement. Our tangent movement has two properties:

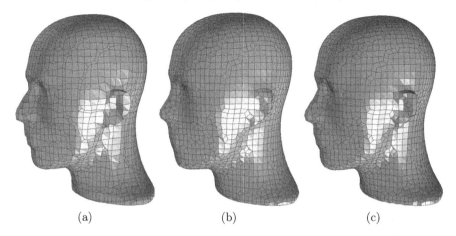

(a)   (b)   (c)

Fig. 6. The quality of a quadrilateral mesh of a human head model is improved (2912 vertices, 2912 quads) after 100 iterations with the time step length 0.01. (a) The original mesh; (b) Each vertex is relocated to its mass center, some facial features are removed; (c) Only tangent movement is applied.

- The tangent movement doesn't change the surface shape ([Sap01], page 72). Figure 6 shows the comparison of the human head model before and after the quality improvement. In Figure 6(b), each vertex is relocated to its mass center, so both normal movement and tangent movement are applied. After some iterations, the facial features, such as the nose, eyes, mouth and ears, are removed. In Figure 6(c), the vertex movement is restricted on the tangent plane, therefore facial features are preserved.
- The tangent movement is an area-weighted averaging method, which is also suitable for adaptive quad meshes as shown in Figure 7 and 8. In Figure 7, there is a cavity in the structure of biomolecule mouse acetylcholinesterase (mAChE), and denser meshes are generated around the cavity while coarser meshes are kept in all other regions. In Figure 8, finer meshes are generated in the region of facial features of the human head.

From Figure 6, 7 and 8, we can observe that after tangent movement, the quadrilateral meshes become more regular and the aspect ratio of the meshes is improved, as well as surface features are preserved.

4 Hexahedral Mesh

There are three steps for surface smoothing and quality improvement of hexahedral meshes, (1) surface vertex normal movement, (2) surface vertex tangent movement and (3) interior vertex relocation.

4.1 Boundary Vertex Movement

The dual contouring hexahedral meshing method [ZBS04] [ZB04] [ZB05] provides a boundary sign for each vertex and each face of a hexahedron, indicating if it lies

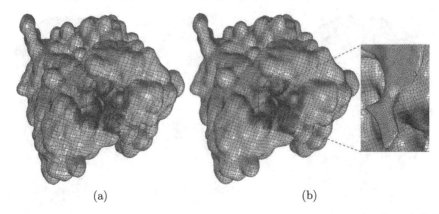

Fig. 7. The quality of an adaptive quadrilateral mesh of a biomolecule mAChE is improved (26720 vertices, 26752 quads). (a) the original mesh; (b) after quality improvement.

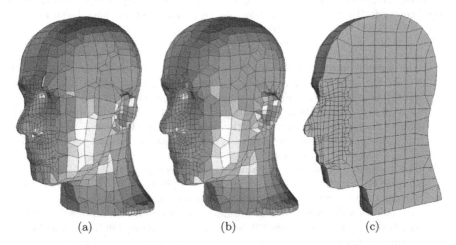

Fig. 8. Adaptive quadrilateral/hexadedral meshes of the human head. (a) the original quad mesh (1828 vertices, 1826 quads); (b) the improved quad mesh; (c) the improved hex mesh (4129 vertices, 3201 hexes), the right part of elements are removed to shown one cross section.

on the boundary surface or not. For example, a vertex or a face is on the surface if its boundary sign is 1, while lies inside the volume if its boundary sign is 0.

The boundary sign for each vertex/face can also be decided by checking the connectivity information of the input hexahedral mesh. If a face is shared by two elements, then this face is not on the boundary; if a face belongs to only one hex, then this face lies on the boundary surface, whose four vertices are also on the boundary surface.

We can use the boundary sign to find the neighboring vertices/faces for a given vertex. For each boundary vertex, we first find all its neighboring vertices and faces lying on the boundary surface by using the boundary sign, then relocate it to its new position calculated from Equation (17). There is a special situation that we need to be careful, a face/edge, whose four/two vertices are on the boundary, may not be a boundary face.

4.2 Interior Vertex Movement

For each interior vertex, we intend to relocate it to the mass center of all its surrounding hexahedra. There are different methods to calculate the volume for a hexahedron. Some people divide a hex into five or six tetrahedra, then the volume of the hex is the summation of the volume of these five or six tetrahedra. This method is not unique since there are various dividing formats. Here we use an trilinear parametric function to calculate the volume of a hex.

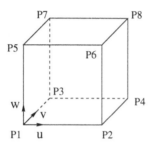

Fig. 9. The trilinear parametric volume V of a hexahedron $[p_1 p_2 \ldots p_8]$.

Volume Calculation: Let $[p_1 p_2 \ldots p_8]$ be a hex in $\mathbb{R}^3$, then we define the trilinear parametric volume $V(u, v, w)$ that interpolates eight vertices of the hex as shown in Figure 9:

$$
\begin{aligned}
V(u, v, w) = {} & (1 - u)(1 - v)(1 - w)p_1 \\
& + u(1 - v)(1 - w)p_2 + (1 - u)v(1 - w)p_3 \\
& + uv(1 - w)p_4 + (1 - u)(1 - v)wp_5 \\
& + u(1 - v)wp_6 + (1 - u)vwp_7 \\
& + uvwp_8.
\end{aligned}
\tag{21}
$$

The tangents of the volume are

$$
\begin{aligned}
V_u(u, v, w) = {} & (1 - v)(1 - w)(p_2 - p_1) + v(1 - w)(p_4 - p_3) \\
& + (1 - v)w(p_6 - p_5) + vw(p_8 - p_7), \\
V_v(u, v, w) = {} & (1 - u)(1 - w)(p_3 - p_1) + u(1 - w)(p_4 - p_2) \\
& + (1 - u)w(p_7 - p_5) + uw(p_8 - p_6), \\
V_w(u, v, w) = {} & (1 - u)(1 - v)(p_5 - p_1) + u(1 - v)(p_6 - p_2) \\
& + (1 - u)v(p_7 - p_3) + uv(p_8 - p_4).
\end{aligned}
$$

Let V denote the volume of $V(u, v, w)$ for $(u, v, w) \in [0, 1]^3$, then we have

$$V = \int_0^1 \int_0^1 \int_0^1 \sqrt{\bar{V}} \, du \, dv \, dw \qquad (22)$$

where

$$\bar{V} = \| (V_u \times V_v) \cdot V_w \|^2 \qquad (23)$$

Numerical integration quadrature could be used. Here we choose the following eight-point Gaussian quadrature rule to compute the integral

$$\int_0^1 \int_0^1 \int_0^1 f(u, v, w) \, du \, dv \, dw \approx \frac{\sum_{j=1}^8 f(q_j)}{8}, \qquad (24)$$

where

$$q^- = \frac{1}{2} - \frac{\sqrt{3}}{6}, \quad q^+ = \frac{1}{2} + \frac{\sqrt{3}}{6},$$

$$
\begin{aligned}
q_1 &= (q^-, \ q^-, \ q^-), & q_2 &= (q^+, \ q^-, \ q^-), \\
q_3 &= (q^-, \ q^+, \ q^-), & q_4 &= (q^+, \ q^+, \ q^-), \\
q_5 &= (q^-, \ q^-, \ q^+), & q_6 &= (q^+, \ q^-, \ q^+), \\
q_7 &= (q^-, \ q^+, \ q^+), & q_8 &= (q^+, \ q^+, \ q^+).
\end{aligned}
$$

The integration rule in Equation (24) is of $O(h^4)$, where h is the radius of the circumscribing sphere.

Mass Center: A mass center p of a region V is defined by finding $p \in V$, such that

$$\int_V \| y - p \|^2 \, d\sigma = min. \qquad (25)$$

V is a piece of volume in $\mathbb{R}^3$, and V consists of hexahedra around vertex x. Then we have

$$\sum (\frac{1}{8} \sum_{j=1}^8 p_j - p_i) V_j = 0, \qquad (26)$$

V_j is the volume of the hex $[p_1 p_2 \ldots p_8]$ calculated from the trilinear function, then we can obtain

$$m(p_i) = \sum_{j \in N(i)} (\frac{1}{8} \sum_{j=1}^8 p_j V_j) / V_{total}^i, \qquad (27)$$

where $N(i)$ is the index set of the one ring neighbors of p_i, and V_{total}^i is the total of hex volume around p_i.

The same Euler scheme is used here for temporal discretization, and the linear system is solved using the conjugate gradient iterative method.

Type	DataSet	MeshSize (Vertex♯, Elem♯)	Scaled Jacobian (best,aver.,worst)	Condition Number (best,aver.,worst)	Oddy Metric (best,aver.,worst)	Inverted Elem♯
quad	Head[1]	(2912, 2912)	(1.0, 0.92, 0.02)	(1.0, 1.13, 64.40)	(0.0, 1.74, 8345.37)	0
	Head[2]	-	(1.0, 0.93, 0.16)	(1.0, 1.11, 6.33)	(0.0, 0.63, 78.22)	
	Head[3]	-	(1.0, 0.96, 0.47)	(1.0, 1.05, 2.12)	(0.0, 0.22, 6.96)	0
	Ribosome 30S[1]	(13705, 13762)	(1.0, 0.90, 0.03)	(1.0, 1.17, 36.90)	(0.0, 1.38, 2721.19)	0
	Ribosome 30S[2]	-	(1.0, 0.90, 0.03)	(1.0, 1.17, 34.60)	(0.0, 1.37, 2392.51)	0
	Ribosome 30S[3]	-	(1.0, 0.93, 0.06)	(1.0, 1.08, 16.14)	(0.0, 0.38, 519.22)	0
hex	Head[1]	(8128, 6587)	(1.0, 0.91, 1.7e-4)	(1.0, 2.99, 6077.33)	(0.0, 29.52, 1.80e5)	2
	Head[2]	-	(1.0, 0.91, 0.005)	(1.0, 1.96, 193.49)	(0.0, 6.34, 5852.23)	0
	Head[3]	-	(1.0, 0.92, 0.007)	(1.0, 1.80, 147.80)	(0.0, 4.50, 1481.69)	0
	Ribosome 30S[1]	(40292, 33313)	(1.0, 0.91, 2.4e-5)	(1.0, 2.63, 4.26e4)	(0.0, 34.15, 2.27e6)	5
	Ribosome 30S[2]	-	(1.0, 0.91, 0.004)	(1.0, 1.74, 263.91)	(0.0, 4.97, 8017.39)	0
	Ribosome 30S[3]	-	(1.0, 0.92, 0.004)	(1.0, 1.59, 237.36)	(0.0, 3.42, 5133.25)	0

Fig. 10. The comparison of the three quality criteria (the scaled Jacobian, the condition number and Oddy metric) before/after the quality improvement for quad/hex meshes of the human head (Figure 12) and Ribosome 30S (Figure 1). DATA[1] – before quality improvement; DATA[2] – after quality improvement using the optimization scheme in [ZB04] [ZB05]; DATA[3] – after quality improvement using the combined geometric flow/optimization-based approach.

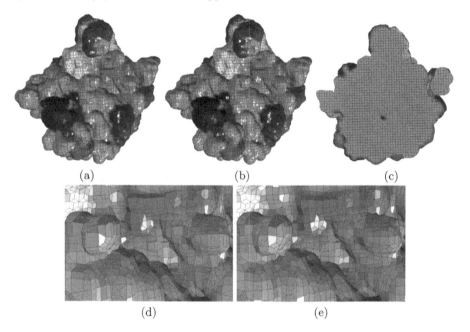

(a)   (b)   (c)

(d)   (e)

Fig. 11. The comparison of mesh quality of Haloarcula Marismortui large Ribosome 50S (1JJ2) crystal subunit. The light yellow and the pink color show 5S and 23S rRNA respectively, the remaining colors are proteins. (a) the original quad mesh (17278 vertices, 17328 quads); (b) the improved quad mesh; (c) the improved hex mesh (57144 vertices, 48405 hexes); (d) the zoom-in picture of the red box in (a); (e) the zoom-in picture of the red box in (b).

5 Results and Applications

There are many different ways to define the aspect ratio for a quad or a hex to measure the mesh quality. Here we choose the scaled Jacobian, the con-

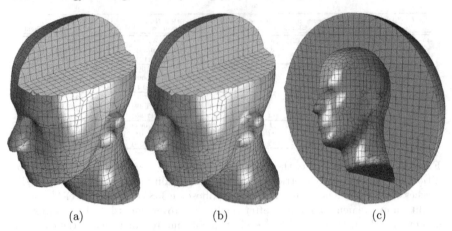

Fig. 12. The comparison of mesh quality of the interior and exterior hexahedral meshes. (a) the original interior hex mesh (8128 vertices, 6587 hexes); (b) the improved interior hex mesh; (c) the improved exterior hex mesh (16521 vertices, 13552 hexes). The mesh quality is measured by three quality metrics as shown in Figure 10.

dition number of the Jacobian matrix and Oddy metric [OGMB88] as our metrics [Knu00a][Knu00b][KM00].

Assume $x \in \mathbb{R}^3$ is the position vector of a vertex in a quad or a hex, and $x_i \in \mathbb{R}^3$ for $i = 1, \ldots, m$ are its neighboring vertices, where $m = 2$ for a quad and $m = 3$ for a hex. Edge vectors are defined as $e_i = x_i - x$ with $i = 1, \ldots, m$, and the Jacobian matrix is $J = [e_1, \ldots, e_m]$. The determinant of the Jacobian matrix is called *Jacobian*, or *scaled Jacobian* if edge vectors are normalized. An element is said to be *inverted* if one of its Jacobians ≤ 0. We use the *Frobenius norm* as a matrix norm, $|J| = (tr(J^T J)^{1/2})$. The condition number of the Jacobian matrix is defined as $\kappa(J) = |J||J^{-1}|$, where $|J^{-1}| = \frac{|J|}{det(J)}$. Therefore, the three quality metrics for a vertex x in a quad or a hex are defined as follows:

$$Jacobian(x) = det(J) \tag{28}$$

$$\kappa(x) = \frac{1}{m}|J^{-1}||J| \tag{29}$$

$$Oddy(x) = \frac{(|J^T J|^2 - \frac{1}{m}|J|^4)}{det(J)^{\frac{4}{m}}} \tag{30}$$

where $m = 2$ for quadrilateral meshes and $m = 3$ for hexahedral meshes.

In [ZB04] [ZB05], an optimization approach was used to improve the quality of quad/hex meshes. The goal is to remove all the inverted elements and improve the worst condition number of the Jacobian matrix. Here we combine our surface smoothing and quality improvement schemes with the optimization-based approach. We use the geometric flow to improve the quality of quad/hex meshes overall and only use the optimization-based smoothing when necessary. Figure 10 shows the comparison of the three quality criteria before and after quality improvement. We can observe that the aspect ratio is improved by using the combined approach.

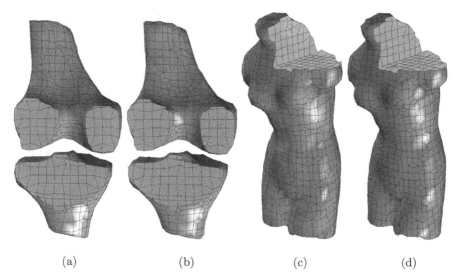

(a)   (b)   (c)   (d)

Fig. 13. The comparison of mesh quality of the human knee and the Venus model. (a) the original hex mesh of the knee (2103 vertices, 1341 hexes); (b) the improved hex mesh of the knee; (c) the original hex mesh of Venus (2983 vertices, 2135 hexes); (d) the improved hex mesh of Venus.

We have applied our surface smoothing and quality improvement technique on some biomolecular meshes. In Figure 4, the surface of a molecule consisting of three amino acids is denoised, the surface quadrilateral mesh becomes more regular and the aspect ratio is improved. The comparison of the quality of quad/hex meshes of Ribosome 30S/50S are shown in Figure 1, Figure 11 and Figure 10. The surface diffusion flow preserves a sphere accurately when the initial mesh is embedded and close to a sphere and the tangent movement of boundary vertices doesn't change the shape, therefore features on the molecular surface are preserved. Our quality improvement scheme also works for adaptive meshes as shown in Figure 7.

From Figure 6 and 8, we can observe that the mesh, especially the surface mesh, becomes more regular and facial features of the human head are preserved as well as the aspect ratio is improved (Figure 10). The interior and exterior hexahedral meshes of the human head as shown in Figure 12 have been used in the electromagnetic scattering simulations. Figure 13 shows the quality improvement of hexahedral meshes, as well as the surface quadrilateral meshes, of the human knee and the Venus model.

Figure 14 shows the quadrilateral meshes for a bubble, which was used in the process of bubble elongation simulation using the boundary element method. First a uniform quad mesh (Figure 14(a)) is extracted from volumetric data for the original state of the bubble, then we use the templates defined in [ZB04] [ZB05] to construct an adaptive mesh as shown in Figure 14(b), the boundary element solutions such as the deformation error are taken as the refinement criteria. Finally we apply our quality improvement techniques to improve the mesh quality. The improved quad mesh is shown in Figure 14(c).

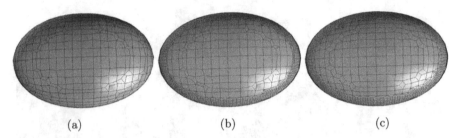

(a)	(b)	(c)

Fig. 14. The comparison of mesh quality of the bubble. (a) a uniform quad mesh (828 vertices, 826 quads); (b) an adaptive quad mesh (5140 vertices, 5138 quads); (c) the improved adaptive quad mesh.

6 Conclusions

We have presented an approach to smooth the surface and improve the quality of quadrilateral and hexahedral meshes. The surface diffusion flow is selected to denoise surface meshes by adjusting each boundary vertex along its normal direction. The surface diffusion flow is volume preserving, and also preserves a sphere accurately when the input mesh is embedded and close to a sphere, therefore it is especially suitable for surface smoothing of biomolecular meshes because biomolecules are usually modelled as a union of hard spheres. The vertex tangent movement doesn't change the surface shape, therefore surface features can be preserved. The interior vertices of hex meshes are relocated to their mass centers in order to improve the aspect ratio. In a summary, our approach has the properties of noise removal, feature preservation and mesh quality improvement. The resulting meshes are extensively used for efficient and accurate finite element calculations.

Acknowledgments

We thank Zeyun Yu for several useful discussions, Jianguang Sun for our system management, Prof. Gregory Rodin for finite element solutions of drop deformation, Prof. Nathan Baker for providing access to the accessibility volume of biomolecule mAChE.

The work at University of Texas was supported in part by NSF grants INT-9987409, ACI-0220037, EIA-0325550 and a grant from the NIH 0P20 RR020647. The work on this paper was done when Prof. Guoliang Xu was visiting Prof. Chandrajit Bajaj at UT-CVC and UT-ICES. Prof. Xu's work was partially supported by the aforementioned grants, the J.T. Oden ICES fellowship and in part by NSFC grant 10371130, National Key Basic Research Project of China (2004CB318000).

[Bak04]   T. Baker. Identification and preservation of surface features. In *13th International Meshing Roundtable*, pages 299–310, 2004.

[BH96]   F. J. Bossen and P. S. Heckbert. A pliant method for anisotropic mesh generation. In *5th International Meshing Roundtable*, pages 63–76, 1996.

[BWX02]   C. Bajaj, J. Warren, and G. Xu. A subdivision scheme for hexahedral meshes. *The Visual Computer*, 18(5-6):343–356, 2002.

[CC78]   C. Charalambous and A. Conn. An efficient method to solve the mini-max problem directly. *SIAM Journal of Numerical Analysis*, 15(1):162–187, 1978.

[CTS98]   S. Canann, J. Tristano, and M. Staten. An approach to combined lapla-cian and optimization-based smoothing for triangular, quadrilateral and quad-dominant meshes. In *7th International Meshing Roundtable*, pages 479–494, 1998.

[EMS98]   J. Escher, U. F. Mayer, and G. Simonett. The surface diffusion flow for immersed hypersurfaces. *SIAM Journal of Mathematical Analysis*, 29(6):1419–1433, 1998.

[Fie88]   D. Field. Laplacian smoothing and delaunay triangulations. *Communications in Applied Numerical Methods*, 4:709–712, 1988.

[FOG97]   L. Freitag and C. Ollivier-Gooch. Tetrahedral mesh improvement using swapping and smoothing. *International Journal Numerical Methathemical Engineeringg*, 40:3979–4002, 1997.

[FP00]   L. Freitag and P. Plassmann. Local optimization-based simplicial mesh untangling and improvement. *International Journal for Numerical Method in Engineering*, 49:109–125, 2000.

[Fre97]   L. Freitag. On combining laplacian and optimization-based mesh smoothing techniqes. *AMD-Vol. 220 Trends in Unstructured Mesh Generation*, pages 37–43, 1997.

[GB98]   P. L. George and H. Borouchaki. *Delaunay Triangulation and Meshing, Application to Finite Elements*. 1998.

[GSK02]   R. V. Garimella, M. J. Shashkov, and P. M. Knupp. Optimization of surface mesh quality using local parametrization. In *11th International Meshing Roundtable*, pages 41–52, 2002.

[Kin97]   P. Kinney. Cleanup: Improving quadrilateral finite element meshes. In *6th International Meshing Roundtable*, pages 437–447, 1997.

[KM00]   C. Kober and M. Matthias. Hexahedral mesh generation for the simula-tion of the human mandible. In *9th International Meshing Roundtable*, pages 423–434, 2000.

[Knu99]   P. Knupp. Winslow smoothing on two dimensional unstructured meshes. *Engineering with Computers*, 15:263–268, 1999.

[Knu00a]   P. Knupp. Achieving finite element mesh quality via optimization of the jacobian matrix norm and associated quantities. part i - a frame-work for surface mesh optimization. *International Journal Numerical Methathemical Engineeringg*, 48:401–420, 2000.

[Knu00b]   P. Knupp. Achieving finite element mesh quality via optimization of the jacobian matrix norm and associated quantities. part ii - a framework for volume mesh optimization and the condition number of the jacobian matrix. *International Journal Numerical Methathemical Engineeringg*, 48:1165–1185, 2000.

[LMZ86]   R. Lohner, K. Morgan, and O. C. Zienkiewicz. Adaptive grid refinement for compressible euler equations. *Accuracy Estimates and Adaptive Re-finements in Finite Element Computations, I. Babuska et. al. eds. Wiley*, pages 281–297, 1986.

[MDSB02]   M. Meyer, M. Desbrun, P. Schröder, and A. Burr. Discrete differential-geometry operators for triangulated 2-manifolds. *VisMath'02, Berlin*, 2002.

[MT95]   S. Mitchell and T. Tautges. Pillowing doublets: Refining a mesh to ensure that faces share at most one edge. In *4th International Meshing Roundtable*, pages 231–240, 1995.

[OGMB88] A. Oddy, J. Goldak, M. McDill, and M. Bibby. A distortion metric for isoparametric finite elements. *Transactions of CSME, No. 38-CSME-32, Accession No. 2161*, 1988.

[Owe98]   S. Owen. A survey of unstructured mesh generation technology. In *7th Internat. Meshing Roundtable*, pages 26–28, 1998.

[Sap01]   G. Sapiro. *Geometric Partial Differential Equations and Image Analysis*. Cambridge, University Press, 2001.

[SC97]   M. Staten and S. Canann. Post refinement element shape improvement for quadrilateral meshes. *AMD-Trends in Unstructured Mesh Generation*, 220:9–16, 1997.

[Sch96]   R. Schneiders. Refining quadrilateral and hexahedral element meshes. In *5th International Conference on Grid Generation in Computational Field Simulations*, pages 679–688, 1996.

[SSH04]   I. B. Semenova, V. V. Savchenko, and I. Hagiwara. Two techniques to improve mesh quality and preserve surface characteristics. In *13th International Meshing Roundtable*, pages 277–288, 2004.

[SV03]   S. M. Shontz and S. A. Vavasis. A mesh warping algorithm based on weighted laplacian smoothing. In *12th International Meshing Roundtable*, pages 147–158, 2003.

[SYI97]   K. Shimada, A. Yamada, and T. Itoh. Anisotropic triangular meshing of parametric surfces via close packing of ellipsoidal bubbles. In *6th International Meshing Roundtable*, pages 375–390, 1997.

[TW00]   S.-H. Teng and C. W. Wong. Unstructured mesh generation: Theory, practice, and perspectives. *International Journal of Computational Geometry and Applications*, 10(3):227–266, 2000.

[Wil93]   T. J. Willmore. Riemannian geometry. *Clareden Press, Oxford*, 1993.

[XPB05]   G. Xu, Q. Pan, and C. Bajaj. Discrete surface modelling using partial differential equations. *Accepted by Computer Aided Geomtric Design*, 2005.

[Xu04]   G. Xu. Discrete laplace-beltrami operators and their convergence. *CAGD*, 21:767–784, 2004.

[ZB04]   Y. Zhang and C. Bajaj. Adaptive and quality quadrilateral/hexahedral meshing from volumetric data. In *13th International Meshing Roundtable*, pages 365–376, 2004.

[ZB05]   Y. Zhang and C. Bajaj. Adaptive and quality quadrilateral/hexahedral meshing from volumetric data. *Computer Methods in Applied Mechanics and Engineering (CMAME), in press, www.ices.utexas.edu/~jessica/paper/quadhex*, 2005.

[ZBS04]   Y. Zhang, C. Bajaj, and B-S Sohn. 3d finite element meshing from imaging data. *Special issue of Computer Methods in Applied Mechanics and Engineering on Unstructured Mesh Generation, in press*, 2004.

[ZS00]   T. Zhou and K. Shimada. An angle-based approach to two-dimensional mesh smoothing. In *9th International Meshing Roundtable*, pages 373–384, 2000.

Quality Improvement of Surface Triangulations

R. Montenegro, J.M. Escobar, G. Montero and E. Rodríguez

Institute for Intelligent Systems and Numerical Applications in Engineering, University of Las Palmas de Gran Canaria, Campus Universitario de Tafira, 35017 Las Palmas de Gran Canaria, Spain, rafa@dma.ulpgc.es

This paper presents a new procedure to improve the quality of triangular meshes defined on surfaces. The improvement is obtained by an iterative process in which each node of the mesh is moved to a new position that minimizes certain objective function. This objective function is derived from an algebraic quality measures of the local mesh (the set of triangles connected to the adjustable or *free node*). The optimization is done in the *parametric mesh*, where the presence of barriers in the objective function maintains the free node inside the *feasible region*. In this way, the original problem on the surface is transformed into a two-dimensional one on the *parametric space*. In our case, the parametric space is a plane, chosen in terms of the local mesh, in such a way that this mesh can be optimally projected performing a *valid* mesh, that is, without *inverted* elements. In order to show the efficiency of this smoothing procedure, its application is presented.

1 Introduction

For 2-D or 3-D meshes the quality improvement [Knu] can be obtained by an iterative process in which each node of the mesh is moved to a new position that minimizes an objective function [Fre]. This function is derived from a quality measure of the local mesh. We have chosen, as a starting point in section 2, a 2-D objective function that presents a barrier in the boundary of the *feasible region* (set of points where the free node could be placed to get a *valid* local mesh, that is, without *inverted elements*). This barrier has an important role because it avoids the optimization algorithm to create a tangled mesh when it starts with a valid one. Nevertheless, objective functions constructed by algebraic quality measures are only directly applicable to inner nodes of 2-D or 3-D meshes, but not to its boundary nodes. To overcome this problem, the local mesh, $M(p)$, sited on a surface Σ, is orthogonally projected on a plane P (the existence and search of this plane will be discuss in section 3) in such a way that it performs a valid local mesh $N(q)$. Therefore, it can be said that $M(p)$ is *geometrically conforming* with respect to P [Fre]. Here p is the free node on Σ and q is its projection on P. The optimization of $M(p)$ is got by the appropriated optimization of $N(q)$. To do this we try to get *ideal* triangles in $N(q)$ that become equilateral in $M(p)$. In general, when the local mesh $M(p)$ is on a surface, each triangle is placed on a different plane and it is not possible to define a feasible region on Σ. Nevertheless, this region is perfectly defined in $N(q)$ as it is analyzed in section 2.1.

To construct the objective function in $N(q)$, it is first necessary to define the objective function in $M(p)$ and, afterward, to establish the connection between them. A crucial aspect for this construction is to keep the barrier of the 2-D objective function. This is done with a suitable approximation in the process that transforms the original problem on Σ into an entirely two-dimensional one on P. We develop this approximation in section 2.2.

The optimization of $N(q)$ becomes a two-dimensional iterative process. The optimal solutions of each two-dimensional problem form a sequence $\{\mathbf{x}^k\}$ of points belonging to P. We have checked in many numerical test that $\{\mathbf{x}^k\}$ is always a convergent sequence. It is important to underline that this iterative process only takes into account the position of the free node in a discrete set of points, the points on Σ corresponding to $\{\mathbf{x}^k\}$ and, therefore, it is not necessary that the surface is smooth. Indeed, the surface determined by the piecewise linear interpolation of the initial mesh is used as a reference to define the geometry of the domain.

If the node movement only responds to an improvement of the quality of the mesh, it can happen that the optimized mesh loses details of the original surface. To avoid this problem, every time the free node p is moved on Σ, the optimization process only allows a small distance between the centroid of the triangles of $M(p)$ and the underlaying surface (the true surface, if it is known, or the piece-wise linear interpolation, if it is not).

There are several alternatives to the previous method. For example, Garimella et al. [Gar] develop a method to optimize meshes in which the nodes of the optimized mesh are kept close to the original positions by imposing the Jacobians of the current and original meshes to be also close. Frey et al. [Fre2] get a control of the gap between the mesh and the surface by modifying the element-size (subdividing the longest edges and collapsing the shortest ones) in terms of an approximation of the smallest principal curvatures radius associated to the nodes. Rassineux et al. [Ras] also use the smallest principal curvatures radius to estimate the element-size compatible with a prescribed gap error. They construct a geometrical model by using the Hermite diffuse interpolation in which local operations like edge swapping, node removing, edge splitting, etc. are made to adapt the mesh size and shape. More accurate approaches, that have into account the directional behavior of the surface, have been considered in by Vigo [Vig] and, recently, by Frey in [Fre3].

Application of our proposed optimization technique is shown in section 4.

2 Construction of the Objective Function

As it is shown in [Fre], [Knu1], and [Knu2] we can derive optimization functions from *algebraic quality measures* of the elements belonging to a local mesh. Let us consider a triangular mesh defined in $\mathbb{R}^2$ and let t be an triangle in the physical space whose vertices are given by $\mathbf{x}_k = (x_k, y_k)^T \in \mathbb{R}^2$, $k = 0, 1, 2$. First, we are going to introduce an algebraic quality measure for t. Let t_R be the reference triangle with vertices $\mathbf{u}_0 = (0,0)^T$, $\mathbf{u}_1 = (1,0)^T$, and $\mathbf{u}_2 = (0,1)^T$. If we choose $\mathbf{x}_0$ as the translation vector, the affine map that takes t_R to t is $\mathbf{x} = A\mathbf{u} + \mathbf{x}_0$, where A is the Jacobian matrix of the affine map referenced to node $\mathbf{x}_0$, given by $A = (\mathbf{x}_1 - \mathbf{x}_0, \mathbf{x}_2 - \mathbf{x}_0)$. We will denote this type of affine maps as $t_R \overset{A}{\to} t$. Let now t_I be an *ideal* triangle (not necessarily equilateral) whose vertices are $\mathbf{w}_k \in \mathbb{R}^2$, $(k = 0, 1, 2)$ and let $W_I = (\mathbf{w}_1 - \mathbf{w}_0, \mathbf{w}_2 - \mathbf{w}_0)$ be the Jacobian matrix, referenced to node $\mathbf{w}_0$, of the affine map $t_R \overset{W_I}{\to} t_I$; then, we define $S = AW_I^{-1}$ as the weighted Jacobian matrix of the affine map $t_I \overset{S}{\to} t$. In the particular case that t_I was the equilateral triangle t_E, the Jacobian matrix $W_I = W_E$ will be defined by $\mathbf{w}_0 = (0,0)^T$, $\mathbf{w}_1 = (1,0)^T$ and $\mathbf{w}_2 = (1/2, \sqrt{3}/2)^T$.

We can use matrix norms, determinant or trace of S to construct algebraic quality measures of t. For example, the Frobenius norm of S, defined by $|S| = \sqrt{tr(S^T S)}$, is specially indicated because it is easily computable. Thus, it is shown in [Knu] that $q_\eta = \frac{2\sigma}{|S|^2}$ is an algebraic quality measure of t , where $\sigma = \det(S)$. We use this quality measure to construct an objective function. Let $\mathbf{x} = (x, y)^T$ be the position vector of the free node, and let S_m be the weighted Jacobian matrix of the m-th triangle of a valid local mesh of M triangles. The objective function associated to m-th triangle is $\eta_m = \frac{|S_m|^2}{2\sigma_m}$, and the corresponding objective function for the local mesh is the n-norm of $(\eta_1, \eta_2, \ldots, \eta_M)$,

$$|K_\eta|_n (\mathbf{x}) = \left[\sum_{m=1}^{M} \eta_m^n (\mathbf{x}) \right]^{\frac{1}{n}} \tag{1}$$

This objective function presents a barrier in the boundary of the feasible region that avoids the optimization algorithm to create a tangled mesh when it starts with a valid one.

Previous considerations and definitions are only directly applicable for 2-D (or 3-D) meshes, but some of them must be properly adapted when the meshes are located on an arbitrary surface. For example, the concept of valid mesh is not clear in this situation because neither the concept of inverted element is. We will deal with these questions in next subsections.

2.1 Similarity Transformation for Surface and Parametric Meshes

Suppose that for each local mesh $M(p)$ placed on the surface Σ, that is, with all its nodes on Σ, it is possible to find a plane P such that the orthogonal projection of $M(p)$ on P is a valid mesh $N(q)$. Moreover, suppose that we define the axes in such a way that the x, y-plane coincide with P. If, in the feasible region of $N(q)$, it is possible to define the surface Σ by the parametrization $\mathbf{s}(x, y) = (x, y, f(x, y))$, where f is a continuous function, then, we can optimize $M(p)$ by an appropriate optimization of $N(q)$. We will refer to $N(q)$ as the *parametric mesh*. The basic idea consists on finding the position $\bar{q}$ in the feasible region of $N(q)$ that makes $M(p)$ be an optimum local mesh. To do this, we search *ideal* elements in $N(q)$ that become equilateral in $M(p)$. Let $\Delta t \in M(p)$ be a triangular element on Σ whose vertices are given by $\mathbf{y}_k = (x_k, y_k, z_k)^T$, $(k = 0, 1, 2)$ and t_R be the reference triangle in P (see Figure 1). If we choose $\mathbf{y}_0$ as the translation vector, the affine map $t_R \overset{A_\pi}{\to} \Delta t$ is $\mathbf{y} = A_\pi \mathbf{u} + \mathbf{y}_0$, where A_π is its Jacobian matrix, given by

$$A_\pi = \begin{pmatrix} x_1 - x_0 & x_2 - x_0 \\ y_1 - y_0 & y_2 - y_0 \\ z_1 - z_0 & z_2 - z_0 \end{pmatrix} \tag{2}$$

Now, consider that $t \in N(q)$ is the orthogonal projection of Δt on P. Then, the vertices of t are $\mathbf{x}_k = \Pi \mathbf{y}_k = (x_k, y_k)^T$, $(k = 0, 1, 2)$, where $\Pi = (\mathbf{e}_1, \mathbf{e}_2)^T$ is 2×3 matrix of the affine map $\Delta t \xrightarrow{\Pi} t$, being $\{\mathbf{e}_1, \mathbf{e}_2, \mathbf{e}_3\}$ the canonical basis in $\mathbb{R}^3$ (the associated projector from $\mathbb{R}^3$ to P, considered as a subspace of $\mathbb{R}^3$, is $\Pi^T \Pi$). Taking $\mathbf{x}_0$ as translation vector, the affine map $t_R \xrightarrow{A_P} t$ is $\mathbf{x} = A_P \mathbf{u} + \mathbf{x}_0$, where $A_P = \Pi A_\pi$ is its Jacobian matrix

$$A_P = \begin{pmatrix} x_1 - x_0 & x_2 - x_0 \\ y_1 - y_0 & y_2 - y_0 \end{pmatrix} \tag{3}$$

Therefore, the 3×2 matrix of the affine map $t \xrightarrow{T} \Delta t$ is

$$T = A_\pi A_P^{-1} \tag{4}$$

Let V_π be the subspace spanned by the column vectors of A_π and let π be the plane defined by V_π and the point $\mathbf{y}_0$. Our goal is to find the *ideal* triangle $t_I \subset P$, moving q on P, such that t_I is mapped by T into an equilateral one, $\Delta t_E \subset \pi$. In general, the strict fulfillment of this requirement is only possible if $N(q)$ is formed by a unique triangle.

Due to $\text{rank}(A_\pi) = \text{rank}(A_P) = 2$, it exists a unique factorization $A_\pi = QR$, where Q is an orthogonal matrix $(Q^T Q = I)$ and R is an upper triangular one with $[R]_{ii} > 0$ $(i = 1, 2)$. The columns of the 3×2 matrix Q define an orthonormal basis $\{\mathbf{q}_1, \mathbf{q}_2\}$ that spans V_π, so we can see Q as the matrix of the affine map $t_R \xrightarrow{Q} \Delta t_R$ and R as the 2×2 Jacobian matrix of the affine map $\Delta t_R \xrightarrow{R} \Delta t$ (see Figure 1). As $t_R \xrightarrow{W_E} t_E$ and Q is an orthogonal matrix that keeps the angles and norms of the vectors, then $t_E \xrightarrow{Q} \Delta t_E$ and, therefore

$$QW_E = A_\pi R^{-1} W_E \tag{5}$$

is the 3×2 Jacobian matrix of affine map $t_R \xrightarrow{QW_E} \Delta t_E$. On the other hand, we define on the plane π

$$S = RW_E^{-1} \tag{6}$$

as the 2×2 weighted Jacobian matrix of the affine map that transforms the equilateral triangle into the physical one, that is, $\Delta t_E \xrightarrow{S} \Delta t$.

We have chosen as ideal triangle in π the equilateral one ($\Delta t_I = \Delta t_E$), then, the Jacobian matrix W_I of the affine map $t_R \xrightarrow{W_I} t_I$ is calculated by imposing the condition $TW_I = QW_E$, because $t_R \xrightarrow{TW_I} \Delta t_I$ and $t_R \xrightarrow{QW_E} \Delta t_E$. Taking into account (5), it yields

$$TW_I = A_\pi R^{-1} W_E \tag{7}$$

and, from (4), we obtain

$$W_I = A_P R^{-1} W_E \tag{8}$$

so we define on P the *ideal-weighted* Jacobian matrix of the affine map $t_I \xrightarrow{S_I} t$ as $S_I = A_P W_I^{-1}$. From (8) it results

$$S_I = A_P W_E^{-1} R A_P^{-1} \tag{9}$$

and, from (6)

$$S_I = A_P W_E^{-1} S W_E A_P^{-1} = A_P W_E^{-1} S \left(A_P W_E^{-1}\right)^{-1} = S_E S S_E^{-1} \tag{10}$$

where $S_E = A_P W_E^{-1}$ is the *equilateral-weighted* Jacobian matrix of the affine map $t_E \overset{S_E}{\to} t$. Finally, from (10), we obtain the next similarity transformation.

$$S = S_E^{-1} S_I S_E \qquad (11)$$

Therefore, it can be said that the matrices S and S_I are *similar*.

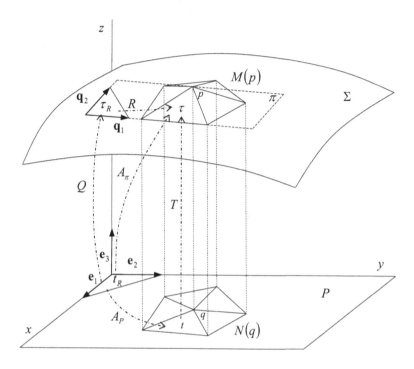

Fig. 1. Local surface mesh $M(p)$ and its associated parametric mesh $N(q)$

2.2 Optimization on the Parametric Space

It might be used S, as it is defined in (6), to construct the objective function and, then, solve the optimization problem. Nevertheless, this procedure has important disadvantages. First, the optimization of $M(p)$, working on the true surface, would require the imposition of the constraint $p \in \Sigma$. It would complicate the resolution of the problem because, in many cases, Σ is not defined by a smooth function. Moreover, when the local mesh $M(p)$ is on a curved surface, each triangle is sited on a different plane and the objective function, constructed from S, lacks barriers. It is impossible to define a feasible region in the same way as it was done at the beginning of this section. Indeed, all the positions of the free node, except those that make $\det(S) = 0$ for any triangle, produce correct triangulations of $M(p)$. However, for many purposes as, for example, to construct a 3-D mesh from the surface triangulation, there are unacceptable positions of the free node.

To overcome these difficulties we propose to carry out the optimization of $M(p)$ in an indirect way, working on $N(q)$. With this approach the movement of the free node will be restricted to the feasible region of $N(q)$, which avoids to construct unacceptable surface triangulations. It all will be carried out using an approximate version of the similarity transformation given in (11).

Let us consider that $\mathbf{x} = (x, y)^T$ is the position vector of the free node q, sited on the plane P. If we suppose that Σ is parametrized by $\mathbf{s}(x, y) = (x, y, f(x, y))$, then, the position of the free node p on the surface is given by $\mathbf{y} = (x, y, f(x, y))^T = (\mathbf{x}, f(\mathbf{x}))^T$.

Note that $S_E = A_P W_E^{-1}$ only depends on $\mathbf{x}$ because W_E is constant and A_P is a function of $\mathbf{x}$. Besides, $S_I = A_P W_I^{-1}$ depends on $\mathbf{y}$, due to $W_I = A_P R^{-1} W_E$, and R is a function of $\mathbf{y}$. Thus, we have $S_E(\mathbf{x})$ and $S_I(\mathbf{y})$. We shall optimize the local mesh $M(p)$ by an iterative procedure maintaining constant $W_I(\mathbf{y})$ in each step. To do this, at the first step, we fix $W_I(\mathbf{y})$ to its initial value, $W_I^0 = W_I(\mathbf{y}^0)$, where $\mathbf{y}^0$ is given by the initial position of p. So, if we define $S_I^0(\mathbf{x}) = A_P(\mathbf{x})(W_I^0)^{-1}$, we approximate the similarity transformation (11) as

$$S^0(\mathbf{x}) = S_E^{-1}(\mathbf{x}) S_I^0(\mathbf{x}) S_E(\mathbf{x}) \tag{12}$$

Now, the construction of the objective function is carried out in a standard way, but using S^0 instead of S. So, we obtain the objective function for a given triangle $\Delta t \subset \pi$

$$\eta^0(\mathbf{x}) = \frac{|S^0(\mathbf{x})|^2}{2\sigma^0(\mathbf{x})} \tag{13}$$

where $\sigma^0(\mathbf{x}) = \det(S^0(\mathbf{x}))$.

With this approach the optimization of the local mesh $M(p)$ is transformed into a two-dimensional problem without constraints, defined on $N(q)$, and, therefore, it can be solved with low computational cost. Furthermore, if we write W_I^0 as $A_P^0(R^0)^{-1}W_E$, where $A_P^0 = A_P(\mathbf{x}^0)$ and $R^0 = R(\mathbf{y}^0)$, it is straightforward to show that S^0 can be simplified as

$$S^0(\mathbf{x}) = R^0 (A_P^0)^{-1} S_E(\mathbf{x}) \tag{14}$$

and our objective function for the local mesh is

$$|K_\eta^0|_n(\mathbf{x}) = \left[\sum_{m=1}^M (\eta_m^0)^n(\mathbf{x}) \right]^{\frac{1}{n}} \tag{15}$$

Let now analyze the behavior of the objective function when the free node crosses the boundary of the feasible region. If we denote $\alpha_P = \det(A_P)$, $\alpha_P^0 = \det(A_P^0)$, $\rho^0 = \det(R^0)$, $\omega_E = \det(W_E)$ and taking into account (14), we can write $\sigma^0 = \rho^0 (\alpha_P^0)^{-1} \alpha_P \omega_E^{-1}$. Note that ρ^0, α_P^0, and ω_E are constants, so η^0 has a singularity when $\alpha_P = 0$, that is, when q is placed on the boundary of the feasible region of

$N(q)$. This singularity determines a barrier in the objective function that prevents the optimization algorithm to take the free node outside this region. This barrier does not appear if we use the exact weighted Jacobian matrix S, given in (6), due to $\det(R) = R_{11} R_{22} > 0$.

Suppose that $\mathbf{x}^1 = \bar{\mathbf{x}}^0$ is the minimizing point of (15). As this objective function has been constructed by keeping $\mathbf{y}$ in its initial position, $\mathbf{y}^0$, then $\mathbf{x}^1$ is only the first approximation to our problem. This result is improved updating the objective function at $\mathbf{y}^1 = (\mathbf{x}^1, f(\mathbf{x}^1))^T$ and, then, computing the new minimizing position, $\mathbf{x}^2 = \bar{\mathbf{x}}^1$. This local optimization process is repeated, obtaining a sequence $\{\mathbf{x}^k\}$ of optimal points, until a convergence criteria is verified. We have experimentally verified in numerous tests, involving continuous functions to define the surface Σ, that this algorithm converges.

Let us consider P as an optimal projection plane (this aspect will be discussed in next section). In order to prevent a loss of the details of the original geometry, our optimization algorithm evaluates the difference of heights ($[\Delta z]$) between the centroid of the triangles of $M(p)$ and the reference surface, every time a new position $\mathbf{x}^k$ is calculated. If this distance exceeds a threshold, $\Delta(p)$, the movement of the node is aborted and the previous position is stored. This threshold $\Delta(p)$ is established attending to the size of the elements of $M(p)$. In concrete, the algorithm evaluates the average distance between the free node and the nodes connected to it, and takes $\Delta(p)$ as percentage of this distance. Other possibility is to fix $\Delta(p)$ as a constant for all local meshes. In the particular case in which we have an explicit representation of the surface by a function $f(x, y)$, $\Delta(p)$ can be established as a percentage of the maximum difference of heights between the original surface and the initial mesh.

3 Search of the Optimal Projection Plane

The former procedure needs a plane in which the local mesh, $M(p)$, is projected conforming a valid mesh, $N(q)$. If this plane exists it is not unique, because a small rotation of the coordinate system produces another valid projection plane, that is, another plane in which $N(q)$ is valid. We have observed that the number of iterations required by our procedure depends on the chosen plane. In general, this number is less if the plane is well *faced* to $M(p)$. We have to find the rotation of reference system x, y, z such that the new x', y'-plane, P', is optimal with respect to a suitable criterion.

We will denote $N(q')$ as the projection of $M(p)$ onto P' and t' the projection of the physical triangle $\Delta t \in M(p)$ onto P'. Let $A'_P = (\mathbf{x}'_1 - \mathbf{x}'_0, \mathbf{x}'_2 - \mathbf{x}'_0)$ be the matrix associated to the affine map that takes the reference element defined on P' to t', then, the area of t' is given by $\frac{1}{2}|\alpha'_P|$ where $\alpha'_P = \det(A'_P)$.

Our goal is to find a coordinate system rotation such that $\sum_{m=1}^{M} \alpha'_{P_m}$ is maximum satisfying the constraints $\alpha'_{P_m} = \det(A'_{P_m}) > 0$ for all the triangles of $N(q')$, that is, $m = 1, ..., M$. In [Ras2] a method to determine a projection plane is considered but without the enforcement of these constraints.

According to Euler's rotation theorem, any rotation may be described using three angles. The so-called *x-convention* is the most common definition. In this convention, the rotation is given by Euler angles (ϕ, θ, ψ), where the first rotation is by an angle $\phi \in [0, 2\pi]$ about the z-axis, the second is by an angle $\theta \in [0, \pi]$ about the x-axis, and the third is by an angle $\psi \in [0, 2\pi]$ about the z-axis (again).

Let $\Phi(\phi, \theta, \psi)$ be the Euler's rotation matrix such that $\mathbf{y}' = \Phi\mathbf{y}$, then, the Jacobian matrix $A_\pi = (\mathbf{y}_1 - \mathbf{y}_0, \mathbf{y}_2 - \mathbf{y}_0)$ associated to the triangle Δt of $M(p)$, defined in (2), can be spanned on the rotated coordinate system as $A'_\pi = (\mathbf{y}'_1 - \mathbf{y}'_0, \mathbf{y}'_2 - \mathbf{y}'_0) = \Phi A_\pi$. Thus, the Jacobian matrix A'_P is written as $A'_P = \Pi A'_\pi = \Pi \Phi A_\pi$. With these considerations it is easy to proof that the value of α'_P is

$$\alpha'_P = \det(\Pi \Phi A_\pi) = m_1 \sin(\phi) \sin(\theta) + m_2 \sin(\theta) \cos(\phi) + m_3 \cos(\theta) \qquad (16)$$

where m_i is the minor obtained by deleting the i-th row of A_π. Note that equation (16) only depends on ϕ and θ angles, as was to be expected.

Although the above maximization problem can be solved taken into account the constraints, we propose an unconstrained approach.

Let us consider, as a first attempt, the objective function $\sum_{m=1}^{M} (\alpha'_{P_m})^{-1}(\phi, \theta)$. The minimization of this function tends to maximize the values of α'_{P_m} and, due to the barrier that appears when $\alpha'_{P_m} = 0$ for some triangle of $N(q')$, the values of α'_{P_m} are maintained positive if the minimization algorithm starts at an interior point, that is, a point (ϕ_0, θ_0) belonging to the set Ψ of angles (ϕ, θ) such that $\alpha'_{P_m}(\phi, \theta) > 0$ for $(m = 1, ..., M)$. On the other hand, if any $\alpha'_{P_m} < 0$ the barrier prevents to reach the required minimum. In next paragraph we propose a method to find an interior point (ϕ_0, θ_0) of Ψ to be used as a starting point in the minimization algorithm.

Let $G = [\mathbf{g}_m]$ be the $3 \times M$ matrix formed by the vectors, $\mathbf{g}_m$, normal to the triangles of $M(p)$. A solution of the inequality system (if it exists) $G^T \mathbf{g} > \mathbf{0}$ provides a direction [Wri], defined by vector $\mathbf{g}$, such that all the triangles of $M(p)$ can be projected on a plane, normal to the unitary vector $\mathbf{n} = \frac{\mathbf{g}}{\|\mathbf{g}\|}$, so that $\alpha'_{P_m} > 0$ for $(m = 1, ..., M)$. Then, it only remains to find the angles ϕ_0 and θ_0 in which the coordinate system needs to be rotated to get the z' axis to point in the direction of $\mathbf{n}$. More precisely, the angles ϕ_0 and θ_0 are the solution of the equation $\Phi^T(\phi_0, \theta_0, 0) \mathbf{e}_3 = \mathbf{n}$, where $\mathbf{e}_3 = (0, 0, 1)^T$. If the inequality system has not solution, then, there is not any valid projection plane for this local mesh, against the premise done in section 2.1. In this case, the local optimization procedure maintains the free node p at its initial position.

We have observed that the previous objective function has computational difficulties as the optimization algorithms use discrete steps to search the optimal point. A step leading outside the region Ψ may indicate a decrease in the value of the objective function and take to a false solution. To overcome this problem we propose a modification of the objective function in such a way that it will be regular all over $\mathbb{R}^3$ and its barrier will be "smoothed". The modification consists of substituting α'_{P_m} by $h(\alpha_{P_m})$, where $h(\alpha)$ is the positive and increasing function given by

$$h(\alpha) = \frac{1}{2}(\alpha + \sqrt{\alpha^2 + 4\delta^2}) \qquad (17)$$

being the parameter $\delta = h(0)$. The behavior of $h(\alpha)$ in function of δ parameter is such that, $\lim_{\delta \to 0} h(\alpha) = \alpha$, $\forall \alpha \geq 0$ and $\lim_{\delta \to 0} h(\alpha) = 0$, $\forall \alpha \leq 0$. The characteristics of h function and its application in the context of mesh untangling and smoothing have been studied in [Gar], [Esc]. Thus, the proposed objective function for searching the projection plane is

$$\Omega(\phi, \theta) = \sum_{m=1}^{M} \frac{1}{h(\alpha'_{P_m}(\phi, \theta))} \tag{18}$$

A crucial property is that the angles that minimize the original and modified objective functions are nearly identical when δ is *small*. Details about the determination of δ value for 3-D triangulations can be found in [Esc].

4 Application to a Scanned Objects

In this section, the proposed technique is applied to smooth the meshes of scanned objects. In particular, we have applied the optimization technique to a pair of meshes obtained from *http://www.cyberware.com/*. The first object, Igea, (see Figure 2) has 67170 triangles and 33587 nodes. The second is a screwdriver (see Figure 5) with 27150 triangles and 13577 nodes. Note the poor quality of these original meshes in several parts.

The projection plane for both surface triangulations have been chosen in terms of the local mesh to be analyzed. We have used the objective function (1) with $n = 2$.

The initial value of the average quality for Igea is 0.794 (measured with the quality metric based on the condition number [Fre]). The optimized mesh, after four iterations of our optimization procedure, is shown in Figure 3. Its average quality has been increased to 0.913. A more significant data is that average quality of the worst 5000 triangles increases from 0.520 to 0.749. Figure 4 shows the quality curves for initial and optimized meshes. These curves are obtained by sorting the elements in increasing order of its quality.

The average quality for the screwdriver is increased from 0.822 to 0.920 in four iterations, see Figure 6. The the worst 500 triangles increases its average quality from 0.486 to 0.704. It is important to remark that the original geometry is almost preserved in the optimization process, as it can be seen by comparing a detail of these meshes in Figures 7 and 8. The quality curves for this application are shown in Figure 9.

We have fixed $\Delta(p)$ for both applications as 10% of average distance between the free node and the nodes connected to it. With this election, 208 nodes has not been modified by the algorithm in the first iteration, 416 in the second, 432 in the third, and 440 in the fourth one, for Igea application. For the screwdriver this number was 85 in the first iteration, 167 in the second, 187 in the third, and 193 in the fourth one.

Finally, we remark that quality curves from the first to the fourth iteration are very close. In particular, the algorithm only needs one iteration to reach an average quality of 0.899 for Igea and 0.907 for the screwdriver.

5 Conclusions and Future Research

We have developed an algebraic method to optimize triangulations defined on surfaces. Its main characteristic is that the original problem is transformed into a fully two-dimensional sequence of approximate problems on the parametric space. This characteristic allows the optimization algorithm to deals with surfaces that only need to be continuous. Moreover, the barrier exhibited by the objective function in the parametric space prevents the algorithm to construct unacceptable meshes.

We have also introduced a procedure to find an optimal projection plane (our parametric space) based on the minimization of a suitable objective function. We have observed that correct choice of this plane plays a relevant role.

The optimization process includes a control on the gap between the optimized mesh and the reference surface that avoids to lose details of the original geometry. In this work we have used a piecewise linear interpolation to define the reference surface when the true surface is not known, but it would be also possible to use a

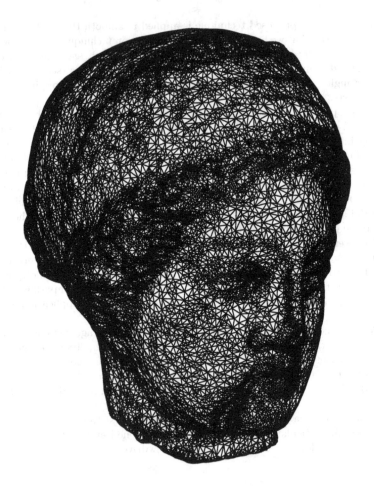

Fig. 2. Original mesh of Igea obtained from *http://www.cyberware.com/*

more regular interpolation, for example, the proposed in [Ras]. Likewise, it would be possible to introduce a more sophisticated criterion for the gap control, by using a local refinement/derefinement techniques, that takes into account the curvature of the surface [Fre2], [Ras], [Vig], [Fre3].

In the present work we have only considered a sole objective function obtained from an isotropic and area independent algebraic quality metric. Nevertheless, the framework that establishes the *algebraic quality measures* [Knu] provides us the possibility to construct anisotropic and area sensitive objective functions by using a suitable metric.

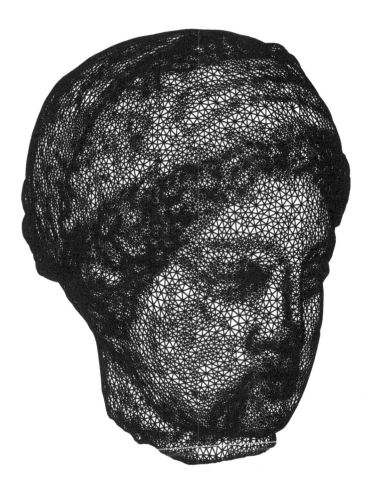

Fig. 3. Optimized mesh of Igea after four iterations of our procedure

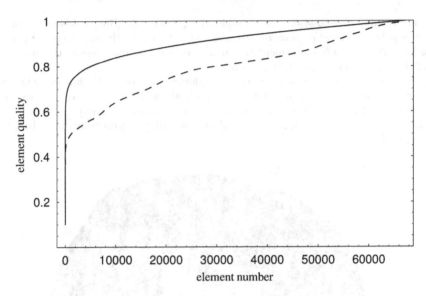

Fig. 4. Quality curves for the initial (dashed line) and optimized (solid line) meshes for Igea

In future works we will use the present smoothing technique for improving the mesh quality of the boundary of 3-D domain triangulations defined over complex terrains [Mon]. A simultaneous smoothing and untangling procedure [Esc] could be applied to inner nodes of the domain after. Authors have developed this tetrahedral mesh generator for wind field simulation in realistic problems [Mon1].

Acknowledgments

This work has been supported by the Spanish Government and FEDER, grant contracts: REN2001-0925-C03-02/CLI and CGL2004-06171-C03-02/CLI.

[Knu] Knupp PM (2001) SIAM J Sci Comp 23:193–218
[Fre] Freitag LA, Knupp PM (2002) Int J Num Meth Eng 53:1377–1391
[Fre] Frey PJ, Borouchaki H (1999) Int J Num Meth Eng 45:101–118
[Gar] Garimella RV, Shaskov MJ, Knupp PM (2004) Comp Meth Appl Mech Eng 9-11:913–928
[Fre2] Frey PJ, Borouchaki H (1998) Comp Vis Sci 1:113–121
[Ras] Rassineux A, Villon P, Savignat JM, Stab O (2000) Int J Num Meth Eng 49:31–49
[Vig] Vigo M, Pla N, Brunet P (1999) Comp Aid Geom Des 16:107–126

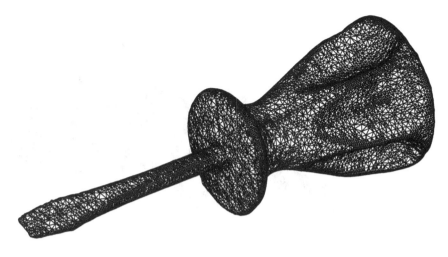

Fig. 5. Original mesh of a screwdriver from *http://www.cyberware.com/*

[Fre3] Frey PJ, Borouchaki H (2003) Int J Num Meth Eng 58:227–245

[Knu1] Knupp PM (2000) Int J Num Meth Eng 48:401–420

[Knu2] Knupp PM (2000) Int J Num Meth Eng 48:1165–1185

[Ras2] Rassineux A, Britkopf P, Villon P (2003) Int J Num Meth Eng 57:371–389

[Wri] Wright S.J. (1997) Primal-dual interior-point methods. SIAM, Philadelphia.

[Gar] Garanzha VA, Kaporin IE (1999) Comp Math Math Phys 39:1426–1440

[Esc] Escobar JM, Rodríguez E, Montenegro R, Montero G, González-Yuste JM (2003) Comp Meth Appl Mech Eng 192:2775–2787

[Mon] Montenegro R, Montero G, Escobar JM, Rodríguez E, González-Yuste JM (2002) Lect N Comp Sci 2329:335–344

[Mon1] Montero G, Rodríguez E, Montenegro R, Escobar JM, González-Yuste JM (2005) Adv Eng Soft 36:3–10

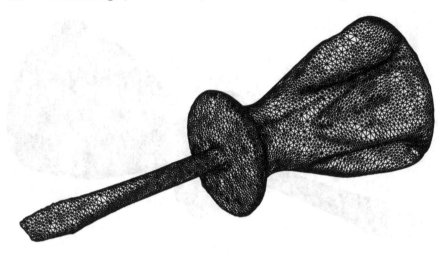

Fig. 6. Optimized mesh of the screwdriver after four iterations

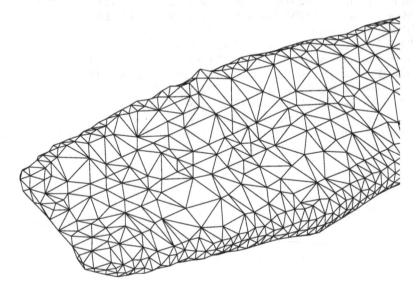

Fig. 7. Detail of the original mesh of the screwdriver end

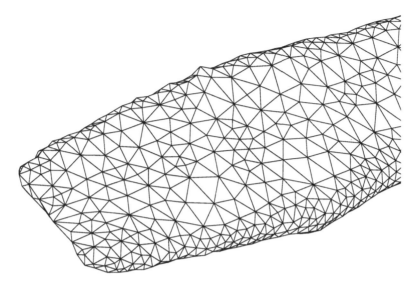

Fig. 8. Detail of the optimized mesh of the screwdriver after four iterations

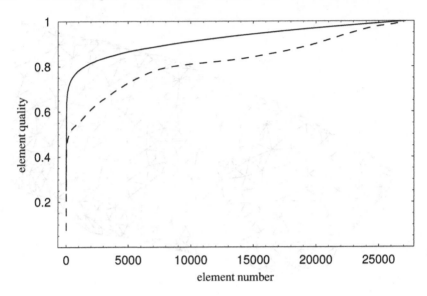

Fig. 9. Quality curves for the initial (dashed line) and optimized (solid line) meshes for the screwdriver

Compact Array-Based Mesh Data Structures

Tyler J. Alumbaugh and Xiangmin Jiao

Center for Simulation of Advanced Rockets
Computational Science and Engineering
University of Illinois at Urbana-Champaign
{talumbau,jiao}@uiuc.edu

Summary. In this paper, we present simple and efficient array-based mesh data structures, including a compact representation of the *half-edge data structure* for surface meshes, and its generalization—a *half-face data structure*—for volume meshes. These array-based structures provide comprehensive and efficient support for querying incidence, adjacency, and boundary classification, but require substantially less memory than pointer-based mesh representations. In addition, they are easy to implement in traditional programming languages (such as in C or Fortran 90) and convenient to exchange across different software packages or different storage media. In a parallel setting, they also support partitioned meshes and hence are particularly appealing for large-scale scientific and engineering applications. We demonstrate the construction and usage of these data structures for various operations, and compare their space and time complexities with alternative structures.

Key words: mesh data structures, half-edge, half-face, parallel computing

1 Introduction

In scientific computing, mesh data structures play an important role in both numerical computations, including finite element and finite volume codes, and geometric algorithms, such as mesh adaptation and enhancement. Historically, the designs of data structures in the numerical and geometric communities have been based on quite different philosophies, due to the diverse requirements of different applications. Specifically, numerical solvers aim at space and time efficiency as well as ease of implementation in traditional programming languages such as Fortran 90, so they tend to use array-based data structures storing minimal information, as exemplified by the CFD General Notation System (CGNS), an AIAA Recommended Practice [The02].

Geometric algorithms, on the other hand, require convenient traversal and modification of mesh entities, and hence tend to use comprehensive pointer-based data structures with substantial memory requirement, and frequently utilize advanced programming features available only in modern programming languages (such as templates in C++), as exemplified by the CGAL library [FGK+00]. Increasingly, modern scientific applications require integrating geometric algorithms with numerical solvers, and this discrepancy in mesh data structures has led to difficulties in integrating different software packages and even more problems when implementing geometric algorithms within engineering codes. In a parallel setting, the necessity of accommodating partitioned meshes introduces additional complexities.

In this paper, we investigate mesh data structures that can serve both numerical and geometric computations on parallel computers. From an applications' point of view, it is desirable that such data structures meet the following requirements:

- Efficient in both time and space, so that mesh entities and their neighborhood information can be queried and modified without performing global search, while requiring a minimal amount of storage.
- Neutral of programming languages, so that it can be implemented conveniently in main-stream languages used in scientific computing, such as C, C++, Fortran 90, and even Matlab.
- Convenient for I/O, so that the data structure can be transferred between different storage (such as between files and main memory) and exchanged across different software modules.
- Easily extensible to support partitioned meshes, and easy to communicate across processors on parallel machines.

Meeting these requirements is decidedly nontrivial. Indeed, none of the pre-existing data structures in the literature appeared to be satisfactory in all these aspects. In particular, the popular data structures used in numerical computations, such as the standard element-vertex connectivity for finite element codes [BCO81], do not support efficient queries (such as whether a vertex is on the boundary) or traversals (such as from neighbor to neighbor) required by many geometric algorithms. The comprehensive pointer-based data structures used in geometric algorithms, such as edge-based data structures for surface meshes [Ket99] and exhaustive incidence-based data structures for volume meshes [BS97], may require a substantial amount of memory, even after optimizing their storage to the bare minimum to store only the required pointers. These pointer-based representations are also difficult to implement in traditional programming languages, and special attention to memory management is needed (even in modern programming languages) to avoid memory fragmentation. In addition, they are inconvenient for I/O and interprocess communication.

In this work, we develop compact array-based data structures for both surface and volume meshes. Our data structures augment, and can be constructed efficiently from, the standard element-vertex connectivity. In the context of parallel computing, only the communication map of shared vertices along partition boundaries is needed as an additional requirement. Our data structures require a minimal amount of storage, primarily composed of an encoding of the incidence relationship of d-dimensional entities along $(d-1)$-dimensional sub-entities, and a mapping from each vertex to one of its incident edges. In two dimensions, our data structure reduces to a compact representation of the well-known *half-edge data structure*. In three dimensions, it delivers a generalization of half-edges to volume meshes, which we refer to as the *half-face data structure*. With additional encoding of adjacency information of $(d-1)$-dimensional entities along partition boundaries, we then obtain a convenient representation of partitioned meshes for parallel computing.

The remainder of the paper is organized as follows. Sec. 2 presents some basic definitions, assumptions, and observations behind our data structures. Sec. 3 describes an array-based data structure for surface meshes, which resembles the well-known half-edge data structure. Sec. 4 introduces a half-face data structure and its array-based implementation. Sec. 5 extends the data structures for partitioned meshes in a parallel setting. Finally, Sec. 6 concludes the paper with a discussion of future work.

2 Preliminaries

Combinatorially, a d-dimensional *mesh* refers to a collection of topological entities of up to d dimensions, along with the *incidence relationship* of entities of different dimensions and *adjacency* of entities of the same dimension, where d is 2 for surface meshes and 3 for volume meshes. In this paper, we refer to the 0-dimensional entities *vertices,* the 1-dimensional entities *edges,* the 2-dimensional entities *faces* (or *facets*), and the 3-dimensional entities *cells*. We use *elements* as a synonym of the d-dimensional entities (i.e., faces in a surface mesh and cells in a volume mesh). We require a mesh to be *conformal*, in the sense that two adjacent cells intersect at a *shared* face, edge, or vertex, and two adjacent faces intersect at a *shared* edge or vertex.

In scientific and engineering applications, in general only a few types of elements are used in a mesh, including triangles and quadrilaterals for surface meshes and tetrahedra, prisms, pyramids, and hexahedra for volume meshes, either linear or quadratic. In this paper, we focus on linear elements. In general, the sub-entities within an element are ordered and assigned local IDs following a given convention, for consistent numerical and geometric computations (such as the calculation of the Jacobian and face normals) and for exchanging data across different software packages. In addition, the sub-entities within a sub-entity (such as the vertices within a face of a tetrahedron) are also ordered consistently. Targeted at these applications, we focus our attention on the meshes composed of the most commonly used element types, and adopt the widely used CGNS conventions [The02] to number sub-entities. Fig. 1 depicts the conventions for the most common elements. In addition, we assume the vertices are assigned consecutive integer IDs ranging from one to the number of vertices, and similarly for elements. In a parallel setting, each partition is assumed to have its own numbering systems for vertices and elements. Given an element, we assume one can determine its type from the element ID in negligible time, by comparing with the minimum and maximum IDs of each type of elements if the elements of the same type are numbered consecutively, or by performing a table lookup.

To simplify presentation, our discussions will mainly focus on *manifold* models with *boundary*. Extension to non-manifold models would involve generalizing the programming interface and tweaking the internal representation. In numerical and geometric computations, the boundary of a mesh frequently plays an important role to impose proper boundary treatments. We classify an entity to be a *border* entity if it is on the boundary, and otherwise to be a *non-border* or *interior* entity. Each

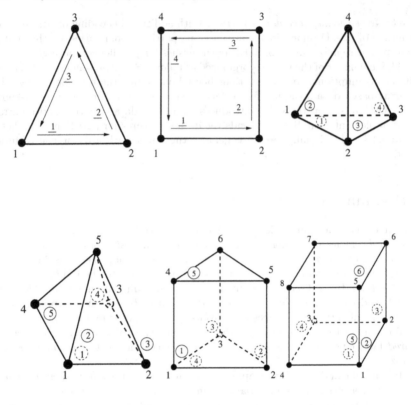

Fig. 1. Local numbering conventions for 2-D and 3-D elements. Underscored numbers correspond to local edge IDs, and circled ones correspond to local face IDs. The vertex next to an edge or face ID is the first vertex of the edge or face.

border entity is said to incident on a *hole*. In our applications, a surface mesh is typically composed of the border entities of a volume mesh, and hence in general is *orientable* with consistent inward and outward surface normals. A manifold surface mesh with boundary has the following useful properties:

- Each *edge* is contained in either *one* or *two faces*.
- There is a *cyclic* sequence of the incident *edges* of each *non-border vertex*.
- There is a *linear* sequence of the incident *edges* of each *border vertex*.

Here, an ordered set of entities is said to be a *sequence* if each pair of consecutive entities are contained in a higher-dimensional entity. Analogously, a manifold volume mesh with boundary has the following properties:

- Each *face* is contained in either one or two *cells*.

- There is a *cyclic* sequence of the incident *border edges* of each *border vertex*.
- There is a *cyclic* sequence of the incident *faces* of each *non-border edge*.
- There is a *linear* sequence of the incident *faces* of each *border edge*.

To manipulate surface and volume meshes, scientific and engineering applications require efficient *mesh data structures* (abbreviated as *MDS* hereafter). An MDS allows iterating through the entities of a mesh, performing queries on incidence, adjacency, and classification (in particular, boundary classification) of entities, and modifying the mesh efficiently. We classify incidence relationships to be either *upward* or *downward*, which map an entity to other higher- or lower-dimensional entities, correspondingly. Furthermore, an entity in general is incident on one or more entities of a given dimension, so we further subdivide the incidence queries into *one-to-any* and *one-to-all*. We assume the *valence* of the mesh (i.e., the maximum number of edges incident on a vertex) is bounded by a small constant. We say an MDS is *comprehensive* if it can perform every incident, adjacent, or classification query in a time independent of mesh size. A data structure is *complete* (or *self-contained*) if it contains all the information necessary to construct a comprehensive MDS. In general, we require an MDS to be complete. Obviously, a comprehensive MDS is complete, but not vice versa. The goal of this paper is to develop complete and comprehensive data structures that require minimal storage.

3 Surface Mesh Data Structures

In this section, we investigate data structures for surface meshes in sequential applications. Issues related to parallelization will be discussed in Sec. 5.

3.1 Traditional Representations

In the literature, two classes of representations of surface meshes have been commonly used: edge-based representations, and connectivity tables.

Edge-Based Representations

Edge-based representations have been studied and used extensively in computational geometry, for their generality and comprehensiveness. Three major variants have been proposed: the winged-edge [Bau75, Gla91], half-edge [Wei85] (or the doubly-connected edge list or DCEL [dvOS00]), and quad-edge [GS85] data structures; see [Ket99] for a comparison and discussion of implementation issues. These data structures allow efficient local traversal and modification of mesh entities, and are designed to handle arbitrary polygons. In practice, their implementations are typically pointer-based, and memory optimization has focused on omitting certain pointers to trade time for space, as in the Computational Geometry Algorithms Library (CGAL) [FGK+00, Ket99].

Among the edge-based structures, the half-edge data structure (abbreviated as HEDS hereafter) has been very popular for orientable manifold surfaces for its intuitiveness and ease-of-use. The HEDS is designed based on the observation that each face is bounded by a loop of directed edges in counterclockwise order, and each hole is bounded by a loop in clockwise order, as illustrated in Fig. 2. Therefore, every edge has two directed half-edges (with opposite directions), which are said to be the *twin* or the *opposite* of each other, one in each incident face (or hole). The programming interface of the HEDS allows a user to query the previous and next half-edges within a face (or hole), the opposite half-edge, an incident vertex or face of a half-edge, and vice versa. By flagging border half-edges or their incident holes, the HEDS also allows querying the boundary classification of any entity.

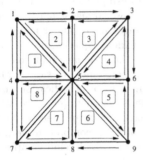

Element	V1	V2	V3
1	1	4	5
2	1	5	2
3	2	5	3
4	3	5	6
5	6	5	9
6	8	9	5
7	7	8	5
8	5	4	7

Fig. 2. Illustration of half-edges.   **Fig. 3.** Connectivity of sample surface mesh.

In a pointer-based implementation of HEDS, an object (or record) is created for each vertex, half-edge, or face. A half-edge object stores pointers to its opposite, previous, and next half-edges, as well as to its origin (or destination) vertex and to an incident face. Each face or vertex stores a pointer to an incident half-edge. For applications involving computations on both faces and vertices, one can slightly reduce the storage by omitting either the previous or next half-edge. Therefore, such an implementation requires at least eight pointers per edge (four per half-edge), and one pointer per vertex or face. The winged-edge and quad-edge structures have comparable storage requirements (six pointers per edge) in this setting.

Element Connectivity

The element connectivity is a classic mesh representation for finite element analysis [BCO81], and is also frequently used for file I/O and for exchanging meshes between different software modules. A connectivity table lists the vertices contained within each face in increasing order of their local IDs. Fig. 3 shows the connectivity table of the sample mesh of Fig. 2. This simple representation is self-contained but not comprehensive, as it does not support queries such as adjacent faces and boundary classification. A geometric software library typically constructs an internal edge-based structure from the connectivity table, and then discards this table.

3.2 Array-Based Half-Edges

Inspired by the HEDS and element connectivity, we design a hybrid data structure that combines the comprehensiveness of the former and the compactness of the latter. Our design is based on the following observations: The traditional HEDS must store many pointers in order to support arbitrary polygons. For surface meshes composed of only triangles and quadrilaterals, we can encode a half-edge based on its location within its incident face. The incident face and the previous or next half-edge can then be obtained with simple arithmetic operations without being stored explicitly. From a local edge ID and the element connectivity table, one can obtain the IDs of the incident vertices of an edge efficiently. Therefore, we only need to store the correspondence between twin half-edges and a mapping from each vertex to any of its incident half-edges.

More specifically, we assign each half-edge an ID composed of a pair of numbers $\langle f, i \rangle$, where f is the ID of its containing face, and i is the index of the edge (starting from 1, as shown in Fig. 1) within the face. To support boundary classification, we assign consecutive integer IDs to border edges (starting from 1), and encode the bth border half-edge as $\langle b, 0 \rangle$, where the zero value of the second part distinguishes a border edge from a non-border one. Since the number of edges per face is at most four, we can encode a half-edge ID in a single integer, using the last three bits to store the second part and the remaining bits the first part. We now define the arrays for the HEDS:

- V2e: Map each vertex to the ID of an incident half-edge originated from the vertex; map a border vertex to a border half-edge.
- E2e: Map each non-border half-edge to the ID of its twin half-edge.
- B2e: Map each border half-edge to the ID of its twin non-border half-edge.

Fig. 4 shows an example of this MDS for the sample mesh in Fig. 2. In terms of memory management, V2e and B2e are dense one-dimensional arrays, whose sizes are equal to the numbers of vertices and border edges, respectively. To allow efficient array indexing, E2e is a two-dimensional array with each row corresponding to a face, and its number of columns is the maximum number of edges per face (3 for triangular meshes and 4 for quadrilateral or mixed meshes). Note that the full array-based HEDS is composed of these three arrays along with the element connectivity (EC), where EC is a two-dimensional array similar to E2e. Let n_i denote the number of i-dimensional entities in a mesh, where i is between 0 and 2. Assume that $n_1 \approx 3n_0$ and $n_2 \approx 2n_0$, then the array-based HEDS requires about $7n_0$ 32-bit integers. Compared to pointer-based HEDS, it reduces memory requirement by about four folds for 32-bit architecture and eight folds for 64-bit architecture.

Given the element connectivity, it is straightforward to construct the three arrays. In particular, we first construct E2e by inserting the half-edges into a hash-table (or map) with their incident vertices as keys and detecting collisions to match twin half-edges. A half-edge without a match is identified as a border edge and assigned a unique border ID, and its and its twin's half-edge IDs are then inserted into the corresponding entries in B2e and E2e, respectively. During the above procedure, we fill in the entry of V2e corresponding to the origin of each half-edge, allowing border half-edge to overwrite non-border ones but not vice versa.

V2e   E2e   B2e

Vertex	Inc. HE
1	$\langle 2,0 \rangle$
2	$\langle 3,0 \rangle$
3	$\langle 4,0 \rangle$
4	$\langle 1,0 \rangle$
5	$\langle 1,3 \rangle$
6	$\langle 5,0 \rangle$
7	$\langle 8,0 \rangle$
8	$\langle 7,0 \rangle$
9	$\langle 6,0 \rangle$

Element	Half-edges		
1	$\langle 1,0 \rangle$	$\langle 8,1 \rangle$	$\langle 2,1 \rangle$
2	$\langle 1,3 \rangle$	$\langle 3,1 \rangle$	$\langle 2,0 \rangle$
3	$\langle 2,2 \rangle$	$\langle 4,1 \rangle$	$\langle 3,0 \rangle$
4	$\langle 3,2 \rangle$	$\langle 5,1 \rangle$	$\langle 4,0 \rangle$
5	$\langle 4,2 \rangle$	$\langle 6,2 \rangle$	$\langle 5,0 \rangle$
6	$\langle 6,0 \rangle$	$\langle 5,2 \rangle$	$\langle 7,2 \rangle$
7	$\langle 7,0 \rangle$	$\langle 6,3 \rangle$	$\langle 8,3 \rangle$
8	$\langle 1,2 \rangle$	$\langle 8,0 \rangle$	$\langle 7,3 \rangle$

Border	Opp. HE
1	$\langle 1,1 \rangle$
2	$\langle 2,3 \rangle$
3	$\langle 3,3 \rangle$
4	$\langle 4,3 \rangle$
5	$\langle 5,3 \rangle$
6	$\langle 6,1 \rangle$
7	$\langle 7,1 \rangle$
8	$\langle 8,2 \rangle$

Fig. 4. Array-based half-edge data structure of sample surface mesh.

3.3 Properties and Operations

The array-based HEDS has the following useful properties:

1. The full array-based HEDS delivers a *comprehensive* MDS.
2. E2e and V2e deliver a *complete* MDS.

To show the comprehensiveness of the full data structure, we summarize the basic queries as follows. Note that all array indices start from 1, and m denotes the number of edges of a given face.

- One-to-any downward incidence
    - ith edge (half-edge) of face f: return $\langle f, i \rangle$
    - ith vertex of face f: return $EC(f, i)$
    - origin of non-border half-edge $\langle f, i \rangle$: return $EC(f, i)$
- One-to-any upward incidence
    - the incident face of a non-border half-edge $\langle f, i \rangle$: return f
    - an incident half-edge of vth vertex: return $V2e(v)$
- Adjacency
    - opposite of non-border half-edge $\langle f, i \rangle$: return $E2e(f, i)$
    - opposite of border half-edge $\langle b, 0 \rangle$: return $B2e(b)$
    - previous of non-border half-edge $\langle f, i \rangle$: return $\langle f, \mathrm{mod}(l + m - 2, m) + 1 \rangle$
    - next of non-border half-edge $\langle f, i \rangle$: return $\langle f, \mathrm{mod}(l, m) + 1 \rangle$
- Boundary classification

- half-edge $\langle f, i \rangle$: return $i = 0$
- vertex v: return V2e(v).second$= 0$

Other types of queries are combinations of the above basic operations. In particular for downward incidence, the destination of a non-border half-edge is the origin of its previous half-edge; the origin and destination of a border half-edge are the destination and origin of its opposite half-edge, respectively; the one-to-all incidences from a face to edges and vertices involves enumerating i from 1 to m. For one-to-all upward incidence, the incident faces of an edge are those incident on its twin half-edges; the incident half-edges and faces at a vertex involve accessing all the half-edges incident on a vertex, enabled by the basic adjacency operations. Starting from a half-edge, we can rotate around its destination vertex in clockwise order following the links of opposite and then previous half-edges; the counterclockwise rotation around the origin vertex of a half-edge follow the links of opposite and then next half-edges. We also use these rotations to get the previous and next of a border half-edge: For the former, we loop around the origin of the border half-edge in counterclockwise order until reaching a border half-edge; for the latter, we loop around its destination clockwise. Since the valence of the mesh is a small constant, these rotations take constant time.

To show E2e and V2e are complete, we simply need to construct a procedure to compute the element connectivity (EC) from these arrays, because EC is a complete MDS itself. At a high level, the procedure goes as follows: First, allocate and fill in the element connectivity with zeros. Then, loop through all vertices, and for each vertex v, visit all the half-edges originated from v as shown above, starting from the half-edge V2e(v) and then rotating using the adjacency information in E2e. When visiting a half-edge $h = \langle f, i \rangle$ with $i > 0$, we assign EC(f, i) to v. After processing all the vertices, we then have the complete EC. We comment that E2e is independent of vertex numbering, but it can also be considered as complete if the vertices are allowed to be renumbered, since we can determine a vertex numbering from E2e using a procedure similar to the above. These complete sub-MDS are useful to reduce the amount of data for I/O and inter-process communications.

3.4 Mesh Modification

A comprehensive mesh data structure allows not only querying but also modifying a mesh efficiently. In this subsection, we demonstrate how to modify the array-based HEDS, using *edge flipping* for triangular meshes as an example. This operation is used in many algorithms, such as Delaunay triangulation and mesh enhancement [FG02]. Fig. 5 illustrates a sample edge-flipping operation, which removes an edge composed of vertices $\{v_1, v_3\}$ and creates a new edge composed of $\{v_2, v_4\}$. As a consequence, the faces $\{v_1, v_2, v_3\}$ and $\{v_1, v_3, v_4\}$ (denoted by f and g) are replaced by $\{v_1, v_2, v_4\}$ and $\{v_2, v_3, v_4\}$, respectively.

Without loss of generality, suppose the fth row of EC is $\{v_3, v_1, v_2\}$ and gth row is $\{v_1, v_3, v_4\}$, so the first half-edges in f and g are the dashed lines in Fig. 5. When flipping the edge $\{v_1, v_3\}$, we need only update the entries associated with v_i, f, and g in E2e, B2e, V2e, and EC. Updating V2e and EC is relatively straightforward. For E2e and B2e, we need to map the half-edges opposite to those in faces f and g to the new half-edges, update mappings between the half-edges within f and g in E2e. In summary, an edge flip involves the following four steps:

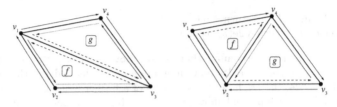

Fig. 5. Illustration of edge flipping.

1. **if** $\langle i, j \rangle \equiv \text{E2e}(f, 1)$ is border **then** $\text{B2e}(i) = \text{E2e}(g, 3)$; **else** $\text{E2e}(i, j) = \text{E2e}(g, 3)$; perform the operation symmetrically by switching f and g;
2. $\text{E2e}(f, 1) = \text{E2e}(g, 3)$; $\text{E2e}(g, 1) = \text{E2e}(f, 3)$;
   $\text{E2e}(f, 3) = \langle g, 3 \rangle$; $\text{E2e}(g, 3) = \langle f, 3 \rangle$;
3. **if** $\text{V2e}(v_1) = \langle g, 1 \rangle$ **then** $\text{V2e}(v_1) = \langle f, 2 \rangle$;
   if $\text{V2e}(v_2) = \langle f, 3 \rangle$ **then** $\text{V2e}(v_2) = \langle g, 1 \rangle$;
   if $\text{V2e}(v_3) = \langle f, 1 \rangle$ **then** $\text{V2e}(v_3) = \langle g, 2 \rangle$;
   if $\text{V2e}(v_4) = \langle g, 3 \rangle$ **then** $\text{V2e}(v_4) = \langle f, 1 \rangle$;
4. $\text{EC}(f, 1) = v_4$; $\text{EC}(g, 1) = v_2$.

In the above, if v_3 was the ith (instead of the first) vertex of face f in EC, then we need to replace half-edge index $\langle f, j \rangle$ (and its corresponding array indices) by $\langle f, \text{mod}(j + i + 2, 3) + 1 \rangle$; similarly for g. Other modification operations, such as edge splitting and edge contraction, are slightly more complex but can be constructed similarly. For edge splitting, since the numbers of vertices and faces are increased by the operation, it is desirable to reserve additional memory for the arrays so that new vertices and faces can be appended to the end. For edge contraction, since the numbers of vertices and faces decrease, we need to swap the IDs of the to-be-removed vertex with the one with the largest ID (and similarly for faces), so that the vertex and element IDs will remain consecutive and the arrays can be shrunk after contraction.

4 Volume Mesh Data Structures

We now extend the array-based half-edge data structure to develop compact representations for volume meshes in serial applications.

4.1 Previous Work

As for surface meshes, the element connectivity is the classic representation for volume meshes in numerical computations, file I/O, and data exchange, but it is not a comprehensive representation. A few alternative mesh representations have been proposed to serve various special purposes, such as mesh generation [DT90,

L88], mesh refinement [CSW88, Riv84], and numerical computations [HTW92]. In [BBCK03], a difference coding was proposed to compress a mesh representation, but vertices may need to be renumbered for its effectiveness. Recently, an index-based mesh representation was developed independently of this work [CPE05], which shares some similarities in mesh-entity representations with our data structures. Another particularly noteworthy MDS is the algorithm oriented mesh database (AOMD) [RS03], which provides unified data structures for numerical and geometric computations.

AOMD is designed based on the observation that a typical application uses only a subset of incidences.[1] It stores the incidences used by an application and omits the unneeded ones. Its design aims at providing a unified programming interface for accessing the mesh database independently of the underlying storage. Unfortunately, the efficiency of AOMD critically depends on a given application. If an application requires nearly all types of incidences, then the efficiency advantage of AOMD diminishes. The implementation of AOMD is fairly complex, extensively utilizing template features of C++ to achieve customizability, and hence its design cannot be easily adopted in engineering applications written in traditional programming language (such as Fortran 90). As coupled applications become more and more commonplace, there are increasing demands for simple volume meshes data structures that are compact and comprehensive.

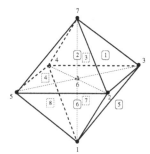

Element	V1	V2	V3	V4
1	2	3	6	7
2	2	6	5	7
3	3	4	6	7
4	4	5	6	7
5	2	6	3	1
6	5	6	2	1
7	6	4	3	1
8	6	5	4	1

Fig. 6. Sample volume mesh.   **Fig. 7.** Connectivity of sample volume mesh.

4.2 Array-Based Half-Faces

To achieve compactness and comprehensiveness for volume meshes, we propose a generalization of the array-based HEDS, called the *half-face data structure* (HFDS). Our generalization is based on the observation that faces in volume meshes play a

---

[1]In AOMD [RS03], the incidence relation are called adjacency.

similar role as edges in surface meshes: Each face is contained in two cells (or a cell and a hole), and the two copies of a face have opposite orientations when its vertices are ordered following the right-hand rule in each cell (i.e., in counterclockwise order with respect to the inward face normal of the cell). We refer to the two copies of a face as *half-faces*, which are said to be the *twin* or the *opposite* of each other. As for half-edges in surface meshes, we encode each half-face by a pair of numbers $\langle c, i \rangle$, where c is the element ID of its containing cell, and i is the face index (starting from 1) within the cell. Furthermore, we assign consecutive IDs to border faces (starting from 1), and encode a half-face with border ID b as $\langle b, 0 \rangle$. Since there are at most six faces per cell, we can encode a half-face ID in one integer, using the last three bits for the second part and the remaining bits for the first part.

The above encoding scheme suggests a straightforward generalization of HEDS with three arrays: V2f, F2f, and B2f, which are the counterparts of V2e, E2e, and B2e, respectively, and in which the half-edge IDs are substituted by half-face IDs. Indeed, this generalization does provide a legitimate data structure. However, it is not an ideal generalization, because unlike E2e and V2e in HEDS, F2f and V2f no longer deliver a complete MDS. The incompleteness is due to the fact that the ordering of vertices in a half-face is cyclic without a designated starting point, so it is not always possible to infer the ordering of the vertices within a cell from this sub-MDS. When supplemented by the element connectivity, this simple generalization can suffer from inefficiency, as it requires comparing the vertex IDs to align the half-faces when performing one-to-all incidence queries.

To overcome these limitations, we define the *anchor* of a half-face as its designated first vertex, and each half-face then has m anchored copies, where m is the number of vertices (or edges) of the face. We encode an *anchored half-face* (AHF) by a three-part ID $\langle c, i, j \rangle$, where the first two parts correspond to the half-face ID, and the third part corresponds to the *anchor index* (staring from 0), which is defined as follows: For a non-border half-face, if the anchor is the kth vertex in the face-vertex list of the face following the CGNS convention, then the anchor index is $k - 1$; for a border half-face, the anchor index is $\mod(m - t, m)$, where t is the anchor index of the vertex in its opposite half-face. Each AHF has one opposite (or twin) AHF, which is its opposite half-face with the same anchor. The last two parts of an AHF ID constitute the *local AHF ID*. The anchor index requires only two bits, and the local AHF ID requires only five bits, so the full AHF ID can be encoded in one integer. With 32-bit unsigned integers, this encoding is sufficient for meshes containing up to 2^{27} (more than 100 million) cells. In addition, we assign the AHF ID to the first edge of the AHF, and then obtain an ID for each edge within each face.

In the CGNS convention, each vertex has a local index within a cell. It is useful to store the mapping from the local AHF ID to the vertex's local index within the cell. We assign a unique ID (between 1 and the number of polyhedron types, 4 in general) to each type of element. To store the mapping for all element types, we introduce a three-dimensional array of size $4 \times 6 \times 4$, denoted by eA2v, whose dimensions correspond to the type IDs, local face IDs, and anchor indices, respectively. In addition, we define an array eAdj of the same size to store the mapping from each AHF to the local AHF ID of its adjacent AHF within a cell along its first edge. We now define the complete representation of the half-face data structure:

- V2f: Map each vertex to the AHF anchored at the vertex; map a border vertex to a border AHF;

- F2f: Map each non-border half-face with anchor index 0 to its twin AHF;
- B2f: Map each border half-face with anchor index 0 to its twin AHF.

Fig. 8 shows an example of this MDS for the sample mesh of Fig. 6. Similar to the array-based HEDS, V2f and B2f are dense one-dimensional arrays, whose sizes are equal to the numbers of vertices and border faces, respectively. F2f is a two-dimensional array with each row corresponding to a cell, and its number of columns is the maximum number of faces in a cell, ranging between four and six. The full HFDS is composed of these three arrays along with the element connectivity (EC). The construction of HFDS follows a procedure similar to that of HEDS, except for the additional operations needed to align the twin half-faces to determine the anchor indices.

V2f   F2f   B2f

Vertex	Inc. HF
1	$\langle 7,0,1 \rangle$
2	$\langle 2,0,2 \rangle$
3	$\langle 3,0,0 \rangle$
4	$\langle 4,0,0 \rangle$
5	$\langle 6,0,2 \rangle$
6	$\langle 1,1,1 \rangle$
7	$\langle 1,0,1 \rangle$

Element	Half-faces			
1	$\langle 5,1,0 \rangle$	$\langle 1,0,0 \rangle$	$\langle 3,4,1 \rangle$	$\langle 2,2,1 \rangle$
2	$\langle 6,1,1 \rangle$	$\langle 1,4,1 \rangle$	$\langle 4,3,1 \rangle$	$\langle 2,0,0 \rangle$
3	$\langle 7,1,1 \rangle$	$\langle 3,0,0 \rangle$	$\langle 4,4,1 \rangle$	$\langle 1,3,1 \rangle$
4	$\langle 8,1,1 \rangle$	$\langle 4,0,0 \rangle$	$\langle 2,3,1 \rangle$	$\langle 3,3,1 \rangle$
5	$\langle 1,1,1 \rangle$	$\langle 6,3,1 \rangle$	$\langle 7,4,1 \rangle$	$\langle 5,0,0 \rangle$
6	$\langle 2,1,1 \rangle$	$\langle 8,2,1 \rangle$	$\langle 5,2,1 \rangle$	$\langle 6,0,0 \rangle$
7	$\langle 3,1,1 \rangle$	$\langle 8,4,1 \rangle$	$\langle 7,0,0 \rangle$	$\langle 5,3,1 \rangle$
8	$\langle 4,1,1 \rangle$	$\langle 6,2,1 \rangle$	$\langle 8,0,0 \rangle$	$\langle 7,2,1 \rangle$

Border	Opp. HF
1	$\langle 1,2,0 \rangle$
2	$\langle 2,4,0 \rangle$
3	$\langle 3,2,0 \rangle$
4	$\langle 4,2,0 \rangle$
5	$\langle 5,4,0 \rangle$
6	$\langle 6,4,0 \rangle$
7	$\langle 7,3,0 \rangle$
8	$\langle 8,3,0 \rangle$

Fig. 8. Array-based half-face data structure for sample volume mesh.

4.3 Properties and Operations

Similar to HEDS, the array-based HFDS has the following useful properties:

1. The full array-based HFDS delivers a *comprehensive* MDS.
2. F2f and V2f deliver a *complete* MDS.

To show the HFDS is comprehensive, we summarize the basic queries as follows. Again, m denotes the number of edges in a given AHF and is either 3 or 4, and all indices (except for anchor index) start from 1.

- One-to-any downward incidence
  - ith face of cell c anchored at jth vertex in the face: return $\langle c, i, j - 1 \rangle$
  - jth edge of AHF $\langle c, i \rangle$: return $\langle c, i, j - 1 \rangle$
  - local index of anchor of non-border AHF $\langle c, i, j \rangle$ within cell c: return $\mathrm{eA2v}(e, i, j + 1)$, where e is type ID of cell c
  - ith vertex of cell c: return $\mathrm{EC}(c, i)$
- One-to-any upward incidence
  - incident cell of AHF (or edge) $\langle c, i, j \rangle$: return c
  - incident AHF (or edge) of vth vertex: return $\mathrm{V2f}(v)$
- Adjacency
  - opposite of non-border AHF $\langle c, i, j \rangle$: return $\langle d, s, \mathrm{mod}(m - j + t, m) \rangle$, where $\langle d, s, t \rangle \equiv \mathrm{F2f}(c, i)$
  - opposite of border AHF $\langle b, 0, j \rangle$: same as above, except that $\langle d, s, t \rangle \equiv \mathrm{B2f}(b, i)$
  - previous of AHF $\langle c, i, j \rangle$ within the face: return $\langle c, i, \mathrm{mod}(j + m - 1, m) \rangle$
  - next of AHF $\langle c, i, j \rangle$ within the face: return $\langle c, i, \mathrm{mod}(j + 1, m) \rangle$
  - in-cell adjacent AHF of non-border AHF $\langle c, i, j \rangle$ along edge: return $\langle c, \mathrm{eAdj}(e, i, j + 1) \rangle$, where e is type ID of cell c
- Boundary classification
  - AHF (edge) $\langle c, i, j \rangle$: return $i = 0$
  - vertex v: return $\mathrm{V2f}(v).\mathrm{second} = 0$

Other types of queries are combinations of the above basic operations. Some queries are straightforward generalization of HEDS, including downward incidences (for border half-faces and one-to-all) and all incident cells of a face. Obtaining incident faces and cells along an edge involves traversing all the half-faces incident on the edge, using the opposite and in-cell adjacent operators, so does determining the border half-face that is adjacent to a border half-face along an edge. The time complexity for traversing the boundary depends on the number of cells incident on a border edge. For applications that frequently traverse the boundary, a separate array B2b can be constructed to save the correspondence of border AHFs, similar to the E2e array in the HEDS, except that border AHF IDs will be stored instead of half-edge IDs.

A more complex query is to enumerate all incident AHFs anchored at a vertex, which is a useful building block for enumerating all incident cells and edges of a vertex. Using the in-cell adjacency and previous (or next) operators, we can iterate through the AHFs around an anchor within a cell. Together with the opposite operator, we can then visit the adjacent cells and their AHFs anchored at the vertex. This process essentially performs a breadth-first traversal over the AHFs around the vertex, and takes time proportional to the output size.

The argument for the completeness of F2f and V2f is similar to that of HEDS: the element connectivity (EC) can be constructed from F2f and V2f by looping through the AHFs around each vertex, starting from the AHF associated with the vertex in V2f. This complete sub-MDS can be used for efficient file I/O and inter-process communication, because EC and B2f can be constructed from them without requiring any additional storage.

Table 1. Memory requirements of mesh data structures. The units I and P stand for the numbers of integers and pointers, respectively.

Mesh type	One-level	Circular	Reduced interior	HFDS
Tetrahedral mesh	$201n_0P$	$153n_0P$	$76n_0P$	$24n_0I$
Hexahedral mesh	$71n_0P$	$55n_0P$	$31n_0P$	$7n_0I$

4.4 Comparison

We now compare the storage requirements of the HFDS with some other MDS. In [BS97], three comprehensive MDS were reported: the "one-level adjacency representation", in which each i-dimensional entity stores points to its incident $(i+1)$- and $(i-1)$-dimensional entities when applicable, the "circular adjacency representation", in which $(i-1)$-dimensional incidence entities are stored for cells, faces, and edges, along with the incidence cells of each vertex, and the "reduced-interior representation", which omit the interior faces and edges in the representation. The storage requirements of these data structures are competitive with other existing mesh data structures [BS97, RKS00].

Let n_i denote the number of i-dimensional entities in a mesh, where i is between 0 and 3. Assume that $23n_0 \approx 4n_3$ for tetrahedral meshes and $n_0 \approx n_3$ for hexahedral meshes [BS97]. For a tetrahedral mesh, the HFDS data structure requires about $24n_0$ integers, of which $23n_0$ are for F2f, in addition to the $23n_0$ integers for EC. For a hexahedral mesh, the HFDS data structure requires about $7n_0$ integers, of which $6n_0$ are for F2f, in addition to the $8n_0$ integers for EC. Table 1 compares the memory requirements of the HFDS (excluding EC) against those reported in [BS97]. For meshes with fewer than 100 million cells, an integer in HFDS requires 4 bytes, whereas a pointer in other data structures requires 4- and 8-bytes on 32- and 64-bit architectures, respectively. Compared to the reduced-interior representation, the HFDS delivers roughly three to four folds of reduction in memory on 32-bit architectures, and six to eight folds on 64-bit architectures. Compared to the one-level and circular representations, the reduction is roughly an order of magnitude.

5 Parallelization

In a parallel environment (especially on distributed memory machines), a mesh must be partitioned so that it can be distributed onto multiple processors [GMT98]. In this context, a border entity on the partition may or may not be on the physical boundary, and it is important for applications to distinguish the two types. Further-more, when a process has multiple partitions, it is desirable to allow an algorithm to traverse across partition boundaries transparently. Our data structures can be extended conveniently to support such queries and traversals. We now describe the extension for meshes partitioned using an element-oriented scheme, which assigns each element to one partition along with its vertices.

Assume that each partition has a unique partition ID, and an efficient mapping exists for querying the owner process of a given partition. A partition typically has its own numbering systems for vertices and elements, from 1 to the numbers of vertices and elements, respectively. For each partition of a surface mesh, we construct an array-based HEDS. Note that a partition may not strictly be a manifold, as a vertex may be incident on more than two border edges. However, the HEDS is still applicable because an edge is always owned by one or two partitions. We define the *counterpart* of a border half-edge (on partition boundaries) to be the non-border half-edge in another partition. Given a mapping between the vertices shared across partitions, we then construct the following arrays to augment the HEDS:

- B2rp: Map each border edge to the partition ID of its counterpart, or map to -1 if the edge is on physical boundary.
- B2re: Map each border edge to the edge ID of its counterpart; undefined if on the physical boundary.

By checking the values in B2rp, we can identify whether entities are on the physical boundary. In addition, when determining the opposite of a half-edge within a parti-tion, if its opposite is on the partition boundary, this extended data structure looks up the counterpart of its opposite border edge and returns the partition ID and the half-edge ID, so that multiple partitions can be traversed seamlessly. From this data structure, one can also easily construct the communication pattern for shared edges across partitions. All the arrays in the extended HEDS are independent of the process mapping of the partitions, so a partition can be migrated easily across processes.

The generalization from surface meshes to volume meshes is straightforward. The *counterpart* of a border AHF is the non-border AHF with the same anchor. We introduce a similar set of arrays to map border faces to the partition and AHF IDs of their counterparts. The construction of these arrays also requires only the vertex mapping for vertices along partition boundaries.

6 Conclusion

In this paper, we introduced a compact array-based representation for the half-edge data structure for surface meshes, and a novel generalization to volume meshes. Our data structures augment the element connectivity by introducing three ad-ditional arrays. These data structures require minimal additional storage, provide comprehensive and efficient support for queries of adjacency, incidence, and bound-ary classification, and can be used to modify a mesh with operations such as edge flipping. Our data structures reduce memory requirement by a factor of between three and eight compared to other comprehensive data structures. A more compact subset of our data structures is also self-contained and can be used for efficient I/O and interprocess communication.

This work so far has mainly focused on two- and three-dimensional conformal manifold meshes, driven by the needs in the coupled parallel simulations at the Center for Simulation of Advanced Rockets [HD00]. These array-based data structures are readily extensible to higher dimensions, and it is also interesting to generalize them to non-manifold and/or non-conforming meshes. Another future direction is to compare the runtime performance of our data structures with other alternative structures, especially in the context of parallel mesh adaptivity.

Acknowledgements

This work was supported by the U.S. Department of Energy through the University of California under subcontract B523819, and in part by NSF and DARPA under CARGO grant #0310446. The first author would like to thank Phillip Alexander of CSAR for help with numerous software issues. We thank anonymous referees for their helpful comments.

[Bau72]   B. Baumgart. Winged-edge polyhedron representation. Technical report, Stanford Artificial Intelligence Report No. CS-320, October 1972.

[Bau75]   B. G. Baumgart. A polyhedron representation for computer vision. In *National Computer Conference*, pages 589–596, 1975.

[BBCK03] Daniel K. Blandford, Guy E. Blelloch, David E. Cardoze, and Clemens Kadow. Compact representations of simplicial meshes in two and three dimensions. In *Proceedings of 12th International Meshing Roundtable*, pages 135–146, 2003.

[BCO81]   E. B. Becker, G. F. Carey, and J. Tinsley Oden. *Finite Elements: An Introduction*, volume 1. Prentice-Hall, 1981.

[Bla03]   J. Blazek. Comparison of two conservative coupling algorithms for structured-unstructured grid interfaces, 2003. AIAA Paper 2003-3536.

[BP91]   J. Bonet and J. Peraire. An alternated digital tree (adt) algorithm for 3d geometric searching and intersection problems. *Int. J. Numer. Meth. Engrg.*, 31:1–17, 1991.

[BS97]   Mark W. Beall and Mark S. Shephard. A general topology-based mesh data structure. *Int. J. Numer. Meth. Engrg.*, 40:1573–1596, 1997.

[BS98]   R. Biswas and R.C. Strawn. Tetrahedral and hexahedral mesh adaptation for cfd problems. *Applied Numerical Mathematics*, 21:135–151, 1998.

[CH94]   S. D. Connell and D. G. Holmes. 3-dimensional unstructured adaptive multigrid scheme for the Euler equations. *AIAA J.*, 32:1626–1632, 1994.

[CPE05]   W. Celes, G.H. Paulino, and R. Espinha. A compact adjacency-based topological data structure for finite element mesh representation. *Int. J. Numer. Meth. Engrg.*, 2005. in press.

[CSW88]   G. F. Carey, M. Sharma, and K.C. Wang. A class of data structures for 2-d and 3-d adaptive mesh refinement. *Int. J. Numer. Meth. Engrg.*, 26:2607–2622, 1988.

[DH02]   W. A. Dick and M.T. Heath. Whole system simulation of solid propellant rockets. In *38th AIAA/ASME/SAE/ASEE Joint Propulsion Conference and Exhibit*, July 2002. 2002-4345.

[DT90]   H. Dannelongue and P. Tanguy. Efficient data structure for adaptive remeshing with fem. *J. Comput. Phys.*, 91:94–109, 1990.

[dvOS00]   Mark de Berg, Marc van Kreveld, Mark Overmars, and Otfried Schwarzkopf. *Computational Geometry: Algorithms and Applications.* Springer, 2nd edition, 2000.

[Ede87]   H. Edelsbrunner. *Algorithms in Combinatorial Geometry.* Springer-Verlag Berlin Heidelberg, Germany, 1987.

[FG02]   Pascal Jean Frey and Paul-Louis George. *Mesh Generation: Application to Finite Elements.* Hermes, 2002.

[FGK+00]   A. Fabri, G.-J. Giezeman, L. Kettner, S. Schirra, and S. Schönherr. On the design of CGAL, a computational geometry algorithms library. *Softw. - Pract. Exp.*, 30:1167–1202, 2000. Special Issue on Discrete Algorithm Engineering.

[Gla91]   A. S. Glassner. Maintaining winged-edge models. In J. Arvo, editor, *Graphics Gems II*, pages 191–201. Academic Press, 1991.

[GMT98]   John. R. Gilbert, Gary L. Miller, and Shang-Hua Teng. Geometric mesh partitioning: Implementation and experiments. *SIAM J. Sci. Comp.*, 19:2091–2110, 1998.

[GS85]   L. J. Guibas and J. Stolfi. Primitives for the manipulation of general subdivisions and the computation of Voronoi diagrams. *ACM Trans. Graphics*, 4:74–123, 1985.

[HD00]   M. T. Heath and W. A. Dick. Virtual prototyping of solid propellant rockets. *Computing in Science & Engineering*, 2:21–32, 2000.

[HTW92]   D. Hawken, P. Townsend, and M. Webster. The use of dynamic data structures in finite element applications. *Int. J. Numer. Meth. Engrg.*, 33:1795–1811, 1992.

[JH03]   X. Jiao and M.T. Heath. Accurate, conservative data transfer between nonmatching meshes in multiphysics simulations. In *7th US National Congress on Computational Mechanics*, July 2003.

[Ket98]   Lutz Kettner. Designing a data structure for polyhedral surfaces. In *Proc. 14th Annu. ACM Sympos. Comput. Geom.*, pages 146–154, 1998.

[Ket99]   Lutz Kettner. Using generic programming for designing a data structure for polyhedral surfaces. *Comput. Geom. Theo. Appl.*, 13:65–90, 1999.

[KV93]   Y. Kallinderis and P. Vijayan. Adaptive refinement-coarsening schemes for three-dimensional unstructured meshes. *AIAA J.*, 31:1440–1447, 1993.

[L88]   R. Löhner. Some useful data structures for the generation of unstructured grids. *Comm. Appl. Numer. Methods*, 4:123–135, 1988.

[M88]   M. Mäntylä. *An Introduction to Solid Modeling.* Computer Science Press, Rockville, MD, 1988.

[Mav00]   D. J. Mavriplis. Adaptive meshing techniques for viscous flow calculations on mixed element unstructured meshes. *Int. J. Numer. Meth. Fluids*, 34:93–111, 2000.

[NLT91]   F. Noel, J.J.C. Leon, and P. Trompette. Data structures dedicated to an integrated free-form surface meshing environment. *Computers and Structures*, 57:345–355, 1991.

[Ove96]   Mark H. Overmars. Designing the computational geometry algorithms library cgal. In *ACM Workshop on Applied Computational Geometry*, May 1996.

[PAM⁺98] D. Poirier, S. R. Allmaras, D. R. McCarthy, M. F. Smith, and F. Y. Enomoto. The CGNS system, 1998. AIAA Paper 98-3007.

[Riv84] M. C. Rivara. Design and data structure of fully adaptive, multigrid, finite-element software. *ACM Trns. Math. Soft.*, 10:242–264, 1984.

[RKFS02] J.-F. Remacle, O. Klaas, J. E. Flaherty, and M. S. Shephard. Parallel algorithm oriented mesh database. *Engineering with Computers*, 18:274–284, 2002.

[RKS00] Jean-Francois. Remacle, B.K. Karamete, and M. Shephard. Algorithm oriented mesh database. In *9th International Meshing Roundtable*, 2000.

[RS03] Jean-François Remacle and Mark S. Shephard. An algorithm oriented mesh database. *Int. J. Numer. Meth. Engrg.*, 58:349–374, 2003.

[The02] The CGNS Steering Sub-committee. *The CFD General Notation System Standard Interface Data Structures.* AIAA, 2002.

[Wei85] K. Weiler. Edge-based data structures for solid modeling in curved-surface environments. *IEEE Computer Graphics and Applications*, 5:21–44, 1985.

[Wei88] K. Weiler. The radial-edge structure: a topological representation for non-manifold geometric boundary representations. In M. Wosny, H. McLaughlin, and J. Encarnacao, editors, *Geometric Modeling for CAD Applications*, pages 3–36. 1988.

Parallel 2D Graded Guaranteed Quality Delaunay Mesh Refinement[*]

Andrey N. Chernikov[1] and Nikos P. Chrisochoides[1,2,3]

[1]Department of Computer Science
College of William and Mary
McGlothlin-Street Hall
Williamsburg, VA 23185, USA
{ancher,nikos}@cs.wm.edu

[2]Department of Mechanical Engineering, MIT
Boston, MA 02139, USA

[3]Department of Radiology
Harvard Medical School
Boston, MA 02115, USA

Summary. We develop a theoretical framework for constructing guaranteed quality Delaunay meshes in parallel for general two-dimensional geometries. This paper presents a new approach for constructing graded meshes, i.e., meshes with element size controlled by a user-defined criterion. The sequential Delaunay refinement algorithms are based on inserting points at the circumcenters of triangles of poor quality or unacceptable size. We call two points Delaunay-independent if they can be inserted concurrently without destroying the conformity and Delaunay properties of the mesh. The contribution of this paper is three-fold. First, we present a number of local conditions of point Delaunay-independence, which do not rely on any global mesh metrics. Our sufficient conditions of point Delaunay-independence allow to select points for concurrent insertion in such a way that the standard sequential guaranteed quality Delaunay refinement procedures can be applied in parallel to attain the required element quality constraints. Second, we prove that a quadtree, constructed in a specific way, can be used to guide the parallel refinement, so that the points, simultaneously inserted in multiple leaves, are Delaunay-independent. Third, by experimental comparison with the well-known guaranteed quality sequential meshing software, we show that our method does not lead to overrefinement, while matching its quality and allowing for code re-use.

1 Introduction

Parallel 2D mesh generation is still important for some 3D simulations like direct numerical simulations of turbulence in cylinder flows with very large Reynolds numbers [8] and coastal ocean modeling for predicting storm surge and beach erosion in

[*]This work was supported by NSF grants: EIA-9972853, EIA-0203974, and ACI-0312980

real-time [23]. In both cases, 2D mesh generation is taking place in the xy-plane and it is replicated in the z-direction in the case of cylinder flows or using bathemetric contours in the case of coastal ocean modeling applications. With the increase of the Reynolds number, the size of the mesh grows in the order of $Re^{9/4}$ [14], which motivates the use of parallel mesh generation algorithms. At the same time, the size of the mesh can be somewhat reduced by employing parallel nonuniform mesh refinement, which is the topic of this paper.

Nave, Chrisochoides, and Chew [17] presented a practical provably-good parallel mesh refinement algorithm for polyhedral domains. The approach in [17] allows rollbacks to occur whenever the simultaneously inserted points can potentially lead to an invalid mesh. It is also labor intensive since it requires changing the sequential meshing kernel in order to accommodate for rollbacks and overlapping of computation with communication. In the present paper, we develop a theoretical framework which allows us to guarantee a priori that concurrently inserted points are Delaunay-independent. The elimination of rollbacks leads to two major benefits: savings in the computation time and the possibility to leverage existing sequential Delaunay meshing libraries like the Triangle [19].

Linardakis and Chrisochoides [15] described a Parallel Domain Decoupling Delaunay method for two-dimensional domains, which is capable of leveraging serial meshing codes. However, it can produce only uniform meshes and is based on the Medial Axis Transform, which is very expensive and difficult to compute for three-dimensional geometries. The approach developed in this paper allows to construct nonuniform meshes and is domain decomposition independent, i.e., it does not require an explicit construction of internal boundaries between the subdomains which will be forced into the final mesh.

Blelloch, Hardwick, Miller, and Talmor [2] describe a divide-and-conquer projection-based algorithm for constructing Delaunay triangulations of pre-defined point sets in parallel. The work by Kadow and Walkington [13, 11, 12] extended [3, 2] for parallel mesh generation and further eliminated the sequential step for constructing an initial mesh, however, all potential conflicts among concurrently inserted points are resolved sequentially by a dedicated processor [12].

Edelsbrunner and Guoy [9] define the points x and y as independent if the closures of their *prestars* (or *cavities* [10]) are disjoint. The approach in [9] does not provide a way to avoid computing the cavities and their intersections for all candidate points, which is very expensive. Spielman, Teng, and Üngör [21] presented the first theoretical analysis of the complexity of parallel Delaunay refinement algorithms. In [22] the authors developed a more practical algorithm.

In [6] we presented a theoretical framework and the experimental evaluation of a parallel algorithm for constructing uniform guaranteed quality Delaunay meshes. We proved a sufficient condition of Delaunay-independence, which is based on a relation of the distance between points and the global circumradius upper bound, and which can be verified very efficiently. We also showed that a coarse-grained mesh decomposition can be used in order to guarantee a priori that the points in certain regions will be Delaunay-independent. In this paper, we build upon the ideas presented in [5] to produce non-uniform (graded) meshes. The non-trivial differences with [6] lie in the introduction of new, local point independence conditions, and in the dynamic construction of a quadtree with leaf size reflecting the local mesh density.

A more extensive review of parallel mesh generation methods can be found in [7].

2 Parallel Refinement Theory

In this section, we develop local Delaunay-independence conditions and show how quadtree leaves can be used to select subsets of circumcenters for concurrent insertion. We extend our previous work [6] by eliminating the use of the global circumradius upper bound and adapting the size of refinement and buffer zones to the user-defined grading function.

2.1 Terminology and Notation

We will denote point number i as p_i and the triangle with vertices p_i, p_j, and p_k as $\triangle(p_ip_jp_k)$. When the vertices of a triangle are irrelevant, we will write simply $\triangle_r$. An edge of a triangle will be denoted as $e(p_ip_j)$ and a line segment connecting two arbitrary points as $\mathcal{L}(p_ip_j)$. Let us call the open disk corresponding to a triangle's circumcircle its *circumdisk*. We will use symbols $\bigcirc(\triangle(p_ip_jp_k))$, $\odot(\triangle(p_ip_jp_k))$, and $r(\triangle(p_ip_jp_k))$ to represent the circumdisk, circumcenter, and circumradius of $\triangle(p_ip_jp_k)$, respectively.

The input to a planar triangular mesh generation algorithm includes a description of *domain* $\Omega \subset \mathbb{R}^2$, which is permitted to contain holes or have more than one connected component. We will use a *Planar Straight Line Graph* (PSLG) [19] to delimit Ω from the rest of the plane. Each segment in the PSLG is considered *constrained* and must appear (possibly as a union of smaller segments) in the final mesh.

The applications that use Delaunay meshes often impose two constraints on the quality of mesh elements: an upper bound on the circumradius-to-shortest edge ratio (which is equivalent to a lower bound on a minimal angle [16, 20]) and an upper bound on the element area. The former is usually fixed and given by a constant value $\bar{\rho}$, while the latter can vary and be controlled by some user-defined grading function $\Delta(x, y) : \mathbb{R}^2 \to \mathbb{R}^1$. As a special case, the grading function can also be constant: $\Delta(x, y) = \bar{\Delta}$.

Typically, a mesh generation procedure starts with constructing an initial mesh, which conforms to the input vertices and segments, and then refines this mesh until the constraints are met. In this paper, we focus on parallelizing the Delaunay refinement stage, which is usually the most memory- and computation-expensive. The general idea of Delaunay refinement is to insert points in the circumcenters of triangles that violate the required bounds, until there are no such triangles left. We will extensively use the notion of *cavity* [10] which is the set of triangles in the mesh whose circumdisks include a given point p_i. We will denote $\mathcal{C}(p_i)$ to be the cavity of p_i and $\partial\mathcal{C}(p_i)$ to be the set of edges which belong to only one triangle in $\mathcal{C}(p_i)$, i.e., external edges.

For our analysis, we will use the Bowyer-Watson (B-W) point insertion algorithm [4, 24], which can be written as

$$
\begin{aligned}
V' &\leftarrow V \cup \{p_i\}, \\
T' &\leftarrow T \setminus \mathcal{C}(p_i) \cup \{\triangle(p_ip_jp_k) \mid e(p_jp_k) \in \partial\mathcal{C}(p_i)\},
\end{aligned}
\tag{1}
$$

where $\mathcal{M} = (V, T)$ and $\mathcal{M}' = (V', T')$ represent the mesh before and after the insertion of p_i, respectively. The set of newly created triangles forms a *ball* [10] of point p_i (denoted $\mathcal{B}(p_i)$), which is the set of triangles in the mesh that have p_i as a vertex.

Sequential Delaunay algorithms treat *constrained* segments differently from triangle edges [20, 18]. A vertex p is said to *encroach upon* a segment s, if it lies within the open diametral disk of s [18]. When a new point is about to be inserted and it happens to encroach upon a constrained segment s, another point is inserted in the middle of s instead [18], and a cavity of the segment's midpoint is constructed and triangulated as before.

We will use the terms *triangulation* and *mesh* interchangeably, depending on the context.

2.2 Delaunay-independent Points

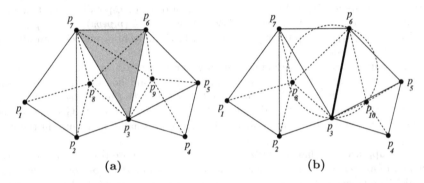

Fig. 1. (a) If $\triangle p_3 p_6 p_7 \in \mathcal{C}(p_8) \cap \mathcal{C}(p_9)$, then concurrent insertion of p_8 and p_9 yields a non-conformal mesh. Solid lines represent edges of the initial triangulation, and dashed lines represent edges created by the insertion of p_8 and p_9. Note that the intersection of edges $p_8 p_6$ and $p_9 p_7$ creates a non-conformity. (b) If edge $p_3 p_6$ is shared by $\mathcal{C}(p_8) = \{\triangle p_1 p_2 p_7, \triangle p_2 p_3 p_7, \triangle p_3 p_6 p_7\}$ and $\mathcal{C}(p_{10}) = \{\triangle p_3 p_5 p_6, \triangle p_3 p_4 p_5\}$, the new triangle $\triangle p_3 p_{10} p_6$ can have point p_8 inside its circumdisk, thus, violating the Delaunay property.

We expect our parallel Delaunay refinement algorithm to insert multiple circumcenters concurrently in such a way that at every iteration the mesh will be both conformal (i.e., simplicial) and Delaunay. Figure 1 illustrates how the concurrently inserted points can violate one of these conditions.

Definition 1 (Delaunay-independence). *Points p and p are Delaunay-independent with respect to mesh $\mathcal{M} = (V, T)$ if their concurrent insertion yields the conformal Delaunay mesh $\mathcal{M}' = (V \cup \{p_i, p_j\}, T')$. Otherwise, p_i and p_j are Delaunay-conflicting.*

Suppose point p_i encroaches upon a constrained segment s_i. Then p_i will not be inserted, and the midpoint p_i' of s_i will be inserted instead (similarly for p_j).

Definition 2 (Strong Delaunay-independence). *Points p_i and p_j are strongly Delaunay-independent with respect to mesh $\mathcal{M} = (V, T)$ iff any pair of points in $\{p_i, p_i'\} \times \{p_j, p_j'\}$ are Delaunay-independent with respect to $\mathcal{M}$.*

2.3 Local Delaunay-Independence Conditions

Lemma 1 (Delaunay-independence criterion I). *Points p_i and p_j are Delaunay-independent iff*

$$C(p_i) \cap C(p_j) = \emptyset, \tag{2}$$

and

$$\forall e(p_m p_n) \in \partial C(p_i) \cap \partial C(p_j) : p_i \notin \bigcirc (\triangle(p_j p_m p_n)). \tag{3}$$

Proof. First, $\mathcal{M}' = (V \cup \{p_i, p_j\}, T')$ is conformal iff (2) holds. Indeed, if (2) holds, then considering (1), the concurrent retriangulation of $C(p_i)$ and $C(p_j)$ will not yield overlapping triangles, and the mesh will be conformal. Conversely, if (2) does not hold, the newly created edges will intersect as shown in Fig. 1a, and $\mathcal{M}'$ will not be conformal.

Now, we will show that $\mathcal{M}'$ is Delaunay iff (3) holds. The Delaunay Lemma [10] states that iff the empty circumdisk criterion holds for every pair of adjacent triangles, then the triangulation is globally Delaunay. Disregarding the symmetric cases, there are three types of pairs of adjacent triangles $\triangle_r$ and $\triangle_s$, where $\triangle_r \in \mathcal{B}(p_i)$, that will be affected: (i) $\triangle_s \in \mathcal{B}(p_i)$, (ii) $\triangle_s \in T' \setminus \mathcal{B}(p_i) \setminus \mathcal{B}(p_j)$, and (iii) $\triangle_s \in \mathcal{B}(p_j)$. The sequential Delaunay refinement algorithm guarantees that $\triangle_r$ and $\triangle_s$ will be locally Delaunay in the first two cases. In addition, condition (3) ensures that they will be locally Delaunay in the third case. Therefore, the mesh will be globally Delaunay. Conversely, if (3) does not hold, triangles $\triangle(p_i p_m p_n)$ and $\triangle(p_j p_m p_n)$ will not be locally Delaunay, and the mesh will not be globally Delaunay.

Corollary 1 (Sufficient condition of Delaunay-independence I [6]). *From Lemma 1 it follows that if (2) holds and $\partial C(p_i) \cap \partial C(p_j) = \emptyset$, then p_i and p_j are Delaunay-independent.*

Lemma 2 (Delaunay-independence criterion II). *Points p_i and p_j are Delaunay-independent with respect to mesh $\mathcal{M} = (V, T)$ iff the edge $e(p_i p_j)$ does not appear in $\mathcal{M}' = (V \cup \{p_i, p_j\}, T')$.*

Proof. To prove the "if" part, let us recall that an edge e exists in a Delaunay triangulation iff there is an empty open disk whose circle passes through the endpoints of e [20]. This means that, in case $e(p_i p_j)$ is not in $\mathcal{M}$, there is no empty open disk whose circle passes through p_i and p_j. This observation has two consequences:

(i) There is no open disk (triangle circumdisk, as a special case), empty of the existing mesh vertices, that includes both p_i and p_j; therefore, condition (2) holds.

(ii) There is no empty open disk, which includes p_i, whose circle passes through p_j. As a special case, there is no such disk whose circle also passes through p_m and p_n; consequently, condition (3) holds.

Thus, p_i and p_j are Delaunay-independent by Lemma 1.

In order to show that the "only if" part of the Lemma holds, we assume that p_i and p_j are Delaunay-independent. Then, by Lemma 1, conditions (2) and (3) hold. Consider Figure 2a. An edge e of a triangulation is either locally Delaunay or is flippable, in which case the edge created by flipping e is locally Delaunay [20]. Since the edge $e(p_m p_n)$ is locally Delaunay, the edge $e(p_i p_j)$ is not locally Delaunay, and, hence, cannot exist in $\mathcal{M}'$.

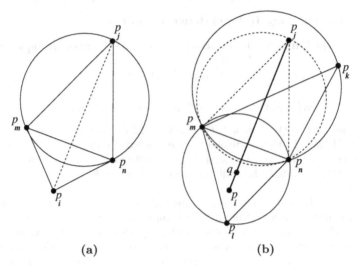

(a)   (b)

Fig. 2. (a) Either $e(p_ip_j)$ or $e(p_mp_n)$ is locally Delaunay. (b) $\bigcirc(\triangle(p_jp_mp_n))$ cannot include p_i and not include q.

Corollary 2 (Sufficient condition of Delaunay-independence II). *Lemma 2 implies that if p_i and p_j are not visible to each other (i.e. the edge $e(p_ip_j)$ cannot exist in a triangulation of Ω, e.g. it would cross a constrained segment), then p_i and p_j are Delaunay-independent.*

Lemma 3 (Sufficient condition of Delaunay-independence III). *Points p_i and p_j are Delaunay-independent if there exists a point $q \in \mathcal{L}(p_ip_j)$ such that*

$$\forall \triangle_s \in T : q \in \bigcirc(\triangle_s) \implies r(\triangle_s) \leqslant \frac{1}{2}\|p_i - p_j\|. \tag{4}$$

Proof. First, condition (4) implies that $\mathcal{C}(p_i) \cap \mathcal{C}(p_j) = \emptyset$. Indeed, if there had been a triangle circumdisk that included p_i and p_j, then this circumdisk would have also included q and had radius greater than $\frac{1}{2}\|p_i - p_j\|$, which contradicts (4).

Now, there are two possibilities:

(i) If $\partial\mathcal{C}(p_i) \cap \partial\mathcal{C}(p_j) = \emptyset$, then, by Corollary 1, p_i and p_j are Delaunay-independent.

(ii) Otherwise, let $\partial\mathcal{C}(p_i) \cap \partial\mathcal{C}(p_j) \neq \emptyset$ and $e(p_mp_n)$ be an arbitrary edge in $\partial\mathcal{C}(p_i) \cap \partial\mathcal{C}(p_j)$ as depicted on Fig. 2b. $\bigcirc(\triangle(p_jp_mp_n))$ cannot include p_i; otherwise, it would have also included q and had radius greater than $\frac{1}{2}\|p_i - p_j\|$, which contradicts (4). Hence, by Lemma 1, p_i and p_j are Delaunay-independent.

2.4 Quadtree Construction

Definition 3 (Quadtree node). *Let a* quadtree node *be an axis-aligned square $S \subset \mathbb{R}^2$. A quadtree node can be either divided into four smaller nodes of equal size or not divided (in this case it is a* leaf*).*

We will denote the length of the side of square S as $\ell(S)$.

Definition 4 (α-neighborhood). *Let the α-neighborhood $\mathcal{N}_\alpha(S_i)$ ($\alpha \in \{Left,$ Right, Top, Bottom\}) of quadtree leaf S_i be the set of quadtree leaves that share a side with S_i and are located in the α direction of S_i. For example, in Fig. 3, $S_k \in \mathcal{N}_{Top}(S_i)$ and $S_l \in \mathcal{N}_{Right}(S_i)$.*

Definition 5 (Orthogonal directions). *Let the orthogonal directions ORT (α) of direction α be*

$$\text{ORT}(\alpha) = \begin{cases} \{Left, Right\} & \text{if } \alpha \in \{Top, Bottom\}, \\ \{Top, Bottom\} & \text{if } \alpha \in \{Left, Right\}. \end{cases}$$

Definition 6 (Buffer zone). *Let the set of leaves*

$$\text{BUF}(S_i) = \bigcup_\alpha \mathcal{N}_\alpha(S_i) \cup \{S_m \in \mathcal{N}_{\text{ORT}(\alpha)}(S_k) \mid S_k \in \bigcup_\alpha \mathcal{N}_\alpha(S_i)\}$$

be called a buffer zone of leaf S_i with respect to mesh $\mathcal{M}$ iff

$$\forall S_n \in \text{BUF}(S_i), \forall \triangle_s \in T : \bigcirc(\triangle_s) \cap S_n \neq \emptyset \implies r(\triangle_s) < \frac{1}{4}\ell(S_n). \quad (5)$$

Equation (5) is the criterion for the dynamic construction of the quadtree. Starting with the root node which covers the entire domain, each node of the quadtree is split into four smaller nodes as soon as all triangles, whose circumdisks intersect this node, have circumradii smaller than one eighth of its side length.

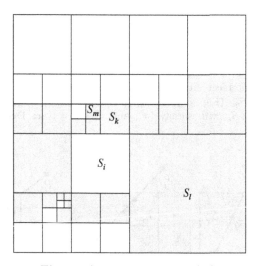

Fig. 3. An example of BUF (S_i).

Definition 7 (Delaunay-separated regions). *Let two regions $R_i \subset \mathbb{R}^2$ and $R_j \subset \mathbb{R}^2$ be called Delaunay-separated with respect to mesh $\mathcal{M}$ iff arbitrary points $p_i \in R_i$ and $p_j \in R_j$ are strongly Delaunay-independent.*

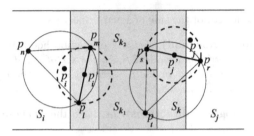

Fig. 4. Splitting constrained segments and strong Delaunay-independence.

Lemma 4 (Sufficient condition of square Delaunay-separateness). *If S_i and S_j are quadtree leaves and $S_j \notin \mathrm{BUF}\,(S_i)$, then S_i and S_j are Delaunay-separated.*

Proof. First, for an arbitrary pair of points $p_i \in S_i$ and $p_j \in S_j \notin \mathrm{BUF}\,(S_i)$, we will prove that p_i and p_j are Delaunay-independent. Then we will extend the proof to show that any pair of points from $\{p_i, p_i'\} \times \{p_j, p_j'\}$ are Delaunay-independent, which will imply that p_i and p_j are strongly Delaunay-independent; hence, S_i and S_j are Delaunay-separated.

By enumerating all possible configurations of leaves in $\mathrm{BUF}\,(S_i)$ and grouping similar cases, w.l.o.g. all arrangements can be accounted for using the following argument.

Suppose $\mathcal{L}\,(p_i p_j)$ intersects the common boundary of S_i and $S_k \in \mathcal{N}_{Top}\,(S_i) \subset \mathrm{BUF}\,(S_i)$ (Fig. 5).

(i) If $\mathcal{L}\,(p_i p_j)$ intersects the upper boundary of S_k (Fig. 5a), then, from (5) and the fact that $\ell\,(S_k) < \|p_i - p_j\|$, any point $q \in \mathcal{L}\,(p_i p_j) \cap S_k$ will satisfy (4). Therefore, by Lemma 3, p_i and p_j are Delaunay-independent.

(ii) Otherwise, let $\mathcal{L}\,(p_i p_j)$ intersect the left boundary of S_k and $S_m \in \mathcal{N}_{Left}\,(S_k) \subset \mathrm{BUF}\,(S_i)$ be the leaf adjacent to this boundary at the point of intersection. $\mathcal{L}\,(p_i p_j)$ can intersect either the upper boundary of S_m (Fig. 5b) or the left boundary of S_m (Fig. 5c). In both cases, $\ell\,(S_m) < \|p_i - p_j\|$, any point $q \in \mathcal{L}\,(p_i p_j) \cap S_m$ will satisfy (4), and p_i and p_j are Delaunay-independent by Lemma 3.

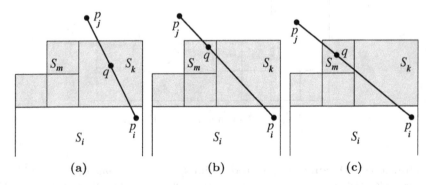

(a)   (b)   (c)

Fig. 5. Some possible positions of points p_i and p_j relative to $\mathrm{BUF}\,(S_i)$.

Now, suppose p_i and p_j encroach upon constrained edges $e\,(p_l p_m)$ and $e\,(p_r p_s)$, respectively (Fig. 4). Then the midpoints p_i' and p_j' of $e\,(p_l p_m)$ and $e\,(p_r p_s)$ will be inserted instead. If p_i' and p_j' lie in the same quadtree leaves as p_i and p_j, then they can be proven Delaunay-independent using the argument above.

Let us analyze the worst case, i.e. $p_i', p_j' \in S_k \in \mathrm{BUF}\,(S_i)$. Since the diametral disk of an edge has the smallest radius among all disks whose circle passes through the endpoints of an edge, then $r\,(e\,(p_l p_m)) \leqslant r\,(\triangle\,(p_l p_m p_n)) < \frac{1}{4}\ell\,(S_k)$ and $r\,(e\,(p_r p_s)) \leqslant r\,(\triangle\,(p_r p_s p_t)) < \frac{1}{4}\ell\,(S_k)$. Therefore, $\|p_i' - p_j'\| > \frac{1}{2}\ell\,(S_k)$. By constructing imaginary buffer squares S_{k_1} and S_{k_2} as shown on Fig. 4, we can still satisfy condition (4), which guarantees that p_i' and p_j' are Delaunay-independent by Lemma 3.

3 Experiments

Figures 6 and 7 compare the meshes produced by our implementation and the Triangle library [19] for a pipe cross-section and a key. Figure 8 also shows the initial geometry and the quadtree produced by our algorithm for the cylinder flow problem, which is similar to the model used in [8]. For all of the quadtree nodes, mesh refinement and node subdivision routines were applied concurrently while preserving the required buffer zones, until the quality constraints were met. The specified grading functions are used as follows. If (x_i, y_i) is the centroid of the triangle $\triangle_i$, then the area of $\triangle_i$ has to be less than $\Delta(x_i, y_i)$. In all experiments we used the same minimal angle bound of $20°$. These tests indicate that while maintaining the required quality of the elements, the number of triangles produced by our method is close, and sometimes is even smaller, than produced by the Triangle [19].

4 Conclusions

We presented a theoretical framework for developing parallel Delaunay meshing codes, which allows to control the size of the elements with a user-defined grading function. We eliminated such disadvantages of the previously proposed methods as the necessity to maintain a cavity (conflict) graph, the rollbacks, the requirement to solve a difficult domain decomposition problem, and the centralized sequential resolution of potential conflicts. Our theory leverages the quality guarantees of the existing sequential Delaunay refinement algorithms. The experimental results confirm that the parallel algorithm produces meshes with the same quality as the sequential Delaunay refinement algorithm and does not lead to overrefinement.

We are currently working on the extension of the proposed approach to three dimensions. While the quadtree immediately generalizes to the octree, the properties of 3D cavities require further study.

5 Acknowledgments

We thank the anonymous reviewers for helpful comments.

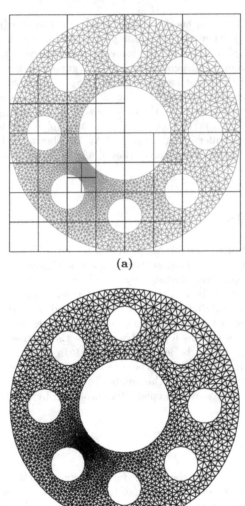

Fig. 6. Pipe cross-section model, $\Delta(x,y) = 0.4\sqrt{(x-200)^2 + (y-200)^2} + 1$. **(a)** Our parallel refinement algorithm, 4166 triangles. **(b)** The Triangle [19], 4126 triangles.

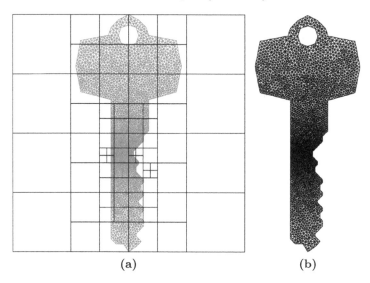

Fig. 7. Jonathan Shewchuk's key model, $\Delta(x, y) = 0.02|y - 46| + 0.1$. **(a)** Our parallel refinement algorithm, 5411 triangles. **(b)** The Triangle [19], 5723 triangles.

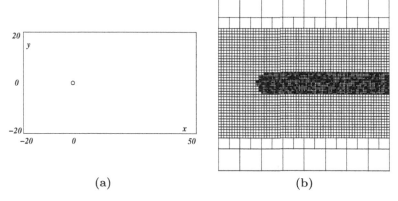

Fig. 8. The cylinder flow model. $\Delta(x, y) = 1.2 \cdot 10^{-3}$ if $((x \geqslant 0) \wedge (y < 5)) \triangledown ((x < 0) \wedge (\sqrt{x^2 + y^2}) < 5)$; $\Delta(x, y) = 10^{-2}$, otherwise. Our parallel refinement algorithm produced 1044756 triangles, and the Triangle [19] produced 1051324 triangles. **(a)** The input model. **(b)** The final quadtree. The complete triangulation is not drawn.

[2] G. E. Blelloch, J. Hardwick, G. L. Miller, and D. Talmor. Design and implementation of a practical parallel Delaunay algorithm. *Algorithmica*, 24:243–269,
1999.

[3] G. E. Blelloch, G. L. Miller, and D. Talmor. Developing a practical projection-
based parallel Delaunay algorithm. In *12th Annual Symposium on Computational Geometry*, pages 186–195, 1996.

[4] A. Bowyer. Computing Dirichlet tesselations. *Computer Journal*, 24:162–166,
1981.

[5] A. N. Chernikov and N. P. Chrisochoides. Parallel guaranteed quality planar
Delaunay mesh generation by concurrent point insertion. In *14th Annual Fall
Workshop on Computational Geometry*, pages 55–56. MIT, Nov. 2004.

[6] A. N. Chernikov and N. P. Chrisochoides. Practical and efficient point insertion scheduling method for parallel guaranteed quality Delaunay refinement.
In *Proceedings of the 18th annual international conference on Supercomputing*,
pages 48–57. ACM Press, 2004.

[7] N. P. Chrisochoides. A survey of parallel mesh generation methods. Technical
Report BrownSC-2005-09, Brown University, 2005. To appear in Numerical So-
lution of Partial Differential Equations on Parallel Computers (eds. Are Magnus
Bruaset, Petter Bjorstad, Aslak Tveito).

[8] S. Dong, D. Lucor, and G. E. Karniadakis. Flow past a stationary and moving
cylinder: DNS at Re=10,000. In *2004 Users Group Conference (DOD_UGC'04)*,
pages 88–95, 2004.

[9] H. Edelsbrunner and D. Guoy. Sink-insertion for mesh improvement. In *Proceedings of the Seventeenth Annual Symposium on Computational Geometry*,
pages 115–123. ACM Press, 2001.

[10] P.-L. George and H. Borouchaki. *Delaunay Triangulation and Meshing. Application to Finite Elements*. HERMES, 1998.

[11] C. Kadow. Adaptive dynamic projection-based partitioning for parallel Delaunay mesh generation algorithms. In *SIAM Workshop on Combinatorial Scientific Computing*, Feb. 2004.

[12] C. Kadow. *Parallel Delaunay Refinement Mesh Generation*. PhD thesis,
Carnegie Mellon University, 2004.

[13] C. Kadow and N. Walkington. Design of a projection-based parallel Delaunay mesh generation and refinement algorithm. In *Fourth
Symposium on Trends in Unstructured Mesh Generation*, July 2003.
http://www.andrew.cmu.edu/user/sowen/usnccm03/agenda.html.

[14] G. Karnriadakis and S. Orszag. Nodes, modes, and flow codes. *Physics
Today*, 46:34–42, 1993.

[15] L. Linardakis and N. Chrisochoides. Parallel domain decoupling Delaunay
method. *SIAM Journal on Scientific Computing*, in print, accepted Nov. 2004.

[16] G. L. Miller, D. Talmor, S.-H. Teng, and N. Walkington. A Delaunay based
numerical method for three dimensions: Generation, formulation, and partition.
In *Proceedings of the Twenty-Seventh Annual ACM Symposium on Theory of
Computing*, pages 683–692. ACM Press, May 1995.

[17] D. Nave, N. Chrisochoides, and L. P. Chew. Guaranteed–quality parallel Delaunay refinement for restricted polyhedral domains. *Computational Geometry:
Theory and Applications*, 28:191–215, 2004.

[18] J. Ruppert. A Delaunay refinement algorithm for quality 2-dimensional mesh
generation. *Journal of Algorithms*, 18(3):548–585, 1995.

[19] J. R. Shewchuk. Triangle: Engineering a 2D quality mesh generator and Delaunay triangulator. In *Proceedings of the First workshop on Applied Computational Geometry*, pages 123–133, Philadelphia, PA, 1996.

[20] J. R. Shewchuk. *Delaunay Refinement Mesh Generation*. PhD thesis, Carnegie Mellon University, 1997.

[21] D. A. Spielman, S.-H. Teng, and A. Üngör. Parallel Delaunay refinement: Algorithms and analyses. In *Proceedings of the Eleventh International Meshing Roundtable*, pages 205–217, 2001.

[22] D. A. Spielman, S.-H. Teng, and A. Üngör. Time complexity of practical parallel Steiner point insertion algorithms. In *Proceedings of the sixteenth annual ACM symposium on Parallelism in algorithms and architectures*, pages 267–268. ACM Press, 2004.

[23] R. A. Walters. Coastal ocean models: Two useful finite element methods. *Recent Developments in Physical Oceanographic Modelling: Part II*, 25:775–793, 2005.

[24] D. F. Watson. Computing the n-dimensional Delaunay tesselation with application to Voronoi polytopes. *Computer Journal*, 24:167–172, 1981.

Index of Authors & Co-Authors

Index by Affiliation

Printed in the United States
By Bookmasters